An Introduction to
STOCHASTIC
PROCESSES

Edward P. C. Kao

Department of Mathematics
University of Houston

DOVER PUBLICATIONS, INC.
Mineola, New York

Bibliographical Note

This Dover edition, first published in 2019, is an unabridged and corrected republication of the work originally published by Wadsworth Publishing Company, Belmont, California, in 1997. A Solutions Manual is available for instructors only by contacting calculus@doverpublications.com.

Library of Congress Cataloging-in-Publication Data

Names: Kao, Edward P. C., author.
Title: An introduction to stochastic processes / Edward P. C. Kao.
Other titles: Stochastic processes
Description: Mineola: Dover Publications, Inc., 2019. | Originally
  published by Wadsworth Publishing Company, Belmont, California in 1997.
  | Includes bibliographical references and index. | Summary: "This Dover
  edition, first published in 2019, is an unabridged and corrected republication
  of the work originally published by Wadsworth Publishing Company,
  Belmont, California, in 1997. A Solutions Manual is available for instructors
  only by contacting calculus@doverpublications.com"—Provided by publisher.
Identifiers: LCCN 2019023975 | ISBN 9780486837925 (paperback)
Subjects: LCSH: Stochastic processes.
Classification: LCC QA274 .K355 2019 | DDC 519.2/3—dc23
LC record available at https://lccn.loc.gov/2019023975

Manufactured in the United States by LSC Communications
83792001
www.doverpublications.com

2 4 6 8 10 9 7 5 3 1

2019

T his is an introductory book on stochastic processes—a subject about modeling and analysis of random phenomena occurring over time or space. Many years ago, we could not *do* stochastic processes in a serious way in the context of real-world problem solving. The rapid advancements in numerical methods and computing facilities have profoundly changed the landscape. This text responds to the challenges of incorporating computer use in the teaching and learning of stochastic process.

This book is written for *students* who are interested in learning concepts, models, and computational approaches in stochastic processes. The intended audience includes upper-level undergraduates and first-year graduate students in operations research, management science, finance, engineering, statistics, computer science, and applied mathematics. The prerequisites for the text are intermediate-level calculus, elementary linear algebra, and an introductory course in probability with an emphasis in operational skills on conditioning.

This book takes an application and computation oriented approach instead of the standard formal and mathematically rigorous approach. The emphasis is on the development of operational skills in stochastic modeling and analysis through a variety of examples drawn from diverse areas while relegating the burden of computation to its rightful master—the computer. Following our approach, we are able to present many topics of practical importance in detail at a very early stage. One such example is the study of a time-dependent service system covered in Chapter 2. There we see that once the model is constructed the time-dependent solutions of the system of differential equations with time-varying parameters can be obtained rather conveniently on a computer.

Organization and Coverage

The book covers standard topics in a first course in stochastic processes. It also includes some additional materials reflecting recent development in computational probability. The first chapter reviews some preliminary materials. They include a brief introduction, transform methods, and some basic concepts in mathematical analysis. Chapters 2 to 6 are organized in a logical sequence. We start with a Poisson process and its variants in Chapter 2, and move to a more general counting process called the renewal process in Chapter 3. To model dependency in random phenomena, we study discrete-time and continuous-time Markov chains in Chapters 4 and 5, respectively. The Markov renewal process presented in Chapter 6 can be considered as the generalization of most models studied earlier. The capstone of all these is the semi-regenerative process in Section 6.4. The last chapter is about Brownian motion, diffusion processes, and Ito's lemmas. The chapter also contains applications of diffusion process in finance.

The book provides a great deal of flexibility for instructors. For students in business and management, Chapters 1–5 should provide a good introduction to stochastic models in management science. For students majoring in finance, the first few sections of Chapters 2–5 along with the last chapter will give them the preliminary background in stochastic processes for further study in continuous-time finance. For students in computer science, electrical and computer engineering, or operations management who want to acquire some knowledge about Markovian service systems necessary for performance evaluations of communication systems, computer networks, or automated manufacturing systems they will find Chapters 2, 4, and 5 useful. In order to reach Section 5.8 on queueing networks in a one-semester course, instructors may choose to skip Sections 5.5–5.7. For students in industrial and systems engineering and operation research who eventually will study queueing theory beyond Markovian models, knowing the materials in Chapter 6 would be helpful. To cover the entire book, a two-semester sequence can be considered with Chapters 1–4 in the first semester and Chapters 5–7 in the second. More difficult examples and problems are marked with an asterisk (*).

A solution manual is available from the publisher for instructors who adopt this text for a course. Readers are welcome to use the perforated card in the back of the book to contact MathWorks, Inc. for a diskette containing the MATLAB programs listed in the appendices.

Notation

We use a capitalized italic letter to denote a matrix, and a bold-italic lower-case letter to denote a vector. The letters I and O denote the identity and zero matrices, respectively. Column and row vectors will not be distinguished but will be stated as such if their types are not clear from the context. We use $exp(\mu)$ to denote an exponential density with parameter μ, $U(0, 1)$ a uniform density over the interval $(0, 1)$, $pos(n; \lambda, t)$ a Poisson probability mass at n whose parameter is λt, $Erlang(n, \mu)$ an Erlang density with parameters n and μ, and $N(\mu, \sigma^2)$ a normal density with mean μ and variance σ^2. We follow the MATLAB notation that "$j: i: k$" denotes $\{j, j+i, j+2i, \ldots, k\}$. For example, $t = 1:1:10$ means $t = 1, 2, \ldots,$ 10. MATLAB commands and function calls are stated in **`bold courier`** fonts.

MATLAB

The software chosen for this text is MATLAB®. MATLAB is easy to learn and numerically reliable. It is most suitable for solving problems involving matrices. In many homework problems, students are expected to experiment with their models and solution procedures with the aid of MATLAB. Of course, Mathematica® or Maple® can also be used to accomplish the same for those who are conversant with and have access to these software.

A brief tutorial on MATLAB is given at the end of the text. For more information about the MATLAB software, readers may contact: The MathWorks, Inc., 24 Prime Park Way, Natick, MA 01760–1500, E-Mail: info@mathworks.com,

WWW: http://www.mathworks.com. We emphasize that the MATLAB programs shown at the end of each chapter were for illustrative purposes and no attempts were made to optimize the codes.

Acknowledgments

I am deeply grateful to Professor Wayne L. Winston, Indiana University, who provided the initial encouragement and a continuing stream of comments and suggestions during the early development of the text. He generously shared his own class notes on Brownian motion and Ito's lemmas with me. In Chapter 7, the part relating to continuous-time finance was greatly influenced by his notes. The feedback from his use of the manuscript in the spring of 1994 in a course on stochastic processes was very helpful. I would like to thank Professor Xiuli Chao, New Jersey Institute of Technology, for his help on queueing networks. My brother Dr. Peichuen Kao, AT&T Bell Labs, read many parts of the original manuscript and whose incisive remarks improved the clarity of a number of arguments. Many of my students at the University of Houston who have read preliminary versions of this text and offered numerous suggestions. In particular, I would like to thank Marvin A. Arostegui, Miguel A. Caceres, Calvin Chen, Jinhu Qian, Meng Rui, Nicola Secomandi, Bradley D. Silver, and Sandra D. Wilson for their many contributions.

Thanks to the reviewers of the manuscript: Professor Apostolos Burnetas, Case Western Reserve University; Professor Ralph L. Disney, Texas A&M University; Professor Halina Frydman, New York University; Professor Carl M. Harris, George Mason University; Professor Vien Nguyen, Massachusetts Institute of Technology; Professor P. Simin Pulat, University of Oklahoma; Professor Shaler Stidham, Jr., University of North Carolina; Professor Wayne L. Winston, Indiana University; and Professor Shelley Zacks, SUNY at Binghamton. Their thoughtful comments and suggestions played an important role in shaping the final version of the manuscript. Finally, I would like to express my appreciation to the staff at Duxbury Press: Editor Curt Hinrichs, Production Editor Jerry Holloway, Editorial Assistant Cynthia Mazow, and Project Development Editors Jennifer Burger and Julie McDonald. Thanks are also due to the staff at Shepherd, Inc. who did the editorial and composition work of the book, in particular, Editor Patricia Noble. Christina Palumbo and Noami Bulock at The MathWorks, Inc. provided excellent support in my use of MATLAB. The MATLAB Tutorial shown at the end of the book benefited by expert feedback from the staff at The MathWorks.

While I was fortunate to receive the help from many people in writing and improving this text, I bear responsibility for any errors and would appreciate hearing about them.

Edward P. C. Kao
Department of Decision and Information Science
University of Houston
Houston, TX 77204-6282
E-mail: ekao@uh.edu
March 1996

To Connie

Introduction

Tips for Chapter 1

■ The details about numerical inversion of transforms (the paragraphs immediately following Examples 1.2.8 and 1.3.4) are there for the curious. These materials are not relevant to subsequent exposition.

■ Motivations of Examples 1.2.6 and 1.3.4: One of the most frequently used approaches in modeling a probabilistic system is to (i) develop a system of difference or differential equations by conditioning on all the possible outcomes of the first step (or the last step) of a sequence of experiments, (ii) obtain the transform of the aforementioned system, (iii) invert the transform algebraically or numerically to solve the problem. In these examples, we consider two somewhat elaborate cases to illustrate this problem-solving approach. A good grasp of the ideas underlying the two examples will give readers a head start for many derivations in subsequent chapters. Readers who find the exposition lengthy or the steps hard to follow may want to consider these two examples as supplementary reading materials for later chapters. For instance, Example 1.3.4 can be read after covering Section 2.1 or 5.2.

■ The Riemann-Stieltjes integral and Riemann-Stieltjes transform discussed in Section 1.4 are useful concepts for handling a continuous random variable that contains discrete components. When a random variable is either continuous or discrete, these concepts *only* provide unified notations. The latter scenario applies to most subjects considered in the text. Examples 1.4.9 and 1.4.10 give two illustrations of applying Leibniz's rule in renewal theory (Section 3.3). Easier examples can be found in most calculus books. The last section (Section 1.4) can be used as quick reference for the various mathematical concepts involved in the text.

■ The moment generating function (shown at the end of Section 1.3) and Taylor-series expansion (given at the end of Section 1.4) are introduced here but will only be needed in Chapter 7.

1.0 Overview

This chapter introduces the subject of stochastic processes, reviews transform techniques to facilitate problem solving and analysis in applied probability, and presents some mathematical background needed in the sequel. In the first section, we define what is meant by a stochastic process and the ideas of stationary and independent increments. The section also gives an overview of the text. The next two sections review generating functions and Laplace transforms. They are quite useful in handling discrete and continuous random variables that we will encounter in the study of stochastic processes. In addition to inversion by algebraic means (manageable only for problems of small size and simple structure), we also present approaches for inverting probability generating functions and Laplace transforms numerically. In this age of computers, numerical inversion enlarges the domain of applicability of transform methods. Readers who have experiences in using generating functions and Laplace transforms in other contexts can go through Sections 1.2 and 1.3 rather quickly. The last section lists a minimal set of results in mathematical analysis that are needed for the text. The section is written primarily for readers who do not have training in mathematics beyond calculus. For others, the section can serve as a source for quick reference. Those who already have had a course in advanced calculus or elementary analysis (say at a level of Rudin [1976] or Bartle [1976]) can skip the last section and go directly to the next chapter.

1.1 Introduction

Let $X(t)$ denote the state of a system at time t. For example, the state $X(t)$ can be the closing price of an IBM stock on day t. The collection of the random variables $X = \{X(t), t \in T\}$ is called a *stochastic process*, in which the set T is called the *index set*. When the index set is countable, X is called a discrete-time process. Thus the daily closing prices of an IBM stock form a discrete-time stochastic process, in which $T = \{0, 1, \ldots\}$. When the index set is an interval of the real line, the stochastic process is called a continuous-time process. If $X(t)$ denotes the price of an IBM stock at time t on a given day, then the process $X = \{X(t), t \in T\}$ is a continuous-time process, in which T is the interval covering a trading day.

If we assume that $X(t)$ takes values in a set S for every $t \in T$, then S is called the state space of the process X. When S is countable, we say that the process has a discrete state space. The two stochastic processes involving the price of an IBM stock both have discrete state spaces whose elements are dollars in increment of

1/8. When S is an interval of a real line, the process has a continuous state space. As an example, if $X(t)$ denotes the temperature at Houston Intercontinental Airport at time t, then, in principle, $X(t)$ can assume any value in an interval S.

A realization of a stochastic process X is called a *sample path* of the process. In Figure 1.1, we depict a sample path associated with a discrete-time process with a discrete state space—namely, the daily closing prices of an IBM stock. In Figure 1.2, we do the same for a continuous-time process with a discrete state space—namely, the price at any time t on a given day. Similarly, in Figure 1.3 we plot a sample path for a continuous-time process with a continuous state space representing the uninterrupted temperature readings at Houston Intercontinental Airport over a given period. If these temperature readings are taken at a set of

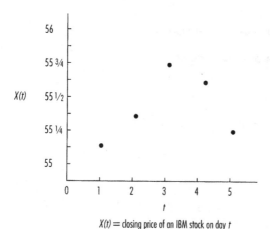

$X(t) =$ closing price of an IBM stock on day t

FIGURE
1.1   A sample path of a discrete-time process with a discrete state space.

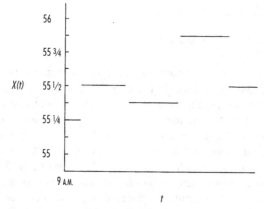

$X(t) =$ price of an IBM stock at time t on a given day

FIGURE
1.2   A sample path of a continuous-time process with a discrete state space.

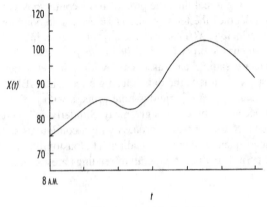

$X(t) =$ temperature at the airport at time t

FIGURE
1.3   A sample path of a continuous-time process with a continuous state space.

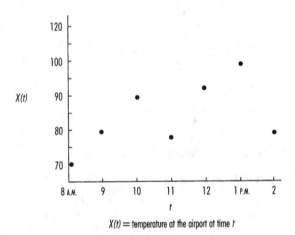

$X(t) =$ temperature at the airport at time t

FIGURE
1.4   A sample path of a discrete-time process with a continuous state space.

distinct epochs, say every hour on the hour, then the process becomes a discrete-time process with a continuous state space. The latter is depicted in Figure 1.4.

Without structural properties, little can be said or done about a stochastic process. Two important properties are the independent-increment and stationary-increment properties of a stochastic process. A process $X = \{X(t), t \geq 0\}$ possesses the *independent-increment* property if for all $t_0 < t_1 < \cdots < t_n$, random variables $X(t_1) - X(t_0), X(t_2) - X(t_1), \ldots, X(t_n) - X(t_{n-1})$ are independent (the time indices will either be discrete or continuous depending on the context). Hence in a process with independent increments, the magnitudes of state change over nonoverlapping intervals are mutually independent. A process possesses the *stationary-increment* property if the random variable $X(t + s) - X(t)$ possesses the same probability distribution for all t and any $s > 0$. In other words, the probability distribution governing the magnitude of state change depends only on the difference in the lengths of the time indices and is independent of the time origin used for the indexing variable.

Let $N(t)$ denote the number of arrivals of a given event by time t (e.g., car arrivals to a toll booth). The stochastic process $N = \{N(t), t \geq 0\}$ is called a *counting* process. The Poisson process studied in Chapter 2 is a counting process in which interarrival times of successive events are independently and identically distributed (i.i.d.) exponential random variables. The process possesses both the independent-increment and stationary-increment properties. Poisson processes are used extensively in modeling arrival processes to service systems and demand processes in inventory systems. There are many useful variants of Poisson processes. An important extension is the nonhomogeneous Poisson process in which we assume that the arrival rate is time dependent. This extension makes Poisson a versatile process for real-world applications.

In a counting process, when interarrival times of successive events follow a probability distribution other than the exponential and yet these times are mutually independent, the resulting process is called a *renewal process*. Chapter 3 is devoted to the study of renewal and related processes. The theory of a renewal process forms a cornerstone for the development of other more complicated stochastic processes, which is accomplished by use of its extension known as the *regenerative process*. At the arrival epoch of a renewal event, the future of the process becomes independent of the past. Therefore the interval between two successive renewals forms a regeneration cycle. The regeneration cycles are probabilistic replica to one another. When we are interested in a long-run property of a stochastic process, studying it over one regeneration cycle will enable us to ascertain its asymptotic value.

In Chapter 4, we introduce Markov chains. In a Markov chain, both the state space and index set are discrete. A change of state depends probabilistically only on the current state of the system and is independent of the past given that the present state is known. A process possessing this property is known to have the *Markovian* property. The successes one can have in employing Markov chains for modeling in applications depend on proper state definitions at selected epochs to maintain the Markovian property at these epochs. When there are rewards associated with state occupancy, the resulting process is called a Markov reward process. Markov chains and Markov reward processes have been used extensively in modeling and analyses of many systems in production, inventory, computers, and communication.

In a Markov chain, we are interested in the state changes over the state space and unconcerned about the sojourn times in each state before a state change takes place. For such a chain, when sojourn times in each state follow exponential distributions with state-dependent parameters, the resulting stochastic process is called a continuous-time Markov process with a discrete state space. For the special case when transitions from a given state will only be made to states other than itself, the resulting process is called the *continuous-time Markov chain*. Various subjects relating to continuous-time Markov chains will be examined in detail in Chapter 5.

A generalization of a Markov chain allows sojourn times in each state to follow probability distributions that depend on the starting and ending states associated with each transition. Stochastic processes resulting from such a generalization are called *Markov renewal processes*. This generalization makes renewal processes and discrete-time and continuous-time Markov chains all special cases of Markov renewal processes. Subjects relating to Markov renewal processes are covered in Chapter 6.

Stochastic processes presented in Chapters 2–6 all have discrete state space. In the last chapter, we will study processes with continuous state space—particularly the Brownian motion process. The mathematics needed to handle Brownian motion and related processes is more demanding. Our coverage of the subjects involved will be relatively limited.

1.2 Discrete Random Variables and Generating Functions

Let $\{a_n\}$ denote a sequence of numbers. We define the *generating function* for the sequence $\{a_n\}$ as

$$a^g(z) = \sum_{n=0}^{\infty} a_n z^n, \tag{1.2.1}$$

where the power series $a^g(z)$ converges in some interval $|z| < R$. $a^g(z)$ is also called the Z-transform or geometric transform for the sequence $\{a_n\}$. To illustrate, consider the case in which $a_n = \alpha^n$, $n = 0, 1, \ldots$. Then we see that $a^g(z) = 1/(1 - \alpha z)$ when $|\alpha z| < 1$. In Table 1.1, we present an abbreviated listing relating some sequences $\{a_n\}$ and their respective generating functions. For the ith pair shown in the table, we use the notation Z-i. The pairs Z-1 and Z-2 imply that the generating function is a linear operator in the sense that if a sequence is a linear combination of two sequences, the linear relation is preserved under the transform by using the generating function. The pair Z-3 implies that the convolution operation of two sequences becomes a multiplication operation if we work with the respective generating functions instead. The sequence $\{b_n\}$ in Z-6 is the sequence $\{a_n\}$ "delayed" by k units, whereas the sequence $\{b_n\}$ in Z-7 is the sequence $\{a_n\}$ "advanced" by k units. The sequences in Z-8 and Z-9 perform respectively the "summing" and "differencing" operations. They are the discrete analogs of integration and differentiation. The two pairs enable us to do these operations when the functions have first been transformed. If A_n is a square matrix with elements $\{a_{ij}(n)\}$, then the (i, j)th element of the matrix generating function $A^g(z)$ is defined as $\sum_{n=0}^{\infty} z^n a_{ij}(n)$. When the elements of matrix A are $\{a_{ij}\}$, Z-10 gives the corresponding matrix generating function.

When $\lim_{n\to\infty} a_n$ exists, we can evaluate this limit by working with the generating function using the *final value property*:

$$\lim_{n\to\infty} a_n = \lim_{z\to 1}(1 - z)a^g(z).$$

TABLE **1.1**
A Table of Generating Functions

The Sequence $\{a_n\}$	Generating Function $a^g(z) = \sum\limits_{n=0}^{\infty} a_n z^n$
1. $\{\alpha a_n\}$	$\alpha a^g(z)$
2. $\{\alpha a_n + \beta b_n\}$	$\alpha a^g(z) + \beta b^g(z)$, where $b^g(z) = \sum\limits_{n=0}^{\infty} b_n z^n$
3. $\left\{ \sum\limits_{m=0}^{n} a_m b_{n-m} \right\}$ Convolution	$a^g(z) b^g(z)$
4. $\{a^n\}$	$\dfrac{1}{1-az}$
5. $\left\{ \dfrac{1}{k!}(n+1)(n+2)\cdots(n+k)a^n \right\}$	$\dfrac{1}{(1-az)^{k+1}}$
6. $\{b_n\}$, where $b_n = 0$ if $n < k$ $\qquad\qquad\quad = a_{n-k}$ if $n \geq k$ and k is a positive integer	$z^k a^g(z)$
7. $\{b_n\}$, where $b_n = 0$ if $n < 0$ $\qquad\qquad\quad = a_{n+k}$ if $n \geq 0$ and k is a positive integer	$\dfrac{1}{z^k}\left[a^g(z) - a_0 - a_1 z - \cdots - a_{k-1} z^{k-1} \right]$
8. $\left\{ \sum\limits_{m=0}^{n} a_m \right\}$	$\dfrac{1}{1-z} a^g(z)$
9. $\{b_n\}$, where $b_n = a_0$ if $n = 0$ $\qquad\qquad\quad = a_n - a_{n-1}$ if $n \geq 1$	$(1-z)a^g(z)$
10. $\{A^n\}$, where A is a square matrix	$\sum\limits_{n=0}^{\infty} (zA)^n = [I - Az]^{-1}$, where I is an identity matrix

A formal proof of the property can be found in a reference cited in the Bibliographic Notes. We leave an alternate proof based on Z-6 as an exercise.

Problem manipulations involving transforms are sometimes referred to as operations in the transform domain. When we invert a transform to its corresponding sequence $\{a_n\}$, we call the procedure an inversion of the transform to the time domain. Generating functions are quite useful in solving systems of difference equations; however, we shall focus our attention on their applications in stochastic modeling.

Let X denote a discrete random variable and $a_n = \text{Prob}\{X = n\}$. Then $P_X(z) = a^g(z) = E[z^X]$ is called the probability generating function for the random variable X. Here we impose the condition $|z| \leq 1$ so as to ensure the uniform convergence of the power series $a^g(z)$. If we know the probability generating function of X, the coefficients of the power series expansion of $a^g(z)$ give the probabilities that X assumes various values. Many times problem solving is somewhat messy in the time domain. We do our manipulations in the transform domain and then make an inversion to obtain the desired result.

We can obtain moments of a random variable X from its probability generating function $P_X(z)$. Define the kth derivative of $P_X(z)$ by

$$P_X^{(k)}(z) = \frac{d^k}{dz^k} P_X(z).$$

Then we see that

$$P_X^{(1)}(z) = \frac{d}{dz} \sum_{n=0}^{\infty} a_n z^n = \sum_{n=0}^{\infty} \frac{d}{dz} a_n z^n = \sum_{n=0}^{\infty} n a_n z^{n-1} \qquad \text{and} \qquad E[X] = P_X^{(1)}(1).$$

Similarly, we have

$$P_X^{(2)}(z) = \frac{d}{dz} P_X^{(1)}(z) = \frac{d}{dz} \sum_{n=1}^{\infty} n a_n z^{n-1} = \sum_{n=1}^{\infty} n(n-1) a_n z^{n-2},$$

and $P_X^{(2)}(1) = E[X(X-1)] = E[X^2] - E[X]$. So the second derivative of $P_X(z)$ with respect to z evaluated at 1 gives the *second factorial moment* of X. The second moment of X is given by

$$E[X^2] = P_X^{(2)}(1) + P_X^{(1)}(1). \tag{1.2.2}$$

Other higher moments of X can be found analogously.

EXAMPLE
1.2.1

The Binomial Random Variable Let X be a binomial random variable with parameters n and p and

$$P\{X = j\} = a_j = \binom{n}{j} p^j q^{n-j} \qquad j = 0, 1, \dots, n,$$

where $q = 1 - p$. The probability generating function is given by

$$P_X(z) = \sum_{j=0}^{n} a_j z^j = \sum_{j=0}^{n} \binom{n}{j} p^j q^{n-j} z^j = \sum_{j=0}^{n} \binom{n}{j} (pz)^j q^{n-j} = (pz + q)^n.$$

With $P_X^{(1)}(z) = n(pz+q)^{n-1} p$ and $P_X^{(2)}(z) = n(n-1)(pz+q)^{n-2} p^2$, we obtain $E[X] = P_X^{(1)}(1) = np$ and $E[X^2] = n(n-1)p^2 + np$ by Equation 1.2.2. This gives $Var[X] = npq$. ∎

EXAMPLE
1.2.2

The Poisson Random Variable Let X be a Poisson random variable with parameter $\lambda > 0$ and

$$P\{X = n\} = a_n = e^{-\lambda} \frac{\lambda^n}{n!} \qquad n = 0, 1, \dots.$$

The probability generating function is given by

$$P_X(z) = \sum_{n=0}^{\infty} e^{-\lambda} \frac{\lambda^n}{n!} z^n = e^{-\lambda} \sum_{n=0}^{\infty} \frac{(\lambda z)^n}{n!} = e^{\lambda(z-1)}.$$

Differentiating $P_X(z)$ with respect to z twice, we obtain

$$P_X^{(1)} = \lambda e^{\lambda(z-1)} \quad \text{and} \quad P_X^{(2)} = \lambda^2 e^{\lambda(z-1)}.$$

So, $E[X] = P_X^{(1)}(1) = \lambda$, and $P_X^{(2)}(1) = \lambda^2$. This gives $E[X^2] = \lambda^2 + \lambda$ and $Var[X] = \lambda$. ∎

EXAMPLE 1.2.3   **The Geometric Random Variable**   Let X be a geometric random variable with parameter p and

$$P\{X = n\} = a_n = pq^n \qquad n = 0, 1, \ldots,$$

where $q = 1 - p$. We can interpret X as the number of failures needed to obtain the first success in a sequence of independent Bernoulli trials with probability of p of finding a success in a single trial. The probability generating function of X is given by

$$P_X(z) = \sum_{n=0}^{\infty} pq^n z^n = \frac{p}{1 - qz}.$$

Finding the first two moments of X will be left as an exercise. ∎

Let $X_1, \ldots, X_k$ denote k independent, nonnegative, and integer-valued random variables where X_i follows probability generating function $P_i(z)$. Let S be the sum of these k random variables. Since S is the convolution of the k independent random variables, by Z-3 we conclude that the probability generation function of S is given by

$$P_S(z) = P_1(z) \cdots P_k(z) \tag{1.2.3}$$

Hence if we work in the transform domain the convolution operations are reduced to multiplication operations.

EXAMPLE 1.2.4   **The Negative Binomial Random Variable**   For a negative binomial random variable S, we have $S = X_1 + \cdots + X_k$, where $X_1, \ldots, X_k$ are i.i.d. random variables with a common geometric distribution $\{pq^n\}$. We note that S can be interpreted as the number of failures needed to obtain the kth success for the first time in a sequence of independent Bernoulli trials with probability p of finding a success in a single trial. In Example 1.2.3, we recall that the probability generating function $P_i(z)$ for X_i is given by $p/(1 - qz)$. Using Equation 1.2.3, we find the probability generating function

$$P_S(z) = \left(\frac{p}{1 - qz} \right)^k = \sum_{n=0}^{\infty} P_n z^n.$$

To invert the previous equation to the time domain, we see that

$$\frac{1}{(1-qz)^k} = \frac{1}{(1-qz)^{(k-1)+1}},$$

and an application of Z-5 shows that the sequence in the time domain reads

$$\frac{1}{(k-1)!}(n+1)\cdots(n+(k-1))q^n.$$

Using Z-1, we conclude that

$$p_n = \frac{p^k}{(k-1)!}(n+1)(n+2)\cdots(n+k-1)q^n = \frac{(n+k-1)\cdots(n+1)n!}{(k-1)!\,n!}p^k q^n$$

$$= \binom{n+k-1}{n}p^k q^n \quad n=0,\,1,\,\dots. \;\blacksquare$$

We now introduce the notion of a *compound* random variable. Let $\{X_i\}$ be a sequence of i.i.d., nonnegative, and integer-valued random variables with a common probability generating function $P_X(z)$. Let N be a nonnegative and integer-valued random variable with a probability generating function $\pi_N(z)$. Assume that N is independent of $\{X_i\}$. The compound random variable S_N is defined as the sum of $X_1, \dots, X_N$. This random variable is often called the *random sum*. We let $H_S(z)$ denote the probability generating function of S_N. Now we see that

$$H_S(z) = E[z^S] = E_N\big[E[z^S|N]\big] = E_N\big[E[z^{X_1+\cdots+X_N}|N]\big]$$

$$= E_N\big[E[z^{X_1+\cdots+X_N}]\big] \qquad \text{(by independence of } N \text{ and } \{X_i\})$$

$$= E_N\big[E[z^{X_1}]\cdots E[z^{X_N}]\big] \qquad \text{(by independence of } X_1, \dots, X_N)$$

$$= E_N\big[(P_X(z))^N\big] = \pi_N(P_X(z)). \tag{1.2.4}$$

Therefore, the probability generating function $H_S(z)$ is obtained by simply using the probability generating function of X (a function in z) as the argument of the probability generating function of N. One way to find the first two moments of the random sum S_N is by using the approach involving differentiation of $H_S(z)$. We leave the application of this approach as an exercise. Another way is by use of conditional expectations. First, we see that $E[S_N|N] = E[X_1 + \cdots + X_N|N] = E[NX_1|N] = NE[X_1]$. This implies that

$$E[S_N] = E\big[E[S_N|N]\big] = E\big[NE[X_1]\big] = E[X_1]E[N]. \tag{1.2.5}$$

For any random variables Y and N, we recall from probability theory that the conditional variance formula is given by

$$Var[Y] = E\big[Var[Y|N]\big] + Var\big[E[Y|N]\big]. \tag{1.2.6}$$

With $Y = S_N$ in Equation 1.2.6 and $Var[S_N|N] = NVar[X_1]$ by independence of $\{X_i\}$, we conclude

$$Var[S_N] = E[NVar[X_1]] + Var[NE[X_1]] = Var[X_1]E[N] + E^2[X_1]\,var[N]. \quad (1.2.7)$$

EXAMPLE 1.2.5   Let N be the number of times a person will visit a store in a year. Assume that N follows the geometric distribution $P\{N = n\} = (1 - \theta)\theta^n, n = 0, 1, \ldots$. From Example 1.2.3, we find the probability generating function $\pi_N(z) = (1 - \theta)/(1 - \theta z)$. During each visit with probability p the person buys something. Purchases over successive visits are probabilistically independent and whether a purchase will be made during a visit is independent of number of times the person visits the store in a year. We are interested in deriving the probability distribution for S, the *number of times* the person will buy something from the store in a year. We let $X_i = 1$ if the person buys something during the ith visit and 0 otherwise. Then we have $S = X_1 + \cdots + X_N$. The probability generating function of X_i is $P_X(z) = q + pz$. Using Equation 1.2.4, we obtain

$$H_S(z) = \pi_N\big(P_X(z)\big) = \frac{1-\theta}{1-\theta P_X(z)} = \frac{1-\theta}{1-\theta[q+pz]} = \frac{1-\theta}{(1-q\theta)-p\theta z}$$

$$= \frac{\dfrac{1-\theta}{1-q\theta}}{1 - \left(\dfrac{p\theta}{1-q\theta}\right)z} = \frac{1-Q}{1-Qz},$$

where we let $Q = p\theta/(1 - q\theta)$. Noting that the previous equation is actually the probability generating function of a geometric distribution, we conclude that $P\{S = k\} = (1 - Q)Q^k, k = 0, 1, \ldots$. ∎

Let $B_1, \ldots, B_k$ be mutually exclusive and collectively exhaustive events. For any event A, we recall from probability theory that

$$P(A) = \sum_{i=1}^{k} P(A|B_i)P(B_i).$$

Moreover, if X and Y are two discrete random variables, we have

$$P(Y = y) = \sum_x P(Y = y|X = x)P(X = x).$$

The preceding formulas are commonly known as the *laws of total probability*. In problem solving in applied probability, sometimes the following three-step approach can be useful: (i) by conditioning on the outcomes of the initial trials and using the law of total probability, write a system of difference equations; (ii) rewrite the system in the transform domain; and (iii) derive desired results from the transform. The next example illustrates the use of such an approach.

EXAMPLE 1.2.6

Consider a biased coin with probability p of obtaining heads and $q = 1 - p$ of getting tails. The coin is tossed repeatedly and stopped when two heads occur in succession for the first time. Let X denote the number of such trials needed and let $a_n = P\{X = n\}$. We want to find the probability generating function and the first two moments of X.

Following the first step of the approach, we note that $a_0 = 0$, $a_1 = 0$, $a_2 = p^2$, and $a_3 = p^2 q$. For $n = 4, 5, \ldots$, the probabilities associated with the three mutually exclusive and collectively exhaustive outcomes $B_1 = \{T\}$, $B_2 = \{H, T\}$, and $B_3 = \{H, H\}$ are q, pq, and p^2. Let H^2 denote the event that two heads occur in succession and A_n denote the event that H^2 occurs for the first time at the nth trial. For $n > 3$, we apply the law of total probability and find

$$P(A_n) = \sum_{i=1}^{3} P(A_n|B_i)P(B_i) = P(A_{n-1})q + P(A_{n-2})pq + (0)p^2.$$

In the previous derivation, to see that $P(A_n|B_1) = P(A_{n-1})$ we observe that given $\{T\}$ occurs at the first trial, then H^2 must occur for the first time at the $(n-1)$st remaining trials so that H^2 indeed occurs at the nth trial of the whole experiment. The term $P(A_n|B_2) = P(A_{n-2})$ can be interpreted similarly. Finally, if $\{H, H\}$ occurs initially, then it is impossible for H^2 to occur for the first time at trial n for $n > 3$. This gives $P(A_n|B_3) = 0$. Since $a_n = P(A_n)$, we obtain

$$a_n = qa_{n-1} + pqa_{n-2} \qquad n = 4, 5, \ldots. \tag{1.2.8}$$

We now move to the second step of the approach. We rewrite Equation 1.2.8 as

$$a_{n+2} = qa_{n+1} + pqa_n \qquad n = 2, 3, \ldots. \tag{1.2.9}$$

We multiply the nth equation of Equation 1.2.9 by z^n and add the resulting equations. This gives

$$\sum_{n=2}^{\infty} z^n a_{n+2} = \sum_{n=2}^{\infty} qz^n a_{n+1} + \sum_{n=2}^{\infty} pqz^n a_n$$

$$\frac{1}{z^2}\sum_{n=2}^{\infty} z^{n+2} a_{n+2} = \frac{1}{z}\sum_{n=2}^{\infty} qz^{n+1} a_{n+1} + pqP_X(z)$$

$$\frac{1}{z^2}\left[P_X(z) - z^3 a_3 - z^2 a_2\right] = \frac{q}{z}\left[P_X(z) - z^2 a_2\right] + pqP_X(z)$$

$$P_X(z) - z^3 a_3 - z^2 a_2 = qz\left[P_X(z) - z^2 a_2\right] + pqz^2 P_X(z).$$

Using the initial conditions a_2 and a_3 and rearranging the terms, we obtain

$$P_X(z) = \frac{p^2 z^2}{1 - qz - pqz^2}. \tag{1.2.10}$$

We now give a slightly simpler way of obtaining Equation 1.2.10. Define the indicator variable $I\{A\} = 1$ if A is true, and 0 otherwise, and define $a_n = 0$ if $n < 0$. Then Equation 1.2.8 and the initial conditions can be combined in a single expression:

$$a_n = qa_{n-1} + pqa_{n-2} + p^2 I\{n = 2\} \qquad n = 0, 1, \dots.$$

It is easy to verify that the above holds for all n. Multiplying the nth equation by z^n and adding the resulting equations give

$$\sum_{n=0}^{\infty} a_n z^n = \sum_{n=0}^{\infty} qa_{n-1} z^n + \sum_{n=0}^{\infty} pqa_{n-2} z^n + p^2 z^2.$$

Changes of indexing variables will produce

$$P_X(z) = \sum_{n=0}^{\infty} qa_n z^{n+1} + \sum_{n=0}^{\infty} pqa_n z^{n+2} + p^2 z^2 = qzP_X(z) + pqz^2 P_X(z) + p^2 z^2$$

and hence Equation 1.2.10.

Having obtained the probability generating function, we are now at the last step. By differentiating Equation 1.2.10 twice with respect to z and proceeding methodically, we will find the first two moments of X—after some cumbersome algebra. A somewhat intriguing alternative is to look at Equation 1.2.10 in the following manner:

$$P_X(z) = \frac{z^2}{\left(\dfrac{1 - qz - pqz^2}{p^2}\right)} \equiv \frac{P_W(z)}{P_Y(z)}$$

and *assume* that W and Y are two legitimate random variables with respective probability generating functions $P_W(z)$ and $P_Y(z)$. If this were the case, then we would have concluded without hesitation that $W = X + Y$ and X and Y are independent. For W, we see that $P\{W = 2\} = 1$ and therefore is a legitimate random variable. While $P_Y(1) = 1$, we see that the coefficients of the power series expansion associated with the terms z and z^2 are both negative. This implies that the "probabilities" that $Y = 1$ and $Y = 2$ are both negative. Fortunately, we can proceed with our computation by assuming that having negative probabilities is acceptable. The reason that this transgression is acceptable in the present context is that the results pertaining to convolution and moments are actually derived for generating functions whose coefficients are not restricted to be numbers in a unit interval as long as generating functions equal 1 when their arguments are set to 1. It is easy to verify that

$$P_Y^{(1)}(1) = \frac{1}{p^2}(-q - 2pq) = 2 - \frac{1}{p} - \frac{1}{p^2} = E[Y] \text{ and } P_Y^{(2)}(1) = \frac{1}{p^2}(-2pq) = 2 - \frac{2}{p}.$$

The variance of Y is then

$$Var[Y] = P_Y^{(2)}(1) + P_Y^{(1)}(1) - \left[P_Y^{(1)}(1)\right]^2 = -\frac{1}{p^4} - \frac{2}{p^3} + \frac{2}{p^2} + \frac{1}{p}.$$

Since $E[W] = 2$ and $Var[W] = 0$, we conclude that

$$E[X] = 2 - E[Y] = \frac{1}{p} + \frac{1}{p^2} \quad \text{and} \quad Var[X] = -Var[Y] = \frac{1}{p^4} + \frac{2}{p^3} - \frac{2}{p^2} - \frac{1}{p}.$$

(The expression $Var[X] = -Var[Y]$ might raise some eyebrows if one forgets that Y is not a legitimate random variable.) ∎

If we have a generating function, we can invert it to the corresponding discrete function in the time domain—namely the sequence $\{a_n\}$—either algebraically or numerically. To do it algebraically, we use *partial fraction expansion*. Typically the transform $A^g(z)$ is written as the ratio of two polynomials. For the method to work, the degree of the numerator polynomial must be at least one less than that of the denominator. If this is not so, we can either factor enough z out of the numerator or divide the two polynomials so that the mentioned condition is met. We then do a partial fraction expansion of the remainder. There are two types of ratio and their respective expansions to consider. To illustrate, one type reads

$$\frac{a + bz}{(1 - cz)(1 - dz)} = \frac{A}{(1 - cz)} + \frac{B}{(1 - dz)},$$

where $c \neq d$. The preceding equation is equivalent to $a + bz = A(1 - dz) + B(1 - cz)$. Setting $z = 1/c$, we find the value of A. The value of B can be found similarly. Another option is to set z at any two distinct values and solve the resulting system of linear equations. Inverting the transform is done by invoking Z-1, Z-2, and Z-4. The other type of ratio is one in which the denominator factors are not distinct. This is illustrated by

$$\frac{a + bz + cz^2}{(1 - dz)^2 (1 - ez)} = \frac{A}{(1 - dz)^2} + \frac{B}{(1 - dz)} + \frac{C}{(1 - ez)},$$

where $d \neq e$. Finding the coefficients A, B, and C can be done in a manner similar to the first case. MATLAB function **residue** will do the aforementioned partial fraction expansion. Since there is a one-to-one correspondence between the probability generating function and the respective probability distribution, the inversion enables us to uncover the functional form of the latter. This is illustrated in the following two examples.

EXAMPLE
1.2.7

Assume that we are given the following probability generating function of random variable X

$$P_X(z) = \frac{4}{(2 - z)(3 - z)^2}.$$

What is the probability distribution $\{p_n\}$ of X? We need to invert $P_X(z)$. First, we write

$$\frac{4}{(2-z)(3-z)^2} = \frac{A}{(2-z)} + \frac{B}{(3-z)} + \frac{C}{(3-z)^2}.$$

Multiplying that by the denominator on the left side gives

$$A(3-z)^2 + B(2-z)(3-z) + C(2-z) = 4.$$

Setting z at 2, 3, and 0 in succession, we find respectively $A = 4$, $C = -4$, and $B = -4$. The partial fraction expansion of $P_X(z)$ is then given by

$$P_X(z) = \frac{4}{(2-z)} - \frac{4}{(3-z)} - \frac{4}{(3-z)^2} = \frac{2}{\left(1-\frac{1}{2}z\right)} - \frac{\frac{4}{3}}{\left(1-\frac{1}{3}z\right)} - \frac{\frac{4}{9}}{\left(1-\frac{1}{3}z\right)^2}.$$

In the Appendix, we illustrate the use of MATLAB to do the partial fraction expansion. Using Z-2, Z-4, and Z-5, we invert the previous equation to the time domain. This gives the probability distribution

$$p_n = 2\left(\frac{1}{2}\right)^n - \left(\frac{4}{3}\right)\left(\frac{1}{3}\right)^n - \left(\frac{4}{9}\right)\left(\frac{1}{3}\right)^n (n+1)$$

$$= 2^{-n+1} - 4(3)^{-(n+1)} - 4(n+1)(3)^{-(n+2)} \qquad n = 0, 1, 2, \dots \ \blacksquare$$

EXAMPLE 1.2.3   We return to Example 1.2.6 with the goal of obtaining a closed-form expression for the distribution of X—the number of trials needed to obtain two heads in succession in a sequence of Bernoulli trials. Recall the probability generating function is given by Equation 1.2.10. Note that

$$\frac{1}{1-qz-pqz^2} = \frac{-1}{pq\left(z^2 + \frac{q}{pq}z - \frac{1}{pq}\right)}.$$

The term in the last parentheses can be factored as $(z - z_1)(z - z_2)$, where

$$z_1 = \frac{-q+\sqrt{q^2+4pq}}{2pq} \qquad \text{and} \qquad z_2 = \frac{-q-\sqrt{q^2+4pq}}{2pq}.$$

Using partial fraction expansion and $z_1 - z_2 = (\sqrt{q^2+4pq})/pq$, we find

$$\frac{1}{1-qz-pqz^2} = \frac{1}{\sqrt{q^2+4pq}}\left[\frac{-1}{z-z_1} + \frac{1}{z-z_2}\right].$$

Hence the probability generating function in a readily invertible form is given by

$$
P_X(z) = \frac{p^2 z^2}{\sqrt{q^2 + 4pq}} \left[\frac{1}{z_1 - z} - \frac{1}{z_2 - z} \right] = \frac{p^2 z^2}{\sqrt{q^2 + 4pq}} \left[\frac{c_1}{1 - c_1 z} - \frac{c_2}{1 - c_2 z} \right]
$$

$$
= dz^2 \left[\frac{c_1}{1 - c_1 z} - \frac{c_2}{1 - c_2 z} \right],
$$

where $c_i = 1/z_i$, for $i = 1$ and 2, and $d = p^2/\sqrt{q^2 + 4pq}$. Because of the term z^2, we use Z-6 along with Z-2 and Z-4 to make the inversion. This gives

$$
p_n = d[c_1(c_1)^{n-2} - c_2(c_2)^{n-2}] = d[c_1^{n-1} - c_2^{n-1}], \qquad n = 2, 3, \ldots
$$

and $p_i = 0$, for $i = 0$, 1. To check our result, it is easy to verify that $p_2 = d[c_1 - c_2]$ $= p^2$. Verifying other results can be done accordingly. For the case of a fair coin, we again illustrate the use of MATLAB for partial fraction expansion in this chapter's Appendix. ∎

In the last two examples, we have seen the inversion of probability generating function algebraically. We now show how it is done numerically. There are times when algebraic inversion of probability generating function is too complicated, and a numerical solution becomes an attractive alternative. For the probability generating function $P_X(z) = \sum_{n=0}^{\infty} p_n z^n$, we let $\tilde{p}_n$ denote the probability p_n obtained from a numerical inversion using

$$
\tilde{p}_n = \frac{1}{2nr^n} \left\{ P_X(r) + (-1)^k P_X(-r) + 2 \sum_{j=1}^{n-1} (-1)^j \, \mathrm{Re}\left(P_X\left(re^{\frac{\pi j i}{n}} \right) \right) \right\},
$$

where $0 < r < 1$, $n \geq 1$, $i = \sqrt{-1}$, and $\mathrm{Re}(z) =$ the real part of z. The derivation of the preceding expression is based on the Fourier-series method. It can be shown that for $0 < r < 1$, and $n \geq 1$, the resulting error bound is given by

$$
|p_n - \tilde{p}_n| \leq \frac{r^{2n}}{1 - r^{2n}}.
$$

For all practical purposes, we can consider the error bound is approximately r^{2n}. To have an accuracy of 10^{-v}, we let $r = 10^{-v/2n}$. The MATLAB program **invt_pgf** given in the Appendix will carry out the numerical inversion. This is illustrated in the next example.

EXAMPLE
1.2.9
Going back to the probability generating function derived in Example 1.2.7, we use the MATLAB program **invt_pgf** in the Appendix to produce the following results for the case $p = 0.6$: $p(1) = 0.0000$, $p(2) = 0.3600$, $p(3) = 0.1440$, $p(4) = 0.1440$, $p(5) = 0.0922$. To verify these results, we see that $p_2 = p^2 = (0.6)^2 = 0.36$, $p_3 = p^2q = (0.36)(0.4) = 0.144$, $p_4 = (0.4)p_3 + (0.6)(0.4)p_2 = (0.4)(0.144) + (0.6)(0.4)(0.36) = 0.144$, and so on. ∎

1.3 Continuous Random Variables and Laplace Transforms

Let f be any real-valued function defined on $[0, \infty)$. The Laplace transform of f is defined as

$$f^e(s) = \int_0^\infty e^{-st} f(t)\, dt,$$

provided that the integrated exists. The Laplace transform is also called the exponential transform. It is a continuous analog of the geometric transform. In Table 1.2, we give an abbreviated listing that shows the frequently used transform pairs. For reference, we use L-i to denote the ith pair in the table. The ten transform pairs are the continuous analogs of their discrete counterparts shown in Table 1.1.

TABLE **1.2**
A Table of
Laplace
Transforms

The Function $f(t)$	Laplace Transform $f^e(s) = \int_0^\infty e^{-st} f(t)\, dt$
1. $\alpha f(t)$	$\alpha f^e(s)$
2. $\alpha f(t) + \beta g(t)$	$\alpha f^e(s) + \beta g^e(s)$ where $g^e(s) = \int_0^\infty e^{-st} g(t)\, dt$
3. $\int_0^\infty f(\tau)g(t-\tau)\, d\tau$	$f^e(s)g^e(s)$
4. e^{-at}	$\dfrac{1}{s+a}$
5. $\dfrac{1}{k!} t^k e^{-at}$	$\dfrac{1}{(s+a)^{k+1}}$
6. $f(t - \tau)$ $(\tau > 0)$	$e^{st} fe(s)$
7. $f(t + \tau)$ $(\tau > 0)$	$e^{st}\left[f^e(s) - \int_0^\tau e^{-st} f(t)\, dt \right]$
8. $\int_0^t f(\tau)\, d\tau$	$\dfrac{1}{s} f^e(s)$
9. $\dfrac{d}{dt} f(t)$	$sf^e(s) - f(0)$
10. e^{At} where A is a square matrix	$\int_0^\infty e^{-st} e^{At}\, dt = [sI - A]^{-1}$, where I is an identity matrix

Thus we omit an elaboration similar to that shown in the last section about the roles played by the various functions.

When $\lim_{t \to \infty} f(t)$ exists, we can use the final-value property to find it by evaluating

$$\lim_{t \to \infty} f(t) = \lim_{s \to 0} s f^e(s).$$

The proof of the preceding property is relegated to Problem 31. The use of the Laplace transform in solving problems in applied probability is demonstrated in the sequel. When f is a probability density of a nonnegative continuous random variable X, we have

$$f_X^e(s) = E[e^{-sX}].$$

The previous integral exists when $\mathrm{Re}(s) \geq 0$, where s is a complex number. When the random variable at stake is clear, we sometimes suppress the subscript X from $f_X^e(s)$. As in the case of probability generating function, we can find moments for random variable X from its Laplace transform $f_X^e(s)$. Define the nth derivative of the Laplace transform $f_X^e(s)$ with respect to s by $f^{(n)}(s)$. Successive differentiation of the transform yields

$$f^{(n)}(s) = \frac{d^n}{ds^n} f_X^e(s) = \frac{d^n}{ds^n} E[e^{-sX}] = E\left[\frac{d^n}{ds^n} e^{-sX} \right]$$

$$= E\left[(-1)^n X^n e^{-sX} \right] = (-1)^n E\left[X^n e^{-sX} \right].$$

From the previous equation, we conclude that

$$E[X^n] = (-1)^n f^{(n)}(0). \tag{1.3.1}$$

There is a one-to-one correspondence between the Laplace transform and the probability density. If $X_1, \ldots, X_k$ are k independent and nonnegative continuous random variables and $f_i^e(s)$ is the Laplace transform of the density f_{X_i}, then the transform pair L-3 implies that the Laplace transform of the random variable $S = X_1 + \cdots + X_k$ is given by

$$f_S^e(s) = f_1^e(s) \cdots f_k^e(s). \tag{1.3.2}$$

Assume that F is a distribution function with density f. A two-fold convolution of F with itself, written commonly as $F_2(\cdot)$, is given by

$$F_2(t) = P(Z \leq t) = \int_0^t P\{X + Y \leq t \mid X = x\} f_X(x) dx \quad \text{by the law of total probability}$$

$$= \int_0^t P\{Y \leq t - x\} f(x) dx = \int_0^t F(t - x) f(x) dx,$$

in which we use random variables X, Y, and Z with $X \sim F$, $Y \sim F$, $Z = X + Y$ and assume that X and Y are independent. Hence we see that $F_2(t) \neq \int_0^t F(t - x)$ $F(x)dx$ even though $f_2(t) = \int_0^\infty f(t - x) f(x) dx$, where f_2 is the density of F_2 (the latter identity can also be established by differentiating $F_2(t)$ with respect to t with an application of Leibnitz's rule, see Section 1.4). The n-fold convolution of F with itself, F_n, is defined similarly.

Let A be an event and X be a continuous random variable with a density f_X on $[0, \infty)$. A version of the law of total probability for this case reads

$$P(A) = \int_0^\infty P(A|X = x) f_X(x) dx.$$

> If Y is a random variable with distribution function F_Y, and $F_{Y|X}(\cdot|x)$ is the conditional distribution function of Y given $X = x$, then another version of the law of total probability is
>
> $$F_Y(y) = \int_0^\infty F_{Y|X}(y|x) f_X(x) dx.$$

EXAMPLE 1.3.1

The Exponential Random Variable   Let X be a random variable having the exponential density with parameter μ, that is,

$$f_X(x) = \mu e^{-\mu x} \qquad x > 0.$$

The Laplace transform of the density is given by

$$f^e(s) = \frac{\mu}{s + \mu}.$$

Since

$$f^{(1)}(s) = (-1) \frac{\mu}{(s + \mu)^2} \qquad \text{and} \qquad f^{(2)}(s) = (-1)(-2) \frac{\mu}{(s + \mu)^3},$$

we find

$$E[X] = -f^{(1)}(0) = \frac{1}{\mu} \qquad \text{and} \qquad E[X^2] = f^{(2)}(0) = \frac{2}{\mu^2}.$$

This gives $Var[X] = 1/\mu^2$. ∎

EXAMPLE 1.3.2

Competing Exponential Random Variables   Let X_1 and X_2 denote the occurrence times of events 1 and 2, respectively, where $X_1 \sim exp(\mu_1)$ and $X_2 \sim exp(\mu_2)$. Assume that X_1 and X_2 are independent. Let X be the first occurrence time, that is, $X = \min\{X_1, X_2\}$. Hence the two events are competing for the first occurrence. There are many applications of the idea of competing exponential random variables in applied probability models—particularly in reliability theory and Markovian queueing models. For example, assume that a piece of equipment contains two key components. Let X_1 and X_2 denote their respective lifetimes and assume that the two lifetimes follow exponential distributions with respective parameters μ_1 and μ_2. If one component fails, then the equipment fails; the equipment lifetime is given by

X. In a queueing example, suppose that the interarrival times of successive customers to a single-server service system follow an exponential distribution with parameter μ_1 and the service time of a customer follows an exponential distribution with parameter μ_2. At time *t*, assume that there is one customer in the system. Then one of the two possible events will occur after time *t*: either a service completion or a new arrival will occur first. In this case, $t + X$ represents the time for a change of the state of the system to occur.

To find the probability distribution for *X*, we note that

$$P\{X > x\} = P\{\min(X_1, X_2) > x\} = P\{X_1 > x, X_2 > x\}$$

$$= e^{-\mu_1 x} e^{-\mu_2 x} = e^{-(\mu_1 + \mu_2)x} \qquad x > 0.$$

Thus we conclude $X \sim exp(\mu_1 + \mu_2)$. The next question about the two competing exponential random variables is who is the winner—which event occurs first? Let $I = 1$ if $X_1 < X_2$, and 0 otherwise. We see that the two events $\{I = 1$ and $X > x\}$ and $\{x < X_1 < X_2\}$ are equivalent. Therefore we write

$$P\{I = 1 \text{ and } X > x\} = P\{x < X_1 < X_2\} = \iint\limits_{x < x_1 < x_2} \mu_1 e^{-\mu_1 x_1} \mu_2 e^{-\mu_2 x_2} \, dx_1 dx_2$$

$$= \int_x^\infty \mu_1 e^{-\mu_1 x_1} \int_{x_1}^\infty \mu_2 e^{-\mu_2 x_2} \, dx_2 dx_1 = \int_x^\infty \mu_1 e^{-\mu_1 x_1} e^{-\mu_2 x_1} \, dx_1$$

$$= \frac{\mu_1}{\mu_1 + \mu_2} e^{-(\mu_1 + \mu_2)x} \qquad x > 0.$$

It is important to note that *I* and *X* are independent: the event type and event time are two independent random variables. The marginal distribution $P(I = 1)$ is obtained by setting *x* equal to 0 in the last expression. This gives

$$P\{I = 1\} = \frac{\mu_1}{\mu_1 + \mu_2}.$$

Since $X_i \sim exp(\mu_i)$, we can envision that for X_i to occur at any time the rate of occurrence is μ_i. In other words, we have $P\{X_i \in (t, t + h) | X_i > t\} = \mu_i h + o(h)$ (for any function *f*, the "little-oh" function $o(h)$ means $\lim_{h \to 0} = f(h)/h = 0$—more will be said about the little $o(h)$ function in Section 1.4). The probability for one of the exponential events to occur first is given by the magnitude of its rate relative to the aggregated rates of occurrence. The result has an intuitively plausible interpretation.

We now generalize these results. Consider $X = \min\{X_1, \ldots, X_n\}$, and assume that $X_i \sim exp(\mu_i)$ for each *i* and the random variables are mutually independent. Then *X* follows an exponential distribution with parameter $\mu_1 + \cdots + \mu_n$ and

$$P(X \in (t, t+h) | X > t) = (\mu_1 + \cdots + \mu_n)h + o(h). \qquad (1.3.3)$$

Let $I_i = 1$ if arg $\min\{X_1, \ldots, X_n\} = i$ and 0 otherwise. Then the probability that random variable X_i occurs first is given by

$$P\{I_i = 1\} = \frac{\mu_i}{\mu_1 + \cdots + \mu_n}, \qquad i = 1, \ldots, n. \qquad (1.3.4)$$

Note that by the memoryless property of the exponential distribution if the n random variables are in progress at some time origin, then Equations 1.3.3 and 1.3.4 still hold. In other words, they need not start at the same epoch. ∎

EXAMPLE
1.3.3

The Erlang Random Variable Let $X_1, \ldots, X_n$ be i.i.d. random variables whose common density is exponential with parameter $\lambda > 0$. Let $S = X_1 + \cdots + X_n$. Then S is called an Erlang random variable with parameters (n, λ). Clearly S is the convolution of n i.i.d. random variables. Using Example 1.3.2, we find the Laplace transform of S:

$$f_S^e(s) = \left(\frac{\lambda}{s+\lambda}\right)^n = \lambda^n \frac{1}{(s+\lambda)^{(n-1)+1}}.$$

Inverting the previous equation by L-5 yields

$$f_S(t) = \lambda^n \frac{1}{(n-1)!} t^{n-1} e^{-\lambda t} = \frac{\lambda e^{-\lambda t}(\lambda t)^{n-1}}{(n-1)!} \qquad t > 0.$$

The moments of S are found by noting

$$f_S^{(1)}(s) = \lambda^n(-n)(s+\lambda)^{-(n+1)} \quad \text{and} \quad f_S^{(2)}(s) = \lambda^n(-n)(-(n+1))(s+\lambda)^{-(n+2)}$$

and

$$E[S] = -f_S^{(1)}(0) = \frac{n}{\lambda}, \quad E[S^2] = f_S^{(2)}(0) = \frac{n(n+1)}{\lambda^2}, \quad \text{and} \quad Var[S] = \frac{n}{\lambda^2}.$$

The preceding results are to be expected since S is the sum of n i.i.d. exponential random variables each with parameter λ. The Erlang density is a special case of the gamma density. In the latter case, its density is identical to $f_S(t)$ except that n is allowed to be any positive real number and the denominator is replaced by the gamma function $\Gamma(n) = \int_0^\infty e^{-x} x^{n-1} dx$ (when n is a positive integer, we have $\Gamma(n) = (n-1)!$ and the density reduces to that of an Erlang). ∎

EXAMPLE
1.3.4

An Insurance Problem This example has two objectives: one is to show a more elaborate use of the law of total probability and the other is to present a nice application of Laplace transforms. The example employs a few important properties of a Poisson process. As such, it serves as a prelude to the next chapter. Some readers may choose to return to this example after reviewing the first two sections of Chapter 2.

Claims arrive at an insurance company with i.i.d. interarrival times. Each follows an exponential distribution with parameter λ. In the next chapter, we will see that the exponential interarrival time assumption implies that the counting process $N = \{N(t), t \geq 0\}$ follows a Poisson process with rate λ, where λ is the instantaneous rate of arrival. Moreover, the process possesses the following properties: (i) $P(N(h) = 1) = \lambda h + o(h)$, (ii) $P(N(h) = 0) = 1 - \lambda h + o(h)$, and (iii) $P(N(h) \geq 2) = o(h)$, where the little-oh function $o(h)$ has been defined in Example 1.3.2. We assume that the successive claim sizes $S_1, S_2, \ldots$ are positive and independent random variables with a common probability distribution B, density b, and mean $E[S]$ and are independent of the arrival process N. Initially the company has a capital reserve of x and its reserve increases at a constant rate of $\sigma > 0$ through the collection of premiums. We assume that $\sigma > \lambda E[S]$. In other words, the "input" rate is greater than the "output" rate. Let $X(t)$ denote the total amount of claims paid by time t. Then we have

$$X(t) = \sum_{i=1}^{N(t)} S_i.$$

Since $x + \sigma t$ represents the company's capital at time t, a ruin occurs when $x + \sigma t < X(t)$. For each $x > 0$, we define the ruin probability as $q(x) = P\{X(t) > x + \sigma t \text{ for some } t \geq 0\}$.

To compute $q(x)$, we focus on $q(x-h)$ (where h is a very small amount of capital), and what happens in the first h/σ time units. If the company's capital is $x - h$ at time 0, then its capital will reach $x = (x - h) + \sigma(h/\sigma)$ at time h/σ. We consider the three scenarios shown in Figure 1.5 (based on Property (iii) of the Poisson process, we do not have to consider cases when there are two or more claims in a small interval).

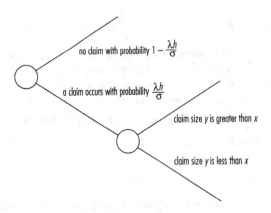

no claim with probability $1 - \dfrac{\lambda h}{\sigma}$

a claim occurs with probability $\dfrac{\lambda h}{\sigma}$

claim size y is greater than x

claim size y is less than x

FIGURE
1.5 The three scenarios to consider for the insurance problem.

Since λ is the claim arrival rate and h/σ is a small interval, $\lambda h/\sigma$ is the probability that there will be a claim occurring in the interval (based on Property (i) of the Poisson process). Define events $B_1 = \{$no claims occur in the next h/σ time units$\}$, $B_2 = \{$a claim occurs in the next h/σ time units, and the claim size is greater than or equal to $x\}$, $B_3 = \{$a claim occurs in the next h/σ time units, and the claim size is less than $x\}$, and $A_x = \{$a ruin will occur some time in the future given that the present capital is $x\}$. An application of the law of total probability yields

$$P(A_{x-h}) = \sum_{i=1}^{3} P(A_{x-h}|B_i)P(B_i).$$

If B_1 occurs, then the probability that a ruin will occur some time in the future becomes $q(x)$. Thus under the first scenario we have $P(A_{x-h}|B_1) = q(x)$ and $P(B_1) = 1 - (\lambda h/\sigma) + o(h)$ (the latter is based on Property (ii) of the Poisson process, and $o(\lambda h/\sigma) = (\lambda/\sigma)o(h) = o(h)$; see Section 1.4 to review little-oh functions). For the second scenario, we see that $P(B_2) = (\lambda h/\sigma)P(S > x) + o(h)$ and $P(A_{x-h}|B_2) = 1$ because in this case a ruin is certain. The last scenario is more delicate. We apply the law of total probability involving the continuous random variable S and find

$$P(A_{x-h}|B_3)P(B_3) = P(A_{x-h} \cap B_3)$$

$$= \frac{\lambda h}{\sigma} \int_0^x P\{\text{a ruin will occur given the capital is } x | S = y\}b(y)dy + o(h)$$

$$= \frac{\lambda h}{\sigma} \int_0^x q(x - y)b(y)dy + o(h).$$

Combining these results, using the fact that the sum of a finite number of little-oh functions is again a little-oh function and $q(x - h) = P\{A_{x-h}\}$, we obtain

$$q(x - h) = \left(1 - \frac{\lambda h}{\sigma}\right)q(x) + \frac{\lambda h}{\sigma} \int_x^\infty b(y)dy + \frac{\lambda h}{\sigma} \int_0^x q(x - y)b(y)dy + o(h).$$

The first three terms on the right side correspond to the outcomes associated with the three scenarios shown in Figure 1.5. The little-oh function $o(h)$ has been defined in Example 1.3.2. Multiplying the preceding equation by -1, adding $q(x)$ to both sides, dividing the resulting equation by h, and taking the limit as $h \to 0$, we find the following

$$q'(x) = \frac{\lambda}{\sigma}q(x) - \frac{\lambda}{\sigma}[1 - B(x)] - \frac{\lambda}{\sigma}\int_0^x q(x - y)b(y)dy$$

$$= \frac{\lambda}{\sigma}\left[q(x) - [1 - B(x)] - \int_0^x q(x - y)b(y)dy\right].$$

Using L-3, L-4, L-8, and L-9, we find the Laplace transform of the previous equation

$$sq^e(s) - q(0) = \frac{\lambda}{\sigma}\left[q^e(s) - \frac{1}{s}\left(1 - b^e(s)\right) - q^e(s)b^e(s)\right].$$

Rearranging the terms yields

$$q^e(s)\left[s - \frac{\lambda}{\sigma}\left(1 - b^e(s)\right)\right] = q(0) - \frac{\lambda}{\sigma s}\left[1 - b^e(s)\right].$$

To find the unknown $q(0)$, we let $s \to 0$ in the preceding equation. The left side goes to zero because $b^e(0) = 1$. The right side is an indeterminacy. We use L'Hôpital's rule and obtain $q(0) = \lambda E[S]/\sigma$. We call this last quantity ρ. Summarizing this we conclude

$$q^e(s) = \frac{\rho - \dfrac{\lambda}{\sigma s}\left[1 - b^e(s)\right]}{s - \dfrac{\lambda}{\sigma}\left[1 - b^e(s)\right]}. \tag{1.3.5}$$

Consider the simple case when the claim size follows an exponential distribution with parameter μ. Hence $E[S] = 1/\mu$ and $\lambda/\sigma = \rho\mu$. We use Equation 1.3.5 to find the Laplace transform of the ruin probability. With $b^e(s) = \mu/(s + \mu)$, we obtain $q^e(s) = \rho/[s + \mu(1 - \rho)]$. After an inversion, we find $q(x) = \rho e^{-\mu(1-\rho)x}$, $x \geq 0$. Consider a numerical example with $\sigma = \$1000$ per day, $\lambda = 5$ and $E[S] = \$190$. The ruin probabilities are shown in Figure 1.6.

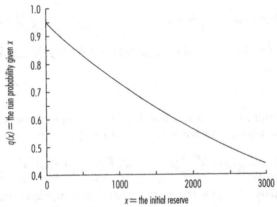

FIGURE
1.6   The ruin probabilities for the numerical example.

The ruin probability has applications not only in insurance but also in other diverse areas. In production and inventory control, consider the case in which demands for a given product arrive according to a Poisson process and successive demands are i.i.d. random variables. The product is produced at a rate of σ per unit time. In this context, the ruin probability $q(x)$ represents the probability that a shortage will occur when the initial inventory is x.

Consider a single server service system in which customers arrive one at a time and are served in order of arrival. If interarrival times of customers follow an exponential distribution with parameter λ, and service times are independent and follow distribution B, we call such a model an $M/G/1$ queue (the first letter M signifies Markovian—specifically it means Poisson arrivals; the second letter indicates that service times are i.i.d. random variables following a common distribution; the third letter shows the number of available servers in the system). Let V denote the amount of work in the system immediately before an arrival. It can be shown that $q(x) = P(V > x)$ if the system has been running for a long time. In other words, the complementary distribution of the virtual waiting time V is given by the ruin probability. A proof of this result can be found in a reference given in the Bibliographic Notes. ▪

Like geometric transforms, we can invert Laplace transforms either algebraically or numerically. To do it algebraically, the approach is similar to that of inverting a geometric transform. Partial fraction expansion often will be needed before we can use the transform pairs shown in Table 1.2 or resort to a more elaborate table. Even when the algebraic approach is feasible, the process of carrying it out could be laborious. A numerical inversion may serve as an alternative. A numerical inversion of the Laplace transform $f^e(s)$ is given by

$$\tilde{f}(t) = \frac{e^{A/2}}{2t} \text{Re}\left(f^e\left(\frac{A}{2t}\right) \right) + \frac{e^{A/2}}{t} \sum_{k=1}^{\infty} (-1)^k \text{Re}\left(f^e\left(\frac{A + 2k\pi i}{2t}\right) \right), \quad (1.3.6)$$

where A is a parameter related to the discretization error. The derivation of Equation 1.3.6 is again based on the Fourier-series method. When $|f(t)| \le 1$ for all $t \ge 0$ (for example, the cases of a distribution function or a complementary distribution function), it can be shown that the discretization error is bounded by

$$|f(t) - \tilde{f}(t)| \le \frac{e^{-A}}{1 - e^{-A}}.$$

When e^{-A} is small, the discretization error is approximately equal to e^{-A}. To have an accuracy of $10^{-\nu}$, we let $A = \nu \ln(10)$. In actual implementation, the infinite sum of Equation 1.3.6 can be replaced by a finite sum. We omit the implementation details. A MATLAB program **invt_lap** given in the Appendix will carry out the numerical inversion. Its use is illustrated in the next example.

EXAMPLE 1.3.5 **The Waiting Time Distribution of an M/H₂/1 Queue** For the $M/G/1$ queue described in Example 1.3.4, we now consider the case when the service time S follows a hyperexponential distribution whose density is given by

$$b(x) = p_1 \mu_1 e^{-\mu_1 x} + p_2 \mu_2 e^{-\mu_2 x} \qquad x \ge 0, \qquad (1.3.7)$$

where $0 \le p_1 \le 1$, $0 \le p_2 \le 1$, and $p_1 + p_2 = 1$. The Laplace transform of the density is given by

$$b^e(s) = \sum_{i=1}^{2} \frac{p_i \mu_i}{s + \mu_i}.$$

Using the Laplace transform, we find the mean and variance of random variable S as follows

$$E[S] = \frac{p_1}{\mu_1} + \frac{p_2}{\mu_2} \qquad Var[S] = \frac{p_1(2 - p_1)}{\mu_1^2} + \frac{p_2(2 - p_2)}{\mu_2^2} - \frac{2 p_1 p_2}{\mu_1 \mu_2}.$$

The hyperexponential density defined by Equation 1.3.7 is sometimes called the H_2 density. The *coefficient of variation* c_S of S defined by $\sqrt{Var(S)}/E[S]$ for the H_2 density is greater than or equal to one. Comparing with the exponential density, we see that the H_2 density has more variability. A special case of H_2 is one with *balanced means* in the sense $p_1/\mu_1 = p_2/\mu_2$. Given $E[S]$ and c_S, the unique H_2 density with balanced means can be found in the following:

$$p_1 = \frac{1}{2}\left(1 + \sqrt{\frac{c_S^2 - 1}{c_S^2 + 1}}\right), \quad p_2 = 1 - p_1, \quad \mu_1 = \frac{2 p_1}{E[S]}, \quad \text{and} \quad \mu_2 = \frac{2 p_2}{E[S]}$$

The H_2 density with balanced means can be rather useful when one performs sensitivity analysis in which the variability of a random variable is to be varied systematically.

In Example 1.3.4, we mentioned that for an $M/G/1$ queue, we have $q(x) = P(V > x)$, where V represents the amount of remaining work immediately before an arrival epoch. When $X(t) = \sum_{i=1}^{N(t)} S_i > t$, the difference between the two terms indeed gives the remaining work at any time. In the context that $q(x) = P\{X(t) > x + \sigma t$ for some $t \geq 0\}$, we set $\sigma = 1$. The remaining work at an arrival epoch is simply the amount of time an arriving customer must wait in line before reaching service. Hence $q(x)$ gives the complementary waiting time distribution for the arriving customer, the integral of $q(x)$ over $(0, \infty)$ produces the mean waiting time in line and $q(0) = \lambda E[S]$ yields the probability that an arriving customer will have to wait. Using Equation 1.3.5 we obtain the Laplace transform of $q(x)$

$$q^e(s) = \frac{\lambda E[S] - \dfrac{\lambda}{s}\left[1 - b^e(s)\right]}{s - \lambda\left[1 - b^e(s)\right]}, \quad \text{where} \quad 1 - b^e(s) = \frac{s(s + \mu_1 + \mu_2 - p_1\mu_1 - p_2\mu_2)}{(s + \mu_1)(s + \mu_2)}.$$

Algebraic inversion of the preceding transform to the time domain is possible but messy. Suppose we are interested in finding the complementary waiting time distribution for the case in which the service time distribution is H_2 with balanced means where $p_1 = 0.72360680$, $\mu_1 = 1.47721360$, $\mu_2 = 0.55278640$. For this H_2, we have $E[S] = 1$ and $c_S = 1.5$. Let $\lambda = 0.8$ so that $\rho = 0.8$. With $q(0) = 0.8$, in the Appendix we use **invt_lap** to invert the Laplace transform $q^e(s)$. The inverted complementary waiting time distribution is shown in Figure 1.7. The results so obtained can be compared to those found from algebraic inversion via partial fraction expansion. Such a comparison will reveal that the results found from numerical inversion are quite accurate. ∎

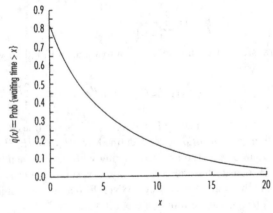

FIGURE
1.7   The complementary waiting time distribution at an arrival epoch.

A rather useful result of the Laplace transform is the *asymptotic-rate theorem.* For example, if a function $f(t)$ represents the total expected rewards received in $(0, t)$, then sometimes we may be interested in the expected average reward per unit time $f(t)/t$ for a large t. The asymptotic-rate theorem provides a means to produce the limit in the transform domain.

The asymptotic-rate theorem: when both sides of limit exist, then

$$\lim_{t \to \infty} \frac{f(t)}{t} = \lim_{s \to 0} s^2 f^e(s). \tag{1.3.8}$$

The proof of this theorem is explored in an exercise.

We conclude this brief exposition on transform methods with a definition of the following combination of probability generating function and Laplace transform for the function $f(n, t)$, where the first argument of the function is defined on the set of nonnegative integers and the second is defined on $[0, \infty)$

$$f^{ge}(z, s) = \int_0^\infty e^{-st} \sum_{n=0}^\infty z^n f(n, t)dt$$

provided the preceding expression is well defined. In studying stochastic processes, we frequently encounter functions of the type $f(n, t)$ where n denotes the number of transitions by time t. The double transforms are used in Chapters 5 and 6. Operations with respect to the previous transform over one argument can be done in the usual manner while treating the other as a constant.

Moment Generating Function

Probability generating functions and Laplace transforms are useful for working with nonnegative random variables. When a random variable can assume negative values, we sometimes employ moment generating functions to play a role similar to that of the two transforms. In this text it is sufficient to restrict our attention to continuous random variables. For random variable X with density f the moment generating function is defined as

$$M(t) = E[e^{tX}] = \int_{-\infty}^\infty e^{tx} f(x)dx,$$

provided that the integral exists for $t \in (-s, s)$ for some $s > 0$. Like Laplace transforms, there is a one-to-one correspondence between a probability density and a moment generating function. By successive differentiating $M(t)$ with respect to t and setting the resulting expressions equal to zero, we find a formula for finding moments similar to Equation 1.3.1:

$$E[X^n] = M^{(n)}(0). \tag{1.3.9}$$

For the standard normal distribution, we have

$$M(t) = \frac{1}{\sqrt{2\pi}} \int_{-\infty}^{\infty} e^{tx} e^{-x^2/2} dx = \frac{1}{\sqrt{2\pi}} \int_{-\infty}^{\infty} e^{-1/2(x^2 - 2tx)} dx$$

$$= \frac{1}{\sqrt{2\pi}} \int_{-\infty}^{\infty} e^{-1/2(x^2 - 2tx + t^2) + (1/2)t^2} dx$$

$$= \frac{1}{\sqrt{2\pi}} e^{(1/2)t^2} \int_{-\infty}^{\infty} e^{-(1/2)(x-t)^2} dx = e^{(1/2)t^2} \left[\frac{1}{\sqrt{2\pi}} \int_{-\infty}^{\infty} e^{-(1/2)u^2} du \right] = e^{t^2/2}$$

In the preceding derivation, the term in the brackets is one because it is the area under a standard normal density.

For any random variable X with moment generating function $M_X(t)$, if we define $Y = a + bX$, where a and b are two constants, we have $M_Y(t) = E[e^{(a+bX)t}]$ $= e^{at} E[e^{(bt)X}] = e^{at} M_X(bt)$. Thus when $X \sim n(0, 1)$ and $Y = \mu + \sigma X$, we know $Y \sim N(\mu, \sigma^2)$ and its moment generating function is given by

$$M_Y(t) = e^{\mu t} M_X(\sigma t) = e^{\mu t} e^{(\sigma t)^2/2} = e^{\mu t + (\sigma^2 t^2/2)}. \tag{1.3.10}$$

To conclude the two sections on generating functions and transforms, remember that the probability generating function, Laplace transform, or moment generation function of a random variable X may not exist. However the characteristic function of random variable X, defined by

$$\phi(t) = E[e^{itX}] \qquad -\infty < t < \infty,$$

always exists. A thorough discussion about characteristic functions can be found in a reference given in the Bibliographic Notes.

1.4 Some Mathematical Background

In this text, we try to keep the level of mathematics needed to cover the various subjects at a minimum. However, there are a few concepts that facilitate the development of the subject matter; these will be reviewed in this section. Readers who have had a course on elementary analysis and/or advanced calculus can skip this section and go directly to subsequent chapters.

Right and Left Continuity and Limits

A function $F(t)$ is defined as *right-continuous* if

$$\lim_{t \downarrow \tau} F(t) = F(\tau) \tag{1.4.1}$$

for all τ ("$t \downarrow \tau$" means that t approaches τ from the right). Recall that when F denotes a distribution function it is defined as right-continuous. The notion of a left-continuous function is defined similarly with $t \uparrow \tau$ replacing $t \downarrow \tau$ in Equation 1.4.1.

A function $F(t)$ has left limits if $\lim_{t \uparrow \tau} F(t)$ exists for all t in the domain of the function. Functions with right limits are defined analogously.

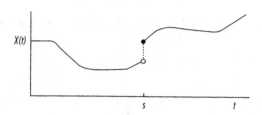

FIGURE
1.8   A right-continuous, left-limits sample path.

A stochastic process $X = \{X(t), t \geq 0\}$ whose sample path is right-continuous and has left limits is depicted in Figure 1.8. Stochastic processes possessing these two properties are sometimes termed *RCLL*. If X is *RCLL*, then $X(t-)$ is the left limit of X at t and $\Delta X(t) = X(t) - X(t-)$ is the amount of jump at t. This is illustrated in the figure at time s.

Riemann-Stieltjes Integrals

In applied probability, a random variable often is neither discrete nor continuous. This is illustrated in the following examples.

EXAMPLE   Let X be the lifetime of a car battery. We assume that X follows an exponential
1.4.1   distribution with parameter μ. Assume that customers replace their batteries at failure or one year of usage, whichever comes first. Let Y denote the *actual* lifetime of a battery. We then see that $Y = \min\{X, 1\}$ and

$$F_Y(y) = \begin{cases} 1 - e^{-\mu y} & 0 \leq y < 1 \\ 1 & y \geq 1. \end{cases}$$

We see that random variable Y is neither continuous nor discrete. It has a discontinuity at $y = 1$ as shown in Figure 1.9.

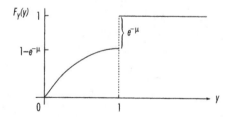

FIGURE
1.9   The distribution function of Y.

The well-known formula for finding the mean $E[Y] = \int_0^\infty y f_Y(y) dy$ does not work because Y only has a density over the interval $(0, 1)$ and has a jump at 1, that is,

$P(Y = 1) = F(1) - F(1-) = e^{-\mu}$. To find the mean, we have to "patch up" the two pieces to obtain

$$E[Y] = \int_0^1 y f_Y(y)dy + P(Y = 1)(1) = \int_0^1 y(\mu e^{-\mu y})dy + e^{-\mu}$$

$$= \left\{ \frac{1}{\mu}[1 - e^{-\mu}] - e^{-\mu} \right\} + e^{-\mu} = \frac{1}{\mu}[1 - e^{-\mu}].$$

An easier way to find the mean in this case is using the identity $E[Y] = \int_0^\infty P(Y > y)dy$ for any nonnegative random variable Y. This gives

$$E[Y] = \int_0^1 e^{-\mu y}dy = \frac{1}{\mu}[1 - e^{-\mu}]. \quad \blacksquare$$

EXAMPLE
1.4.2
Consider Example 1.4.1 again. Assume now that a fraction p of customers replaces their batteries at failure or one year of usage, whichever comes first and the remaining fraction $q = 1 - p$ replaces their batteries at failure. Let random variable Z denote the *actual* battery life. Then random variable Z is a probability mixture of random variables Y and X. Its distribution function is given by

$$F_Z(z) = pF_Y(z) + qF_X(z)$$

$$= \begin{cases} p[1 - e^{-\mu z}] + q[1 - e^{-\mu z}] = 1 - e^{-\mu z} & 0 \le z < 1 \\[2mm] p[1] + q[1 - e^{-\mu z}] = 1 - qe^{-\mu z} & z \ge 1. \end{cases}$$

In Figure 1.10, we see that the discontinuity occurs at $z = 1$. In fact, we have $P\{Z = 1\} = pe^{-\mu}$. In contrast to Figure 1.9, in this case there are still probability densities lingering after $z \ge 1$.

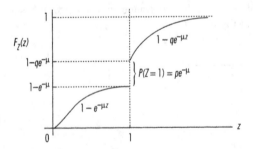

FIGURE
1.10   The distribution function of Z.

Finding the mean of Z can again be done on a "piecemeal" approach or using the distribution function directly as shown in Example 1.4.1. A simple way is to take the weighted average of $E[Y]$ and $E[X]$. $\blacksquare$

EXAMPLE
1.4.3
Consider a customer who goes to his barber for a haircut once every two weeks. Through years of experience, the customer observes that with probability $1 - \rho$ the barber will be able to serve him immediately, with probability ρ there will be one customer already in service, and the time for each haircut follows an exponential distribution with parameter μ. Let W denote the waiting time of the customer needing a haircut. Clearly, $P(W = 0) = 1 - \rho$. Because of the memoryless property of the exponential distribution at the arrival epoch of the customer, if there is one customer already in service the remaining service time of the customer again follows an exponential distribution with the same parameter. The distribution function for random variable W is given by

$$F_W(w) = 1 - \rho + \rho[1 - e^{-\mu w}] \qquad w \geq 0.$$

In this case the discontinuity of the distribution function occurs at the starting point of the interval $[0, \infty)$. Random variables having distribution functions of the type shown occur frequently in queueing theory. The waiting time distribution of Example 1.3.5 is one such example. ▪

The existence of random variables that are neither continuous nor discrete demands the introduction of the Riemann-Stieltjes integral. Let g be a continuous function and F a nondecreasing function. A subdivision of interval (a, b) is a set of numbers $\{x_0, x_1, \ldots, x_n\}$ such that $a = x_0 < x_1 < \cdots < x_n = b$. The subdivision divides the interval into n disjoint subintervals $(x_0, x_1), \ldots, (x_{n-1}, x_n)$. The Riemann-Stieltjes integral of g with respect to F from a to b is

$$\int_a^b g(x)dF(x) = \lim_{\|\Delta\| \to 0} \sum_{k=1}^n g(\zeta_k)[F(x_k) - F(x_{k-1})], \qquad (1.4.2)$$

where $x_{k-1} < \zeta_k \leq x_k$, $k = 1, \ldots, n$, and $\|\Delta\| = \max\{x_1 - x_0, \ldots, x_n - x_{n-1}\}$. The requirement that F be nondecreasing can be relaxed. If we can write $F(x) = F_1(x) + F_2(x)$, where F_1 is nondecreasing and F_2 is nonincreasing (in this case, F is called a function of *bounded variation*), then Equation 1.4.2 exists and

$$\int_a^b f(x)dF(x) = \int_a^b f(x)dF_1(x) + \int_a^b f(x)dF_2(x).$$

The Laplace-Stieltjes integral version of integration by parts is

$$\int_a^b g(x)dF(x) + \int_a^b F(x)dg(x) = g(b)F(b) - g(a)F(a). \qquad (1.4.3)$$

In the preceding expression, g and F appear symmetrically. Hence equation 1.4.3 also works when g is nondecreasing and F is continuous. In summary, if either g or F is of bounded variation and the other is continuous, then Equation 1.4.2 exists and Equation 1.4.3 works.

Two special cases of interest are: (i) F is differentiable and (ii) F is a step function. In the first case, we let $f(x) = dF(x)/dx$. Then Equation 1.4.2 reduces to

$$\int_a^b g(x)dF(x) = \int_a^b g(x)f(x)dx.$$

In other words, Riemann-Stieltjes integration becomes Riemann integration. When f is a density of random variable X, the preceding is simply $E[g(X)]$. In the second case, we consider the situation in which $F(x)$ is a step function with jumps of magnitude p_k at x_k, where $a < x_1 < \cdots < x_n < b$. This implies $F(x_k+) - F(x_k-) = p_k$, $k = 1, \ldots, n$. Then Equation 1.4.2 becomes

$$\int_a^b g(x)dF(x) = \sum_{k=1}^n g(x_k)p_k.$$

When F is the distribution function of a discrete random variable X, the preceding is again $E[g(X)]$. This implies that if we write

$$E[g(X)] = \int g(x)dF(x),\tag{1.4.4}$$

the expression works for random variables that are continuous, discrete, or neither. While the expression works for any one of these cases, the operations involved reduce to those that are context specific. An application of Equation 1.4.4 is where $g(X) = e^{-sX}$, and X is a random variable with distribution function F. This gives the Laplace-Stieltjes transform of F

$$\tilde{F}(s) = \int_0^\infty e^{-st}dF(t).\tag{1.4.5}$$

When F is a distribution function, we note that

$$\tilde{F}(0) = \int_0^\infty dF(t) = 1.\tag{1.4.6}$$

The previous expression provides a simple accuracy check of the Laplace-Stieltjes transform of F. Let $\tilde{F}^{(n)}(s)$ denote the nth derivative of $\tilde{F}(s)$ with respect to s. Similar to the derivation of Equation 1.3.1, we have

$$E[X^n] = (-1)^n \tilde{F}^{(n)}(0).\tag{1.4.7}$$

EXAMPLE 1.4.4 Consider a discrete random variable X with $P(X = d_1) = 1/3$ and $P(X = d_2) = 2/3$. The Laplace-Stieltjes transform of F is given by

$$\tilde{F}(s) = \frac{1}{3}e^{-sd_1} + \frac{2}{3}e^{-sd_2}. \quad \blacksquare$$

EXAMPLE 1.4.5 Returning to Example 1.4.1, we find the Laplace-Stieltjes transform of the distribution function F

$$\tilde{F}(s) = \int_0^1 e^{-st}(\mu e^{-\mu t})dt + e^{-s(1)}[F(1) - F(1-)]$$

$$= \frac{\mu}{(s+\mu)}[1 - e^{-(s+\mu)}] + e^{-s}[1 - (1 - e^{-\mu})]$$

$$= \frac{\mu}{s+\mu} + \frac{s}{s+\mu}e^{-(s+\mu)}.$$

Letting $s = 0$, we find $\tilde{F}(0) = 1$. Using Equation 1.4.7, it is easy to verify that the mean of the random variable is equal to the $E[Y]$ given in Example 1.4.1. ∎

EXAMPLE
1.4.6

Returning to Example 1.4.2, we find the Laplace-Stieltjes transform of the distribution function F

$$\tilde{F}(s) = \int_0^1 e^{-st}(\mu e^{-\mu t})dt + e^{-s(1)}[F(1) - F(1-)] + \int_1^\infty e^{-st}(q\mu e^{-\mu t})dt$$

$$= \frac{\mu}{s+\mu}[1 - e^{-(s+\mu)}] + e^{-s}[pe^{-\mu}] + \frac{q\mu}{s+\mu}[e^{-(s+\mu)}].$$

Setting s to 1 in the preceding equation, we find

$$\tilde{F}(0) = [1 - e^{-\mu}] + pe^{-\mu} + q[e^{-\mu}] = 1.$$

Using Equation 1.4.7 with $n = 1$, we obtain the mean of Z

$$E[Z] = -e^{-\mu} + [1 - e^{-\mu}]\frac{1}{\mu} + pe^{-\mu} + qe^{-\mu} + \frac{q}{\mu}e^{-\mu} = \frac{1}{\mu} - \frac{p}{\mu}e^{-\mu}.$$

If we are only interested in $E[Z]$, then an easy way other than taking the weighted averages is to use the formula. $E[Z] = \int_0^\infty [1 - F_Z(z)]dz$. This yields

$$E[Z] = \int_0^1 e^{-\mu z}dz + \int_1^\infty qe^{-\mu z}dz = \frac{1}{\mu}[1 - e^{-\mu}] + \frac{q}{\mu}e^{-\mu} = \frac{1}{\mu} - \frac{p}{\mu}e^{-\mu}. ∎$$

EXAMPLE
1.4.7

Returning to Example 1.4.3, we find the Laplace-Stieltjes transform of the distribution function F_W

$$\tilde{F}_W(s) = (1-\rho)e^{-s(0)} + \int_0^\infty e^{-st}(\rho\mu e^{-\mu t})dt = (1-\rho) + \frac{\rho\mu}{s+\mu}.$$

The mean waiting time is therefore ρ/μ. ∎

EXAMPLE
1.4.8

Let Y be the sum of n i.i.d. uniform random variables over the interval $(0, 1)$. We now show how to use the Laplace-Stieltjes transform to find the density of Y. Let X be a standard uniform variate. Then the Laplace transform of the density of X is given by

$$E[e^{-sX}] = \int_0^1 e^{-sx}dx = \frac{1}{s}[1 - e^{-s}].$$

Consequently, the Laplace transform of Y is given by

$$E[e^{-sY}] = \left[\frac{1}{s}[1 - e^{-s}]\right]^n = s^{-n}\sum_{k=0}^n \binom{n}{k}(-e^{-s})^k(1)^{n-k} = \sum_{k=0}^n \binom{n}{k}(-1)^k e^{-sk}s^{-n}.$$

We will next show that the last two terms, e^{-sk} and s^{-n}, can be stated as two transforms and hence the function in the time domain is actually a convolution of two functions. Using the gamma function given in Example 1.3.3, we can derive the following identity

$$s^{-n} = \int_0^\infty e^{-sx} \frac{x^{n-1}}{(n-1)!} dx.$$

If we define $f(x) = x^{n-1}/(n-1)!$, we see that $f^e(s) = s^{-n}$. Define a distribution function $G(x) = 1$ if $x \geq k$ and 0 if $x < k$. The Laplace-Stieltjes transform of G is given by

$$\tilde{G}(s) = \int_0^\infty e^{-sx} dG(x) = e^{-sk}.$$

It is clear now that

$$s^{-n} e^{-sk} = f^e(s)\tilde{G}(s) = \int_0^\infty e^{-sx}\left[\int_0^x f(x-y)dG(y)\right]dx,$$

where the last equality is due to the fact that the function in the time domain of the product of two transforms is the convolution of the two functions yielding the product. The term in the last brackets reduces to

$$\int_0^x f(x-y)dG(y) = \begin{cases} f(x-k) & \text{if } x \geq k \\ 0 & \text{if } x < k. \end{cases}$$

So we define

$$h(x, n, k) = \begin{cases} \dfrac{(x-k)^{n-1}}{(n-1)!} & \text{if } x \geq k \\ 0 & \text{if } x < k. \end{cases}$$

Summarizing these results, we obtain

$$E[e^{-sY}] = \int_0^\infty e^{-sx}\left[\sum_{k=0}^n \binom{n}{k}(-1)^k h(x, n, k)\right]dx,$$

and the density of Y is given by the term inside of the last brackets. ∎

Little-oh Functions

In Examples 1.3.2 and 1.3.4, we have already used the idea of a little-oh function $o(h)$. Little-oh functions will be used frequently in the sequel. For any function f, we say that it is a little-oh function if f possesses the following property:

$$\lim_{h \to 0} \frac{f(h)}{h} = 0.$$

The little-oh function, abbreviated as $o(h)$, is used to describe a function that goes to zero faster than its argument. For instance, if $f(h) = h$ then f is not $o(h)$, whereas if $f(h) = h^2$ then f is $o(h)$. A constant multiple of a little-oh function is $o(h)$. The sum of a finite number of little-oh functions is $o(h)$. While the sum of a countable number of little-oh functions is not always $o(h)$, the situations considered in this text are such that they meet the sufficiency condition so that the resulting function is again $o(h)$.

Limits and Averages

Let $h(t)$ be a function defined in $[0, \infty)$. Consider a limit of the function h, $\lim_{t\to\infty} h(t)$. This limit is called a *pointwise* limit. It is the limiting value of the function h when the point t is "large." The limit may not exist. For example, if $h(t) = \sin(t)$ for $t \geq 0$, then the function oscillates between $+1$ and -1 without reaching a pointwise limit. Another quantity related to the pointwise limit is called the time average, or the Cesàro limit. It is defined as

$$\lim_{t\to\infty} \frac{\int_0^t h(\tau)d\tau}{t}.$$

The time average is simply the "area under the curve from 0 to t" divided by t. As an example, when $h(t) = \sin(t)$, the time average is 0 by symmetry. If the pointwise limit of a function exists, the time average of the function always exists; however, the converse is not always true as illustrated by the sine function example. When $\lim_{t\to\infty} h(t) = a < \infty$, then the final value property of Laplace transform introduced in Section 1.3 gives

$$\lim_{s\downarrow 0} s h^e(s) = a, \tag{1.4.8}$$

where $h^e(s)$ is the Laplace transform of h.

The discrete counterparts of the previous limits follow similarly. Let $\{a_n\}$ denote a sequence of real numbers. The limit of the sequence is defined as $\lim_{n\to\infty} a_n$. If $\lim_{n\to\infty} a_n = a < \infty$, then the time average of $\{a_n\}$ exists and

$$\lim_{n\to\infty} \frac{\sum_{k=1}^n a_k}{n} = a,$$

and moreover the final value property of generating function introduced in Section 1.2 gives

$$\lim_{z\uparrow 1}(1 - z)a^g(z) = a, \tag{1.4.9}$$

where $a^g(z)$ is the generating function of $\{a_n\}$.

Leibnitz's Rule

Leibnitz's rule is about differentiating (with respect to x) an integral that has the following form

$$\int_{a(x)}^{b(x)} f(x, y)\,dy.$$

Assuming the $a(x)$, $b(x)$, and $f(x, y)$ are all differentiable with respect to x, then the rule reads

$$\frac{d}{dx}\int_{a(x)}^{b(x)} f(x, y)\,dy = \int_{a(x)}^{b(x)} \frac{\partial f(x,y)}{\partial x}\,dy + f(x, b(x))\frac{db(x)}{dx} - f(x, a(x))\frac{da(x)}{dx}.$$

EXAMPLE
1.4.9

For a fixed $t > 0$, we consider the following function in x

$$U_t(x) = F(t) - \int_x^t [1 - F(y)]m(t - y)dy.$$

(In Section 3.3, we will see that this is the distribution function for the current life at time t of a renewal process for the case when $x < t$.) Define $u_t(x) = dU_t(x)/dx$. Let

$$a(x) = x, \qquad b(x) = t, \qquad \text{and} \qquad f(x, y) = [1 - F(y)]m(t - y).$$

Then it is easy to see that

$$\frac{da(x)}{dx} = 1, \qquad \frac{db(x)}{dx} = 0, \qquad \text{and} \qquad \frac{\partial f(x, y)}{\partial x} = 0.$$

Since $f(x, a(x)) = [1 - F(x)]m(t - x)$, an application of Leibnitz's rule gives

$$u_t(x) = [1 - F(x)]m(t - x). \quad \blacksquare$$

EXAMPLE
1.4.10

For a fixed $t > 0$, we consider the following function in x

$$L_t(x) = \int_{t-x}^t [F(x) - F(t - y)]m(y)dy,$$

where $dF(x)/dx = f(x)$ and $dM(y)/dy = m(y)$. (In Section 3.3, we will see that this is the distribution function for the total life at time t of a renewal process for the case when $x < t$.) Define $l_t(x) = dL_t(x)/dx$. Let

$$a(x) = t - x, \qquad b(x) = t, \qquad \text{and} \qquad g(x, y) = [F(x) - F(t - y)]m(y).$$

We see that

$$\frac{da(x)}{dx} = -1, \qquad \frac{db(x)}{dx} = 0, \qquad \text{and} \qquad \frac{\partial g(x, y)}{\partial x} = f(x)m(y).$$

Since $g(x, a(x)) = [F(x) - F(t - (t - x))]m(t - x) = [F(x) - F(x)]m(t - x) = 0$, we have

$$l_t(x) = \int_{t-x}^t \frac{\partial}{\partial x} g(x, y)dy = \int_{t-x}^t f(x)m(y)dy = f(x)[M(t) - M(t - x)]. \quad \blacksquare$$

Taylor Series Expansion

Let $f(x)$ be a continuous function possessing $n + 1$ derivatives for all x in the interval $[a, b]$. For any $0 \le h \le b - a$, we have, for some s between a and $a + h$,

$$f(a + h) = f(a) + \sum_{i=1}^{n} \frac{f^{(i)}(a)}{i!}h^i + \frac{f^{(n+1)}(s)}{(n+1)!}h^{n+1}, \qquad (1.4.10)$$

where $f^{(i)}(a)$ denotes the ith derivative of $f(x)$ with respect to x evaluated at a. The last term of Equation 1.4.10 is known as the Lagrange's form of the remainder.

Equation 1.4.10 is called the nth order Taylor series expansion of f around a. Another way to state Equation 1.4.10 is by letting $x = a + h$. This gives

$$f(x) = f(a) + \sum_{i=1}^{n} \frac{f^{(i)}(a)}{i!}(x - h)^i + \text{remainder.} \tag{1.4.11}$$

When f has derivatives of *all* orders, then we have

$$f(a + h) = \sum_{i=0}^{\infty} \frac{f^{(i)}(a)}{i!} h^i. \tag{1.4.12}$$

This representation is called the Taylor series expansion of f around a.

We now give a multidimensional generalization of Taylor series expansion. Let $x = \{x_1, \ldots, x_n\}$ and $a = \{a_1, \ldots, a_n\}$. Assume that the required conditions for a single variable Taylor series expansion to hold are all generalized to multiple dimension. Then we have

$$f(x) = f(a) + \sum_{i=1}^{n} f_i(a)(x_i - a_i) + \frac{1}{2!} \sum_{i,j=1}^{n} f_{ij}(a)(x_i - a_i)(x_j - a_j) + \cdots + \text{remainder,}$$

$$\tag{1.4.13}$$

where $f_i(a)$ is the partial derivative of f with respect to x_i evaluated at a and

$$f_{ij}(a) = \frac{\partial^2}{\partial x_i \partial x_j} f(x)\Big|_{x=a}.$$

Taylor series expansion is used extensively in Chapter 7.

Problems

1 Consider Example 1.2.7 again. (a) Use the "trick" introduced in Example 1.2.6 to find $E[X]$ and $Var[X]$. (b) Do the same by using differentiation; see Equation 1.2.2.

2 Consider random variable S defined by $S = X_1 + \cdots + X_N$, where $\{X_i\}$ are i.i.d. Bernoulli random variables with $P(X_i = 1) = p$ and $P(X_i = 0) = 1 - p \equiv q$, and N is a Poisson random variable with parameter λ and is independent of $\{X_i\}$. Find the probability generating function of S and identify the probability distribution of S from the form of the probability generating function.

3 The number of aircraft arriving at a maintenance depot for repairs in a month follows a Poisson distribution with parameter λ. The number of repair hours for repairing an arriving aircraft follows another independent Poisson distribution with parameter α. Let Y denote the number of repair hours needed in a month. Find the probability generating function of Y. Use the generating function to obtain $E[Y]$ and $Var[Y]$.

4 A will throw a six-sided fair die repeatedly until he obtains a 2. B will throw the same die repeatedly until she obtains a 2 or 3. We assume that successive throws are independent, and A and B are throwing the die independently of one another. Let X be the sum of the number of throws required by A and B. Find the probability mass function for X.

5 Three six-sided fair dice are thrown. Use the probability generating function approach to find the probability of obtaining a total of 9.

6 An urn contains a white and b black balls. Balls are randomly drawn from the urn one at a time. After a ball is drawn, it is returned to the urn if it is white; if it is black, it is replaced by a white ball from another urn. Let M_n denote the expected number of white balls in the urn after drawing from the urn n times. (a) By conditioning on the composition of the urn after the nth drawing, derive a difference equation for M_{n+1}, $n = 0, 1, 2, \ldots$. (b) Use the geometric transform approach to find a closed-form expression for M_n.

7 Let $X_1, \ldots, X_N$ be i.i.d. continuous and positive-valued random variables with a common density f_X. Let $f_X^e(s)$ denote the Laplace transform of f_X. Define $S = X_1 + \cdots + X_N$, where N is a positive, integer-valued random variable with probability generating function $\pi_N(z)$. We assume that N is independent of $\{X_i\}$. Let $f_S^e(s)$ denote the Laplace transform of random variable S. Find $f_S^e(s)$ in terms of $\pi_N(z)$ and $f_X^e(s)$.

8 Use the result obtained from Problem 7 to find $f_S^e(s)$ for the problem in which $X_i \sim exp(\lambda)$ and N follows the geometric distribution where $P\{N = n\} = (1 - \rho)\rho^{n-1}$, $n = 1, 2, \ldots$. What is the probability density of random variable S?

9 At time 0, Mr. Smith puts two steel-belted radial tires at the front of his car for extra safety and two nonradial tires at the back of his car to save some money. There is a spare in the trunk of his car. Let X_i be the lifetime of his ith steel-belted tire ($i = 1, 2$), Y_i the lifetime of his ith nonradial tire ($i = 1, 2$), and Z the lifetime of the spare. Assume the five random variables are mutually independent and $X_i \sim exp(\lambda)$ for $i = 1, 2$, $Y_i \sim exp(\mu)$ for $i = 1, 2$, and $Z \sim exp(\sigma)$. As soon as one of the four tires in use fails, Mr. Smith replaces the failed tire with the spare and continues to drive his car until one of the four fails again. When this occurs, he considers the tire usage completes a *cycle*. Let W denote the length of a cycle. Find the Laplace transform $f_W^e(s)$ of the density of W, $E[W]$, and $Var[W]$.

10 A personal computer has two important components: a memory board populated with memory chips and a hard disk drive for storing data and programs. Ms. Jones embarks on a mission requiring the use of such a personal computer. To prepare for possible equipment failures, she carries three extra memory boards and two extra disk drives as spares. When a part fails, it is immediately replaced by a spare—if one is still available. Assume that lifetimes of memory boards are i.i.d. with a common exponential density with rate μ_1 and lifetimes of hard drives are i.i.d. with a common exponential density with rate μ_2. Let Y denote the total length during which the computer is operational. Find $E[Y]$. Compute $E[Y]$ for the case in which the mean lifetime of memory board is 1,000 hours and the mean lifetime of hard drive is 500 hours.

11 A project consists of three independent tasks that can be done concurrently. For $i = 1, 2$, and 3, the time to complete task i follows an exponential distribution with parameter μ_i. Find the Laplace transform for the earliest project completion time X. What is $E[X]$?

12 Consider a single-server service facility with one server and an unlimited number of waiting spaces. Customers are served one at a time. Interarrival times of successive customers follow a common distribution G. Service times are i.i.d. random variables following an exponential density with parameter μ. Assume that service times and the arrival process are independent. If there are three customers already in the facility as the fourth one arrives, what is the probability that the facility will be empty when the fifth customer arrives?

13 Consider a service facility with three independent and identical servers. The facility has an unlimited number of waiting spaces. Customers are served one at a time. Interarrival times of successive customers follow a common distribution G. Service times are i.i.d. random variables following an exponential density with parameter μ. Let random variable X denote the number of customers already in the system at an arrival epoch and Y those already in the system at the next arrival epoch.

Find the conditional probabilities $P\{Y=j\mid X=i\}$ for the following four cases: **(a)** $i=2$ and $j=1$, **(b)** $i=4$ and $j=2$, **(c)** $i=1$ and $j=4$, **(d)** $i=5$ and $j=4$.

14 **Tail Probabilities of a Discrete Random Variable**   Let X be a nonnegative random variable with $P\{X=i\}=p_i$, $i=0,1,2,\dots$. Define the tail probabilities of X as $P\{X>i\}=q_i$, where $q_i=p_{i+1}+p_{i+2}+\cdots$, $i\geq 0$. Let $P(z)\equiv p^g(z)$ be the probability generating function of X and define the probability generating function of tail probabilities by

$$Q(z)\equiv q^g(z)=\sum_{n=0}^{\infty}q_n z^n.$$

(a) Show that $Q(z)=\dfrac{1-P(z)}{1-z}$ (Hint: Examine the individual terms of $(1-z)Q(z)$.) **(b)** Establish that $E[X]=Q(1)$ and $E[X(X-1)]=2Q^{(1)}(1)$. (In other words, $Q(1)=P^{(1)}(1)$ and $Q^{(1)}(1)=(1/2)P^{(2)}(1)$.)

15 Consider a clock powered by a single AAA battery. When a battery fails, the owner replaces the failed battery by one chosen randomly from two different brands. With probability p one of brand 1 is chosen, and with probability $1-p$ one of brand 2 is chosen. Let X_i denote the lifetime of a brand i battery. For $i=1$ and 2, we assume that $E[X_i]=\mu_i$ and $Var[X_i]=\sigma_i^2$. Let X denote the lifetime of the battery in use. Find $E[X]$ and $Var[X]$.

16 Find $E[X]$ and $Var[X]$ of the geometric random variable X, defined in Example 1.2.3, using the given probability generating function.

17 Find $E[X]$ and $Var[X]$ of the negative binomial random variable X, defined in Example 1.2.4, using the given probability generating function,

18 Consider the random sum $S_N=X_1+\cdots+X_N$ whose probability generating function is given by Equation 1.2.4. Establish the expressions 1.2.5 and 1.2.7 by the use of Equation 1.2.4.

19 Let $P_X(z)$ denote the probability generating function of a nonnegative, integer-valued random variable X. Use the Taylor series expansion to show that

$$P\{X=i\}=\frac{1}{i!}P_X^{(i)}(z)\Big|_{z=0}.$$

The preceding result provides another means of finding the probability mass distribution of X.

20 Let $f_X^e(s)$ denote the Laplace transform of a nonnegative continuous random variable X and $f_X^e(s)$ denote the nth derivative of $f_X^e(s)$ with respect to s. Show that the nth coefficient of the Taylor series expansion of $f_X^e(s)$ gives $(-1)^n E[X^n]/n!$.

21 For a given movie theater, customers arrive in groups of varying sizes. Let X_i denote the size of the ith group and assume that $\{X_i\}$ are i.i.d. random variables with a common probability mass function: $p_1 = 0.2$, $p_2 = 0.5$, $p_3 = 0.2$, and $p_4 = 0.1$, where $p_k = P\{X_i = k\}$. For a given show, let S denote the total number of customers present. For a given show, if we know that there are twenty-five groups present, compute the probability mass function of S and graph it. Also, find $E[S]$ and $Var[S]$.

22 Consider Problem 21 again. Assume now that the number of groups present for a given show is a random variable N. If N follows a Poisson distribution with mean twenty-five, compute the probability mass function of S, graph it, and find $E[S]$ and $Var[S]$.

23 Consider a first-come-first-served (FCFS) service system serving one customer at a time. In the service of a customer, the probability that there are k new arrivals is p_k. Let $P_X(z)$ represent the probability generating function associated with $\{p_k\}$, where X is the number of arrivals in a service interval. Assume that the arrival process and service times are independent, and the service times themselves are mutually independent. We call the first arrival to an empty system "the zeroth generation customer." All arrivals during the service of the zeroth generation customer are called "the first generation customers." Similarly, all arrivals during the services of the first generation customers are called "the second generation customers." For modeling purposes, we may view that all arrivals in the service of the ith customer of the nth generation as the $(n + 1)$st generation "offspring" produced by the customer. Let S_n denote the total size of the nth generation customers and let $S_{n,i}$ denote the size of the nth generation customers who are actually the offspring of the ith "parent" of the $(n - 1)$st generation customer. In other words, we have $S_n = S_{n,1} + \cdots + S_{n,S_{n-1}}$. Moreover $\{S_{n,1}, \ldots, S_{n,S_{n-1}}\}$ are i.i.d. random variables with a common distribution $\{p_k\}$, and S_{n-1} is independent of $\{S_{n,1}, \ldots, S_{n,S_{n-1}}\}$. Let $P_{S_n}(z)$ denote the probability generating function of S_n. **(a)** Find $P_{S_n}(z)$ as a function of $P_X(z)$ for $n \geq 0$. **(b)** For the case with $p_0 = 0.2$, $p_1 = 0.3$, $p_2 = 0.2$, $p_3 = 0.2$, and $p_4 = 0.1$, find the probability distribution of the random variable S_2. **(c)** What is the probability that the system will be empty again after serving two extra generations of customers?

24 The stochastic process $\{S_n, n \geq 0\}$ described in Problem 23 is called a *branching process*. Stated formally, we have $S_0 = 1$, and

$$S_n = \begin{cases} S_{n,1} + \cdots + S_{n,S_{n-1}} & \text{if } S_{n-1} > 0 \\ 0 & \text{otherwise,} \end{cases}$$

where $\{S_{n,1}, \ldots, S_{n,S_{n-1}}\}$ are i.i.d. random variables with a common distribution $\{p_k\}$, mean μ, and variance σ^2. Find $E[S_n]$ and $Var[S_n]$ in terms of μ, σ^2, and n. (Hint: Use Equation 1.2.7.)

25 Establish the generating function pair Z-7.

26 Establish the generating function pair Z-8.

27 Establish the Laplace transform pair L-5 using the gamma function given in Example 1.3.3.

28 Let $\{X_i\}$ be i.i.d. random variables with the common probability mass function

$$P\{X_1 = i\} = \begin{cases} p & \text{if } i = +1 \\ q & \text{if } i = -1, \end{cases}$$

where $p + q = 1$. Define $S_0 = 0$ and $S_n = X_1 + \cdots + X_n$. The stochastic process $S = \{S_n, n \geq 0\}$ is called a *simple random walk*. Let h_n denote the probability that $S_n = 1$ for the first time. We call the number of steps (the first n such that $S_n = 1$) needed to reach 1 the *first passage time*. (a) Give an argument that supports the following system of recursive equations: $h_0 = 0$, $h_1 = p$, and

$$h_n = \sum_{j=1}^{n-2} qh_j h_{n-j-1} \qquad n = 2, 3, \ldots.$$

(b) Compute the cumulative distributions $\left\{ \sum_{k=0}^{n} h_k \right\}$ for $p = 0.45, 0.50,$ and 0.55, over the range $n = 0{:}20$. (c) Speculate what will happen to these cumulative distributions when $n \to \infty$—specifically, whether they will approach to 1.

29 Consider Problem 28 again. Let $H(z)$ denote the probability generating function of $\{h_n\}$. (a) Show that $H(z)$ satisfies the quadratic equation in $H(z)$: $qz(H(z))^2 - H(z) + pz = 0$. (b) Solve the equation for $H(z)$. (c) Let N denote the first passage time. Note that $P(N < \infty) = \sum_{n=1}^{\infty} h_n = H(z)\big|_{z=1}$. Find $P(N < \infty)$ for the three cases: $p > q$, $p = q$, and $p < q$. (d) Compute the cumulative distribution $\left\{ \sum_{k=0}^{n} h_k \right\}$ for $p = 0.45$ over the range $n = 0{:}20$ by inverting $H(z)$ numerically using MATLAB M-function **invt_pgf**.

30 Use integration by parts to establish the Laplace transform pair L-9.

31 Assume that $\lim_{t \to \infty} f(t)$ exists. Use the result obtained in Problem 30 to establish the final value property

$$\lim_{t \to \infty} f(t) = \lim_{s \to 0} s f^e(s),$$

where $f^e(s)$ is the Laplace transform of f.

32 Let $\{a_n\}$ be a sequence of functions defined on the set of nonnegative integers. Assume that $\lim_{n \to \infty} a_n$ exists. Use the generating function pair Z-9 to establish the final value property

$$\lim_{n \to \infty} a_n = \lim_{z \to \infty} (1 - z) a^g(z),$$

where $a^g(z)$ is the generating function of $\{a_n\}$.

33 Let $g(t) = f(t)/t$. Show that

$$g^e(s) = \int_s^\infty f^e(x)dx,$$

where $f^e(\cdot)$ and $g^e(\cdot)$ are the Laplace transforms of $f(\cdot)$ and $g(\cdot)$, respectively.

34 Use the result obtained in Problem 33 to establish the asymptotic-rate theorem of the Laplace transform

$$\lim_{t\to\infty} \frac{f(t)}{t} = \lim_{s\to 0} s^2 f^e(s),$$

where we assume that the limits on both sides of the preceding equation exist.

35 Let X and Y be two independent exponential random variables with parameters μ_1 and μ_2, respectively. Define $Z = X + Y$. Use Laplace transforms to find the distribution function $F_Z(z)$.

36 Consider an experiment involving a fair coin and two biased dice, one with a probability of 1/3 of coming up with an ace and the other with a probability of 1/5 of coming up with an ace. We start with a coin flip to determine which die should be used to perform subsequent trials involving repeated tosses of the die chosen. Let T denote the number of tosses of the die needed to produce an ace for the first time. **(a)** Find the probability generating function of T, $P_T^g(z)$. **(b)** Find $E[T]$. **(c)** Find $Var[T]$.

37 Let p_n be the probability of an even number of successes in a sequence of n Bernoulli trials with constant probability p of a success at any trial. (For simplicity, you may let $q = 1 - p$.) **(a)** Derive a system of recursive equations for, p_n, $n = 0, 1, \ldots$. **(b)** Derive a closed form expression for p_n, $n = 0, 1, \ldots$.

Bibliographic Notes

A detailed account of geometric and Laplace transforms and their uses in applied probability can be found in Howard (1971). Many results relating to probability generating functions are based on the results given in Hunter (1983). The use of "illegitimate" random variables in Example 1.2.7 is given in Graham, et. al. (1989). A formal proof of the final value property for the generating function can be found as Theorem 2.3.3 in Hunter (1983). The term of competing exponential random variables in Example 1.3.2 was used in Howard (1971). In Section 1.5.2 of Taylor and Karlin (1994), the authors give a clear exposition about issues involving exponential distributions. The insurance example given as Example 1.3.4 is taken from Tijms (1986). A proof that $q(x) = P(V > x)$ was given there. Example 1.4.8 is from Resnick (1992). Problems 4 and 5 can be found in Hunter (1983) and Problem 6 is from Ross (1983). Problems 28 and 29 are based on Resnick (1992). A thorough discussion of characteristics function is given in Chapter 15 of Feller (1971).

References

Abate, J., and W. Whitt. 1992. Numerical Inversion of Probability Generating Functions. *Operations Research Letters* 12(4):245–51.

Abate, J., and W. Whitt. 1995. Numerical Inversion of Laplace Transforms of Probability Distributions. *ORSA Journal on Computing* 7(1):36–43.

Bartle, R. G. 1976. *The Element of Real Analysis*. 2nd ed. New York: John Wiley & Sons.

Feller, W. 1971. *An Introduction to Probability Theory and Its Applications, Vol. II.* New York: John Wiley & Sons.

Graham, R. L., D. E. Knuth, and O. Patashnik. 1989. *Concrete Mathematics.* Reading, MA: Addison-Wesley.

Howard, R. A. 1971. *Dynamic Probabilistic Systems, Vol. I.* New York: John Wiley & Sons.

Hunter, J. J. 1983. *Mathematical Techniques of Applied Probability, Vol. 1.* New York: Academic Press.

Resnick, S. I. 1992. *Adventures in Stochastic Processes.* Boston: Birkhäuser.

Ross, S. M. 1983. *Stochastic Processes.* New York: John Wiley & Sons.

Rudin, W. 1976. *Principles of Mathematical Analysis.* 3rd ed. New York: McGraw-Hill.

Taylor, H. M., and S. Karlin. 1994. *An Introduction to Stochastic Modeling.* Revised ed. New York: Academic Press.

Tijms, H. C. 1986. *Stochastic Modelling and Analysis: A Computational Approach.* New York: John Wiley & Sons.

Appendix

This appendix reproduces MATLAB runs cited in the examples given in Chapter 1. It also lists two M-files, **invt_pgf** and **invt_lap** for numerically inverting the probability generating function and Laplace transform, respectively.

Chapter 1: Section 2

Example 1.2.7 This example performs partial fraction expansion using MATLAB function **residue**. Before invoking the function, we first express the denominator as a polynomial by using MATLAB function **conv**.

```
a=[-1 2]; c=[-1 3]; a=conv(conv[a,c],c); b=[4];
[r,p,k]=residue(b,a)

r =
    4.0000
   -4.0000
   -4.0000
p =
    3.0000
    3.0000
    2.0000
k =
   []
```

The preceding shows that

$$\frac{4}{z-3} + \frac{-4}{(z-3)^2} + \frac{-4}{(z-2)}.$$

This is exactly what we found in the text. Note that in this example, we need to convert $(2-z)(3-z)^2$ into a polynomial using MATLAB function **conv** before we do the partial fraction expansion by **residue**.

Example 1.2.8   This example gives a numerical illustration for $p = 0.5$, the case of a fair coin. The result obtained from using MATLAB function **residue** is then compared with that of the closed form solution with $p = 0.5$.

```
p=0.5; q=0.5; b=[1]; a=[-p*q -q 1]; [r,p,k]=residue(b,a)

r =
    0.8944
   -0.8944
p =
   -3.2361
    1.2361
k =
    []
```

The preceding shows that, for $p = 0.5$,

$$\frac{1}{1 - qz - pqz^2} = \frac{0.8944}{z + 3.2361} + \frac{-0.8944}{z - 1.2361}.$$

The following run uses the results given in the text.

```
p=0.5; q=0.5; sx=sqrt(q^2 + 4*p*q); z1=(-q+sx)/(2*p*q);
z2=(-q-sx)/(2*p*q);
r1=-1/sx; r2=1/sx;
fprintf('  %8.4f  %8.4f  %8.4f  %8.4f ',z1,r1,z2,r2);

    1.2361   -0.8944   -3.2361    0.8944
```

Therefore, the two sets of results coincide.

Example 1.2.9   The MATLAB function **invt_pgf** is a general purpose function for inverting probability generating function. To achieve the error bound stated in the Abate and Whitt (1992) paper, it is best to invert the complementary distribution function instead. For the present example, since we already know the answers for the first few $\{p_n\}$, we invert the probability generating function of the mass function directly. The function **invt_pgf** calls the user-supplied MATLAB function **e129pgf**, which contains the probability generating function. If readers have a different probability generating function under a different function name, the three occurrences of **e129pgf** must be replaced accordingly.

```
function  invt_pgf
%
%   Inverting PGF to the time domain
%   Algorithm based on Abate and Whitt
%   "Numerical inversion of probability generation function"
%   OR Letter, Vol 12 (1992) 245-251
%
pmf=[];
for n=1:5
ga=8; r=10^(ga/(2*n)); r=1/r; h=pi/n; u=1/(2*n*r^n); sum=0;
for k=1:n-1
   z=r*exp(i*h*k); sum=sum+((-1)^k)*e129pgf(z);
end
pn=2*sum+e129pgf(r)+(-1)^n*e129pgf(-r); pn=u*pn; pmf=[pmf pn];
end
for n=1:5
   fprintf('  p(%2.0f) = %8.4f  \n',n,pmf(n));
end

function [y]=e129pgf(z)
%
%   pgf of Example 1.2.7
%
p=0.6; q=1-p; up=(p^2)*(z^2); down=(1-q*z-p*q*z^2); y=up/down;
y=real(y);
```

invt_pgf

```
   p( 1) =    0.0000
   p( 2) =    0.3600
   p( 3) =    0.1440
   p( 4) =    0.1440
   p( 5) =    0.0922
```

Chapter 1: Section 3

Example 1.3.5   The MATLAB function **invt_lap** is a general purpose function for inverting Laplace transforms. The function calls the user-supplied MATLAB function **e135**, which contains the Laplace transform. Again, if readers have a different Laplace transform, the corresponding MATLAB function must be created and referenced in **invt_lap** (again three times). In this illustration, the output are two vectors "tx" and "qx," the vector "tx" contains the x values, with $x = 0.5:0.5:20$, and the vector "qx" contains the respective $q(x)$ values. The initial value at $x = 0$ is given to the function with $q(0) = 0.8$. Finally, the pairs of $(x, q(x))$ can be plotted using MATLAB function **plot**.

```
function [tx,qx]=invt_lap
%
%   Numerical Inversion of Laplace transform
%   Ref: "Numerical Inversion of Laplace Transform of Probability
%        Distribution," by Abate and Whitt, ORSA J on Computing,
%        Vol 7(1995) 36-43.
%   For example 1.3.5
%
rho=0.8;
qx=[0.8]; tx=[0];
m=11; c=[]; ga=8; A=ga*log(10); mm=2^m;
```

```
for k=0:m
  d=binomial(m,k); c=[c d];
end
for t=0.5:0.5:20
tx=[tx t];
ntr=15; u=exp(A/2)/t; x=A/(2*t); h=pi/t; su=zeros(m+2);
sm=e135(x,0)/2;
for k=1:ntr
   y=k*h; sm=sm+((-1)^k)*e135(x,y);
end
su(1)=sm;
for k=1:12
   n=ntr+k; y=n*h; su(k+1)=su(k)+((-1)^n)*e135(x,y);
end
av1=0; av2=0;
for k=1:12
   av1=av1+c(k)*su(k);   av2=av2+c(k)*su(k+1);
end
f1=u*av1/mm; f2=u*av2/mm; qx=[qx f2];
end

function [w]=binomial(n,m)
%
if m <= n
x=1:n; x=cumprod(x); x=x(1,n);
if m==0
y=1;
else
y=1:m; y=cumprod(y); y=y(1,m);
end
if (n==m)
z=1;
else
z=1:n-m; z=cumprod(z); z=z(1,n-m);
end
w=x/(y*z);
else
w=0;
end

function [z]=e135(x,y)
%
%   The Laplace transform of  q(x)
%   fo  M/H2/1   queue
%
s=x+y*i; mu1=1.44721360; mu2=.55278640; p1=0.72360680; p2=1-p1;
lm=.8; ES=1;
up=s*(s+mu1+mu2-(p1*mu1)-(p2*mu2)); dm=(s+mu1)*(s+mu2); x=up/dm;
up=(lm*ES)-(lm/s)*x; dm=s-(lm*x); qs=up/dm; z=real(qs);

[tx,qx]=invt_lap; plot(tx,qx);
```

Poisson Processes

Tips for Chapter 2

■ The part relating to the derivation of Equation 2.3.2 is similar to the process of deriving Equation 2.1.1 except it is done in a somewhat more intricate manner. The part can be used as supplementary reading.

■ The section on nonhomogeneous Poisson processes Section 2.3 is relatively long by design, recognizing that most random phenomena in real-world applications exhibit time dependency.

■ If readers have more than a casual interest in queueing applications, Examples 2.3.8–2.3.11 can be quite helpful in understanding the important role played by time dependency in modeling arrival processes. You will also find Section 2.7 about PASTA indispensable.

■ Readers with an interest in stochastic modeling of repairable inventory systems will find Section 2.4, particularly Examples 2.4.1–2.4.2, and all parts relating to the Palm theorem relevant.

2.0 Overview

A Poisson process is a counting process in which the interarrival times of successive events are i.i.d. exponential random variables. If events occur in a purely random manner, then a Poisson process will generally be an adequate representation of the arrival process. In Section 2.1, we give a precise definition of a Poisson process; in Section 2.2, we survey many salient properties of the process. An important generalization to include time dependency is given in Section 2.3. The resulting process is called a nonhomogeneous Poisson process. Under the generalization, a variety of arrival processes occurring in practice can be approximated satisfactorily—for example, jobs arriving at a job shop, messages received by a transmitting station, emergency patients arriving to a hospital, and so on. In Section 2.3, we also describe means to simulate nonhomogeneous Poisson arrivals, illustrate the use of the process in building analytical models of service systems, and give numerical examples. Sections 2.4, 2.5, and 2.6 present other generalizations of a Poisson process. They include compound, filtered, two-dimensional, and marked Poisson processes. The last section introduces the idea of Poisson arrivals see time averages (PASTA). Based on the structure of a system, often the system's long-run fraction of time (the time average) at a given state can be computed. However, at an epoch a customer arrives to the system, the probability that the system is in a given state may not be the same as the corresponding time average. This is because the customer's arrival may interact with the state of the system. PASTA says that when the arrival process is Poisson, then the arriving customer indeed sees the time averages. Using the time averages, we can then compute many performance measures pertaining to the arriving customer, for example, the waiting time in a service system or the lead time for an order to arrive in an inventory system.

2.1 Introduction

Consider a counting process $N = \{N(t), t \geq 0\}$, where $N(t)$ denotes the number of arrivals in the interval $(0, t]$. In this chapter we will be looking at the counting process in which the interarrival time of successive arrivals follows an exponential distribution with parameter λ. Such a counting process is called a Poisson process. For a Poisson process, the rate of arrival remains constant. Poisson processes are frequently used stochastic processes that model arrival processes to service systems. In many cases, Poisson processes are adequate representations of real-world systems. When this is not so, a generalization of the process known as the nonhomogeneous Poisson process can extend its range of applicability. In such an extension, we allow the arrival rate function to be time dependent. We start with a set of assumptions underlying the Poisson process and derive the probability distribution for $N(t)$. The underlying assumptions are amenable to intuitive interpretations, and hence potential users of the process can decide the adequacy of using a Poisson process to approximate an arrival process based on behavior observations of the actual system. We will then study various salient properties of Poisson processes and several variants of the Poisson process.

> A counting process $N = \{N(t), t \geq 0\}$ is a Poisson process with rate $\lambda > 0$, if it possesses the following properties: (i) $N(0) = 0$, (ii) it satisfies the stationary and independent increment properties, (iii) $P\{N(h) = 1\} = \lambda h + o(h)$, and (iv) $P\{N(h) \geq 2\} = o(h)$.

The notions of stationary and independent properties were introduced in Section 1.1, and the little-oh function was defined in Example 1.3.3. Property (i) simply starts the clock of counting from time 0. The stationary-increment property says that if we change the time origin of the counting process it will not alter the probabilistic behaviors of the process; it is the *length* of the interval at stake that matters. The independent-increment property requires that the number of arrivals in two disjoint intervals be independent. As an example, arrivals to a system that experiences a "rush-hour" phenomenon during known time segments will not satisfy the independent-increment property. Property (iii) implies that in a small interval the probability of one arrival is approximately equal to the arrival rate λ times the length of the interval, and Property (iv) says that in a small interval the probability of more than one arrival is negligible. Hence if arrivals can occur in batches of more than one, the number of arrivals in an interval will not follow a Poisson process, whereas the number of *arrival occurrences* in the same interval could well follow a Poisson process. Using some of the preceding properties as yardsticks, Poisson processes can be excluded outright as candidates for modeling some arrival processes. Many times, however, we still find that Poisson processes are useful in serving as approximations for arrival processes occurring in practice.

Based on the aforementioned four properties, we will show in the following that

$$P\{N(t) = n\} = e^{-\lambda t} \frac{(\lambda t)^n}{n!} \qquad n = 0, 1, 2, \ldots. \qquad (2.1.1)$$

Let $P_n(t) = P\{N(t) = n\}$. We observe that

$$P_0(t + h) = P\{N(t + h) = 0\} = P\{N(t) = 0, N(t + h) - N(t) = 0\}$$

$$= P\{N(t) = 0\}P\{N(t + h) - N(t) = 0\} \qquad \text{by the independent-increment property}$$

$$= P_0(t)P\{N(h) = 0\} \qquad \text{by the stationary-increment property}$$

$$= P_0(t)[1 - \lambda h + o(h)] \qquad \text{by Property (iii).}$$

Subtracting $P_0(t)$ from both sides of the previous equation and dividing the resulting expression by h, we obtain

$$\frac{P_0(t + h) - P_0(t)}{h} = -\lambda P_0(t) + \frac{o(h)}{h}$$

Taking the limit as $h \to 0$ yields

$$P_0'(t) = -\lambda P_0(t). \qquad (2.1.2)$$

For $n \geq 1$, we condition on the number of arrivals by time t and write

$$P_n(t + h) = P\{N(t + h) = n\}$$

$$= \sum_{i=0}^{n} P_{n-i}(t)P_i(h) \quad \text{by independent- and stationary-increment properties}$$

$$= P_n(t)P_0(h) + P_{n-1}(t)P_1(h) + \sum_{i=2}^{n} P_{n-i}(t)P_i(h)$$

$$= P_n(t)[1 - \lambda h + o(h)] + P_{n-1}(t)[\lambda h + o(h)] + o(h)$$

$$= (1 - \lambda h)P_n(t) + \lambda h P_{n-1}(t) + o(h).$$

Subtracting $P_n(t)$ from both sides of the previous equation and dividing the resulting expression by h, we obtain

$$\frac{P_n(t+h) - P_n(t)}{h} = -\lambda P_n(t) + \lambda P_{n-1}(t) + \frac{o(h)}{h}.$$

Taking the limit as $h \to 0$, we obtain

$$P_n'(t) = -\lambda P_n(t) + \lambda P_{n-1}(t) \qquad n = 1, 2, \ldots. \qquad (2.1.3)$$

In fact, Equation 2.1.3 also works for $n = 0$ if we define $P_n(t) = 0$ when $n < 0$. Equations 2.1.2 and 2.1.3 form a system of linear differential-difference equations. We use transform methods to solve for $\{P_n(t), n \geq 0\}$. For a fixed t, we define the probability generating function for random variable $N(t)$:

$$P^g(z, t) = \sum_{n=0}^{\infty} z^n P_n(t), \qquad |z| < 1.$$

Let $P^{(1)}(z, t) = \partial P^g(z, t)/\partial t$. Differentiating the preceding equation with respect to t and using Equations 2.1.2 and 2.1.3, we obtain

$$P^{(1)}(z, t) = \sum_{n=0}^{\infty} z^n P_n'(t) = -\lambda \sum_{n=0}^{\infty} z^n P_n(t) + \lambda z \sum_{n=0}^{\infty} z^n P_n(t)$$

$$= -\lambda P^g(z, t) + \lambda z P^g(x, t) = \lambda(z-1) P^g(z, t).$$

For a fixed z, the previous equation is a first-order linear differential equation with a constant coefficient. The boundary condition of the equation is given by

$$P^g(z, 0) = \sum_{n=0}^{\infty} z^n P\{N(0) = n\} = z^0 P\{N(0) = 0\} = 1,$$

where we invoke Property (i) underlying the Poisson process to establish the last identity. For notational convenience, we let $f(t) = P^g(z, t)$ and $a = \lambda(z - 1)$. Then the preceding system reduces to: $f'(t) = af(t)$ and $f(0) = 1$. Using L-9, we find the Laplace transform of the differential equation: $sf^e(s) - f(0) = af^e(s)$. With the boundary condition we obtain $f^e(s) = 1/(s - a)$. Inverting the transform gives $f(t) = P^g(z, t) = e^{at} = e^{\lambda(z-1)t}$, $t \geq 0$. Consequently, we conclude that

$$P^g(z, t) = e^{-\lambda t} e^{\lambda z t} = e^{-\lambda t} \sum_{n=0}^{\infty} \frac{(\lambda t z)^n}{n!} = \sum_{n=0}^{\infty} e^{-\lambda t} \frac{(\lambda t)^n}{n!} z^n.$$

Extracting the coefficients of the previous probability generating function, we obtain the desired result, Equation 2.1.1.

The four properties stated earlier provide one characterization of a Poisson process $N = \{N(t), t \geq 0\}$ with rate λ. A simpler but equivalent characterization of the process N contains the following two properties: (i) each jump of N is of size 1 and (ii) for any $t, s \geq 0$,

$$E\big[N(t + s) - N(t)|N(u), u \leq t\big] = \lambda s. \qquad (2.1.4)$$

In the next section, we explore various important results pertaining to the Poisson process.

2.2   Properties of Poisson Processes

A Poisson process $N = \{N(t), t \geq 0\}$ with rate λ has many interesting and useful properties; we will now study some of the more important ones.

Interarrival Time Distribution

Let S_n denote the epoch of the nth arrival of N and define $S_0 = 0$. The interarrival time X_n is then given by $S_n - S_{n-1}$. In other words, we have

$$S_n = \sum_{k=1}^{n} X_k \qquad n = 1, 2, \ldots. \qquad (2.2.1)$$

We now show that $\{X_n\}$ are i.i.d. random variables, each following an exponential distribution with parameter λ. Since $\{X_1 > t\} \Leftrightarrow \{N(t) = 0\}$ and we know the probability of the right side from Equation 2.1.1, we have $P\{X_1 > t\} = P\{N(t) = 0\} = e^{-\lambda t}$. This shows that X_1 follows an exponential distribution with parameter λ. For any $s > 0$ and $t > 0$, we see that

$$P\{X_2 > t | X_1 = s\} = P\{0 \text{ events in } (s, \ s+t] | X_1 = s\}$$

$$= P\{0 \text{ events in } (s, \ s+t]\}$$

by the independent-increment property

$$= P\{0 \text{ events in } (0, \ t]\}$$

by the stationary-increment property

$$= e^{-\lambda t}.$$

We conclude that X_1 and X_2 are independent and X_2 also follows an exponential distribution with parameter λ. Repeating the previous argument yields the conclusion that the adjacent pairs of $\{X_n\}$ are i.i.d. random variables, each following an exponential distribution with parameter λ. Using a similar conditioning argument, we can also show that the nonadjacent pairs of $\{X_n\}$ are also independent.

A consequence of the preceding observation is that the arrival time of the nth event S_n follows an Erlang distribution with parameters (n, λ). This is due to Equation 2.2.1 and the result obtained in Example 1.3.2 (the sum of i.i.d. exponential random variables with parameter λ follows an Erlang distribution with parameters (n, λ)). Another instructive way to derive the result is by using the identity $\{N(t) \geq n\} \Leftrightarrow \{S_n \leq t\}$ and writing

$$P\{S_n \leq t\} = P\{N(t) \geq n\} = \sum_{k=n}^{\infty} e^{-\lambda t} \frac{(\lambda t)^k}{k!}.$$

An insightful interpretation can be given to the density of S_n. We see that

$$f_{S_n}(t)dt \approx P\{t < S_n \leq t+dt\} = P\{n-1 \text{ arrivals in } (0, \ t] \text{ and } 1 \text{ arrival in } (t, \ t+dt)\}$$

$$= P\{n-1 \text{ arrivals in } (0, \ t]\} \times P\{1 \text{ arrival in } (t, \ t+dt)\} = e^{-\lambda t} \frac{(\lambda t)^{n-1}}{(n-1)!} \lambda dt.$$

The preceding implies that for the nth arrival to occur around t, there must be $n - 1$ Poisson arrivals in $(0, t]$ and one arrival around time t. The probability for the latter to occur is approximately λdt.

EXAMPLE **2.2.1**

Generating Arrival Times of a Poisson Process by Computer Simulation Assume that the Poisson process has a rate λ. To generate arrival times $\{S_n\}$, we can successively generate the exponential interarrival times $\{X_n\}$ and use Equation 2.2.1 to find the arrival times. The generation of an exponential variate X with parameter λ can be done by the inverse transform method: (i) generate $U \sim U(0, 1)$ and (ii) let $X = F^{-1}(U)$, where $X \sim F$. Since $F_X(x) = 1 - e^{-\lambda x}$, $x \geq 0$, it is easy to verify that $X = -(1/\lambda)\log(1 - U)$. Furthermore, since $1 - U \sim U(0, 1)$, we also have $X = -(1/\lambda)\log(U)$. It is more efficient computationally to use the latter because it avoids subtraction. ∎

EXAMPLE **2.2.2**

Generating the Poisson Arrival Count by Computer Simulation For a fixed t, we may at times want to simulate the random variable $N(t)$, the number of arrivals by time t. By definition, we have $N(t) = \max\{n|S_n \leq t\}$, where S_n is defined by Equation 2.2.1. Using the result given in Example 2.2.1, we see that

$$N(t) = \max\left\{n \left| \sum_{k=1}^{n} -\frac{1}{\lambda}\log(U_k) \leq t \right.\right\} = \max\left\{n \left| \sum_{k=1}^{n} \log(U_k) \geq -\lambda t \right.\right\}$$

$$= \max\left\{n|\log(U_1 \cdots U_n) \geq -\lambda t\right\} = \max\left\{n|U_1 \cdots U_n \geq e^{-\lambda t}\right\}.$$

where U_k denotes the kth standard uniform variate generated. In simulation, we generate successive $\{U_k\}$ until the last condition is violated for the first time. Let U_N be the last uniform variate so obtained; the simulated $N(t)$ is then given by $N - 1$. ∎

Past Arrival Times Given $N(t)$

First we review a result from elementary probability theory. Let $Y_1, Y_2, \ldots, Y_n$ be i.i.d. random variables with common density f, and $Y_{(1)}, Y_{(2)}, \ldots, Y_{(n)}$ are the corresponding n order statistics ($Y_{(i)}$ is the ith smallest of $\{Y_i\}$). Then the joint density of $\{Y_{(i)}\}$ is given by

$$f_{Y_{(1)}, \ldots, Y_{(n)}}(y_1, \ldots, y_n) = n! \prod_{i=1}^{n} f(y_i) \qquad\qquad y_1 < \cdots < y_n. \qquad (2.2.2)$$

If f follows the uniform density over $(0, t)$, then the preceding reduces to

$$f_{Y_{(1)}, \ldots, Y_{(n)}}(y_1, \ldots, y_n) = \frac{n!}{t^n} \qquad\qquad y_1 < \cdots < y_n < t.$$

Given that $N(t) = n$, we next show that the n arrival times $S_1, \ldots, S_n$ have the same distribution as the order statistics corresponding to the n i.i.d. samples from $U(0, t)$. That is,

$$f_{S_1, \ldots, S_n | N(t)}(t_1, \ldots, t_n | n) = \frac{n!}{t^n} \qquad 0 < t_1 < \cdots < t_n < t. \qquad (2.2.3)$$

To establish Equation 2.2.3, we see that

$$P\{t_i \le S_i \le t_i + h_i, \ i = 1, \ldots, n | N(t) = n\}$$

$$= \frac{P\{\text{one event in } (t_i, \ t_i + h_i], \ 1 \le i \le n, \ \text{no events elsewhere in } (0, \ t]\}}{P\{N(t) = n\}}$$

$$= \frac{\lambda h_1 e^{-\lambda h_1} \cdots \lambda h_n e^{-\lambda h_n} \ e^{-\lambda(t - h_1 - \cdots - h_n)}}{\dfrac{e^{-\lambda t}(\lambda t)^n}{n!}} = \frac{n!}{t^n} h_1 \cdots h_n.$$

Dividing the last equality by $h_1, \ldots, h_n$ yields

$$\frac{P\{t_i \le S_i \le t_i + h_i, \ i = 1, \ldots, n | N(t) = n\}}{h_1 \cdots h_n} = \frac{n!}{t^n}.$$

Taking the limits as $h_i \to 0$ for all i, we obtain Equation 2.2.3.

EXAMPLE
2.2.3   A cable TV company collects $1/unit time from each subscriber. Subscribers sign up in accordance with a Poisson process with rate λ. What is the expected total revenue received in $(0, t]$?

Let S_i denote the arrival time of the ith customer. The revenue generated by this customer in $(0, t]$ is $t - S_i$. Adding the revenues generated by all arrivals in $(0, t]$, we obtain the expected total revenue received in $(0, t]$

$$E\left[\sum_{i=1}^{N(t)}(t - S_i)\right].$$

We first find the previous expectation by conditioning on $N(t)$

$$E\left[\sum_{i=1}^{N(t)}(t - S_i)\Big|N(t) = n\right] = E\left[\sum_{i=1}^{n}(t - S_i)\Big|N(t) = n\right] = nt - E\left[\sum_{i=1}^{n}S_i\Big|N(t) = n\right].$$

Let $U_1, \ldots, U_n$ be i.i.d. $\sim U(0, t)$. Applying Equation 2.2.3, the last term of the preceding expression can be stated as

$$E\left[\sum_{i=1}^{n}S_i\Big|N(t) = n\right] = E\left[\sum_{i=1}^{n}U_{(i)}\right] = E\left[\sum_{i=1}^{n}U_i\right] = n\left(\frac{t}{2}\right),$$

where the second equality holds because we are summing over all i in both cases. Thus, we can write

$$E\left[\sum_{i=1}^{N(t)}(t-S_i)\big|N(t)\right] = N(t)t - N(t)\frac{t}{2} = \frac{N(t)t}{2}.$$

Taking expectation with respect to $N(t)$, we find

$$E\left[\sum_{i=1}^{N(t)}(t-S_i)\right] = \frac{E[N(t)]t}{2} = \frac{1}{2}\lambda t^2. \quad \blacksquare$$

Decomposition of Poisson Process

Consider a Poisson process $N = \{N(t), t \geq 0\}$ with rate λ. We consider the case in which if an arrival occurs at time s, it is a type-1 arrival with probability $P(s)$ and a type 2 arrival with probability $1 - P(s)$. Hence the type of arrival depends probabilistically on the epoch of arrival. This way of decomposing an arrival stream includes the time-invariant decomposition (that is, an arrival is of type 1 with probability p and of type 2 with probability $1 - p$) as a special case. Let $N_i = \{N_i(t), t \geq 0\}$, $i = 1$ and 2, where $N_i(t)$ denotes the number of type-i arrivals in $(0, t]$. We now show that

$$P\{N_1(t) = n, N_2(t) = m\} = \left[e^{-\lambda pt}\frac{(\lambda pt)^n}{n!}\right]\left[e^{-\lambda qt}\frac{(\lambda qt)^m}{m!}\right], \qquad (2.2.4)$$

where

$$p = \frac{1}{t}\int_0^t P(s)ds \qquad \text{and} \qquad q = 1 - p.$$

We note that p is the time-average fraction of type-1 arrivals, and Equation 2.2.4 implies that $N_1(t)$ and $N_2(t)$ are two independent Poisson random variables with means λpt and λqt, respectively.

Applying the law of total probability, we find

$$P\{N_1(t) = n, N_2(t) = m\} = \sum_{k=0}^{\infty} P\{N_1(t) = n, N_2(t) = m|N(t) = k\}P\{N(t) = k\}$$

$$= P\{N_1(t) = n, N_2(t) = m|N(t) = n+m\}P\{N(t) = n+m\}. \qquad (2.2.5)$$

From Equation 2.2.3, we know that given $N(t) = n + m$, each arrival time is a sample from $U(0, t)$ and the arrival times are independent. We now consider one such arrival and let S be the arrival time. Then Equation 2.2.3 implies that $f_s(s|N(t) = 1) = 1/t$, $0 < s < t$. This in turn implies

$$P\{\text{the arrival is of type 1}| N(t) = 1\}$$

$$= \int_0^t P\{\text{the arrival is of type 1}|S = s, N(t) = 1\}f_s(s|N(t) = 1)ds = \int_0^t P(s)\frac{1}{t}ds = p.$$

Since the $(n + m)$ arrival times are mutually independent and each has a probability of p being of type 1, the first term of Equation 2.2.5 is given by the binomial probability with parameters $(n + m, p)$. Therefore Equation 2.2.5 becomes

$$P\{N_1(t) = n, \ N_2(t) = m\} = \frac{(n+m)!}{n!m!} p^n q^m e^{-\lambda t} \frac{(\lambda t)^{n+m}}{(n+m)!} = e^{-\lambda(p+q)t} \frac{(\lambda p t)^n}{n!} \frac{(\lambda q t)^m}{m!}.$$

Thus we have established the desired result, Equation 2.2.4. Generalization of Equation 2.2.4 to the case of n types of arrivals is straightforward. Let $P_i(s)$ denote the probability that an arrival at time s is of type i and define

$$p_i = \frac{1}{t}\int_0^t P_i(s)ds \qquad i = 1, \ldots, n,$$

where $p_1 + \cdots + p_n = 1$. Then $N_i(t)$ follows a Poisson distribution with parameter λp_i, and $\{N_i(t)\}$ are mutually independent.

EXAMPLE
2.2.4

Customers arriving at a department store form a Poisson process with rate λ. A customer will buy i items with probability p_i, where $i = 0, 1, \ldots, n$. Let $N_i(t)$ denote the number of customers, each of whom have purchased i items by time t. This is a special case in which the probability that an arrival of type i (namely, one who purchases i items during a visit) is a constant p_i. Based on the generalization of Equation 2.2.4 to the n arrival types, we conclude that $N_i(t)$ follows a Poisson distribution with parameter λp_i, and $\{N_i(t)\}$ are mutually independent. ■

EXAMPLE
2.2.5

Cars arrive at Galveston Beach during spring break. Assume that the interarrival time of cars follows an exponential distribution with parameter λ and the sojourn time of a car on the beach follows a probability distribution G. Also, we assume that the sojourn times are independent of each other and the arrival process, the beach can hold an unlimited number of cars, and at time 0 there are no cars on the beach. Let $N_1(t)$ denote the number of cars that have left the beach at time t and $N_2(t)$ the number of cars still at the beach at time t. What can be said about the two random variables $N_1(t)$ and $N_2(t)$?

Note that at the arrival time s of a car, if its sojourn time is less than $t - s$ the car will be classified as a type-1 arrival: thus the probability that an arriving car at time s will be a type-1 arrival with probability $G(t - s)$. We have $P(s) = G(t - s)$. Based on Equation 2.2.4, we conclude that $N_1(t)$ follows a Poisson distribution with parameter $\lambda t p$, where

$$p = \frac{1}{t}\int_0^t G(t-s)ds = \frac{1}{t}\int_0^t G(x)dx$$

and $N_2(t)$ follows a Poisson distribution with parameter $\lambda t q$, where $q = 1 - p$. Moreover, the two random variables are independent.

In the parlance of queueing theory, the preceding model is known as the $M/G/\infty$ queue. In the model, the arrival process (the beach goers) is Poisson with rate λ, the service-time (the sojourn time at the beach) distribution is G, and there are an infinite number of servers (the beach can accommodate everybody). In the model, we also assume that service times are mutually independent and independent of the arrival process. Let $X(t)$ denote the number of customers in the system at time t (the $N_2(t)$ equivalent). Then the previous exposition implies that $X(t)$ follows a Poisson distribution with parameter $\lambda \int_0^t \overline{G}(x)dx$ (where $\overline{G}(t) \equiv 1 - G(t)$). Let $E[Y]$ denote the mean of service time Y_i.

When $t \to \infty$, the parameter becomes $\lambda E[Y]$ and

$$h(x) = \lim_{t \to \infty} P\{X(t) = x\} = e^{-\lambda E[Y]} \frac{(\lambda E[Y])^x}{x!} \qquad x = 0, 1, 2, \dots. \quad (2.2.6)$$

This well-known result has been called a *Palm theorem*. It has its origin in the study of telephone traffic. ∎

2.3 Nonhomogeneous Poisson Processes

Nonhomogeneous Poisson processes are quite useful in modeling arrival processes to service systems because the processes enable model builders to capture the time-varying aspect of arrival streams in an intuitively appealing and operationally convenient manner. Consider a counting process $N = \{N(t), t \geq 0\}$.

The process N is called a *nonhomogeneous Poisson process* with *intensity function* $\{\lambda(t), t \geq 0\}$ if it possesses the following properties: (i) $N(0) = 0$, (ii) it satisfies the independent-increment property, (iii) $P\{N(t + h) - N(t) \geq 2\} = o(h)$, and (iv) $P\{N(t + h) - N(t) = 1\} = \lambda(t)h + o(h)$.

Comparing with a Poisson process, we note that in a nonhomogeneous Poisson process the stationary-increment assumption is absent, and the constant arrival rate of a Poisson process is replaced by a time-varying intensity function. Sometimes such a process is also called a *nonstationary* Poisson process. Define the *integrated intensity function*

$$m(t) = \int_0^t \lambda(u)du. \quad (2.3.1)$$

We first show that

$$P\{N(t + s) - N(t) = n\} = e^{-[m(t+s) - m(t)]} \frac{[m(t + s) - m(t)]^n}{n!} \qquad n = 0, 1, \dots. \quad (2.3.2)$$

The preceding equation indicates that the number of arrivals in interval $(t, t+s]$ follows a Poisson distribution with parameter $m(t+s) - m(t)$. The expected number of arrivals in the interval is then $m(t+s) - m(t)$. Setting $t = 0$, the number of arrivals in interval $(0, s]$ is Poisson with mean $m(s)$. Hence $E[N(s)] = m(s)$, and Equation 2.3.1 is sometimes called the *mean value function*. To derive Equation 2.3.2, we define $P_n(s) = P\{N(t+s) - N(t) = n\}$ and divide the interval $(0, t+s+h]$ into three segments shown in Figure 2.1.

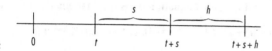

FIGURE
2.1   Three parts of an interval.

For $n = 0$, we use the arguments similar to the derivation of Equation 2.1.1 to write the following

$$P_0(s+h) = P\{N(t+s+h) - N(t) = 0\}$$

$$= P\{0 \text{ events in } (t, t+s], 0 \text{ events in } (t+s, t+s+h]\}$$

$$= P\{0 \text{ events in } (t, t+s]\} \times P\{0 \text{ events in } (t+s, t+s+h]\}$$

$$= P_0(s)[1 - \lambda(t+s)h] + o(h).$$

We see that the second equality of the previous formula holds because of the independent-increment property, and the third equality results from Properties (iii) and (iv). This leads to

$$P_0'(s) = -\lambda(t+s)P_0(s). \tag{2.3.3}$$

For $n \geq 1$, we use the law of total probability, Properties (i)–(iii), and the nature of little-oh functions to obtain

$$P_n(s+h) = P\{N(t+s+h) - N(t) = n\}$$

$$= P\{n-1 \text{ events in } (t, t+s]\} \times P\{1 \text{ event in } (t+s, t+s+h]\}$$

$$+ P\{n \text{ events in } (t, t+s]\} \times P\{0 \text{ event in } (t+s, t+s+h]\} + o(h)$$

$$= P_{n-1}(s)[\lambda(t+s)h + o(h)] + P_n(s)[1 - \lambda(t+s)h + o(h)] + o(h).$$

Similarly, this leads to

$$P_n'(s) = -\lambda(t+s)P_n(s) + \lambda(t+s)P_{n-1}(s) \qquad n \geq 1. \tag{2.3.4}$$

Again, we note that if we define $P_n(s) = 0$ for $n < 0$, then Equation 2.3.4 also covers the case when $n = 0$. We want to solve Equations 2.3.3 and 2.3.4 for $P_n(s)$. For a fixed s, we define the probability generating function

$$P^g(z, s) = \sum_{n=0}^{\infty} z^n P_n(s).$$

Multiplying the nth equation of Equations 2.3.3 and 2.3.4 by z^n and summing the resulting equations, we find

$$P^{(1)}(z, s) = -\lambda(t+s)P^g(z, s) + z\lambda(t+s)P^g(z, s)$$

$$= [-\lambda(t+s) + z\lambda(t+s)]P^g(z, s), \tag{2.3.5}$$

where $P^{(1)}(z, s) \equiv \partial P^g(z, s)/\partial s$. For simplicity, we let $a(s) = [-\lambda(t + s) + z\lambda(t + s)]$ and $f(s) = P^g(z, s)$. Then Equation 2.3.5 reduces to $f'(s) = a(s)f(s)$. Recall that

$$\frac{\partial}{\partial s} \log f(s) = a(s)$$

Integrating the preceding yields

$$\log f(s) = \int_0^s a(u)du.$$

Hence the solution of the differential equation in its original notations is given by

$$\log P^g(z, s) = \int_0^s [-\lambda(t + u) + z\lambda(t + u)]du = -[m(t + s) - m(t)] + z[m(t + s) - m(t)].$$

Exponentiating the preceding expression gives

$$P^g(z, s) = e^{-[m(t+s)-m(t)]+z[m(t+s)-m(t)]} = e^{-[m(t+s)-m(t)]}e^{z[m(t+s)-m(t)]}$$

$$= e^{-[m(t+s)-m(t)]} \sum_{n=0}^{\infty} \frac{[m(t+s) - m(t)]^n}{n!} z^n.$$

An examination of the coefficient associated with z^n establishes Equation 2.3.2. Observe from Equation 2.3.2 that when we lump all arrivals in the interval $(t, t + s]$ together, the number of arrivals in the interval follows a Poisson distribution even though the underlying process is actually nonstationary. Hence one should not jump too quickly to the conclusion that the arrival process is indeed Poisson by only looking at the aggregated counts.

Let $\{S_n\}$ be the arrival times of the nonhomogeneous Poisson process with the intensity function $\lambda(t)$ and the integrated intensity function from Equation 2.3.1. A consequence of Equation 2.3.2 is that the interarrival time $S_{n+1} - S_n$ conditional on $\{S_1, \ldots, S_n\}$ has the following complementary cumulative distribution

$$P\{S_{n+1} - S_n > t | S_1, \ldots, S_n\} = e^{-[m(S_n+t)-m(S_n)]} \qquad t > 0. \qquad (2.3.6)$$

The previous equation holds after an application of Equation 2.3.1 to the last event of the following equivalent events:

$$\{S_{n+1} - S_n > t | S_1, \ldots, S_n\} \Leftrightarrow \{\text{no arrivals in } (S_n, S_n + t] | S_1, \ldots, S_n\}$$

$$\Leftrightarrow \{\text{no arrivals in } (S_n, S_n + t] | S_n\}.$$

The density of the interarrival time $S_{n+1} - S_n$ conditioning on $\{S_1, \ldots, S_n\}$ is

$$f_{S_{n+1}-S_n}(t | S_1, \ldots, S_n) = f_{S_{n+1}-S_n}(t | S_n) = \lambda(S_n + t)e^{-[m(S_n+t)-m(S_n)]} \qquad t > 0. \qquad (2.3.7)$$

For the nonhomogeneous Poisson process, given $N(t) = n$, Equations 2.3.6 and 2.3.7 enable us to show that the arrival times are the ordered statistics of n i.i.d. samples from the following distribution

$$F(x) = \begin{cases} \dfrac{m(x)}{m(t)} & x \leq t \\ 1 & x > t. \end{cases} \qquad (2.3.8)$$

To establish the preceding result, we let $S_1 = s_1, \ldots, S_n = s_n$, and note that there are no arrivals in $(s_n, t]$. From Equations 2.3.6 and 2.3.7, we see that

$$f_{S_1 \cdots S_n}(s_1, \ldots, s_n \text{ and no arrivals in } (s_n, t])$$

$$= \lambda(s_1)e^{-[m(s_1)-m(0)]}\lambda(s_2)e^{-[m(s_2)-m(s_1)]} \cdots \lambda(s_n)e^{-[m(s_n)-m(s_{n-1})]}e^{-[m(t)-m(s_n)]}$$

$$= \prod_{i=1}^{n} \lambda(s_i)e^{-m(t)}.$$

The conditional joint density is then given by

$$f_{S_1 \cdots S_n}\left(s_1, \ldots, s_n \big| N(t) = n\right) = \frac{\displaystyle\prod_{i=1}^{n} \lambda(s_i)e^{-m(t)}}{\dfrac{(m(t))^n}{n!}e^{-m(t)}} = n!\prod_{i=1}^{n} \frac{\lambda(s_i)}{m(t)}.$$

From (2.2.2), we conclude that $S_1 = s_1, \ldots, S_n = s_n$ are ordered samples from a density $\lambda(x)/m(t)$ for $0 \le x \le t$. Upon integrating the last density, we obtain Equation 2.3.8.

EXAMPLE 2.3.1

Generating Arrival Times of a Nonhomogeneous Poisson Process by Computer Simulation—Method 1

Consider a nonhomogeneous Poisson process N with intensity function $\lambda(t)$. Assume that $\lambda(t) \le \lambda$ for all $t \ge 0$. We use the scheme developed in Example 2.2.1 to generate a Poisson arrival sequence $\{S_i\}$. The arrival at S_i will be counted as an arrival of N with probability $\lambda(S_i)/\lambda$. Such a process is called *thinning* in the sense that the newly created arrival stream has been "thinned out" from the original arrival stream. The resulting arrival times are the simulated observations from the nonhomogeneous Poisson process N. This is true because the thinned sequence inherits all the properties of a Poisson process except the stationary-increment assumption. To check whether Property (iv) of a nonhomogeneous Poisson process in the current situation is satisfied, we define $A = \{$one arrival of N in $(t, t + h]\}$ and $B = \{$a Poisson arrival in $(t, t + h]\}$. Then we see that

$$P\{A \cap B\} = P\{A|B\}P\{B\} = \frac{\lambda(t)}{\lambda}[\lambda h + o(h)] = \lambda(t)h + o(h).$$

So the results obtained from this sampling procedure will indeed produce simulated arrival times from the nonhomogeneous Poisson process. ∎

EXAMPLE 2.3.2

Generating Arrival Times of a Nonhomogeneous Poisson Process by Computer Simulation—Method 2

A second method to generate arrival times from a nonhomogeneous Poisson process is by exploiting Equation 2.3.7. Suppose we have already obtained n samples $S_1 = s_1, \ldots, S_n = s_n$, we are about to generate the next arrival time S_{n+1}.

Let $\tau = S_{n+1} - S_n$. Then Equation 2.3.7 implies that τ follows the following distribution

$$F_\tau(x|S_n = t) = 1 - e^{-[m(t+x)-m(t)]} \qquad x > 0.$$

So, to obtain the next sample $S_{n+1} = s_{n+1}$, we simply take a sample from the preceding distribution. The process is then repeated to generate additional observations. ∎

EXAMPLE **Generating Arrival Times of a Nonhomogeneous Poisson Process by Computer Simulation—Method 3**
2.3.3 If t is given and we want to generate arrivals for an eight-hour period, instead of using the earlier methods to generate arrival times and then discarding those exceeding t, we can use an approach based on Equation 2.3.8. First we use the procedure introduced in Example 2.2.2 to take a sample of $N(t)$. Given $N(t) = n$, we take n independent samples from the distribution defined by Equation 2.3.8. We sort the n observations in ascending order. These ordered samples are the arrival times of the nonhomogeneous Poisson process. ∎

If we have a set of arrival times $\{S_i\}$ of a nonhomogeneous Poisson process with intensity function $\lambda(t)$, we can convert them to a corresponding set of arrival times $\{Z_i\}$, where the latter are samples from a Poisson process with parameter $\lambda = 1$. This is done by setting

$$Z_i = m(S_i), \tag{2.3.9}$$

where $m(t)$ is the integrated intensity function defined by Equation 2.3.1. To prove the validity of the transformation defined by Equation 2.3.9, we use the second characterization of a Poisson process defined at the end of Section 2.1. The first property of the characterization is clearly satisfied by the transformation. To see whether the second property holds for the transformed data $\{Z_i\}$, we define the respective counting process $M = \{M(u), u \ge 0\}$, where $M(u)$ denotes the number of arrivals $\{Z_i\}$ in interval $(0, u]$, and need to show that

$$E[M(u+s) - M(u)|M(v), v \le u] = s \tag{2.3.10}$$

for all $u, s \ge 0$. In Figure 2.2, we depict the effect of the transformation.

With Figure 2.2 as a reference, we see that

$$E[M(u+s) - M(u)|M(v), v \le u]$$

$$= E\left[N(m^{-1}(u+s)) - N(m^{-1}(u)) \,\middle|\, N(m^{-1}(v)), \; m^{-1}(v) \le m^{-1}(u)\right]$$

$$= E\left[N(m^{-1}(u+s)) - N(m^{-1}(u))\right] \text{ by the independent-increment property of } N$$

$$= m(m^{-1}(u+s)) - m(m^{-1}(u)) = u + s - u = s.$$

In stating the first equality in the previous derivation, we assume that the integrated intensity function is such that the transformation is one to one.

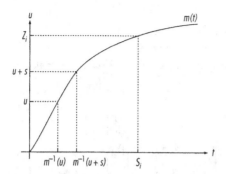

FIGURE
2.2   The transformation through the integrated rate function $m(t)$.

The transformation defined by Equation 2.3.9 is not only useful in *generating samples* from a nonhomogeneous Poisson process but also in *hypothesis testing*. Through the transformation, the arrival times of the nonhomogeneous Poisson process can be converted to a set of arrival times of a Poisson process with an arrival rate 1. Testing whether these arrival times are order statistics from a uniform distribution is a standard subject in statistical analysis.

EXAMPLE   **Generating Arrival Times of a Nonhomogeneous Poisson Process by Computer Simulation—Method 4**
2.3.4   This method is based on the transformation of Equation 2.3.9. First, we generate the arrival times from a Poisson process with rate 1. We then do the *inverse* transformation of Equation 2.3.9 to obtain the corresponding arrival times of the non-homogeneous Poisson process. ∎

In the next three examples we show how arrival processes occurring in practice are actually modeled by nonhomogeneous Poisson processes. In the first two examples, arrivals are calls for the use of a specialized computer for online analysis of electrocardiograms in the Veteran Administration Hospital located in Houston. In the third example, arrivals are emergency 911 calls to a precinct of the New York City Police Department. In these cases, the uses of nonhomogeneous Poisson process have been found adequate to model the arrival processes.

EXAMPLE   **Modeling Arrivals to a Computer System**   In a study of the use patterns of a Hewlett-
2.3.5   Packard computer designed for online analysis of electrocardiograms, arrival data have been analyzed for developing an input process for subsequent uses in computer simulation and analytical model building. In Figure 2.3, a plot of hourly arrival counts for a one-week period is shown. In the figure, we see that each regular work day in a week has its own pattern of variation with single peaks in

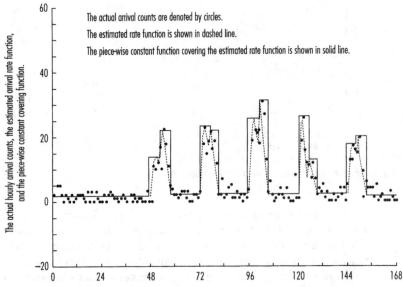

2.3   Plots of hourly arrival counts, the arrival rate function estimated from arrival times and the covering function.

Source: Reprinted by permission, Edward P. C. Kao and S. L. Chang, "Modeling Time-Dependent Arrivals to Service Systems: A Case in Using a Piecewise-Polynomial Rate Function in a Nonhomogeneous Poisson Process," *Management Science,* Vol. 34, No. 11, November 1988, © 1988, The Institute of Management Sciences (currently INFORMS), 290 Westminster Street, Providence, RI 02903.

the morning and afternoon, and variations over evening and weekends are random around constant means. The day-of-week effect on a regular work day is mostly influenced by scheduled surgeries occurring at Veterans Administration hospitals and clinics located in various parts of the country. Calls come in via local as well as long-distance phone lines. The peak load occurs on Wednesday, coinciding with the peak of surgical activities. The calls in evenings and weekends are mostly of emergency nature.

In modeling the arrival process, a piece-wise polynomial is used to approximate the intensity function $\lambda(t)$. For the morning and afternoon periods of each workday, low-degree polynomials are used to model the patterns of variation of the intensity function. For weekends and evenings, piece-wise constant functions are used instead. Breakpoints of these functions are connected to form a continuous intensity function over a one-week period. Using a maximum likelihood estimator based on actual arrival times of individual calls, the coefficients of the piece-wise polynomial are estimated. The estimated intensity function $\lambda(t)$ is shown in dashed line in Figure 2.3. The actual arrival times are then transformed through Equation 2.3.4 to a set of arrival times *presumably* of a Poisson process with rate 1—if the original arrival process can indeed be represented by a nonhomogeneous Poisson process. Statistical tests of this transformed data set show that there is no reason to reject the Poisson hypothesis. Hence it is concluded that the nonhomogeneous Poisson process representation of the arrival process is adequate. We remark that the piece-wise constant function covering the estimated intensity function shown in Figure 2.3 is for generating simulated arrival times using a method in the spirit of Example 2.3.1. The *integrated* covering function plays the role of $m(x)$ in Equation 2.3.9 for each time interval of interest (that is, interval between two successive breakpoints). ∎

EXAMPLE
2.3.6

Modeling Arrivals to a Computer System—Another Study   In Example 2.3.5, the arrival stream represents the aggregate of many substreams emanating from different sources. In a different study under the same backdrop, calls from different sources were analyzed separately. We now look at the calls generated during a workday from Houston Veterans Administration Hospital specifically and model the arrival process by a nonhomogeneous Poisson process. In this particular study only hourly arrival counts are available over a six-week period. A simple regression approach has been used to approximate the parameters of the intensity function $\lambda(t)$ whose form is assumed to be harmonic. Only those coefficients that are statistically significant are selected in the representation. This gives

$$\lambda(t) = 8.924 - 1.584\cos\frac{\pi t}{1.51} + 7.897\sin\frac{\pi t}{3.02} - 10.434\cos\frac{\pi t}{4.53} + 4.293\cos\frac{\pi t}{6.04}.$$

In Figure 2.4, we can compare the actual hourly arrival counts against the estimated average hourly arrival counts used in the preceding harmonic function. ∎

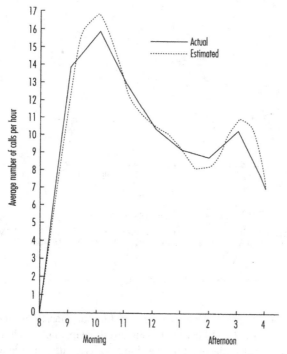

FIGURE
2.4   The intensity function for Houston calls, actual versus estimated.

Source: Reprinted by permission, Edward P. C. Kao, "Simulating a Computer System for Processing Electrocardiograms," *IEEE Transactions on Systems, Man, and Cybernetics*, SMC-10, No. 6, © 1980 IEEE.

EXAMPLE
2.3.7

Modeling Arrivals of 911 Calls to a Police Department   In an effort to validate a multiple-car dispatch queueing model of police patrol operations for the New York City Police Department, the data relating to 911 calls have been analyzed to establish a representation of the arrival process for subsequent studies of the resulting queueing models. In Figure 2.5, the total numbers of calls to Precinct 77 in the 96 fifteen-minute intervals are shown for the study period. Figure 2.5 shows a strong time-dependent arrival pattern for these emergency calls for police assistance. Using the interarrival time data, an estimated intensity function $\lambda(t)$ was constructed and tested for modeling adequacy. To supplement the hypothesis testing based on interarrival times, tests based on arrival counts during selected intervals were also used. The conclusion of the study showed that the nonhomogeneous Poisson process gave an adequate description of the overall pattern of calls for service. ▪

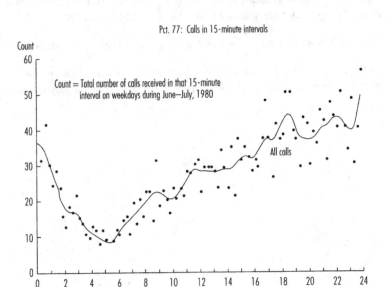

FIGURE
2.5

Emergency 911 calls to Precinct 77 of the New York City Police Department in a day.

Source: Reprinted by permission, Linda Green and Peter Kolesar, "Testing the Validity of a Queueing Model of Police Control," *Management Science*, Vol. 35, No. 2, February 1989, © 1989, The Institute of Management Sciences (currently INFORMS), 290 Westminster Street, Providence, RI 02903.

Nonhomogeneous Poisson processes are useful not merely in modeling arrival processes *empirically* as illustrated in the last three examples. They can be incorporated in analytical studies and used as input for computer simulation. The next example shows how this is done for a multiserver queueing system.

EXAMPLE
2.3.8

A Multiserver Queue with Nonhomogeneous Poisson Arrivals and Exponential Service Times—the $M(t)/M/s$ **Queue** Consider a service system with s identical servers. The arrivals to the system follow a nonhomogeneous Poisson process with intensity function $\lambda(t)$. The service time of each server follows an exponential distribution with parameter μ. When k ($\leq s$) servers are busy at the same time, we assume that the k service times are mutually independent and independent of the arrival process. Moreover, when k servers are busy simultaneously at any epoch, the time S for the first service completion to occur follows an exponential distribution with parameter $k\mu$—this is due to the memoryless property of the exponential distribution and the fact that S is the minimum of k exponential random variables each with parameter μ (see Example 1.3.2).

Let $X(t)$ denote the number of customers in the system at time t and assume $X(0) = 0$. Define $P_n(t) = P\{X(t) = n\}$. In the following, we derive a system of differential-difference equations characterizing $\{P_n(t)\}$. The general approach is similar to the derivations of Equations 2.1.1 and 2.3.2 for the probabilities $\{P_n(t)\}$ corresponding to the Poisson and nonhomogeneous Poisson processes, respectively. Recall that $N(t)$ denotes the number of arrivals in $(0, t]$. We define $N(t, t + h)$ as the number of arrivals in $(t, t + h]$. Hence $N(t, t + h) = N(t + h) - N(t)$, $P\{N(t, t + h) = 1\} = \lambda(t)h + o(h)$, and $P\{N(t, t + h) = 0\} = 1 - \lambda(t)h + o(h)$. Let $S(t, t + h|k)$ denote the number of service completions in $(t, t + h]$ given that at time t there are k busy servers. From Example 1.3.2, we recall that $P\{S(t, t + h|k) = 1\} = k\mu h + o(h)$ and $P\{S(t, t + h|k) = 0\} = 1 - k\mu h + o(h)$. The "headcount" balance equation is $X(t + h) = X(t) + N(t, t + h) - S(t, t + h|k)$. For $n = 0$, an application of the law of total probability gives

$$P_0(t+h) = P\{X(t+h) = 0\} = \sum_{k=0}^{\infty} P\{X(t) = k, X(t+h) = 0\}.$$

In Table 2.1, we give an enumeration of the scenarios that may lead to $X(t + h) = 0$ from various values of $X(t)$ and ignore some that are of $o(h)$ variety. By the independence assumptions relating to the arrival process and service times, we give the joint probabilities $P\{X(t) = k, X(t + h) = 0\}$ in the last column of the table for the cases with $X(t + h) = 0$. Using these results, we find

$$P_0(t+h) = P_0(t)[1 - \lambda(t)h + o(h)] + P_1(t)[\mu h + o(h)] + o(h)$$

$$= P_0(t)[1 - \lambda(t)h] + P_1(t)\mu h + o(h).$$

Subtracting both sides of the preceding expression by $P_0(t)$, dividing the resulting equation by h, and taking the limit as $h \to 0$ will yield

$$P_0'(t) = -\lambda(t)P_0(t) + \mu P_1(t). \qquad (2.3.11)$$

TABLE **2.1**
A Head Count Table for $X(t+h) = 0$

$X(t)$	$N(t, t + h)$	$S(t, t + h\|k)$	$X(t + h)$	$P\{X(t) = k, X(t + h) = 0\}$
0	1	—	1	
0	0	—	0	$P_0(t)[1 - \lambda(t)h + o(h)]$
1	0	1	0	$P_1(t)[1 - \lambda(t)h + o(h)][\mu h + o(h)]$
1	0	0	1	

For $1 \le n < s$, again we apply the law of total probability to write

$$P_n(t+h) = P\{X(t+h) = n\} = \sum_{k=0}^{\infty} P\{X(t) = k,\ X(t+h) = n\}.$$

In Table 2.2, we give an enumeration of the scenarios that may lead to $X(t+h) = n$ from various values of $X(t)$ and ignore some that are of $o(h)$ variety. Again, we use the various independence assumptions to compute the joint probabilities shown in the last column of Table 2.2. Using these results, we obtain

$$P_n(t+h) = P_{n-1}(t)[\lambda(t)h + o(h)] + P_n(t)[1 - \lambda(t)h - n\mu h + o(h)]$$
$$+ P_{n+1}(t)[(n+1)\mu h + o(h)] + o(h).$$

By the approach identical to the derivation of Equation 2.3.11, we find

$$P_n'(t) = \lambda(t)P_{n-1}(t) + (n+1)\mu P_{n+1}(t) - [\lambda(t) + n\mu]P_n(t) \qquad 1 \le n < s. \quad \text{(2.3.12)}$$

Similarly, we can derive the remaining equation

$$P_n'(t) = \lambda(t)P_{n-1}(t) + s\mu P_{n+1}(t) - (\lambda(t) + s\mu)P_n(t) \qquad n \ge s. \quad \text{(2.3.13)}$$

The system of differential-difference equations given by Equations 2.3.11–2.3.13 can be stated succinctly in matrix notation. For the case when $s = 2$, the system is displayed in Equation 2.3.14,

$$\begin{bmatrix} P_0'(t) \\ P_1'(t) \\ P_2'(t) \\ \cdot \\ \cdot \end{bmatrix} = \begin{bmatrix} -\lambda(t) & \mu & & & \\ \lambda(t) & -(\lambda(t)+\mu) & 2\mu & & \\ & \lambda(t) & -(\lambda(t)+2\mu) & 2\mu & \\ & & \lambda(t) & -(\lambda(t)+2\mu) & \cdot \\ & & & \cdot & \cdot \end{bmatrix} \begin{bmatrix} P_0(t) \\ P_1(t) \\ P_2(t) \\ \cdot \\ \cdot \end{bmatrix}. \quad \text{(2.3.14)}$$

In other words, Equation 2.3.14 can be stated simply as $P'(t) = Q(t)P(t)$, where $P'(t)$ and $P(t)$ are two column vectors and $Q(t)$ is a square matrix. In the next

TABLE 2.2	$X(t)$	$N(t, t+h)$	$S(t, t+h\|k)$	$X(t+h)$	$P\{X(t) = k,\ X(t+h) = n\}$
A Head Count Table for	$n-1$	1	1	$n-1$	
$X(t+h) = n$	$n-1$	1	0	n	$P_{n-1}(t)[\lambda(t)h + o(h)][1 - (n-1)\mu h + o(h)]$
	$n-1$	0	1	$n-2$	
	$n-1$	0	0	$n-1$	
	n	1	1	n	$P_n(t)[\lambda(t)h + o(h)][n\mu h + o(h)] = o(h)$
	n	1	0	$n+1$	
	n	0	1	$n-1$	
	n	0	0	n	$P_n(t)[1 - \lambda(t)h + o(h)][1 - n\mu h + o(h)]$
	$n+1$	1	1	$n+1$	
	$n+1$	1	0	$n+2$	
	$n+1$	0	1	n	$P_{n+1}(t)[1 - \lambda(t)h + o(h)][(n+1)\mu h + o(h)]$
	$n+1$	0	0	$n+1$	

example, we will show how Equation 2.3.14 is solved numerically for time-dependent system behaviors. ∎

EXAMPLE
2.3.9 **An M(t)/M/1/4 Queue**   Consider the computer system for processing electrocardiograms presented in Examples 2.3.5 and 2.3.6. We start with the case in which the arrival process follows the nonhomogeneous Poisson process with the intensity function $\lambda(t)$ shown in Example 2.3.6. Integrating $\lambda(t)$ over (0, 8) and dividing the result by 8 show that the mean hourly arrival rate is 10.76. Assume that the system has three waiting spaces and the exponential service time distribution has a rate of 7 per hour. An arriving person seeing all waiting spaces are occupied will leave and have no influence on the future of the system. Since the arrival rate exceeds the service rate, the system will be full from time to time. We are interested in obtaining the probability that an arrival will be lost as a function of time of day.

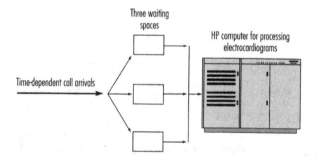

FIGURE
2.6   The computer system for processing electrocardiograms.

The service system shown in Figure 2.6 is known as the $M(t)/M/1/4$ queue. As in Example 2.3.8, we let $P_n(t)$ denote the probability that at time t there are n customers in the system. For example, when $n = 1$ then this means that one customer is in service at time t and no one is waiting. To be specific, we assume the system is empty at time 0 (8 A.M.) and therefore $P_0(0) = 1$ and $P_n(0) = 0$, for $n > 0$. The system of difference-differential equations for this queue is slightly different from Equation 2.3.14. The dimension of the $Q(t)$ matrix is finite and of size 5×5 and there is only one server so the service rates shown in the matrix must be modified. We present the $Q(t)$ matrix in the *transposed* form as follows

$$\left[P_0'(t),\ P_2'(t),\ \dots,\ P_4'(t) \right] =$$

$$\left[P_0(t),\ P_1(t),\ \dots,\ P_4(t) \right]
\begin{bmatrix}
-\lambda(t) & \lambda(t) & & & \\
\mu & -(\lambda(t)+\mu) & \lambda(t) & & \\
& \mu & -(\lambda(t)+\mu) & \lambda(t) & \\
& & \mu & -(\lambda(t)+\mu) & \lambda(t) \\
& & & \mu & -\mu
\end{bmatrix}.$$

We show Equation 2.3.14 in the transposed form because in Chapter 5 we will see many systems of difference-differential equations used that are similar in form to the one shown previously. The $Q(t)$ matrix displayed has suggestive features: the $(i, i+1)$st element gives the arrival rate at time t, the $(i, i-1)$st element gives the service rate when the number of customers in the system is n, and the diagonal elements have values such that the corresponding row sums are 0. The system can be solved readily using ordinary differential equations, for example, the MAT-LAB subroutines **ode23** or **ode45**. The details of the computer implementation are given in the Appendix.

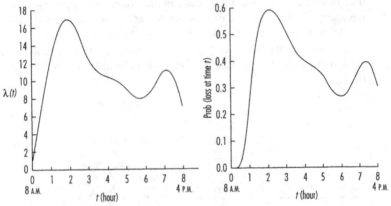

FIGURE
2.7   The intensity function versus the loss probabilities over time.

   In Figure 2.7, we plot the loss probabilities $\{P_4(t)\}$ as a function of t over the workday. The two curves show that the variation of loss probability over time resembles that of the intensity function $\lambda(t)$—as we would expect. ∎

EXAMPLE   **The Departure Process from an M/G/∞ Queue**   In an $M/G/\infty$ queue, the arrival process is
2.3.10   Poisson with rate λ, the service time distribution is given by G, and there is an infinite number of servers in the system (see Example 2.2.5). We assume that service times are mutually independent and independent of the arrival process. Let $M(t)$ denote the number of service completions in $(0, t]$. In this example we show that the departure process $M = \{M(t), t \geq 0\}$ from this queue is a nonhomogeneous Poisson process with intensity function $\lambda(t) = \lambda G(t)$. This result is useful in applications because the departure process forms the arrival process of the next service system if the two service systems are organized in tandem.

   Let $D(s, s+r)$ denote the number of service completions in the interval $(s, s+r]$ in $(0, t]$. If we can show that (i) $D(s, s+r)$ follows a Poisson distribution with mean $\lambda \int_{s}^{s+r} G(y)dy$ and (ii) the numbers of service completions in disjoint intervals are independent, then we are finished by definition of a nonhomogeneous Poisson process.

To establish the first assertion, the approach is identical to that employed in Example 2.2.5. An arrival at time y is called a type-1 arrival if its service completion occurs in $(s, s + r]$. To find the probability $P(y)$ that an arrival at time y is a type-1 arrival, there are three cases to consider. They are shown in Figure 2.8, in which we mark the first case where the arrival time $y \le s$. There, if the service time S is at least of length $s - y$ and no more than $s + r - y$, clearly the arrival is of type 1 and $P(y) = P\{s - y < S < s + r - y\} = G(s + r - y) - G(s - y)$. In the second case, the arrival time y is in $(s, s + r]$ and if the service time S is less than $s + r - y$, the arrival is of type 1. Then we have $P(y) = G(s + r - y)$ if $s < y \le s + r$. For the last case with $s + r < y \le t$, clearly $P(y) = 0$.

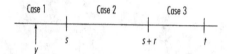

2.8   Three cases for finding $P(y)$.

Based on the result in Equation 2.2.4 about the decomposition of a Poisson process, we conclude that $D(s, s + r)$ follows a Poisson distribution with mean λpt, where $p = (1/t)\int_0^t P(y)dy$. The integral of the last expression is equal to

$$\int_0^t P(y)dy = \int_0^s [G(s + r - y) - G(s - y)]dy + \int_s^{s+r} G(s + r - y)dy + \int_{s+r}^t (0)dy$$

$$= \int_0^{s+r} G(s + r - y)dy - \int_0^s G(s - y)dy$$

$$= \int_0^{s+r} G(z)dz - \int_0^s G(z)dz.$$

(change of variable by letting $z = s + r - y$ and $z = s - y$, respectively)

This completes the first part of the proof. The second part of the proof again uses Equation 2.2.4. Consider any two disjoint intervals I_1 and I_2 in $(0, t]$. All arrivals in $(0, t]$ who depart in I_1 are called type-1 arrivals and all arrivals in $(0, t]$ who depart in I_2 are called type-2 arrivals. Since types 1 and 2 arrivals originated from two disjoint sets, each comprising a collection of the respective arrival epochs of the Poisson process, we conclude that the departures in the two disjoint intervals are independent. An alternate proof for the result given in this example that uses the concept of a filtered Poisson process (to be introduced in Section 2.5) will be left as an exercise. As $t \to \infty$, we have $G(t) \to 1$ and $\lambda(t) = \lambda$. In fact, as soon as $G(t) \to 1$, the departure process M turns into a Poisson process with rate λ. In the next example, we apply the result obtained in this example to a queueing network involving two queues in tandem. ∎

EXAMPLE
2.3.11 **A Small Cafeteria in a Big University**   Consider a small cafeteria in a big university that serves lunch between 11:30 A.M. and 1:00 P.M. The cafeteria is self-service, with food and drinks spread over many tables where people can pick their choices

without having to wait in line. During the ninety-minute period, arrivals to the cafeteria can be modeled by a Poisson process with parameter λ. Each customer's total sojourn time at these food and drink stalls follows a distribution function G. Once a customer finishes the selection, the customer joins a *single* waiting line leading to three cashiers. Any cashier who becomes free will serve whoever is at the head of the line. The service time at each cashier follows an exponential distribution with parameter μ. Service times at the cashiers are mutually independent and are independent of the arrival process leading to the cashiers. The complete system is depicted in Figure 2.9.

FIGURE
2.9   A small cafeteria in a big university.

The cafeteria operation can be modeled as two queues in tandem. The first queue is of $M/G/\infty$ type. For this queue, the term *queue* is really a misnomer since there is no waiting involved when a customer is in the first system. Hence the term queue is used more in the tradition of queueing theory. From the last example, we know that the departure process M is a nonhomogeneous Poisson process with intensity function $\lambda(t) = \lambda G(t)$. The departure process becomes the arrival process to the three-server exponential queue with an infinite number of waiting spaces. Since the departure process M is a nonhomogeneous Poisson process, the second system is an $M(t)/M/3$ queue discussed in Example 2.3.8. Using the queueing theory notations, this queueing network is displayed in Figure 2.10.

We consider a numerical example in which the arrival rate to the cafeteria is two customers per minute (hence $\lambda = 2$), the mean sojourn time (food selection time) is eight minutes per customer, the mean check-out time is one minute (hence $\mu = 1$). Let S denote the sojourn time and G its distribution. We assume that S follows a two-stage Erlang distribution with parameters $(n = 2, v)$. From Example 1.3.3, we find that $E[S] = n/v$ and $Var[S] = n/v^2$. Since $E[S] = 8$, this gives $v = 1/4$, $Var[S] = 32$, and coefficient of variation $\sqrt{32}/8 = 0.7071$. The

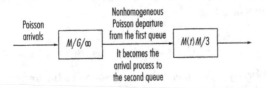

FIGURE
2.10   Two queues in tandem.

distribution function of an Erlang random variable is given in Section 2.2. Therefore, we have

$$G(t) = P\{S \le t\} = \sum_{k=2}^{\infty} e^{-0.25t} \frac{(0.25t)^k}{k!} = 1 - e^{-0.25t}[1 + 0.25t].$$

The arrival process is then a nonhomogeneous Poisson process with intensity
function

$$\lambda(t) = 2[1 - e^{-0.25t}(1 + 0.25t)].$$

In Figure 2.11, we plot the preceding intensity function and observe that the function reaches its asymptotic value of 2 after about thirty minutes. So after the cafeteria opens for about a half hour the arrival process to the cashiers is close to Poisson
with a constant rate of 2 per minute and the second queue turns into an $M/M/3$ queue.

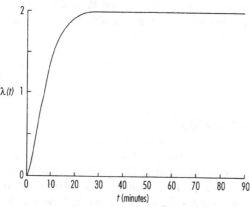

FIGURE 2.11 The intensity function of the arrival process to the second queue.

To find the performance characteristics of the second queue, we compute the
time-dependent probabilities $\{P_n(t)\}$ for $0 \le t \le 90$. The procedure is identical to
that introduced in Example 2.3.8 and demonstrated in Example 2.3.9 except that
the number of possible values $X(t)$ can range in principle from 0 to ∞ (of course
we need to make the changes of $Q(t)$ defined by Equation 2.3.14 so that it reflects
that $s = 3$). As a result, the $Q(t)$ matrix is infinite in dimension. However, in this
numerical example for large t, the traffic intensity of the second queue is
$\lambda/3\mu = 2/3 < 1$. The queue size will not grow without bound. In actual computation, the dimension of $Q(t)$, say N, can be set at a reasonably large value and
checked empirically to determine whether $P_N(t) < \varepsilon$ for a small tolerance ε. In the
MATLAB run shown in the Appendix, we let $N = 20$ and find that the value of ε
is less than 10^{-14} for all t of interest. Hence we know the truncation at $N = 20$ is
quite adequate. Let $L(t)$ denote the number of customers in the system at time t.
For each t, once $\{P_n(t)\}$ are found, we compute the first two moments of $L(t)$ using

$$E[L(t)] = \sum_{k=0}^{N} nP_n(t) \quad \text{and} \quad E[L^2(t)] = \sum_{k=0}^{N} n^2 P_n(t).$$

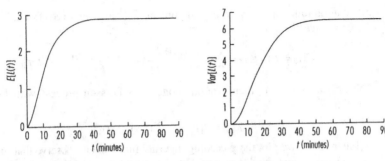

FIGURE
2.12 The means and variances of time-dependent queue length of the second queue.

In Figure 2.12, the means and variances of $L(t)$ are shown for $0 \le t \le 90$. Remember that the patterns of convergence of $E[L(t)]$ and $Var[L(t)]$ to their respective asymptotic values (those of the $M/M/3$ queue) resemble that of the intensity function displayed in Figure 2.11. ∎

2.4 Compound Poisson Processes

Consider a Poisson process $N = \{N(t), t \ge 0\}$ with rate λ. Let $\{Y_n\}$ be i.i.d. random variables with a common mean and variance $E[Y]$ and $Var[Y]$, respectively. The process N and the sequence $\{Y_n\}$ are assumed to be independent. While $\{Y_n\}$ can be continuous random variables, we only consider the case in which they are nonnegative and integer-valued and with a probability generating function $P_Y(z)$.

Define $X(t) = \sum_{n=1}^{N(t)} Y_n$, then the stochastic process $X = \{X(t), t \ge 0\}$ is called a *compound Poisson process.*

The compound Poisson process is used in many applications. For an example, in the maintenance depot of a fleet of company cars, assume that the interarrival time between successive cars for service follows an exponential distribution with parameter λ. The number of labor hours spent for repairing the nth car is Y_n. Then $X(t)$ represents the number of labor hours spent up to time t. Assuming that the independence assumptions are met, we see that $\{X(t), t \ge 0\}$ is a compound Poisson process. Consider another example involving an insurance company. We assume that the number of claims occurring in $(0, t]$ follows a Poisson process, the claim sizes $\{Y_n\}$ are i.i.d. random variables, and $\{Y_n\}$ are independent of the Poisson process. Then $X(t)$ represents the total amount of claims accumulated up to time t, and the resulting process $\{X(t), t \geq 0\}$ is a compound Poisson process. This last example has been studied at length in Example 1.3.4.

Following Equation 1.2.3, we know that the probability generating function of $X(t)$ is

$$H_t(z) = E[z^{X(t)}] = \pi_N(P_Y(z)) = e^{\lambda t[P_Y(z)-1]}, \tag{2.4.1}$$

where $\pi_N(z)$ has been found in Example 1.2.2. Let $d^n H_t(z)/dz^n = H_t^{(n)}(z)$ and $d^n P_Y(z)/dz^n = P_Y^{(n)}(z)$. After differentiation, we obtain

$$H_t^{(1)}(z) = e^{\lambda t[P_Y(z)-1]}\lambda t\left[P_Y^{(1)}(z)\right]$$

$$H_t^{(2)}(z) = e^{\lambda t[P_Y(z)-1]}\lambda t\left[P_Y^{(2)}(z)\right] + e^{\lambda t[P_Y(z)-1]}\left(\lambda t\left[P_Y^{(1)}(z)\right]\right)^2.$$

Hence

$$E[X(t)] = H_t^{(1)}(1) = \lambda t E[Y] \tag{2.4.2}$$

and $H_t^{(2)}(1) = \lambda t\left[P_Y^{(2)}(1)\right] + (\lambda t E[Y])^2$. From Equation 1.2.4, we write

$$E\left[(X(t))^2\right] = H_t^{(2)}(1) + H_t^{(1)}(1) = \lambda t\left[E[Y^2] - E[Y]\right] + (\lambda t E[Y])^2 + \lambda t E[Y]$$

$$= \lambda t E[Y^2] + (\lambda t E[Y])^2.$$

This gives

$$Var[X(t)] = \lambda t E[Y^2]. \tag{2.4.3}$$

For a compound Poisson process, the variance to mean ratio, defined as q, is given by

$$q = \frac{Var[X(t)]}{E[X(t)]} = \frac{E[Y^2]}{E[Y]}. \tag{2.4.4}$$

This ratio is a function only of Y. It is a convenient measure of variability of $X(t)$ relative to its mean (see the other measure—coefficient of variation). Since $Var[Y] = E[Y^2] - E^2[Y] \geq 0$, we have $E[Y^2] \geq E^2[Y] \geq E[Y]$ and $q \geq 1$. Thus, for any compound Poisson process, its variance exceeds or equals its mean. When the two are the same, the process reduces to a Poisson.

Since the Poisson process N has independent increments and $\{Y_n\}$ are i.i.d. random variables, it is clear that the compound Poisson process X has independent increments. To show that X has stationary increments, it is sufficient to establish that for any $t > s \geq 0$ the probability generating function depends only on the time increment $t - s$. In other words, we need to show that $E[z^{X(t)-X(s)}] = e^{\lambda(t-s)[P_Y(z)-1]}$. Conditioning on $N(t) - N(s)$, we write

$$E\left[z^{X(t)-X(s)}|N(t) - N(s) = n\right] = \left[P_Y(z)\right]^n.$$

Using the law of total probability, we obtain the desired result

$$E[z^{X(t)-X(s)}] = \sum_{n=0}^{\infty}[P_Y(z)]^n e^{-\lambda(t-s)}\frac{(\lambda(t-s))^n}{n!} = e^{-\lambda(t-s)}\sum_{n=0}^{\infty}\frac{(\lambda(t-s)P_Y(z))^n}{n!}$$

$$= e^{-\lambda(t-s)}e^{\lambda(t-s)P_Y(z)} = e^{\lambda(t-s)[P_Y(z)-1]}.$$

The next two examples present two compound Poisson processes used frequently in modeling repairable inventory systems by the U.S. Air Force where the interarrival time of successive demand occurrences follows a Poisson distribution, and the number of units (for example, aircraft engines) demanded at each demand occurrence is the random variable Y.

EXAMPLE
2.4.1

The Stuttering Poisson Process   For a compound Poisson process, consider the situation in which $\{Y_n\}$ follow a geometric distribution with

$$P\{Y = y\} = (1-\rho)\rho^{y-1} \qquad y = 1, 2, \dots$$

and probability generating function

$$P_Y(z) = \frac{(1-\rho)z}{1-\rho z}, \tag{2.4.5}$$

where $0 < \rho < 1$. It is easy to find that

$$E[Y] = \frac{1}{1-\rho}, \quad E[Y^2] = \frac{1+\rho}{(1-\rho)^2}, \quad Var[Y] = \frac{\rho}{(1-\rho)^2}, \quad \text{and} \quad q = \frac{1+\rho}{1-\rho}.$$

Moreover, from Equations 2.4.2 and 2.4.3 we obtain

$$E[X(t)] = \frac{\lambda t}{1-\rho} \qquad \text{and} \qquad Var[X(t)] = \frac{\lambda t(1+\rho)}{(1-\rho)^2}.$$

For the probability distribution of $X(t)$, we note that the probability generating function is given by Equation 2.4.1 with $P_Y(z)$ defined by Equation 2.4.5. To compute the probability distribution of $X(t)$ directly, we use the law of total probability to write

$$P\{X(t) = 0\} = e^{-\lambda t}$$

$$P\{X(t) = x\} = \sum_{k=1}^{x} e^{-\lambda t} \frac{(\lambda t)^k}{k!} P\{Y_1 + \dots + Y_k = x\} \qquad x = 1, 2, \dots. \tag{2.4.6}$$

Let $S = Y_1 + \dots + Y_k$. The probability generating function of S is

$$P_S(z) = \left[\frac{(1-\rho)z}{1-\rho z}\right]^k = (1-\rho)^k z^k \left[\frac{1}{(1-\rho z)^{(k-1)+1}}\right].$$

The term in the last pair of brackets has appeared in Example 1.2.4. The coefficient associated with z^n of the power series expansion of the term (that is, the discrete function in the time domain) is given in that example, namely,

$$\binom{n+k-1}{n} \rho^n (1-\rho)^k.$$

Combining with $(1-\rho)^k z^k$, we conclude that

$$P\{S = n + k\} = \binom{n+k-1}{n} \rho^n (1-\rho)^k.$$

Letting $x = n + k$, the preceding reads

$$P\{S = x\} = \binom{x-1}{x-k} \rho^{x-k}(1-\rho)^k.$$

Putting the expression in Equation 2.4.6, it becomes

$$P\{X(t) = x\} = \sum_{k=1}^{x} e^{-\lambda t} \frac{(\lambda t)^k}{k!} \binom{x-1}{x-k} \rho^{x-k}(1-\rho)^k \qquad x = 1, 2, \dots.$$

The previous probability distribution can be computed recursively. However, we omit the detail.

Why is the stochastic process "stuttering"? In the context of an inventory system, at a given demand occurrence after each demand, there is a probability ρ of a stutter (and hence generating another demand), and a probability $1 - \rho$ that the next demand occurrence will occur after an exponential sojourn. This compound Poisson process sometimes is called the geometric Poisson process. ∎

EXAMPLE 2.4.2   **The Logarithmic Poisson Process**   For a compound Poisson process, consider the situation in which $\{Y_n\}$ follow the following distribution

$$P\{Y = y\} = \frac{1}{y(\ln q)} \left(\frac{p}{q} \right)^y \qquad y = 1, 2, \dots,$$

where $q = p + 1 > 1$. The probability generating function of Y is given by

$$P_Y(z) = \frac{1}{\ln q} \sum_{y=1}^{\infty} \frac{1}{y} \left(\frac{pz}{q} \right)^y = \frac{1}{\ln q} \left[\ln \left(1 - \frac{pz}{q} \right)^{-1} \right] = \frac{1}{\ln q} \left[\ln \left(\frac{q}{q - pz} \right) \right]$$

$$= \frac{1}{\ln q} [\ln q - \ln(q - pz)] = 1 - \frac{1}{\ln q} [\ln(q - pz)].$$

In establishing the second equality, we have used the identity $\ln(1 - x)^{-1} = \sum_{n=1}^{\infty} x^n/n$, $|x| < 1$. As expected, we have $P_Y(1) = 1$. Define $\lambda t = k(\ln q)$. We see that

$$P_Y^{(1)} = \frac{p}{\ln q} \left[\frac{1}{q - pz} \right] \quad \text{and} \quad P_Y^{(2)} = \frac{p^2}{\ln q} \left[\frac{1}{q - pz} \right]^2.$$

Thus the first two moments of Y are given by

$$E[Y] = \frac{p}{\ln q} \quad \text{and} \quad E[Y^2] = P_Y^{(2)}(1) + P_Y^{(1)}(1) = \frac{pq}{\ln q}.$$

From Equations 2.4.2 and 2.4.3, we in turn find

$$E[X(t)] = \frac{\lambda t p}{\ln q} = kp \quad \text{and} \quad Var[X(t)] = \frac{\lambda t p q}{\ln q} = kpq.$$

Thus the variance-to-mean ratio of the compound Poisson process X is exactly equal to q. Using Equation 2.4.1, we find the probability generating function of $X(t)$

$$H_t(z) = e^{-\frac{\lambda t \ln[q-pz]}{\ln q}} = [e^{\ln[q-pz]}]^{-k} = \left[\frac{1/q}{1-(p/q)z}\right]^k.$$

Comparing the right side of the preceding equation with the result given in Example 1.2.4, we conclude that $X(t)$ follows a negative binomial distribution with

$$P\{X(t) = x\} = \binom{x+k-1}{x}\left(\frac{1}{q}\right)^k\left(\frac{p}{q}\right)^x \qquad x = 0, 1, \dots.$$

With $P\{X(t) = 0\} = 1/q^k$, the previous probability can be computed recursively by

$$P\{X(t) = x+1\} = \left(\frac{x+k}{x+1}\right)\left(\frac{p}{q}\right)P\{X(t) = x\} \qquad x = 0, 1, \dots.$$

The logarithmic Poisson distribution is convenient for modeling demand arrival processes of repairable items. When the variance-to-mean ratio of $X(t)$ is small, multiple demands by a single customer occur infrequently (as in the case of aircraft engines). For example, when $q = 2$, the distribution of Y is shown in Figure 2.13.

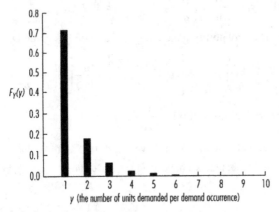

FIGURE
2.13 The distribution of random variable Y.

2.5 Filtered Poisson Processes

Consider a Poisson process $N = \{N(t), t \geq 0\}$ with rate λ. Let $\{Y_n\}$ be i.i.d. continuous random variables distributed as a random variable Y, where Y_n is associated with the nth arrival of N. The process N and the sequence $\{Y_n\}$ are assumed to be independent.

A stochastic process $X = \{X(t), t \geq 0\}$ is called a *filtered Poisson process* if we define

$$X(t) = \sum_{n=1}^{N(t)} \omega(t, S_n, Y_n),$$ (2.5.1)

where $\{S_n\}$ are the arrival times defined by Equation 2.2.1, and ω is called the *response* function, which is defined by the three arguments shown.

A frequently used response function assumes the following form:

$$\omega(t, \tau, y) = \omega_0(t - \tau, y).$$ (2.5.2)

For $t - \tau > 0$, the difference represents the length of the interval from the arrival time before t to the current time t. If we define $\omega_0(s, y) = 1$ if $s \geq 0$, and 0 otherwise, the resulting process X is a Poisson process; if we define $\omega_0(s, y) = y$, if $s \geq 0$, and 0 otherwise, the resulting process X is a compound Poisson process in which the compounding random variables $\{Y_n\}$ are continuous.

**EXAMPLE
2.5.1**

The M/G/∞ Queue Consider Example 2.2.5 again. In the $M/G/\infty$ queue, $X(t)$ represents the number of customers in the system at time t, and Y_n represents the service time of the nth customer. Let $s = t - \tau$, where τ is an arrival time before t, and let y be the length of the service time. We define the response function:

$$\omega_0(s, y) = \begin{cases} 1 & \text{if } 0 < s < y \\ 0 & \text{otherwise.} \end{cases}$$

Under the preceding response function, $X(t)$ gives the number of customers in the system at time t. As an example, in Figure 2.14, we see that $s_1 > Y_1$ and $s_4 < 0$. Hence the two arrivals at times S_1 and S_4 are not included in $X(t)$. For each of the other two arrivals, the response function registers a 1 so that $X(t) = 2$.

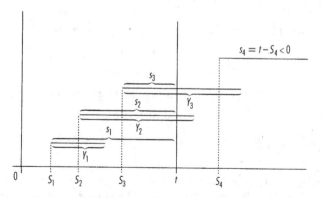

FIGURE
2.14 The $M/G/\infty$ queue from the vantage point of a filtered Poisson process.

If we define $X(t)$ as the total amount of remaining service time for the customers who are present at time t, the following response function will give $X(t)$ the corresponding quantity

$$\omega_0(s, y) = \begin{cases} y - s & \text{if } 0 < s < y \\ 0 & \text{otherwise.} \end{cases}$$

The correspondence is easily seen from Figure 2.14. On the other hand, if we define a response function $\omega_0(s, y) = s$ if $0 < s < y$, and 0 otherwise, then the resulting $X(t)$ represents the total amount of work being performed on those who are present at time t. ▪

EXAMPLE
2.5.2

The Number of Criminals Incarcerated at Time t Let $\{S_i\}$ denote times at which criminals are incarcerated and $\{Y_i\}$ lengths of sentence. All other assumptions underlying a filtered Poisson process stay the same. Using the first response function shown in Example 2.5.1, $X(t)$ represents the number of criminals incarcerated at time t. ▪

EXAMPLE
2.5.3

The Number of Claims on a Workers' Compensation Insurance Let $\{S_i\}$ denote times at which accidents occur at work and $\{Y_i\}$ the lengths during which a worker is disabled. All other assumptions underlying a filtered Poisson process stay the same. Using the first response function shown in Example 2.5.1, $X(t)$ represents the number of workers who at time t are on compensation insurance. ▪

EXAMPLE
2.5.4

The Cable TV Company Revisited Returning to Example 2.2.3, we define $X(t)$ as the total revenue received in $(0, t]$. With S_i defined as the arrival time of the ith customer, we have

$$X(t) = \sum_{i=1}^{N(t)} (t - S_i).$$

In modeling $\{X(t), t \geq 0\}$ as a filtered Poisson process, we define

$$\omega_0(s, y) = \begin{cases} s & \text{if } s \geq 0 \\ 0 & \text{otherwise.} \end{cases}$$

A glance at Figure 2.14 will convince us that the given response function produces the desired result. ▪

The principal result we give in this section is the probability generating function of $X(t)$ as defined by Equation 2.5.1. Let $\{U_i\}$ be i.i.d. random variables with

$U_i \sim U(0, t)$, and $U_{(i)}$ denote the ith order statistic of $\{U_i\}$ with its corresponding $Y_{(i)}$ (it is important to observe that $Y_{(i)}$ is *not* the ith order statistic of $\{Y_i\}$). Conditioning on $N(t)$ and using Equation 2.2.3, we write

$$E\left[z^{X(t)}|N(t) = n\right] = E\left[z^{\sum_{i=1}^{n}\omega(t,S_i,Y_i)}|N(t) = n\right] = E\left[z^{\sum_{i=1}^{n}\omega(t,U_{(i)},Y_{(i)})}|N(t) = n\right]$$

$$= E\left[z^{\sum_{i=1}^{n}\omega(t,U_i,Y_i)}\right] = \left(E[z^{\omega(t,U,Y)}]\right)^n = \left(\frac{1}{t}\int_0^t E[z^{\omega(t,\tau,Y)}]d\tau\right)^n.$$

Using the law of total probability, we obtain the probability generating function of $X(t)$

$$E[z^{X(t)}] = \sum_{n=0}^{\infty} e^{-\lambda t}\frac{(\lambda t)^n}{n!}\left(\frac{1}{t}\int_0^t E[z^{\omega(t,\tau,Y)}]d\tau\right)^n$$

$$= e^{-\lambda t}\sum_{n=0}^{\infty}\frac{1}{n!}\left(\lambda\int_0^t E[z^{\omega(t,\tau,Y)}]d\tau\right)^n = \exp\left[\lambda\int_0^t E[z^{\omega(t,\tau,Y)} - 1]d\tau\right]. \quad (2.5.3)$$

The mean and variance of $X(t)$ can be found from Equation 2.5.3. They are

$$E[X(t)] = \lambda\int_0^t E[\omega(t, \tau, Y)]d\tau \qquad (2.5.4)$$

$$Var[X(t)] = \lambda\int_0^t E[\omega^2(t, \tau, Y)]d\tau. \qquad (2.5.5)$$

EXAMPLE **The M/G/∞ Queue**   We now return to Example 2.5.1 with $X(t)$ denoting the sys-
2.5.5   tem size at time t. Based on Equation 2.5.3, we have the probability generating function

$$P_{X(t)}(z) = \exp\left[\lambda\int_0^t E[z^{\omega_0(t-\tau,Y)} - 1]d\tau\right].$$

By a change of variable with $s = t - \tau$, we rewrite the preceding as

$$P_{X(t)}(z) = \exp\left[\lambda\int_0^t E[z^{\omega_0(s,Y)} - 1]ds\right].$$

To find $E[z^{\omega_0(s,Y)} - 1]$, we condition on Y. If $Y > s$, then $\omega_0(s, Y) = 1$; if $Y \le s$, then $\omega_0(s, Y) = 0$. Hence $E[z^{\omega_0(s,Y)} - 1] = (z-1)P\{Y > s\} = (z-1)\overline{G}(s)$, and

$$P_{X(t)}(z) = e^{[\lambda\int_0^t \overline{G}(s)ds](z-1)}.$$

It is obvious now that $X(t)$ follows a Poisson distribution with parameter $\lambda\int_0^t \overline{G}(s)ds$—a result established in Example 2.2.5. ■

EXAMPLE   **The Cable TV Company Revisited**   In Example 2.2.3, we find $X(t)$, the expected total
2.5.6   revenue received in $(0, t]$. Using the response function defined in Example 2.5.4,
from Equation 2.5.4 we obtain

$$E[X(t)] = \lambda \int_0^t E[\omega_0(t - \tau, Y)]d\tau = \lambda \int_0^t E[\omega_0(s, Y)]ds = \lambda \int_0^t s\,ds = \frac{\lambda t^2}{2}.$$

This agrees with the result found in Example 2.2.3. ∎

The idea of a filtered Poisson process can be generalized to that of a filtered
nonhomogeneous Poisson process. We leave this as an exercise.

2.6 Two-Dimensional and Marked Poisson Processes

Consider a two-dimensional plane S. Let A be a subset of plane S. We envision
points being scattered randomly over S and let $N(A)$ denote the number of points
in A. The stochastic process $N(A)$ is called a *point process* in S. Let $|A|$ denote the
size of the set A. In this case $|A|$ represents the area of A.

Stochastic process $N = \{N(A), A \subset S\}$ is a *two-dimensional Poisson
process* if (i) $N(A)$ follows a Poisson distribution with mean $\lambda|A|$, and
(ii) the numbers of points occurring in disjoint subsets of S are mutually
independent.

The two-dimensional Poisson process can be generalized to a two-dimensional
nonhomogeneous Poisson process. Let $\lambda(x, y)$ be the intensity function of the
point process N. The process is a two-dimensional nonhomogeneous Poisson
process if (i) for each $A \subset S$, $N(A)$ follows a Poisson distribution with mean

$$\iint_A \lambda(x, y)dxdy, \tag{2.6.1}$$

and (ii) the numbers of points occurring in disjoint subsets of S are mutually
independent.

Consider a Poisson process $N = \{N(t), t \geq 0\}$ with parameter λ. Let $\{Y_n\}$ be
i.i.d. random variables with a common distribution function G, where Y_n is associated with the nth arrival of N. The process N and $\{Y_n\}$ are assumed to be independent. The stochastic process $(S, Y) = \{(S_n, Y_n), n \geq 0\}$ defined in the (t, y) plane
is called a *marked Poisson process*, where $\{S_n\}$ are the arrival times of N defined
by Equation 2.2.1. A sample path of the process is shown in Figure 2.15.

The marked Poisson process (S, Y) is a two-dimensional nonhomogeneous
Poisson process in the (t, y) plane with intensity function $\lambda(t, y) = \lambda g(y)$, where

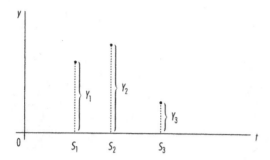

FIGURE
2.15   A marked Poisson process.

$g(y)$ is the density of Y. To see why the intensity function is given by $\lambda g(y)$, we observe that on the t-axis, the rate of occurrence of a Poisson arrival is λ and on the y-axis we have $P\{Y \in (y, y+h)\} = g(y)h + o(h)$. Along with the independence assumption, the two rates multiply. Since $N(A)$ denotes the number of points in a region A in the (t, y) plane, $N(A)$ follows the two-dimensional nonhomogeneous Poisson with mean

$$\iint_A \lambda g(y)\,dy\,dt. \tag{2.6.2}$$

To allow the possibility that Y may be a discrete random variable, we can write Equation 2.6.2 in a more general form using the Riemann-Stieltjes integral

$$\iint_A \lambda\,dG(y)\,dt. \tag{2.6.3}$$

As an example, when G assumes the form of a multinomial distribution with $P\{Y = i\} = p_i$, $i = 1, \ldots, n$ and $p_1 + \cdots + p_n = 1$, Equation 2.6.3 implies that the Poisson process is decomposed into n independent Poisson streams, each with respective density λp_i.

EXAMPLE
2.6.1   A tour boat that makes trips through the Houston ship channel leaves every T minutes from Pier 1. Tourists arrive at the pier according to a Poisson process with rate λ. A tourist waits until the next boat departure with probability $e^{-\mu s}$ if s is the length of the interval from the tourist's arrival epoch and the next boat departure epoch. In other words, with probability $1 - e^{-\mu s}$, the tourist may leave without taking the tour perhaps because of impatience. What is the expected number of tourists present at each boat departure epoch?

Consider a cycle starting right after the departure of a tour boat. Let $t = 0$ be the start of the cycle, $\{S_i\}$ the successive tourists' arrival times, and $\{Y_i\}$ their corresponding willingness-to-wait measures. Upon reflection, we see that $Y_i \sim exp(\mu)$ for each i. A sample path of $\{(S_i, Y_i)\}$ is shown in Figure 2.16. In the figure, we observe that the points staying outside of the triangle enclosed by the line $S + Y = T$ correspond to tourists making the forthcoming boat trip.

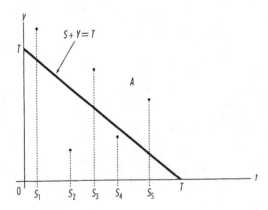

FIGURE
2.16   Tourist arrival times versus waiting times.

Let A be the region representing the first quadrant in the (t, y) plane excluding the triangle enclosed by the line $S + Y = T$. The expected number of tourists making the boat trip is given by $E[N(A)]$. Following Equation 2.6.2, we find

$$E[N(A)] = \int_0^T \int_{T-t}^\infty \lambda \mu e^{-\mu y} \, dy \, dt = \int_0^T \lambda e^{-\mu t} \, dt = \frac{\lambda}{\mu}[1 - e^{-\mu T}]. \quad \blacksquare$$

EXAMPLE   **The Departure Process from an M/G/∞ Queue Revisited**   From Example 2.3.10, we recall
2.6.2   that random variable $D(s, s + r)$ denotes the number of service completions in the interval. In the example, the key step is to show that $D(s, s + r)$ follows a Poisson process with mean $\lambda \int_s^{s+r} G(y) \, dy$. Using the concept of a marked Poisson process we now show that the proof is transparent. Let $\{Y_n\}$ be service times; we know that $Y_n \sim G$. In Figure 2.17, we plot a sample path in the spirit of a marked Poisson process. An arrival at time S_n has a service time of magnitude Y_n going

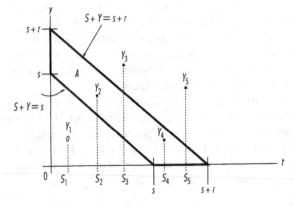

FIGURE
2.17   The sample path of a marked Poisson process for an M/G/∞ queue.

upward in the figure. If the point is below the line $S + Y = s$, then the corresponding arrival at time S_n has completed its service by time s; if the point is below the line $S + Y = s + r$, then the corresponding arrival at time S_n has completed its service by time $s + r$. Moreover, the points between the two parallel lines $S + Y = s$ and $S + Y = s + r$ correspond to those arrivals who have completed their services in the interval $(s, s + r]$. The region A of interest is then the area in the first quadrant enclosed by the two parallel lines shown in the figure. For the realization shown, we see that the two points associated with the arrivals at times S_2 and S_4, displayed as solid dots, are in the region A. Following Equation 2.6.3, we conclude that $N(A)$ follows a Poisson distribution with its mean given by

$$\int_0^{s+r} \int_{s-t}^{s+r-t} \lambda dG(y) dt = \int_0^{s+r} \lambda[G(s+r-t) - G(s-t)]dt$$

$$= \int_0^{s+r} \lambda[G(y) - G(y-r)]dy \quad \text{let } y = s + r - t$$

$$= \int_0^{s+r} \lambda G(y) dy - \int_{-r}^{s} \lambda G(z) dz \quad \text{let } z = y - r$$

$$= \int_0^{s+r} \lambda G(y) dy - \int_0^{s} \lambda G(z) dz \quad \text{since } G(z) = 0 \text{ for } z < 0$$

$$= \lambda \int_s^{s+r} G(y) dy.$$

Hence we establish the desired result. ∎

2.7 Poisson Arrivals See Time Averages (PASTA)

In many applications—particularly in inventory management and queueing—when an arrival to a system occurs, it is important to know probabilistically the state of the system so that the "fate" of the arriving customer can be assessed accordingly. For example, if at an arrival epoch we know the number of customers already in the system (or pieces of equipment already waiting to be repaired), then we can compute the waiting time distribution of the arriving customer (or the repair time for the arriving piece of equipment). If the fraction of arrivals who see the system in a given state upon arrival is equal to the fraction of time the system is in the state, it is simple because the latter generally can be found from knowing the dynamics of the system and without having to worry about the interacting relationship between the arrival process and the system itself. In the next example we demonstrate that the two fractions may not be the same.

EXAMPLE Consider a service system that serves one customer at a time. A customer arrives
2.7.1   every hour on the hour and the service time is exactly thirty minutes. Thus the fraction of time the system is idle is $1/2$. On the other hand, at an arrival epoch with probability 1, the arrival will see an empty system. The two fractions are not the same. ∎

In this section, we will see that when the arrival process is Poisson, the two fractions assume the same value for large t. This important property is known as *Poisson arrivals see time average (PASTA)*. This property occupies a prominent position in stochastic modeling. We now discuss various issues relating to PASTA in some detail.

Let $X = \{X(t), t \geq 0\}$ be a stochastic process with a discrete state space S. Let B be a proper subset of S. Define an indicator random variable

$$U(t) = \begin{cases} 1 & \text{if } X(t) \in B \\ 0 & \text{otherwise,} \end{cases}$$

where we assume that $U(t)$ is defined as a left-continuous function. Define random variable

$$V(t) = \frac{\int_0^t U(s)ds}{t}, \tag{2.7.1}$$

where $V(t)$ is the fraction of time the process X is in the set B in $[0, t]$. As an example, if $X(s)$ represents the number of customers in a service facility at time $0 < s < t$ and B denotes the event that the facility is empty, then random variable $V(t)$ is the fraction of time in the interval $[0, t]$ that the facility is idle. For the sample path shown in Figure 2.18, we project the line segments for which $X(t) \in B$ down to the s-axis. We see that the integral of $U(s)$ over $[0, t]$ is just the sum of the lengths of the darkened line segments. In Figure 2.19, the sample path of U is explicitly displayed, and a and b are the lengths of interval during which $U(s) = 1$. To highlight the left-continuity of U, in the figure the values of U at points of discontinuity are marked by the solid dots. Let $N = \{N(t), t \geq 0\}$ be a Poisson process with rate λ denoting the arrival process to the system whose status is being tracked by the stochastic process X. Typically, the arrival process N interacts with the process X, so the two processes are dependent. For example, at an arrival epoch, the random variable $X(t)$ is immediately incremented by one.

Define random variable

$$Y(t) = \int_0^t U(s)dN(s). \tag{2.7.2}$$

We see that $Y(t)$ is the number of arrivals in $(0, t]$ who find X in B. The Riemann-Stieltjes integral of Equation 2.7.2 is not the type described within the context of

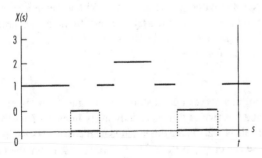

2.18 A sample path of $X(t)$.

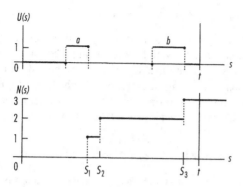

FIGURE
2.19   Sample paths of U and N.

Equations 1.4.1 and 1.4.2 in that both are step functions. However, we can circumvent the difficulty by splitting the interval $[0, t]$ at the arrival epochs. The definition of left continuity of U makes the evaluation of the integral easier. The left-continuity assumption is reasonable in that when an arrival occurs one sees the state of the system *immediately before* one's appearance. For the sample path shown in Figure 2.19, we have three arrivals in $[0, t]$. For the arrivals at S_1 and S_3, each sees that the facility is idle at the respective arrival epoch and hence the function U assumes a value of 1 at each of the two jump points of the function N. This implies that $N(t) = 3$ and $Y(t) = 2$. Define random variable

$$Z(t) = \frac{Y(t)}{N(t)}.$$

We see that $Z(t)$ is the fraction of arrivals in $[0, t]$ who find X in B. The ratio is said to be a *customer average* in that the denominator is the customer count (as opposed to a *time average* in which the denominator is the length of an interval). Since $Z(t) = \frac{2}{3}$, the fraction of time the system is idle as seen by an arrival is two thirds. The fraction of time the system is idle is given $V(t) = (a + b)/t$, where a and b are the lengths of the line segments shown in Figure 2.19. It is obvious that $V(t) < Z(t)$.

For PASTA to hold, we need the *lack of anticipation assumption (LAA)*: for each $t \geq 0$, the arrival process $\{N(t + u) - N(t), u \geq 0\}$ is independent of $\{X(s), 0 \leq s \leq t\}$ and $\{N(s), 0 \leq s \leq t\}$. Since the Poisson arrival process does not interfere with the state of the system before each arrival, it satisfies LAA. Under LAA, it can be shown that $Z(t) \rightarrow V(t)$ with probability 1 as $t \rightarrow \infty$. Thus, the fraction of arrivals that sees the process in some state is equal to the fraction of the time the process is in that state. Under LAA, it also can be shown that

$$E[Y(t)] = \lambda t E[V(t)] = \lambda \left\{ \int_0^t U(s)ds \right\}. \qquad (2.7.3)$$

The preceding implies that the expected number of arrivals who *see* the system in the set B in the interval $[0, t]$ is equal to the arrival rate multiplied by the length of time the system has been in B in $[0, t]$.

EXAMPLE **The M/G/c Queue** In an *M/G/c* queue, arrivals to the system follow a Poisson
2.7.2   process, service times are i.i.d. random variables following a common distribution *G*, and there are *c* servers. Service times are assumed to be independent of
the arrival process. Let π_n denote the long-run fraction of arrivals that find *n* customers in the system and p_n the long-run fraction of time the system has *n* customers. Following PASTA, we have $p_n = \pi_n$ for $n = 0, 1, \ldots$. ∎

EXAMPLE **The GI/M/c Queue** In a *GI/M/c* queue, there are *c* servers. Each customer is served
2.7.3   by one server. Service times are i.i.d. random variables and follow an exponential
distribution with mean $1/\mu$. The times between successive arrivals are i.i.d. random variables and follow a common distribution function *G*. Interarrival times
are assumed to be independent of service times. Define π_n and p_n exactly as in
Example 2.7.2. Let $\mu_n = \mu \min(c, n)$. Hence μ_n is the service rate when there are
n customers in the system. We now present the following well-known result from
queueing theory

$$\mu_n p_n = \lambda \pi_{n-1} \qquad n = 1, 2, \ldots \qquad (2.7.4)$$

In establishing the previous identity, we look at the two transitions shown in
Figure 2.20. The transition going from left to right corresponds to an arrival causing a change of state of the system from $n - 1$ to n. The long-run fraction of time
an arrival sees that the system already has $n - 1$ customers is given by π_{n-1}. The
rate of such an arrival is λ. Therefore the transition rate of moving across the
demarcation line to its right side is $\lambda \pi_{n-1}$. To give an interpretation to the term on
the left of Equation 2.7.4, we consider service completions as "arrivals." When the
system has *n* customers, the "arrival" process $N = \{N(t), t \geq 0\}$ is Poisson with rate
μ_n. Let $Y(t)$ denote the number of service completions in $(0, t)$ who see that the system has *n* customers and that it is about to make a change to having $n - 1$ customers because of a service completion. Then LAA implies that $Y(t)/N(t) \to p_n$ for
large *t*, where p_n is the long-run fraction of time the system has *n* customers. Moreover, $Y(t)/t$ represents the rate of making such a transition. Hence we conclude that

$$\frac{Y(t)}{t} = \frac{Y(t)}{N(t)} \frac{N(t)}{t} \to \mu_n p_n.$$

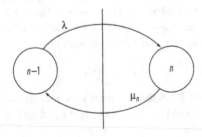

FIGURE
2.20   A transition diagram for the *GI/M/c* queue.

When the system has been running for a long time, the two rates should be the same. This establishes the desired result. ▪

One question of interest is whether there are instances in which non-Poisson arrival see time averages. The answer is yes. One such example is the case of an $M/M/1$ queue with feedback. There, each departure from the queue will join the queue again with a positive probability and the composite arrival stream is no longer Poisson. However, in this case the arrivals from the composite stream do see time averages. This result has implications in queueing network applications to be explored in Chapter 5. Recent works concerning research in this direction are given in the References.

Problems

1 The Bayou City Symphony accepts season ticket subscriptions during an interval $(0, t]$ starting time 0. We assume that each subscription entails the purchase of one season ticket costing c dollars payable at the time of purchase. Assume that the arrivals of subscription orders follow a Poisson process with a rate of λ and that the symphony is able to accommodate all purchase requests. Let β be the discount factor. A dollar received s time units from now will have the present value of $e^{-\beta s}$. Compute the expected present value at time 0 of all revenues received from the ticket sales in $(0, t]$. Consider a variant of the problem. All assumptions remain unchanged except that with probability p_i a subscription is for i tickets, where $i = 1, \ldots, 4$. Compute the expected present value at time 0 of all revenues received from the ticket sales in $(0, t]$.

2 A gas station is providing the state auto inspection service for the general public. Arrivals of cars for inspection follow a Poisson process with rate λ. Each inspection takes a constant of c minutes (about thirty minutes in Texas). The gas station opens for business at 7 A.M. Find the probability that the second arriving car will not have to wait and also find the mean waiting time. Do the problem also for the case in which the inspection time follows an exponential distribution with parameter μ.

3 Consider a single-server service facility with one server and unlimited number of waiting spaces. Customers are served one at a time. Arrivals follow a Poisson process with rate λ. Service times are i.i.d. random variables following a distribution G. Assume that service times and the arrival process are independent. If there are three customers left behind in the facility at a service completion epoch, what is the probability that there will be seven customers still in the facility when the next service completion occurs?

4 Let $\{T_n\}$ be the arrival times of a nonhomogeneous Poisson process with intensity function $\lambda(t)$. Define interarrival time $X_n = T_n - T_{n-1}$, $n \geq 1$, with $T_0 = 0$. **(a)** Find $P\{X_1 > x\}$ for $x \geq 0$. **(b)** Find $P\{X_2 > x\}$ for $x \geq 0$. **(c)** Are $\{X_n\}$ i.i.d.?

5 The arrival of buses (from remote park and ride sites) carrying visitors to the Houston Livestock Show follows a Poisson process with rate λ. A bus carries i visitors with probability α_i, $i = 1, 2, \ldots$. The sojourn times of each visitor at Astrohall, which houses the show, are i.i.d. random variables with a common distribution G. Assume that sojourn times are independent of the arrival process and Astrohall can accommodate all visitors who arrive at the show. Let $X(t)$ denote the number of visitors who arrive during $(0, t]$ that are still in Astrohall at time t. Find $E[X(t)]$. Does $\{X(t), t \geq 0\}$ follow a Poisson process?

6 The arrival of taxis at a taxi stand follows a Poisson process with rate λ. The arrival of customers at the stand follows another Poisson process with parameter μ. Taxis will not stop at the stand if another taxi is already waiting there. Customers will not wait at the stand if another customer is already waiting there. Therefore, the stand is either empty or is occupied by a taxi or customer. Assume that at time 0, the stand is empty. (a) Let S denote the time until the stand is occupied. What is the probability distribution of S? (b) Let T denote the time until the first customer leaves in a taxi. Find $E[T]$. (c) Let Z denote the interval between two successive epochs at which a customer leaves in a taxi. Find $E[Z]$.

7 Consider an inventory system in which demand for an item follows a Poisson process with rate λ. At each demand occurrence only one unit of the item is needed. The system starts with S units on hand and will replenish when the inventory becomes empty. Each replenishment brings the inventory back to level S. Let $X(t)$ denote the inventory level at time t and T the length of the interval starting with the epoch at which the inventory moves up to S and ending with the epoch when the inventory reaches zero. The area under the sample path $X(t)$ is given by $A = \int_0^T X(t)dt$. Find $E[A]$. Note that if h is the cost of one unit inventory per unit time then $hE[A]$ gives the cost of holding inventory per replenishment cycle.

8 The owner of Gallery Mattress hands out "Save You Money" certificates to every other visitor visiting the store starting with the first arrival who receives the first certificate. Assume that arrivals follow a Poisson process with rate λ. Find the expected number of certificates given by time t, $E[X(t)]$. Use your result to compute $E[X(t)]$ when $\lambda = 5$ arrivals per hour and $t = 4$ hours.

9 Solve the Houston ship channel tour boat problem described in Example 2.6.1 by an approach that uses the idea of a filtered Poisson process. What is the response function that does the trick?

10 Consider the nonhomogeneous Poisson process with its intensity function specified in Example 2.3.6. (a) Write a MATLAB program to generate (simulate) the first eighty arrival times. (b) Given $t = 8$ (hours), write a MATLAB program to generate $N(8)$ and then the arrival times in the interval $(0, 8]$. For both parts, draw the respective histograms showing the hourly arrival counts. (Hint for Part b: Using Equation 2.3.8 requires first integrating $\lambda(t)$ to obtain $m(t)$. Another way is by sampling from the density of Equation 2.3.8. Generating random samples from a density can be accomplished by the rejection

method described as follows. We want to take a sample X from the density f, but it is not easy to do it directly. Suppose that taking a sample Y from the density g is easy, for example, g is uniform. Let c be a constant such that $f(x)/g(x) \leq c$ for all x. Then we generate a Y and a standard uniform deviate U. If $U \leq f(Y)/cg(Y)$, then we set $X \leftarrow Y$; otherwise, we repeat the procedure.)

11 Consider the problem described in Example 2.3.9. Suppose now that we have two identical HP computers to handle the incoming traffic. Assume that the service time of each computer is exponential with a rate of 3.5 per hour (so the aggregate total service rate is still 7 per hour). Again we assume that there are three waiting spaces. A waiting customer will be served by the first computer that becomes free on a first-come-first-served basis. Compute the loss probabilities as a function of time t over the interval $[0, 8]$. Plot your results and compare them against those shown in Figure 2.7.

12 Database Maintenance   A time 0, a file in a database is loaded with b records. Additional records will be added in the future in accordance with a Poisson process with rate r. For each record in the file, its useful life follows an exponential distribution with parameter L. At the end of an interval of length t, there are two types of records—dead and live records. Let $M(t)$ and $N(t)$ denote them, respectively. The total number of records in the file at time t is then given by $X(t) = M(t) + N(t)$. (We remark that dead records are flagged as logically deleted and physical removals are done only when the file is compacted at database maintenance time.) The ratio $f(t) = E[X(t)]/E[N(t)]$ gives the mean search overhead per live record reference at time t and reveals the extent of file deterioration due to the existence of dead records (in accessing a live record the computer must search through a fraction of dead records as well). **(a)** Find $f(t)$ in terms of the known parameters. **(b)** Do Part a for the case in which the record life follows a distribution function $B(\cdot)$. Assume that the b records loaded at time 0 also have the same lifetime distribution. **(c)** Do Part b for the case in which record arrivals follow a nonhomogeneous Poisson process with an intensity function $\lambda(\cdot)$. **(d)** Consider a numerical example of Part a with $b = 500$, $r = 5.0508$ per day, and $L = 10$ days. Compute $f(t)$, the average number of record accesses per reference, for $t = 0{:}10{:}150$ (assume that file compaction occurs after 150 days). Do the same when record arrivals follow a nonhomogeneous Poisson process with intensity function

$$\lambda(t) = \exp\left[1.6094 + 0.000135t + 0.000489 \sin\left(\frac{\pi t}{15}\right) - 0.00275 \cos\left(\frac{\pi t}{15}\right)\right].$$

In the preceding intensity function, we note that the arrival rate starts from five per day, increases at a rate of 0.000135 per day, and has a month cyclic variation. The average arrival rate over the 150-day interval is again 5.0508. Plot the intensity function over the interval $[0, 150]$. Plot $f(t)$ over the interval $[0, 150]$ for the two arrival patterns in one graph. What can you say about the effect of arrival pattern on $f(t)$ based on this set of data?

13 Consider a Poisson process with rate λ. Let T be the time for the first arrival to occur and $N(t)$ denote the number of arrivals by time t of the Poisson process. Clearly, for a constant $k > 0$, $N(T/k)$ represents the number of arrivals in the time interval of length T/k. Find the first two moments of random variable $N(T/k)T$.

14 Time-Dependent Probabilities of an *M/M/∞* Queue   In an *M/M/∞* queue, we have Poisson arrivals of individual customers, exponential service times, and an infinite number of servers. Let λ be the arrival rate and μ be the service rate. Define $p_n(t)$ as the probability that at time t there are n customers in the system. Find the probabilities $\{p_n(t)\}$ for all n for any $t \geq 0$. (Hint: Example 2.2.5.)

15 Consider an aircraft maintenance depot that keeps a stock of S spare engines for a fleet of aircrafts. The demand for spare engines can be approximated by a Poisson process with rate λ. When a demand for a spare engine occurs, an order to replenish the item demanded is issued immediately. The time to receive the replenishment follows an arbitrary distribution G with mean $1/\mu$. This inventory replenishment scheme is called one-for-one ordering policy. It is also known as the $(S - 1, S)$ policy, where $S - 1$ denotes the reorder point and S the order-up-to quantity (in other words, the order size is always one). When a demand occurs and there are no spare engines in stock, the unmet demand is back ordered. Let random variable $X(t)$ denote the number of spare engines on hand at time t. When $X(t)$ is positive, it represents the number of spare engines in stock; when $X(t)$ is negative, it represents the number of units back ordered. Let $Y(t)$ denote the number of replenishment orders outstanding at time t. Then we see that $X(t) = S - Y(t)$. **(a)** Describe how the stochastic process $\{Y(t), t \geq 0\}$ can be modeled as an *M/G/∞* queue. **(b)** Let $p(x) = \lim_{t \to \infty} P\{X(t) = x\}$. Find the probability mass function $\{p(x)\}$. **(c)** Let h be the cost of carrying one engine per unit time and r the cost per engine back ordered. For a given S, derive an expression for the total expected costs per unit time. **(d)** Find the optimal S that minimizes the total expected cost when the average demand rate for spare engines is 0.8 per month, the average replenishment lead time is one month, the unit holding cost is $60 per month, and the unit cost of back ordering is $2000.

16 Consider the one-for-one ordering policy described in Problem 15 again. Define *fill rate* as the long-run fraction of time that a demand for a spare engine can be met at the time it is requested and *back orders* as the number of unfilled demands that exist in steady state. For a given S, let $FR(S)$ denote the fill rate and $EBO(S)$ the expected number of back orders. Use the data given in Problem 15d to compute $FR(S)$ and $EBO(S)$ for $S = 0{:}10$.

17 Consider a Poisson process with rate λ. For $i = 1, 2$, we let S_i denote the time of the ith arrival. Use the fact that the two events $\{S_i \leq t_1, S_2 \leq t_2\}$ and $\{N(t_1) \geq 1, N(t_2) \geq 2\}$ are equivalent to find the joint distribution $F_{S_1, S_2}(t_1, t_2)$ of S_1 and S_2, where $0 < t_1 < t_2$.

18 Consider an earthquake that caused serious damage in a given locality. Assume that the locations of fatality follow a two-dimensional homogeneous Poisson process with rate $\lambda = 0.001$ per square mile. Let $A(r)$ be a circular area of radius r with its center located at the city hall. What is the probability that there are no fatalities within ten miles radius of the city hall?

19 Filtered Nonhomogeneous Poisson Processes   Consider a nonhomogeneous Poisson process $N = \{N(t), t \geq 0\}$. Let $\{Y_n\}$ be i.i.d. continuous random variables, where Y_n is associated with the nth arrival of N. The process N and the sequence $\{Y_n\}$ are assumed to be independent. A stochastic process $X = \{X(t), t \geq 0\}$ is called a filtered nonhomogeneous Poisson process if we define

$$X(t) = \sum_{n=1}^{N(t)} \omega(t, S_n, Y_n),$$

where $\{S_n\}$ are the arrival times of N and ω is a response function defined by the three arguments. Show that the probability generating function for $X(t)$ is given by

$$E[z^{X(t)}] = \exp\left\{ \int_0^t \lambda(\tau) E[z^{\omega(t,\tau,Y)} - 1] d\tau \right\}.$$

(Hint: Mimic the derivation of Equation 2.5.3 and use Equation 2.3.8.)

20 The $M(t)/G/\infty$ Queue: A Palm Theorem for Nonhomogeneous Processes
Consider a service system with an infinite server. The service times of individual customers are i.i.d. random variables with a common distribution G. The arrivals to the system form a nonhomogeneous Poisson process with intensity function $\lambda(t)$. The arrival process and service times are mutually independent. Let $X(t)$ denote the number of customers in the system at time t. Use the result given in Problem 19 to show that $X(t)$ follows a Poisson distribution with mean

$$\alpha(t) \equiv \int_0^t \lambda(u) \overline{G}(t - u) du,$$

where $\alpha(t)$ is the mean number of customers still in service at time t.

21 Consider the $M(t)/G/\infty$ queue described in Problem 20. Find the mean number of customers still in service at time t, $\alpha(t)$, when **(a)** the service time is a constant T, and **(b)** the service time follows an exponential distribution with mean $1/\mu$.

22 Demands for aircraft spare engines needed to replace failed ones at an Air Force base follow a Poisson process with a rate of 0.8 units per day. Upon the onset of hostilities, the rate jumps to 3.16 and then tapers off exponentially. Let $t = 0$ denote the epoch when hostilities start. Then the demand process is a nonhomogeneous Poisson with intensity function

$$\lambda(t) = \begin{cases} 0.8 & t < 0 \\ 3.16e^{-0.1t} & t \geq 0. \end{cases}$$

Assume that all repairs of failed engines are done at the base. The repair time is a constant of five days and the repair facility can accommodate all requests at the same time (in other words, there is no waiting). Let $X(t)$ denote the number of engines in repair at time t. Then the facility operates like an

$M(t)/G/\infty$ queue. **(a)** For the next twenty days from the onset of hostilities, compute the expected numbers of engines in repair at the end of each day, that is, find $\alpha(t)$ for $t = 1{:}20$. **(b)** Assume that the stock level S_t of spares on day t must meet the performance goal that $P\{D_t \le S_t\} \ge 0.80$, where D_t is the demand for spare engines on day t. For $t = 1{:}20$, compute S_t.

23 Do Problem 22 for the case in which the repair time follows an exponential distribution with mean $1/\mu = 5$ days. Also, give a closed-form expression of $\alpha(t)$.

24 Do Problem 22 for the case in which the repair time follows a hyperexponential distribution with balanced means (see Example 1.3.5) where the mean is again five days but the coefficient of variation is 5. Also, give a closed-form expression of $\alpha(t)$. By comparing the plots of $\alpha(t)$ for the three cases studied in Problems 22, 23, and 24 where the coefficients of variations of the repair time are respectively 0, 1, and 5, what can you say about the effect of repair time variability on the mean numbers of engines in repair? What about its effect on the spare stock levels?

25 **Departures from an $M/G/\infty$ Queue** We revisit Example 2.3.10. Use the concept of a filtered Poisson process to show that the departure process $M = \{M(t), t \ge 0\}$ from the queue follows a nonhomogeneous Poisson process with intensity function $\lambda G(t)$.

26 Boats arrive at a drawbridge waiting for passage in accordance with a Poisson process with rate λ. The bridge is raised to permit passage beneath it only when there are five boats waiting and is lowered as soon as the boats have passed. Assume that the interval between bridge open and close times is rather short and thus negligible. **(a)** What is the probability that an arriving boat will see that there are i boats already waiting for passage? **(b)** For an arriving boat, let W denote the waiting time before the bridge is raised. Find the expressions for computing the distribution F_W and $E[W]$. **(c)** For $\lambda = 5$ boats per hour, compute $F_W(t)$, with $t = 0{:}0.1{:}3$, and $E[W]$.

27 We return to Example 2.5.1. Let $X(t)$ denote the total amount of remaining service times for customers who are present at time t. Use the idea of a filtered Poisson process to find $\lim_{t \to \infty} E[X(t)]$.

28 We return to Example 2.6.1. Use the idea of a filtered Poisson process to find the expected number of tourists at each boat departure epoch.

29 Use the idea of a filtered Poisson process to obtain the probability generating function of Equation 2.4.1 for the compound Poisson process.

30 Packages arrive at a post office in accordance with a Poisson process with rate λ. Given that seven packages arrive in the interval $[3, 8]$, what is the expected arrival time of the first of these packages?

31 Mail orders for sportswear arrive at L.L. Corns in accordance with a Poisson process with mean λ. The time needed to process each order is a random variable that follows a distribution G with mean $1/\mu$. There are sufficient workers available to handle incoming orders so that all orders receive immediate attention

when they are received. Let $W(t)$ denote the number of orders in process at time t. **(a)** Find the limiting distribution of $W(t)$ as $t \to \infty$. **(b)** Let $V(t)$ denote the length of time needed to process all orders in process at time t given that at time 0 there are no unprocessed orders. Find $P\{V(t) < y\}$.

32 Consider the game of tossing a fair die repeatedly. Let $n_1, \ldots, n_6$ be a set of given numbers. Let random variable N_i denote the number of tosses needed to obtain side i for a total of n_i times. **(a)** What is the probability distribution of N_i? **(b)** Are $\{N_i\}$ independent? Now we assume that the tosses are done at random times generated by a Poisson process with mean one. Let T_i denote the time until side i has appeared a total of n_i times. **(c)** What is the distribution of T_i? **(d)** Are $\{T_i\}$ independent? **(e)** Let $T = \min\{T_1, \ldots, T_6\}$. Find $E[T]$.

Bibliographic Notes

An excellent introduction to Poisson processes can be found in Chapter 4 Çinlar (1975) and Chapter 4 of Parzen (1962). The alternate characterization of Poisson process involving Equation 2.1.4 was given as Theorem (1.17) in Chapter 4 of Çinlar (1975). Many examples given in Sections 2.2 and 2.3 are based on Ross (1983). They include Examples 2.2.3, 2.2.5, 2.3.1, and 2.3.3. Example 2.2.2 is from Ross (1990, 61–62). A useful reference regarding statistical analysis of Poisson processes is Cox and Lewis (1978). An important application of Poisson process in operations research is the work done by Sherbrooke (1968) in modeling repairable inventory systems for the U.S. Air Force. A summary of his work is contained in a book by Sherbrooke (1992). P. A. W. Lewis has made substantial contributions to nonhomogeneous Poisson processes—particularly in statistical analysis and simulation. The papers cited in Lewis (1966, 1972, 1979) provided additional references. There are several examples based on real data demonstrating the use of nonhomogeneous Poisson process: Examples 2.3.5 and 2.3.6 are from Kao (1980) and Kao and Chang (1988), respectively, and Example 2.3.7 is from Green and Kolesar (1989). Analytical modeling of service systems with nonhomogeneous Poisson arrivals has a long history. Earlier papers include those of Luchak (1956), Galliher and Wheeler (1958), and Koopman (1972). The last two papers modeled time-dependent aircraft arrivals to an airport by nonhomogeneous Poisson processes. The works of Green, Kolesar, and Svornos (1991) and Green and Kolesar (1995) are two applications of recent vintage. Example 2.3.11 represents a numerical implementation of the idea of Ross's (1983) Example 2.4(b). Problem 12 is based on Leung (1986). The two examples of compound Poisson process are based on the papers of Sherbrooke (1968a, b). Parzen (1962) has a nice section on filtered Poisson processes. Some examples given in Section 2.5 have their origins in Parzen (1962). The generalization of the Palm theorem to admit nonhomogeneous Poisson arrivals was discussed in Crawford (1981). This generalization led to Dyna-METRIC—an inventory control system in use by the U.S. Air Force for multiechelon systems of repairable items (Hillestad 1982). In Problems 20 and 21, we consider an alternate yet simple proof of the Palm theorem for nonstationary processes using the idea of a filtered nonhomogeneous Poisson process. Problem 22 is based on Muckstadt (1980). Taylor and Karlin (1994) give a readable account of the two-dimensional and marked Poisson processes. The last section on PASTA is based on Wolff (1982, 1989). Additional work in this direction and references can be found in Melamed and Whitt (1990a, b).

References

Çinlar, E. 1975. *Introduction to Stochastic Processes.* Englewood Cliffs, NJ: Prentice-Hall.

Cox, D. R., and P. A. W. Lewis. 1978. *The Statistical Analysis of Series of Events.* London: Chapman and Hall.

Crawford, G. B. 1981. *Palm's Theorem for Nonstationary Processes.* R-2750-RC, The Rand Corporation.

Galliher, H. P., and R. C. Wheeler. 1958. Nonstationary Queueing Probabilities for Landing Congested Aircraft. *Operations Research* 6:264–75.

Green, L., and P. Kolesar. 1989. Testing the Validity of a Queueing Model of Police Patrol. *Management Science* 35(2):127–48.

Green, L., P. Kolesar, and A. Svoronos. 1991. Some Effects of Nonstationarity on Multiserver Markovian Queueing Systems. *Operations Research* 39(3):502–11.

Green, L., and P. Kolesar. 1995. On the Accuracy of the Simple Peak Hour Approximation for Markovian Queues. *Management Science* 41(8):1353–70.

Hillestad, R. J. 1982. Dyna-METRIC: Dynamic Multi-Echelon Technique for Recoverable Item Control. R-2785-AF, The Rand Corporation.

Kao, E. P. C. 1980. Simulating a Computer System for Processing Electrocardiograms. *IEEE Transactions on Systems, Man, and Cybernetics* SMC-10, 6.

Kao, E. P. C., and S. L. Chang. 1988. Modeling Time-Dependent Arrivals to Service Systems: A Case in Using a Piecewise-Polynomial Rate Function in a Nonhomogeneous Poisson Process. *Management Science* 34(11):1367–79.

Koopman, B. O. 1972. Air-Terminal Queues under Time-Dependent Conditions. *Operations Research* 20(6):1089–114.

Leung, C. H. C. 1986. Dynamic Storage Fragmentation and File Deterioration. *IEEE Transactions on Software Engineering* SE-12, 3.

Lewis, P. A. W. 1966. A Computer Program for the Statistical Analysis of Series of Events. *IBM Systems Journal* 5:4.

Lewis, P. A. W. 1972. Recent Advances in the Statistical Analysis of Univariate Point Processes. In *Stochastic Point Processes,* edited by P. A. W. Lewis, 1–54. New York: John Wiley & Sons.

Lewis, P. A. W. 1979. Simulation of Nonhomogeneous Poisson Processes by Thinning. *Naval Research Logistics Quarterly* 26(3):403–13.

Luchak, G. 1956. The Solution of the Single-Channel Queueing Equations Characterized by a Time-Dependent Poisson-Distributed Arrival Rate and a General Class of Holding Times. *Operations Research* 4(4):711–32.

Melamed, B., and W. Whitt. 1990a. On Arrivals That See Time Averages. *Operations Research* 38(1):156–72.

Melamed, B., and W. Whitt. 1990b. On Arrivals That See Time Averages: A Martingale Approach. *Journal of Applied Probability* 27, 376–84.

Muckstadt, J. A. 1980. Comparative Adequacy of Steady-State Versus Dynamic Models for Calculating Stockage Requirements. R-2636-AF, The Rand Corporation.

Parzen, E. 1962. *Stochastic Processes.* San Francisco: Holden-Day.

Ross, S. M. 1983. *Stochastic Processes.* New York: John Wiley & Sons.

Ross, S. M. 1990. *A Course in Simulation.* New York: Macmillan.

Sherbrooke, C. C. 1968a. METRIC: A Multi-Echelon Technique for Recoverable Item Control. *Operations Research* 16:1.

Sherbrooke, C. C. 1968b. Discrete Compound Poisson Processes and Tables of the Geometric Poisson Distribution. *Naval Research Logistics Quarterly* 15(2):189–204.

Sherbrooke, C. C. 1992. *Optimal Inventory Modeling of Systems: Multi-Echelon Techniques.* New York: John Wiley & Sons.

Taylor, H. M., and S. Karlin. 1994. *An Introduction to Stochastic Modeling.* Revised ed. New York: Academic Press.

Wolff, R. W. 1982. Poisson Arrivals See Time Averages. *Operations Research* 30(2):223–31.

Wolff, R. W. 1989. *Stochastic Modeling and the Theory of Queues.* Englewood Cliffs, NJ: Prentice-Hall.

Appendix

Chapter 2: Section 3

Example 2.3.9 This example uses MATLAB function **ode45** to solve the following time-dependent differential equation $P'(t) = Q(t)P(t)$, where $P'(t)$ and $P(t)$ are two column vectors and $Q(t)$ is a square matrix. First we list the function **e236** that generates the intensity $\lambda(t)$. This function is based on the parameters given in Example 2.3.6. The next program **e239a** provides the derivative function $P'(t)$. The last function **e239b** does the work of solving the time-dependent differential equation for $P(t)$. The output from **e239b** is a matrix, called X, with two columns. The first column gives the values of t and the second column the corresponding values of $P_4(t)$. This output matrix X is used for plotting the graph on the right side of Figure 2.7.

```
function [lm]=e236(t)
%
%   Generate the arrival rate for Example 2.3.6
%
a1=8.924; b1=-1.584; b2=7.897; b3=-10.434; b4=4.293;
d1=1.51; d2=3.02; d3=4.53; d4=6.04;
lm=a1+b1*cos(pi*t/d1)+b2*sin(pi*t/d2)+b3*cos(pi*t/d3)+b4*cos(pi*t/d4);

function [xdot]=e239a(t,x)
%
%   M(t)/m/1/4   queue
%
mu=7; u=mu*(ones(1,4)); M=diag(u,-1); u=[0 u]; N=diag(-u,0);
Q=M+N; [lm]=e236(t);
lmt=lm*(ones(1,4)); M=diag(lmt,1);
u=[lmt 0]; N=diag(-u,0); Qt=Q+M+N; Qt=Qt'; xdot=Qt*x;

function [X]=e239b
%
%   M(t)/m/1/4   Queue
%
t0=0; tf=8; x0=zeros(1,5); x0(1,1)=1;
[t,x]=ode45('e239a',[t0 tf],x0);
[n]=length(t); X=zeros(n,2); X(:,1)=t; X(:,2)=x(:,5);
```

Example 2.3.11: Two Queues in Tandem   This example is similar to Example 2.3.9. The main difference is that here the problem in theory has an infinite dimension; however, we do finite-state truncation at 20. Program **e2311** calls **e2311d** for $\{P'(t)\}$ for a given t, and **e2311b** gives the corresponding $\lambda(t)$ for use in **e2311d**. The output from the main program **e2311** is a matrix containing three columns giving t, $E[L(t)]$, and $Var[L(t)]$.

```
function [X]=e2311
%
%   Example 2.3.11  - M(t)/M/3   Queue
%   Differential Equation Approach
%
t0=0; tf=90; NS=20; NS1=NS+1;
x0=zeros(1,NS1); x0(1,1)=1;
[t,x]=ode45('e2311d',t0,tf,x0);
[n,m]=size(x); a=ones(1,NS); b=cumsum(a); b=[0 b];
r1=[]; r2=[];
for k=1:n
pi=x(k,:);
L=sum(b.*pi); b2=b.^2; L2=sum(b2.*pi); V=L2-L^2;
r1=[r1 L]; r2=[r2 V];
end
X=zeros(n,3); X(:,1)=t; X(:,2)=r1'; X(:,3)=r2';

function [xdot]=e2311d(t,x)
%
%   M(t)/m/3 Queue - The small cafeteria in a Big University
%   Differential Equation approach
%
NS=20; NS1=NS+1; a=[1 2]; mu=1; a=a*mu;
u=3*mu*(ones(1,NS)); u(1,1:2)=a;
M=diag(u,-1);
u=[0 u]; N=diag(-u,0); Q=M+N;
[lm]=e2311b(t);
lmt=lm*(ones(1,NS)); M=diag(lmt,1);
u=[lmt 0]; N=diag(-u,0); Qt=Q+M+N;
Qt=Qt'; xdot=Qt*x ;

function [lm]=e2311b(t)
%
%   Generate the arrival rate for Example 2.3.11
%
lm=2*(1-(exp(-0.25*t))*(1+0.25*t));
```

3

Renewal Processes

Tips for Chapter 3

- Understanding the parts relating to the derivations of the distribution functions for the excess life, current life, and total life given in Section 3.3 requires some maturity in probabilistic reasoning. Some may want to skim these parts initially and return to them when the need arises.
- Examples 3.6.3 and 3.6.4 make clever uses of regenerative theory. These examples, while lengthy, give those who are interested in queueing applications some meaty materials on which to ponder.
- The first six sections of this chapter address renewal processes whose interarrival times are continuous random variables. The last section considers discrete renewal processes in which interarrival times are discrete random variables. Although the treatment of the various topics involving both types of interarrival times could have been presented in a unified manner using Riemann-Stieltjes integrals, we choose to give two separate presentations. The last section (3.7), which can be read independently of the other six sections, provides some results needed for discrete-time Markov chains (Chapter 4).

3.0 Overview

In a Poisson process the times between successive occurrences of a given event are i.i.d. exponential random variables. When the interarrival time distribution follows a general distribution, the resulting counting process is a renewal process.

In the first six sections, we assume that the interarrival time is a continuous random variable. In the last section, we briefly look at the case in which the interarrival time is a discrete random variable. Presenting these two cases separately seems to ease the exposition in the sense there is no need to "translate" our results for the continuous case and the discrete case separately. An alternative will be to derive all results in a unified manner through the use of Riemann-Stieltjes integral.

In Section 3.2, we present the heart of renewal theory—the renewal-type equation, 3.2.1, which is typically obtained by conditioning on the time of first renewal before a specified epoch. Its time-dependent solution is given by Equation 3.2.2 and its limiting solution by Equation 3.2.3. The focus of the chapter is the construction of renewal-type equations—the modeling part—and their solutions—the analysis part. This is illustrated in Section 3.3 where we study three related random variables: the excess life, the current life, and the total life. In Section 3.4, we introduce the concept of a renewal reward process. The renewal reward paradigm is a handy tool for problem solving. Additional variants of renewal process along with some limit theorems are given in Section 3.5. The regenerative process presented in Section 3.6 makes the renewal theory an indispensable part of the study of stochastic processes. In a regenerative process, the stochastic process starts afresh when it reaches a regeneration epoch. The interval between two successive regeneration epochs forms a regeneration cycle. Random phenomena occurring in different regenerative cycles are probabilistically equivalent to one another. To analyze the process, we can restrict our attention to the study of events occurring in one regeneration cycle. The examples given in Section 3.6 amply illustrate the power of regenerative theory. The last section focuses on discrete renewal process in which the interarrival times are discrete random variables. The section primarily serves as a prelude to Chapter 4—the discrete-time Markov chain.

3.1 Introduction

We let nonnegative continuous random variable X_n denote the interarrival time between the nth and $(n-1)$st events. Assume that $\{X_n\}$ are i.i.d. random variables with a common distribution F, mean μ, variance σ^2, and that $F(0) < 1$ so its mean $\mu > 0$. Let S_n denote the arrival time of the nth event, then

$$S_n = \sum_{i=1}^{n} X_i, \qquad n = 1, 2, \dots.$$

Assume $S_0 = 0$. This implies that there is an arrival at time 0. Let $N(t) = \max\{n | S_n \leq t\}$, so $N(t)$ represents the number of renewals in $(0, t]$. We call the stochastic process $\{N(t), t \geq 0\}$ a *renewal process*. The Poisson process described in Chapter 2 supplies one example of a renewal process. There, the interarrival time distribution is exponential with parameter μ.

A key identity that enables us to obtain the distribution of $N(t)$ based on our knowledge of F is

$$\{N(t) \geq n\} \Leftrightarrow \{S_n \leq t\}.$$

Two immediate consequences of the preceding identity are

$$P\{N(t) \geq n\} = P\{S_n \leq t\} = F_n(t)$$

and

$$P\{N(t) = n\} = P\{N(t) \geq n\} - P\{N(t) \geq n+1\}$$
$$= P\{S_n \leq t\} - P\{S_{n+1} \leq t\} = F_n(t) - F_{n+1}(t),$$

where F_n is the n-fold convolution of F with itself. Let $M(t) = E[N(t)]$. We call $M(t)$ the *renewal function*.

The renewal function is related to F_n as follows:

$$M(t) = E[N(t)] = \sum_{n=1}^{\infty} P\{N(t) \geq n\} = \sum_{n=1}^{\infty} F_n(t).$$

The first derivative of $M(t)$ with respect to t is called the *renewal density* and denoted by $m(t)$. In such a case, we have

$$m(t) = \sum_{n=1}^{\infty} f_n(t), \tag{3.1.1}$$

where f_n is the density of F_n. We note that $f_n(t)dt$ can be interpreted as the probability that the nth renewal occurs in $(t, t+dt)$. Their sum, $m(t)dt$, corresponds to the probability that one renewal occurs in $(t, t+dt)$.

Define the Laplace transforms

$$f^e(s) = \int_0^{\infty} e^{-st} f(t) dt,$$

$$m^e(s) = \int_0^{\infty} e^{-st} m(t) dt,$$

and

$$M^e(s) = \int_0^{\infty} e^{-st} M(t) dt.$$

Multiplying both sides of Equation 3.1.1 by e^{-st} and integrating over $(0, \infty)$, we obtain a useful relation:

$$m^e(s) = \sum_{n=1}^{\infty} \left[f^e(s) \right]^n = \frac{f^e(s)}{1 - f^e(s)}. \tag{3.1.2}$$

EXAMPLE
3.1.1   In a Poisson process with arrival rate λ, the interarrival time of successive arrivals follows the exponential distribution with rate λ. Then $f^e(s) = \lambda/(s+\lambda)$, which yields $m^e(s) = \lambda/s$. After an inversion, we find the renewal density $m(t) = \lambda$ for all $t \geq 0$. Since $M^e(s) = (1/s)m^e(s)$, we find $M^e(s) = \lambda/s^2$. This gives a familiar expression for the Poisson process $E[N(t)] = M(t) = \lambda t$. ∎

EXAMPLE
3.1.2

Consider a renewal process with an Erlang $(2,1)$ interarrival time distribution

$$f(t) = te^{-t}, \qquad t \ge 0.$$

The Laplace transform of f is given by $f^e(s) = 1/(s+1)^2$. The transform suggests that the interarrival time is the sum of two independent exponential random variables, each with parameter 1. Hence the interarrival time has a mean $\mu = 2$ and a variance $\sigma^2 = 2$. Using Equation 3.1.2, we obtain

$$m^e(s) = \frac{1}{s(s+2)}.$$

After a partial fraction expansion, we find

$$M^e(s) = \frac{1}{s^2(s+2)} = \left(-\frac{1}{4}\right)\left(\frac{1}{s}\right) + \left(\frac{1}{2}\right)\left(\frac{1}{s^2}\right) + \left(\frac{1}{4}\right)\left(\frac{1}{s+2}\right).$$

Inverting the previous expression yields the renewal function

$$M(t) = \left(-\frac{1}{4}\right) + \left(\frac{1}{2}\right)t + \left(\frac{1}{4}\right)e^{-2t}, \qquad t \ge 0.$$

We note that the preceding renewal function contains three components: a constant term (called the *bias*), a *linear asymptote*, and a *transient* term that tapers off to zero as $t \to \infty$. The function is depicted in Figure 3.1. ∎

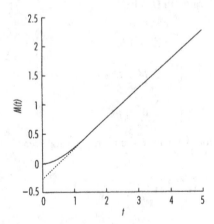

FIGURE
3.1   The renewal function with Erlang $(2,1)$ interarrival times.

Applying the asymptotic-rate theorem (Equation 1.3.8) of the Laplace transform, we find

$$\lim_{t \to \infty} \frac{M(t)}{t} = \lim_{s \to 0} s^2 M^e(s) = \lim_{s \to 0} \frac{1}{(s+2)} = \frac{1}{2}.$$

The last identity

$$\lim_{t \to \infty} \frac{M(t)}{t} = \frac{1}{\mu},$$ (3.1.3)

holds in a general setting and is known as the *elementary renewal theorem*.

Applying the final-value theorem of the Laplace transform, we obtain

$$\lim_{t \to \infty} m(t) = \lim_{s \to 0} sm^e(s) = \lim_{s \to 0} \left(\frac{1}{s+2} \right) = \frac{1}{2}.$$

The preceding identity suggests

$$\lim_{t \to \infty} \frac{dM(t)}{dt} = \frac{1}{\mu},$$ (3.1.4)

namely, the expected asymptotic rate of renewal is equal to the rate of arrival. Finally, we note that

$$\lim_{t \to \infty} \left\{ M(t) - \left(\frac{1}{2}t \right) \right\} = -\frac{1}{4}.$$

We will see later that the last identity can be found by applying the following expression

$$\lim_{t \to \infty} \left\{ M(t) - \frac{1}{\mu}t \right\} = \frac{\sigma^2 - \mu^2}{2\mu^2}.$$ (3.1.5)

The expression gives the constant term in the asymptotic expansion of $M(t)$.

3.2 Renewal-Type Equations

Many problems in renewal theory can be solved methodically by applying the following approach.

First, by conditioning on the first renewal epoch X_1, we construct a *renewal-type* equation

$$g(t) = h(t) + \int_0^t g(t-x)f(x)dx.$$ (3.2.1)

In Equation 3.2.1 the functions h and f are known, and g is unknown. To illustrate the use of the renewal-type equation, we consider a simple equipment replacement problem in which the equipment is replaced upon failure and lifetimes are i.i.d. random variables following a common density f. Let $g(t)$ denote the failure rate of the equipment in use at time t. Conditioning on the time of the first failure X_1, the conditional failure rate at time t is then given by

$$g(t|X_1 = x) = \begin{cases} \dfrac{f(t)}{1 - F(t)} & \text{if } x > t \\ g(t - x) & \text{if } x \le t. \end{cases}$$

Using the law of total probability, we obtain

$$g(t) = \frac{f(t)}{1 - F(t)} \int_t^\infty f(x)dx + \int_0^t g(t - x)f(x)dx = f(t) + \int_0^t g(t - x)f(x)dx.$$

Therefore, the failure rate $g(t)$ at time t is characterized by the preceding renewal-type equation with $h(t) = f(t)$.

The solution of the renewal-type equation is given by

$$g(t) = h(t) + \int_0^t h(t - x)m(x)dx. \tag{3.2.2}$$

One way to obtain Equation 3.2.2 is by taking the Laplace transform of Equation 3.2.1 and inverting the corresponding result. Specifically, we let

$$g^e(s) = \int_0^\infty e^{-st} g(t)dt \quad \text{and} \quad h^e(s) = \int_0^\infty e^{-st} h(t)dt.$$

Taking the Laplace transform of Equation 3.2.1, we find that

$$g^e(s) = h^e(s) + g^e(s)f^e(s)$$

or

$$g^e(s) = \frac{h^e(s)}{1 - f^e(s)} = h^e(s)\left[1 + f^e(s) + \left(f^e(s)\right)^2 + \cdots\right] = h^e(s) + h^e(s)m^e(s).$$

Inverting the last expression we obtain Equation 3.2.2. In Equation 3.2.2, we see that the terms on the right side are all known (the renewal density $m(x)$ is given by Equation 3.1.1). For a finite t, we can compute $g(t)$ by using numerical methods. In some simple cases, $g(t)$ may even be found in closed form.

For large t, we can apply the following *key renewal theorem*

$$\lim_{t \to \infty} g(t) = \frac{\int_0^\infty h(t)dt}{\mu}, \tag{3.2.3}$$

where μ is the mean interarrival time.

For the key renewal theorem to hold, the function h must be *directly Riemann integrable*. A sufficient condition for the function to be directly Riemann integrable is that (i) $h(t) \ge 0$, (ii) $h(t)$ is nonincreasing, and (iii) $\int_0^\infty h(t)dt < \infty$. We now give some examples illustrating the use of the approach.

EXAMPLE
3.2.1

Suppose we are interested in studying the renewal function $M(t)$ in more detail. Since the process starts afresh at the epoch X_1, we can write

$$E\big[N(t)|X_1 = x\big] = \begin{cases} 1 + M(t-x) & \text{if } x \le t \\ 0 & \text{otherwise.} \end{cases}$$

Using the law of total probability, we obtain

$$M(t) = \int_0^t E\big[N(t)|X_1 = x\big]f(x)dx$$

$$= \int_0^t [1 + M(t-x)]f(x)dx = F(t) + \int_0^t M(t-x)f(x)dx.$$

Letting $g(t) = M(t)$ and $h(t) = F(t)$, the preceding expression can be stated in the form of Equation 3.2.1. An application of Equation 3.2.2 gives

$$M(t) = F(t) + \int_0^t F(t-x)m(x)dx.$$

Since $h(t)$ is nondecreasing, it is not directly Riemann integrable. Thus Equation 3.2.3 is not applicable. In fact, in this case the numerator in Equation 3.2.3 is unbounded. ▪

EXAMPLE
3.2.2

If $\{X_i\}$ are identically distributed random variables with a common mean μ, and N is independent of $\{X_i\}$, then we know that $E[S_N] = \mu E[N]$, where $S_N = X_1 + \cdots + X_N$. In a renewal process, we see that $S_{N(t)} = X_1 + \cdots + X_{N(t)}$ denotes the time of the last renewal before t. Unfortunately, it is not true that $E[S_{N(t)}] = \mu E[N(t)] = \mu M(t)$ because $N(t)$ depends on $\{X_i\}$. However the following related result holds

$$E[S_{N(t)+1}] = \mu(M(t)+1). \tag{3.2.4}$$

To establish this result, we let $g(t) = E[S_{N(t)+1}]$. Conditioning on the epoch of the first arrival, we write

$$E\big[S_{N(t)+1}|X_1 = x\big] = \begin{cases} x & x > t \\ x + g(t-x) & x \le t. \end{cases}$$

Applying the law of total probability, we find

$$g(t) = \int_0^\infty E\big[S_{N(t)+1}|X_1 = x\big]f(x)dx$$

$$= \int_0^\infty xf(x)dx + \int_0^t g(t-x)f(x)dx$$

$$= \mu + \int_0^t g(t-x)f(x)dx.$$

With $h(t) = \mu$ for all $t \ge 0$, we use Equation 3.2.2 to obtain $g(t) = \mu + \int_0^t \mu m(x)dx = \mu[1 + M(t)]$. ▪

In the previous example, we established Equation 3.2.4. If we define random variable $N = N(t) + 1$, then Equation 3.2.4 can be stated as

$$E[S_N] = E[X]E[N], \tag{3.2.5}$$

where $S_N = X_1 + \cdots + X_N$ and $E[X] = \mu$. We stated in the example that Equation 3.2.5 does not hold when $N = N(t)$. What makes $N(t) + 1$ so special such that Equation 3.2.5 holds? The answer lies with a special property that the random variable $N(t) + 1$ possesses—namely, it is a *stopping time*.

An integer-valued random variable N is a *stopping time* with respect to i.i.d. random variables $\{X_n\}$ if the occurrence or nonoccurrence of the event $\{N = n\}$ is independent of $X_{n+1}, X_{n+2}, \ldots$.

To show that $N(t)$ is *not* a stopping time, we observe that $\{N(t) = n\}$ if and only if $\{X_1 + \cdots + X_n \le t\}$ *and* $\{X_1 + \cdots + X_{n+1} > t\}$. The event $\{N(t) = n\}$ also depends on X_{n+1} and consequently it is not a stopping time. For the event $\{N(t) + 1 = n\}$, we see that it is equivalent to the event $\{N(t) = n - 1\}$. The latter is equivalent to the event $\{X_1 + \cdots + X_{n-1} \le t\}$ *and* $\{X_1 + \cdots + X_n > t\}$. Hence, it is indeed a stopping time. When N is a stopping time with respect to $\{X_n\}$, $E[N] < \infty$, and $E[|X_1|] < \infty$, Equation 3.2.5 holds and is called *Wald's equation*. To establish the equation, we write

$$S_N = \sum_{n=1}^{N} X_n = \sum_{n=1}^{\infty} X_n I\{N \ge n\}, \tag{3.2.6}$$

where we let indicator random variable $I\{A\} = 1$ if A occurs and 0 otherwise. Now the occurrence of the event $\{N \ge n\}$ means that we have not stopped after observing $X_1, \ldots, X_{n-1}$. This implies that $I\{N \ge n\}$ depends on $X_1, \ldots, X_{n-1}$ and is independent of X_n. Taking the expectation of Equation 3.2.6, interchanging the order of expectation and summation, and using the fact that the expectation of the product of two *independent* random variables is equal to the product of the two respective expectations, we obtain Wald's equation, 3.2.5. Having noted that $\{N(t) + 1\}$ is a stopping time with respect to $\{X_n\}$, we could have found Equation 3.2.4 by an application of Wald's equation. When N is a stopping time with respect to $\{X_n\}$, the generating function of Equation 1.2.4 for the random sum S_N and in particular the variance formula of Equation 1.2.7 for the random sum hold as well.

EXAMPLE
3.2.3   Let $\{X_i\}$ be i.i.d. random variables with $P\{X_1 = 1\} = p$, $P\{X_1 = -1\} = q$, and hence a mean $E[X_1] = p - q$. Consider a simple random walk defined by $S_n = X_1 + \cdots + X_n$. Let $N = \min\{n | S_n = 1\}$. Thus N is the first time the random sum reaches 1. We see that N is a stopping time with respect to $\{X_i\}$ since the event $\{N = n\}$ is determined solely by the sequence $\{X_1, \ldots, X_n\}$. Since $E[|X_1|] < \infty$, we write Wald's equation

$$E[S_N] = (p - q)E[N]. \tag{3.2.7}$$

We observe that $S_N = 1$ for all N and $E[S_N] = 1$. The right side of Equation 3.2.7 deserves some close scrutiny. When $p = q$ and $E(N) < \infty$, the right side is zero, which does not make sense. This must imply $E[N] = \infty$, and Wald's equation is not applicable. Consider now the case $q > p$: the right side of Equation 3.2.7 is negative while the left side is 1. Thus we conclude that $E[N] = \infty$, and Wald's equation does not apply. For the last case where $p > q$, we obtain $E[N] = 1/(p - q)$. In Part b of Problem 29 in Chapter 1, we have established that for this simple random walk, $P\{N < \infty\} = 1$ if $p \geq q$. This reveals a curious situation in which $P\{N < \infty\} = 1$ does not imply $E[N] < \infty$ (the case in which $p = q$).

Define M as the last time before the random sum S_n reaches 1 for the first time. Since $M = N - 1$, S_M must be 0 for all M and $E[S_M] = 0$. Wald's equation becomes $E[S_M] = (p - q)E[M]$. Consider the case $p > q$, in which $E[N] < \infty$ and consequently $E[M] < \infty$. While the right side of the equation is finite and positive, the left side is zero—a contradiction. If we look at the random variable M a little closer, we see that M is not a stopping time because the occurrence or non-occurrence of the event $\{M = n\}$ depends on what happens to X_{n+1}. ∎

EXAMPLE 3.2.4   Consider a single round of offensive assault in a basketball game. At time 0, the offensive team attempts a shot and fails to score. With probability p the offensive team retains control of the ball. If so, the team waits a random time X_1 before making the next shot. This shot is successful with probability $1 - q$. If it fails to score, then the process repeats itself. Assume that $\{X_i\}$ are i.i.d. random variables with a common density f.

Each shot other than the first one made at time 0 is called a reattempt. Let $R(t)$ denote the expected number of reattempts made by time t. We now derive a closed-form expression for $R(t)$. Conditioning on X_1, we write

$$R\left(t \mid X_1 = x\right) = \begin{cases} [1 + qR(t - x)]p & \text{if } x \leq t \\ 0 & \text{otherwise.} \end{cases}$$

Applying the law of total probability, we obtain

$$R(t) = \int_0^t p[1 + qR(t - x)]f(x)dx = pF(t) + pq\int_0^t R(t - x)f(x)dx.$$

Due to the appearance of the term pq, the preceding integral equation is not in the form of Equation 3.2.1; we can use the transform approach to solve the problem. Define

$$r^e(s) = \int_0^\infty e^{-st}r(t)dt \quad \text{and} \quad R^e(s) = \int_0^\infty e^{-st}R(t)dt,$$

where $r(t) = dR(t)/dt$. Taking the derivative of $R(t)$ with respect to t, we obtain

$$r(t) = pf(t) + pq\int_0^t r(t - x)f(x)dx.$$

The Laplace transform of the previous equation is $r^e(s) = pf^e(s) + pqr^e(s)f^e(s)$, or

$$r^e(s) = \frac{pf^e(s)}{1 - pqf^e(s)}.$$

We now consider the case in which f is exponential with parameter λ. Since $f^e(s) = \lambda/(s + \lambda)$. The preceding expression reduces to

$$r^e(s) = \frac{p\lambda}{s + \lambda(1 - pq)}.$$

This gives

$$R^e(s) = \frac{1}{s} \cdot \frac{p\lambda}{s + \lambda(1 - pq)} = \frac{p}{1 - pq}\left[\frac{1}{s} - \frac{1}{s + \lambda(1 - pq)}\right],$$

where the last equality is obtained from a partial fraction expansion. Inverting the preceding expression, we find

$$R(t) = \left[\frac{p}{1 - pq}\right]\left[1 - e^{-\lambda(1 - pq)t}\right] \qquad t \geq 0.$$

Consider a team that is capable of having offensive rebounds, say with a $p = 0.7$, but a poor shooting record of $1 - q = 0.3$. Assume that the team on average holds the ball for five seconds before making a shot. This gives $\lambda = 1/\mu = 1/5 = 0.2$. In Figure 3.2, we plot $R(t)$ over the first minute since the team takes the first shot. The figure shows that $R(t)$ reaches its asymptotic value of about 0.88 when t goes beyond thirty seconds. ∎

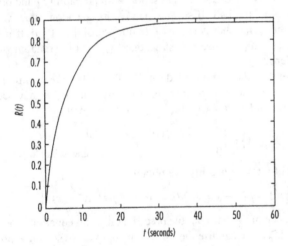

FIGURE
3.2 The expected number of reattempts in a basketball game.

EXAMPLE
3.2.5 Consider a renewal process with a Weibull interarrival time distribution given by

$$F_W(w) = 1 - \exp\left\{-\left(\frac{w}{\alpha}\right)^\beta\right\} \qquad w \geq 0,$$

where α is the scale parameter and β the shape parameter. For a given distribution function F, MATLAB program `c3_mt_f` shown in the Appendix computes the renewal function over the interval of interest. In Figure 3.3, we display the renewal function over the interval $(0, 1]$ for a renewal process that has a Weibull interarrival time distribution with $\alpha = 1$ and $\beta = 2$. For this Weibull distribution, we have $\mu = E[W] = 0.8862$ and hence the slope of the linear asymptote of the renewal function is $1/0.8862 = 1.1284$. ∎

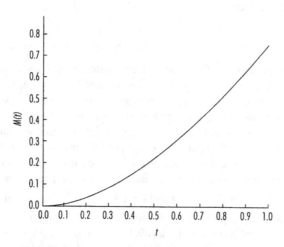

FIGURE
3.3   The renewal function for the case of Weibull interarrival times.

3.3   Excess Life, Current Life, and Total Life

Several random variables involving a given t play important roles in renewal theory and its applications. They are defined by

$$Y(t) = S_{N(t)+1} - t$$
$$A(t) = t - S_{N(t)}$$
$$T(t) = Y(t) + A(t) = X_{N(t)+1}.$$

$Y(t)$ is called the *excess life, residual life,* or the *forward recurrence time* at time t, $A(t)$ is called the *current life, age,* and *backward recurrence time* at t, and $T(t)$ is called the *total life, spread,* and *recurrence time* at t. The three random variables are depicted graphically in Figure 3.4.

One immediate reaction to the given definitions is: should the total life at time t follow the interarrival time distribution F? The answer is no in most cases. The following example gives a numerical illustration.

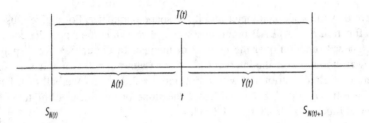

FIGURE
3.4   The excess life, current life, and total life at t.

EXAMPLE
3.3.1   Consider that we have a socket with a light bulb on continuously. Assume that the time-to-failure distribution F of the light bulb is exponential with mean 1. Whenever a light bulb goes out, we immediately replace it with a good one. When a total of t time units is accumulated, the light bulb that is in use at the time is called the marked bulb. After the marked bulb fails, we record its lifetime and start the whole experiment afresh. Let $T_j(t)$ denote the observed lifetime of the jth marked bulb. Clearly, $T_j(t)$ is the jth sampled total life of a renewal process with inter-arrival time F. If we do the experiment n times, we can compute the average total life by $\sum_{j=1}^{n} T_j(t)/n$. With $n = 100$ and $t = 100:100:1000$, we simulate the process on a computer and report the results in Figure 3.5. The figure suggests that the sampled mean total life fluctuates around 2.0. In this section, we will see that when F is exponential the mean lifetime covering the sampled epoch is asymptotically twice as long as the mean lifetime of the item. This is called the *length biased sampling*. The MATLAB program that did the simulation is given in the Appendix. ∎

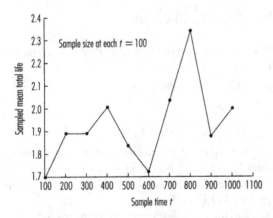

FIGURE
3.5   Sampled mean total life.

Excess-Life Distribution

Let V_t denote the distribution function of the excess-life random variable $Y(t)$, that is, $V_t(x) = P\{Y(t) \le x\}$. We now define the complementary distribution $\overline{V}_t(x) = P\{Y(t) > x\}$. Conditioning on the epoch of the first arrival X_1, we can write

$$P\{Y(t) > x | X_1 = z\} = \begin{cases} 1 & \text{if } z > t + x \\ 0 & \text{if } t < z \le t + x \\ \overline{V}_{t-z}(x) & \text{if } 0 < z \le t. \end{cases}$$

To gain a better understanding of the preceding expressions, we consider three cases shown in Figure 3.6.

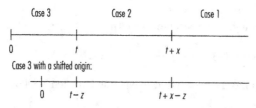

FIGURE
3.6   The three scenarios for $P\{Y(t) > x | X_1 = z\}$.

In the first case, $X_1 = z > t + x$, thus certainly $Y(t) > x$. In the second case, $t < X_1 = z \le t + x$, it is impossible for $Y(t) > x$. In fact, $Y(t)$ can at most be $z - t$. For the last case, since $X_1 = z \le t$, the process starts afresh at z with its origin shifted to z. So the original problem is identical to the new one with a new interval length of $t - z$. Applying the law of total probability, we obtain

$$\overline{V}_t(x) = \int_{t+x}^{\infty} f(z)dz + \int_0^t \overline{V}_{t-z}(x)f(z)dz$$

$$= 1 - F(t + x) + \int_0^t \overline{V}_{t-z}(x)f(z)dz.$$

Let $g(t) = \overline{V}_t(x)$ and $h(t) = 1 - F(t + x)$. We now apply Equation 3.2.2 to the preceding and find

$$\overline{V}_t(x) = 1 - F(t + x) + \int_0^t \{1 - F(t + x - y)\}m(y)dy. \tag{3.3.1}$$

The distribution function of the excess life at time t is then given by

$$V_t(x) = F(t + x) - \int_0^t \{1 - F(t + x - y)\}m(y)dy \qquad x \ge 0. \tag{3.3.2}$$

Differentiating with respect to x yields the density for $Y(t)$

$$v_t(x) = f(t + x) + \int_0^t f(t + x - y)m(y)dy \qquad x \ge 0. \tag{3.3.3}$$

Let $h(t) = 1 - F(t + x)$ in Equation 3.3.1. Then the resulting expression is in the form of Equation 3.2.2. The function $h(t)$ is clearly directly Riemann integrable and Equation 3.2.3 applies accordingly. This gives

$$\lim_{t \to \infty} \overline{V}_t(x) = \frac{\int_0^\infty [1 - F(t + x)] dt}{\mu} = \frac{\int_x^\infty [1 - F(y)] dy}{\mu}. \tag{3.3.4}$$

The limiting excess-life distribution is then given by

$$\lim_{t \to \infty} V_t(x) = 1 - \frac{\int_x^\infty [1 - F(y)] dy}{\mu} = \frac{\int_0^\infty [1 - F(y)] dy - \int_x^\infty [1 - F(y)] dy}{\mu}$$

$$= \frac{\int_0^x [1 - F(y)] dy}{\mu}. \tag{3.3.5}$$

Differentiating the preceding with respect to x gives the limiting excess-life density

$$\lim_{t \to \infty} v_t(x) = \frac{1 - F(x)}{\mu} \qquad x \ge 0. \tag{3.3.6}$$

The limiting mean excess life can be found by integrating Equation 3.3.4 over $(0, \infty)$:

$$\lim_{t \to \infty} E[Y(t)] = \frac{1}{\mu} \int_0^\infty \int_x^\infty [1 - F(y)] dy \, dx = \frac{1}{\mu} \int_0^\infty [1 - F(y)] \left[\int_0^y dx \right] dy$$

$$= \frac{1}{\mu} \int_0^\infty y [1 - F(y)] dy = \frac{1}{\mu} \int_0^\infty y \left[\int_y^\infty f(z) dz \right] dy$$

$$= \frac{1}{\mu} \int_0^\infty f(z) \left[\int_0^z y \, dy \right] dz = \frac{1}{2\mu} \int_0^\infty z^2 f(z) dz = \frac{1}{2\mu} E[X^2]. \tag{3.3.7}$$

Using the previous result, we are now ready to find the constant term of Equation 3.1.5 in the asymptotic expansion of $M(t)$. We recall that in Example 3.2.2, $E[S_{N(t)+1}] = \mu(M(t) + 1)$. Since $S_{N(t)+1} = t + Y(t)$, we conclude that $t + E[Y(t)] = \mu(M(t) + 1)$, or equivalently,

$$M(t) - \frac{t}{\mu} = \frac{E[Y(t)]}{\mu} - 1.$$

Taking the limits on both sides of the preceding equation and applying Equation 3.2.7, we obtain

$$\lim_{t \to \infty} M(t) - \frac{t}{\mu} = \frac{1}{2\mu^2} E[X^2] - 1 = \frac{E[X^2] - 2\mu^2}{2\mu^2} = \frac{\sigma^2 - \mu^2}{2\mu^2}.$$

Equation 3.1.5 is the result.

EXAMPLE 3.3.2 **A Renewal Process with Uniform Interarrival Times** Let $\{X_i\}$ be the interarrival times of a renewal process, where $X_i \sim U(0, 1)$. For this renewal process, the Laplace transform of the interarrival time density is given by

$$f^e(s) = \int_0^1 e^{-st} dt = \frac{1 - e^{-s}}{s}.$$

Using Equation 3.1.2, we find the Laplace transforms of the renewal density and the renewal function

$$m^e(s) = \frac{1 - e^{-s}}{s - (1 - e^{-s})} \quad \text{and} \quad M^e(s) = \frac{1 - e^{-s}}{s\left(s - (1 - e^{-s})\right)}.$$

With $\mu = E[X] = 1/2$ and $\sigma^2 = Var[X] = 1/12$, we use Equation 3.1.5 to find

$$\lim_{t \to \infty}\left\{M(t) - \frac{t}{\mu}\right\} = \frac{\sigma^2 - \mu^2}{2\mu^2} = \frac{\frac{1}{12} - \frac{1}{4}}{2\left(\frac{1}{2}\right)^2} = -\frac{1}{3}.$$

By numerically inverting $M^e(s)$, we compute the renewal function over $[0, 10]$. The MATLAB programs for doing the inversion are given in the Appendix. We plot $M(t) - t/\mu$ in Figure 3.7. The figure shows that the convergence of the difference to the asymptotic value $-1/3$ occurs around $t = 2$.

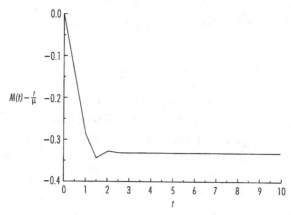

FIGURE
3.7 A plot of $M(t) - t/\mu$.

Our next task is to compute the mean excess life at time $t = 1$, namely, $E[Y(1)]$. The distribution function of X is given by $F(x) = 0$, x, and 1, for $x < 0$, $0 < x < 1$, and $x > 0$. The complementary excess-life distribution of the random variable $Y(1)$ is given by

$$\overline{V}_1(x) = 1 - F(1 + x) + \int_0^1 \{1 - F(1 + x - y)\}m(y)dy.$$

Since $0 < X < 1$, we have $0 < Y(1) < 1$ and $1 - F(1 + x) = 0$ for all $x > 0$. Now

$$F(1 + x - y) = \begin{cases} 1 & 1 + x - y > 1 \\ 1 + x - y & 0 \le 1 + x - y \le 1 \\ 0 & 1 + x - y < 0 \end{cases}$$

$$= \begin{cases} 1 & y < x \\ 1 + x - y & y \ge x \text{ and } y \le 1 \\ 0 & y > 1 \end{cases}.$$

The preceding results enable us to simply write the complementary excess-life distribution as follows

$$\bar{V}_t(x) = \int_x^1 (y - x)m(y)dy \qquad 0 < x < 1.$$

We compute the mean time-dependent excess life at time 1 using

$$E[Y(1)] = \int_0^1 \bar{V}_t(x)dx = \int_0^1 \int_x^1 (y - x)m(y)dydx.$$

Applying the MATLAB programs given in the Appendix, we find $E[Y(1)]$ = 0.3568. Define $S_n = X_1 + \cdots + X_n$ and $N(1) = \min\{n|S_n > 1\}$. Thus $S_{N(1)}$ is the epoch of the first arrival after time 1 and $S_{N(1)} = 1 + Y(1)$. Consequently we have numerically obtained $E[S_{N(1)}] = 1 + E[Y(1)] = 1.3568$. Next we will show that this result can easily be found by an application of Wald's equation and without resorting to numerical integration.

First, we observe that $N(1)$ is a stopping time with respect to $\{X_n\}$ and $E[|X_n|] = E[X_n] < \infty$. Since $N(1) \geq n$ if and only if $S_n \leq 1$, we find

$$P\{N(1) \geq n\} = P\{S_n \leq 1\} = \int_0^1 \int_0^{s_n} \cdots \int_0^{s_2} ds_1 \cdots ds_{n-1}ds_n = \frac{1}{n!}$$

and

$$E[N(1)] = \sum_{n=0}^{\infty} P\{N \geq n\} = \sum_{n=0}^{\infty} \frac{1}{n!} = e.$$

Hence $E[N(1)] < \infty$ and Wald's equation is applicable with

$$E[S_{N(1)}] = E[X_1 + \cdots + X_{N(1)}] = E[X]E[N(1)] = (1/2) \times E[N(1)].$$

This in turn gives $E[S_{N(1)}] = e/2 = 1.3591$. We see that the earlier result obtained by numerical integration and inversion is accurate only to two decimal points. ∎

EXAMPLE **A Free-Replacement Product Warranty Policy** Let $\{X_i\}$ denote product lifetimes. Assume
3.3.3 that $\{X_i\}$ are i.i.d. random variables with a common mean μ. Let t denote the warranty length of a given product. Under a free-replacement warranty, a consumer can obtain replacements of failed items so long as failures occur within the warranty period $(0, t]$. When a product fails outside of the warranty period, the consumer pays for another replacement with a new warranty, again of length t, and the process repeats afresh. A typical scenario is shown in Figure 3.8.

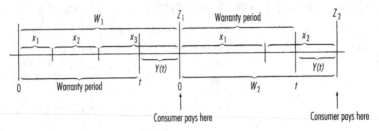

FIGURE
3.8 Two renewal processes under a free-replacement product warranty policy.

We now cast the process of a consumer's purchases as a renewal process. Let Z_i denote the ith time the consumer pays for the item (ignoring the initial one) and $W_i = Z_i - Z_{i-1}$. Define $N_1(s)$ as the number of consumer purchases in $(0, s]$. Then we see that $\{N_1(s), s \geq 0\}$ is a renewal process with interarrival times $\{W_i\}$.

Nested in each W_i is another renewal process $\{N_2(u), u \geq 0\}$ with interarrival times $\{X_i\}$, where $N_2(u)$ denotes the number of failures in $(0, u]$. Let $M_2(u)$ denote its renewal function and $Y(t)$ the excess life at t. We note that $E[W_i] = \mu(M_2(t) + 1)$ and $W_i = t + Y(t)$. Let $B(x) = P\{W_i - t \leq x\}$. Then $B(x) = P\{Y(t) \leq x\}$. We write

$$Z_n = W_1 + \cdots + W_n = nt + (W_1 - t) + \cdots + (W_n - t) = nt + H,$$

where H is the sum of the n excess lives. The distribution of H is given by $B_n(x)$, the n-fold convolution of B with itself. Hence

$$P\{Z_n \leq s\} = P\{nt + H \leq s\} = P\{H \leq s - nt\} = B_n(s - nt).$$

Now we find

$$P\{N_1(s) = n\} = P\{Z_n \leq s\} - P\{Z_{n+1} \leq s\} = B_n(s - nt) - B_{n+1}(s - (n+1)t). \quad \blacksquare$$

EXAMPLE 3.3.4   **A Periodic Review (s, S) Inventory Policy**   Assume that $\{X_i\}$ are demands for a given item in week i. $\{X_i\}$ are i.i.d. random variables with density f and mean μ. Inventory levels are reviewed at the end of each week. At the end of each week, if the inventory level is below s, we order up to S. Assume that deliveries are instantaneous. When stock is depleted, unmet demands are back ordered and met at the end of the week. We depict a typical realization in Figure 3.9.

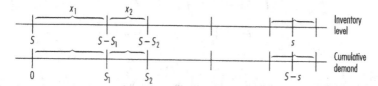

FIGURE 3.9   Inventory depletions under an (s, S) policy.

In Figure 3.9 the graph on the top displays the process of inventory depletion from S to a level below s, where $S_n = \sum_{i=1}^{n} X_i$ represents the cumulative demand in a total of n weeks since the last replenishment. In the lower graph, we relabel the axis of the top graph and the resulting graph depicting a realization of a renewal process with interarrival "times" $\{X_i\}$. Indeed we now call such a process a renewal process $N = \{N(t), t \geq 0\}$, where $N(t)$ denotes the number of weeks elapsed since the last replenishment, while during these weeks the cumulative demand is less than or equal to t, that is, $N(t) = \max\{n | S_n \leq t\}$. The expected number of weeks elapsed between two successive reordering epochs is then $1 + M(S - s)$, where $M(t)$ is the renewal function of N.

Let Z_i denote the amount demanded between the ith replenishment and the $(i - 1)$th replenishment. Then random variable $Z_i = S - s + Y(S - s)$, where $Y(t)$

denotes the excess at t of the renewal process N. The average order size is given by $(S - s) + E[Y(S - s)]$.

For large t, we recall from Equation 3.3.7 that

$$E[Y(t)] = \frac{E[X_1^2]}{2\mu} \qquad t \to \infty.$$

Hence the average order size is given by $(S - s) + \dfrac{E[X_1^2]}{2\mu}$. ∎

Current-Life Distribution

Recall that the current life at t is defined as $A(t) = t - S_{N(t)}$. Let $U_t(x) = P\{A(t) \le x\}$ denote the distribution function of $A(t)$. One way to obtain the distribution function is to use what we know about $Y(t)$, the excess life at t. This is done by noting that

$$A(t) > x \qquad \Leftrightarrow \qquad Y(t - x) > x,$$

where $t > x$. In other words, the length of current life at t is the same as the length of excess life at $t - x$. Using this relation and Equation 3.3.1, we write

$$P\{A(t) > x\} = P\{Y(t - x) > x\} = \overline{V}_{t-x}(x)$$

$$= 1 - F(t - x + x) + \int_0^{t-x} \{1 - F(t - x + x - y)\} m(y) dy$$

$$= 1 - F(t) + \int_0^{t-x} \{1 - F(t - y)\} m(y) dy.$$

Thus we find the distribution function of the current life at time t

$$U_t(x) = \begin{cases} F(t) - \int_0^{t-x} \left[1 - F(t - y)\right] m(y) dy & \text{if } x < t \\ 1 & \text{if } x \ge t. \end{cases} \qquad (3.3.8)$$

In Figure 3.10 we plot the distribution function U_t and note that $P\{A(t) = t\} = 1 - F(t)$. The last identity is expected because the current life must be t if no renewals have occurred by t.

To find the density $u_t(x)$ of $A(t)$ for $x < t$, we first let $z = t - y$ and rewrite Equation 3.3.8 as

$$U_t(x) = F(t) - \int_x^t \left[1 - F(z)\right] m(t - z) dz.$$

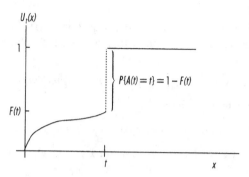

An application of Leibnitz's rule gives

$$u_t(x) = m(t - x)[1 - F(x)]. \tag{3.3.9}$$

We can give an intuitive interpretation of Equation 3.3.9. Recall that $m(t - x)\Delta x$ gives the probability that a renewal will occur in $(t - x, t - x + \Delta x)$, and $1 - F(x)$ is the probability that no arrivals occur in an interval of length x. Combining the two we can view $u_t(x)\Delta x$ as the likelihood that the current life at time t is about x.

Total-Life Distribution

Let L_t be the distribution function of the total life $T(t)$ and $\bar{L}_t(x) = P\{T(t) > x\}$. Conditioning on X_1, we can write

$$P\{T(t) > x \mid X_1 = z\} = \begin{cases} 1 & \text{if } z > \max(x, t) \\ \bar{L}_{t-z}(x) & \text{if } z < t \\ 0 & \text{otherwise.} \end{cases}$$

For the case when $z > \max(x, t)$, we see in Figure 3.11 the probability that $T(t) > x$ is clearly 1. For the other two cases, arguments similar to those used in the derivations of the excess-life distribution apply. Applying the law of total probability, we obtain

$$\bar{L}_t(x) = \int_{\max(x,t)}^{\infty} f(z)dz + \int_0^t \bar{L}_{t-z}(x)f(z)dz$$

$$= \left[1 - F(\max(x, t))\right] + \int_0^t \bar{L}_{t-z}(x)f(z)dz. \tag{3.3.10}$$

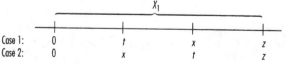

Let $g(t) = \bar{L}_t(x)$ and $h(t) = 1 - F(\max(x, t))$ in Equation 3.3.10. We see that the resulting expression is again in the form of Equation 3.2.1. Applying Equation 3.2.2, we find

$$\bar{L}_t(x) = 1 - F(\max(x, t)) + \int_0^t [1 - F(\max(x, t-y))] m(y) dy. \qquad (3.3.11)$$

The distribution function is then given by

$$L_t(x) = F(\max(x, t)) - \int_0^t [1 - F(\max(x, t-y))] m(y) dy.$$

For the previous equation, we consider two cases. For $x \geq t$, the equation simplifies to

$$L_t(x) = F(x) - \int_0^t [1 - F(x)] m(y) dy = F(x) - [1 - F(x)] M(t).$$

For $x < t$, from Figure 3.12 we can see that

$$F(\max(x, t-y)) = \begin{cases} F(t-y) & \text{if } 0 < y \leq t-x \\ F(x) & \text{if } t-x < y \leq t. \end{cases}$$

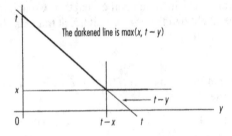

The darkened line is max($x, t-y$)

$t-y$

FIGURE
3.12   The max($x, t-y$) for y in (0, t).

Hence we have, for $x < t$,

$$L_t(x) = F(t) - \int_0^{t-x} [1 - F(t-y)] m(y) dy - \int_{t-x}^t [1 - F(x)] m(y) dy$$

$$= F(t) + \int_0^{t-x} F(t-y) m(y) dy - M(t) + \int_{t-x}^t F(x) m(y) dy.$$

Since $M(t) = F(t) + \int_0^t F(t-y) m(y) dy$ (Example 3.2.1), we rewrite the preceding expression as

$$L_t(x) = \int_{t-x}^t [F(x) - F(t-y)] m(y) dy.$$

Summarizing the two results, the total-life distribution at time t is given by

$$L_t(x) = \begin{cases} \int_{t-x}^t [F(x) - F(t-y)] m(y) dy & x < t \\ F(x) + M(t)[F(x) - 1] & x \geq t. \end{cases} \qquad (3.3.12)$$

Applying Leibnitz's rule to Equation 3.3.12 (see Example 1.4.10), we find the total-life density

$$l_t(x) = \begin{cases} f(x)[M(t) - M(t-x)] & x < t \\ f(x)[1 + M(t)] & x \geq t. \end{cases} \tag{3.3.13}$$

With $g(t) = \bar{L}_t(x)$, we apply Equation 3.2.3 to Equation 3.3.11 to obtain the limiting complementary total-life distribution as follows:

$$\lim_{t \to \infty} P\{T(t) > x\} = \frac{1}{\mu} \int_0^\infty [1 - F(\max(x, t))]dt$$

$$= \frac{1}{\mu} \left[\int_0^x [1 - F(x)]dt + \int_x^\infty [1 - F(t)]dt \right]$$

$$= \frac{1}{\mu} \left[x[1 - F(x)] + \int_x^\infty \int_t^\infty f(z)dzdt \right]$$

$$= \frac{1}{\mu} \left[x[1 - F(x)] + \int_x^\infty f(z) \int_x^z dt dz \right]$$

$$= \frac{1}{\mu} \left[x[1 - F(x)] + \int_x^\infty f(z)(z - x)dz \right] = \frac{1}{\mu} \int_x^\infty z f(z)dz.$$

The limiting total-life distribution is now given by

$$\lim_{t \to \infty} P\{T(t) \leq x\} = \frac{1}{\mu} \int_0^x z f(z)dz. \tag{3.3.14}$$

Differentiating the preceding equation with respect to x, we find the limiting total-life density

$$\lim_{t \to \infty} l_t(x) = \frac{x f(x)}{\mu} \qquad x \geq 0. \tag{3.3.15}$$

EXAMPLE 3.3.5   Consider the renewal process with an exponential interarrival time distribution $F(x) = 1 - \exp(-\lambda x)$. The counting process $N = \{N(t), t \geq 0\}$ is a Poisson process with rate λ and $M(t) = E[N(t)] = \lambda t$. Using Equation 3.3.13, we find that for a given t the total-life density is given by

$$l_t(x) = \begin{cases} \lambda^2 x e^{-\lambda x} & x < t \\ \lambda e^{-\lambda x}(1 + \lambda t) & x \geq t. \end{cases}$$

To confirm the preceding density, we find

$$\int_0^\infty l_t(x)dx = \int_0^t \lambda^2 x e^{-\lambda x}dx + \int_t^\infty \lambda e^{-\lambda x}[1 + \lambda t]dx$$

$$= [1 - e^{-\lambda t} - \lambda t e^{-\lambda t}] + e^{-\lambda t}[1 + \lambda t] = 1$$

as expected. Using Equation 3.3.12, we also obtain the total-life distribution

$$L_t(x) = 1 - [1 + \lambda \min(t, x)]e^{-\lambda x} \qquad x > 0.$$

The mean total life of a Poisson process can then easily be found. It is given by

$$E[T(t)] = \frac{1}{\lambda} + \frac{1}{\lambda}\left[1 - e^{-\lambda t}\right].$$

For the limiting total-life density, from Equation 3.3.15 we conclude that the density is given by $\lambda^2 x e^{-\lambda x}$. However, this is just the density of an Erlang $(2, \lambda)$. Therefore, for a large t the total life at t is the sum of two independent exponential distributions each with parameter λ (see Example 3.1.2). ▪

3.4 Renewal Reward Processes

Consider a renewal process $\{N(t), t \geq 0\}$ with interarrival distribution F. Let R_n denote the reward earned at the time of the nth renewal. Assume that nonnegative rewards $\{R_n\}$ are i.i.d. random variables but R_n may depend on the length of the nth renewal interval X_n, the pairs $\{(X_n, R_n), n = 1, 2, \ldots\}$ are i.i.d, random variables and $E[R_n] < \infty$ for all n.

$$\text{Let } R(t) = \sum_{n=1}^{N(t)} R_n = \text{the total reward earned in } (0, t].$$

It can be shown that

$$\lim_{t \to \infty} \frac{E[R(t)]}{t} = \frac{E[R_1]}{E[X_1]}. \tag{3.4.1}$$

If we call a renewal a *cycle*, then Equation 3.4.1 can be stated in words as

$$\text{The long-run average reward per unit time}$$
$$= \frac{\text{Expected reward received per cycle}}{\text{Expected cycle length}}.$$

This intuitively reasonable relation is quite useful and holds in fairly general conditions. For instance, the timing of the reward received in a cycle is not important and the mode of the reward received in a cycle can be arbitrary (whether in a lump sum or continuously at a constant or varying rate).

EXAMPLE
3.4.1
Alternating Renewal Processes Consider a car that can be in one of two states at any time: working or idle. Initially it is working and stays working for a length Y_1, after that it stays idle for a length Z_1, it then works for a length Y_2, and then stays idle for a length Z_2. The process continues *infinitely*. Assume $\{Y_n\}$ are i.i.d. random variables with a common mean $E[Y_n]$, $\{Z_n\}$ are i.i.d. random variables with a common mean $E[Z_n]$, but Y_n and Z_n can be dependent.

Let $X_n = Y_n + Z_n$. Then we have a renewal process $N = \{N(t), t \geq 0\}$ with inter-arrival times $\{X_n\}$. Let f be the density of X_n and μ its mean. We define the indicator random variable $I(t) = 1$ if the car is working at time t and 0 otherwise. Our objective is to find the long-run average fraction of time the car is working.

Let $g(t) = P\{I(t) = 1\}$. One way to find the time average of $g(t)$ is by using Equation 3.4.1. We envision that every time the car is working we earn a reward of one unit per unit time. Then we have $E[R_1] = E[Y_1]$, that is, the expected reward received in a cycle is equal to the expected length when the car is working in a cycle. Hence, we find

$$\lim_{t \to \infty} \frac{\int_0^t g(u)du}{t} = \frac{E[Y_1]}{\mu} = \frac{E[Y_1]}{E[Y_1] + E[Z_1]}. \tag{3.4.2}$$

Equation 3.4.2 gives the time-average probability that the car is working at any time. To derive the limiting value of $g(t)$, we condition on X_1 and write

$$g(t|X_1 = x) = \begin{cases} P\{Y_1 > t | X_1 > t\} & \text{if } x > t \\ g(t - x) & \text{if } x \leq t. \end{cases}$$

Applying the law of total probability, we find

$$g(t) = \int_t^\infty P\{Y_1 > t | X_1 > t\} f(x)dx + \int_0^t g(t - x)f(x)dx.$$

We see that the preceding equation is of the form of Equation 3.2.1 with $h(t) = \int_t^\infty P\{Y_1 > t | x > t\} f(x)dx$. Applying the key renewal theorem of Equation 3.2.3, we obtain

$$\lim_{t \to \infty} g(t) = \frac{1}{\mu} \int_0^\infty \int_t^\infty P\{Y_1 > t | x_1 > t\} f(x)dxdt$$

$$= \frac{1}{\mu} \int_0^\infty \int_t^\infty \frac{P\{Y_1 > t, X_1 > t\}}{P\{X_1 > t\}} f(x)dxdt$$

$$= \frac{1}{\mu} \int_0^\infty \int_t^\infty \frac{P\{Y_1 > t\}}{P\{X_1 > t\}} f(x)dxdt = \frac{1}{\mu} \int_0^\infty \frac{P\{Y_1 > t\}}{P\{X_1 > t\}} \int_t^\infty f(x)dxdt$$

$$= \frac{1}{\mu} \int_0^\infty P\{Y_1 > t\}dt = \frac{E[Y_1]}{\mu}.$$

We remark that in the previous derivation the third equality holds because $\{Y_1 > t\} \cap \{X_1 > t\} = \{Y_1 > t\}$.

An alternative to this approach is to condition on the time of the last renewal by time t. Recall that $m(u)du$ gives the probability that a renewal is about to occur in $(u, u + du)$. When this occurs, the renewal process starts afresh at u and Prob $\{I(t) = 1\}$ is given by $g(t - u)$. On the other hand, if no renewals occur in $(0, t]$, then Prob $\{I(t) = 1\}$ is given by $P\{Y_1 > t\}$. Combining these observations, we write

$$g(t) = P\{Y_1 > t\} + \int_0^t g(t - u)m(u)du.$$

Let $h(t) = P\{Y_1 > t\}$. It is easy to see that $h(t)$ is directly Riemann integrable, and the key renewal theorem is applicable. The limiting value of $g(t)$ can be obtained readily. ∎

EXAMPLE **A Continuous Review (s, S) Inventory Policy** Customers arrive at a store for a given item
3.4.2 in accordance with a renewal process having interarrival time $\{T_n\}$. Let D_n be the amount demanded by customer n (the nth demand occurrence). We assume that $\{T_n\}$ are i.i.d. random variables with a common distribution F, $\{D_n\}$ are i.i.d. random variables with a common distribution G, and $\{D_n\}$ are continuous random variables. A scenario is depicted in Figure 3.13.

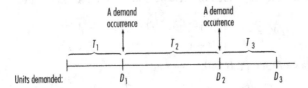

FIGURE
3.13 A sample demand occurrence scenario.

The store uses a continuous-review (s, S) ordering policy: If the inventory level after a demand occurrence is below s, then an order is placed to bring it up to S. Otherwise no order is placed. Thus if the inventory level after serving a customer is x, then the amount ordered is $S - x$ if $x < s$ and 0 otherwise. We assume instantaneous deliveries. Let $X(t) =$ the inventory level at time t (after replenishment if any) and assume $X(0) = S$. Our objective is to obtain the limiting distribution for the inventory position. Following many of our earlier examples, we first consider the complementary distribution $P\{X(t) \geq x\}$. Let $I(t) = 1$ if $X(t) \geq x$ and 0 otherwise.

In Figure 3.14, we first study the top graph. The vertical axis of the graph can be viewed as the "time" axis of a renewal process with "interarrival time" $\{D_n\}$. The horizontal axis is the time axis of another renewal process with interarrival time $\{T_n\}$. Define $N_x = \min\{n | D_1 + \cdots + D_n > S - x\}$. Customer N_x causes the inventory level to drop below x since the last replenishment. It is clear that during the interval of length $T_1 + \cdots + T_{N_x}$ the inventory level is greater than or equal to x, and during the interval of length $T_1 + \cdots + T_{N_s}$ the inventory level is greater than or equal to s. It follows that we can consider the first interval an "on" interval of an alternating renewal process; the cycle length is the latter interval. During an "on" interval we have $I(t) = 1$. This is depicted in the bottom graph of Figure 3.14. Summarizing these observations, we have

$$E[Y_1] = E\left[\sum_{i=1}^{N_x} T_i\right] \quad \text{and} \quad E[X_1] = E\left[\sum_{i=1}^{N_s} T_i\right].$$

Assume that $\{T_n\}$ and $\{D_n\}$ are independent. Hence N_x and $\{T_n\}$ are independent. So, we have $E[Y_1] = E[N_x]E[T_1]$ and similarly, $E[X_1] = E[N_s]E[T_1]$. Applying the result of Equation 3.4.2 of alternating renewal processes and Equation 3.2.4, we obtain

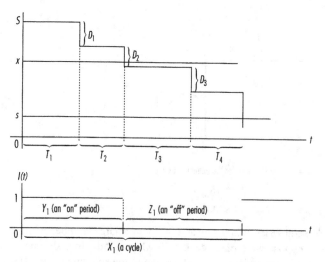

3.14   The continuous review (s, S) inventory policy.

$$\lim_{t \to \infty} P\{I(t) = 1\} = \lim_{t \to \infty} P\{X(t) \geq x\} = \frac{E[N_x]}{E[N_s]} = \frac{M_G(S-x)+1}{M_G(S-s)+1} \qquad x \leq S,$$

where $M_G(t)$ is the renewal function with "interarrival time" distribution G. In the preceding expression, it is interesting to observe that the limiting distribution has nothing to do with F—the interarrival time distribution of customers.

Consider the case in which the interarrival time distribution of customers is arbitrary and the amount demanded by a customer follows an exponential distribution with parameter λ, that is, $G(t) = 1 - e^{-\lambda t}$, $t \geq 0$. Hence $M_G(t) = \lambda t$. This gives

$$\lim_{t \to \infty} P\{X(t) \geq x\} = \frac{M_G(S-x)+1}{M_G(S-s)+1} = \frac{1+\lambda(S-x)}{1+\lambda(S-s)} \qquad s \leq x \leq S.$$

So, for large t and $s \leq x < S$, we have

$$P\{X(t) \leq x\} = \frac{\lambda(x-s)}{1+\lambda(S-s)} \qquad \text{and} \qquad f_{X(t)}(x) = \frac{\lambda}{1+\lambda(S-s)}.$$

Also,

$$P\{X(t) \geq S\} = P\{X(t) = S\} = \frac{1}{1+\lambda(S-s)}.$$

Consider the case with $s = 3$, $S = 16$, and $\lambda = 3$ items per each demand occurrence. Then for large t, we have $P\{X(t) = 16\} = 1/(1 + 3(13)) = 1/40$ and

$f_{X(t)}(x) = 3/40$. The distribution for the limiting inventory level is shown in Figure 3.15. We note that the limiting inventory level is neither continuous nor discrete. It has a uniform density over (3, 16) plus a probability mass at S. ∎

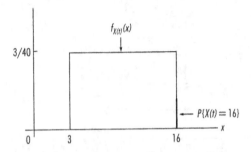

FIGURE
3.15 The limiting distribution of inventory level for the numerical example.

EXAMPLE **A Block Replacement Policy**   Consider a stage with a large number of high-intensity
3.4.3 light bulbs. Suppose we replace each bulb when it fails. In addition, all bulbs are replaced every T time units to take advantage of the economies of scale. This type of replacement policy is called the block replacement policy. Let c_1 be the unit cost of replacement at the block replacement time and c_2 the unit cost of replacement at failure.

Under this cost structure, we can simply look at the costs of each socket holding a bulb independently. If replacements were done only at failure, then they would have formed a renewal process with interarrival time distribution F, the time-to-failure distribution of the bulb. The renewal function $M(t)$ would give the expected number of failures by time t. However under the block replacement policy the aforementioned renewal process is terminated and restarted every T time units. Hence it is natural to introduce another renewal process with a constant interarrival time T. The relation between the two renewal processes is depicted in Figure 3.16.

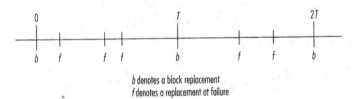

b denotes a block replacement
f denotes a replacement at failure

FIGURE
3.16 The two renewal processes—with one nested under the other.

The expected cost per renewal for this second renewal process is then $c_1 + c_2 M(t)$. Let $C(T)$ be the long-run expected average cost per unit time. For this renewal reward process, using Equation 3.4.1, we obtain

$$C(T) = \frac{c_1 + c_2 M(T)}{T}.$$

Setting $dC(T)/dT = 0$, we get the necessary condition for T that minimizes $C(T)$:

$$Tm(T) - M(T) = \frac{c_1}{c_2}. \tag{3.4.3}$$

We now consider the case in which F follows the gamma distribution

$$f(t) = te^{-t} \qquad t \geq 0.$$

Then, after some algebra, Equation 3.4.3 reduces to

$$e^{-2T}(1 + 2T) = \left\{ 1 - \left(\frac{4c_1}{c_2} \right) \right\}. \tag{3.4.4}$$

If $c_1 \geq (1/4)c_2$, then the right side of Equation 3.4.4 will always be nonpositive. However, the left side of Equation 3.4.4 will always be nonnegative for $T \geq 0$. Thus no finite T will satisfy Equation 3.4.4. This implies that the optimal T will be infinite and replacements at failure will be the minimum cost solution. On the other hand if $c_1 < (1/4)c_2$, then the left side of Equation 3.4.4 is strictly decreasing in T from an initial value of 1. Since the right side is a constant, a unique value of T exists.

We consider a numerical example with $c_1 = 1$ and $c_2 = 5$. In Figure 3.17, we plot the two sides of Equation 3.4.4 as a function of T and remark that the cost-minimizing block replacement time is around 1.5 with $c(1.5) = 2.3748$. A plot of the asymptotic cost rate would show that $C(T)$ is rather flat for $T > 0.75$ even though it is slowly trending upward after 1.5. ∎

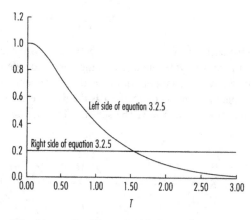

Left side of equation 3.2.5

Right side of equation 3.2.5

T

FIGURE

3.17 A plot of the two sides of Equation 3.4.4 for the numerical example.

EXAMPLE
3.4.4

An Age Replacement Policy We assume that $X_1, X_2, \ldots$ are lifetimes of items that are successively placed in service (say, the batteries successively placed in your car). Assume that these random variables are independent and identically distributed with a common distribution F and mean μ. An age replacement policy calls for replacing an item upon its failure or upon reaching a prescribed age, say T, whichever comes first. Such a policy can be economically advantageous over one calling for replacement only at failure when the item ages and/or the cost of replacement at failure is higher than that of before failure.

Similar to Example 3.4.3, we let c_1 denote the cost of a planned replacement and c_2 the cost of an unplanned replacement (imagine that your car battery fails and you are on a desert road). For a fixed T, we present three different ways to find the asymptotic cost rate. These alternate approaches demonstrate that in stochastic modeling there is often more than one way to tackle a problem; some, of course, are more intuitively appealing than others. In this example, we refer to X_n as X for notational convenience.

Approach 1

Consider the case in which we define a cycle (renewal) to be the time between two successive replacements—planned or otherwise. Let Z be the cycle length and R the cost associated with a cycle. Then $Z = \min\{X, T\}$, and $R = c_1$ with probability $[1 - F(T)]$ and c_2 with probability $F(T)$. Hence, we have

$$E[Z] = \int_0^T [1 - F(x)]dx$$

and $E[R] = c_1[1 - F(T)] + c_2 F(T)$. We call the last integral μ_T. Applying the result of Equation 3.4.1 regarding the renewal reward process, we obtain the long-run expected average cost rate $C(T)$ as follows:

$$C(T) = \frac{c_1[1 - F(T)] + c_2 F(T)}{\mu_T}. \tag{3.4.5}$$

Approach 2

Consider the case in which we define a cycle as the time between two successive unplanned replacements. Let Z be the cycle length and R the cost associated with a cycle. Then $Z = NT + Y$, where N is the number of planned replacements before the occurrence of the next unplanned replacement and Y is the lifetime associated with the unplanned replacement. We have $Y < T$. The scenario for one such cycle is depicted in Figure 3.18.

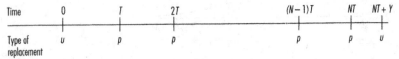

u denotes an unplanned replacement and *p* a planned replacement

FIGURE
3.18   A typical cycle under the second approach.

Couched in terms of a geometric random variable, we can view N as the number of failures before the first success, where a failure corresponds to a planned replacement and occurs with probability $q = 1 - F(T)$. For this geometric random variable, we know $E[N] = [1 - F(T)]/F(T)$. To find $E[Y]$, we see that

$$F_Y(y) = \begin{cases} 1 & y \geq T \\ \dfrac{F(y)}{F(T)} & 0 \leq y \leq T \\ 0 & \text{otherwise.} \end{cases} \tag{3.4.6}$$

So, we have

$$E[Y] = \int_0^T [1 - F_Y(y)] dy$$

$$= \left(\frac{1}{F(T)}\right) \left[TF(T) - \int_0^T F(y) dy \right]. \tag{3.4.7}$$

Since $E[Z] = E[N]T + E[Y]$, we obtain, after some cancellations, $E[Z] = \mu_T/F(T)$. For the expected cost per cycle, we see that $E[R] = c_1 E[N] + c_2 = \{c_1[1 - F(T)]/F(T)]\} + c_2$. Therefore, the long-run expected average cost per unit time is again

$$C(T) = \frac{E[R]}{E[Z]} = \frac{c_1[1 - F(T)] + c_2 F(T)}{\mu_T}.$$

Approach 3

Consider the case in which we define a cycle as the time between two successive planned replacements. Let Z be the cycle length and R the cost associated with a cycle. Then we can write

$$Z = \sum_{n=1}^{N} Y_n + T,$$

where Y_n is the length of the nth unplanned replacement since the last planned replacement, and N is the number of unplanned replacements before the occurrence of the next planned replacement. By symmetry, we conclude $E[N] = F(T)/[1 - F(T)]$. Clearly Y_n has distribution Equation 3.4.6. Even though $Y_1, Y_2, \ldots$ are independent and identically distributed random variables, N depends on $Y_1, Y_2, \ldots$. However, N is a stopping time with respect to $Y_1, Y_2, \ldots$. Thus Wald's equation applies, and $E[Z] = E[N]E[Y] + T$, where $E[Y]$ is given by Equation 3.4.7. After some substitutions and cancellations, we obtain $E[Z] = \mu_T/[1 - F(T)]$. The expected cost per cycle is given by

$$E[R] = c_2 E[N] + c_1 = c_2 \frac{F(T)}{1 - F(T)} + c_1. \tag{3.4.8}$$

Using the relation $C(T) = E[R]/E[Z]$ again produces Equation 3.4.5.

A more direct approach for obtaining Equation 3.4.5 using the third approach can be found by conditioning on X. For the denominator $E[Z]$ of the long-run expected average cost function, we see that

$$E[Z|X=x] = \begin{cases} x + E[Z] & \text{if } x < T \\ T & \text{if } x \geq T. \end{cases}$$

Then we have

$$E[Z] = \int_0^T [x + E[Z]] dF(x) + \int_T^\infty T \, dF(x)$$

$$= \int_0^T x \, dF(x) + E[Z]F(T) + T[1 - F(T)].$$

Rearranging terms yields

$$E[Z] = \frac{\int_0^T x \, dF(x) + T[1 - F(T)]}{[1 - F(T)]} = \frac{\int_0^T [1 - F(y)] dy}{[1 - F(T)]} = \frac{\mu_T}{1 - F(T)}.$$

For the numerator $E[R]$ of the long-run expected average cost function, define

$$E[R|X=x] = \begin{cases} c_2 + E[R] & \text{if } x < T. \\ c_1 & \text{if } x \geq T \end{cases}$$

Taking the expectation with respect to X, we find

$$E[R] = c_1[1 - F(T)] + [c_2 + E[R]]F(T) = c_1[1 - F(T)] + c_2 F(T) + E[R]F(T).$$

Rearranging the terms gives

$$E[R] = \frac{c_1[1 - F(T)] + c_2 F(T)}{[1 - F(T)]} = c_2 \frac{F(T)}{1 - F(T)} + c_1.$$

This is conceptually a more straightforward approach than the earlier scheme, but it is somewhat tedious in its derivations.

Consider a numerical example with its data identical to those given in Example 3.4.3. The gamma lifetime distribution is then given by $F(x) = 1 - e^{-x}(1 + x)$, $x \geq 0$. We use Equation 3.4.5 to compute $C(T)$ and plot the result in Figure 3.19. The cost-minimizing T occurs at about $T = 1.3$ with $C(1.3) = 2.2648$. ∎

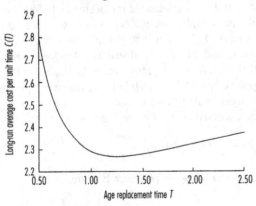

FIGURE

3.19   A numerical example of an age replacement policy.

The notion of a renewal reward process is not only useful in applications, as shown in the given examples, but also in obtaining results about renewal processes themselves. The next example illustrates this issue.

EXAMPLE **The Time Average of Expected Excess Life**   Earlier we derived the limiting mean excess
3.4.5   life a renewal process (Equation 3.3.7)

$$\lim_{t \to \infty} E[Y(t)] = \frac{1}{2\mu} E[X_1^2],$$

where μ is the mean interarrival time and $E[X_1^2]$ the second moment of inter-arrival time. Using the idea of a renewal reward process, we can easily derive a *weaker* form of the limit

$$\lim_{t \to \infty} \frac{E\int_0^t Y(u)du}{t},$$

namely, the time average of expected excess life. Assume that at any time t we are earning a reward at a rate equal to the excess life at time t. A sample path of the reward process is shown in Figure 3.20. We consider a cycle of the reward process to coincide with a renewal interval. Then the reward received in a cycle is simply the area under the triangle shown in Figure 3.20. The expected reward received in a cycle is then given by

$$E[R_1] = E\int_0^{X_1} (X_1 - t)dt = E\int_0^{X_1} u\,du = \frac{E[X_1^2]}{2}.$$

Applying Equation 3.4.1, we conclude that

$$\lim_{t \to \infty} \frac{E\int_0^t Y(u)du}{t} = \frac{E[R_1]}{E[X_1]} = \frac{E[X_1^2]}{2\mu},$$

as expected. ∎

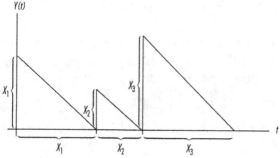

FIGURE
3.20   A sample path of the reward process.

3.5 Limiting Theorems, Stationary and Transient Renewal Processes

In the last two sections, we studied the limiting distributions of excess, current, and total lives of a renewal process. We now present a few more important asymptotic results about renewal theory.

> The *elementary renewal theorem* reads
>
> $$\lim_{t \to \infty} \frac{M(t)}{t} = \lim_{t \to \infty} \frac{E[N(t)]}{t} = \frac{1}{\mu}. \qquad (3.5.1)$$

The theorem implies that asymptotically the expected number of renewals per unit time grows linearly at a rate of $1/\mu$. We omit the formal proof. However, the result can be anticipated from the vantage point of a renewal reward process in which the expected reward per renewal is 1. When the variance σ^2 of the interarrival time is finite, it can be shown that the asymptotic variance of $N(t)$ is given by

$$\lim_{t \to \infty} \frac{Var[N(t)]}{t} = \frac{\sigma^2}{\mu^3}. \qquad (3.5.2)$$

For large t the random variable $N(t)$ is approximately normal with mean t/μ and variance $\sigma^2 t/\mu^3$—a result that can be established by the central limit theorem.

For the renewal-type Equation 3.2.1, we now let $h(y) = 1$ if $0 < y < a$ and 0 otherwise. Applying Equation 3.2.2, we see that its solution is given by

$$g(t) = h(t) + \int_0^t h(t-x)m(x)dx. \qquad (3.5.3)$$

In Equation 3.5.3, we see that when $0 < t - x < a$, or equivalently when $t - a < x < t$, then $h(t-x) = 1$. Hence for $t > a$, Equation 3.5.3 becomes

$$g(t) = \int_{t-a}^t m(x)dx = M(t) - M(t-a).$$

> Applying Equation 3.2.3, we obtain *Blackwell's renewal theorem* for the case in which interarrival times are continuous random variables:
>
> $$\lim_{t \to \infty} M(t) - M(t-a) = \frac{1}{\mu} \int_0^a dt = \frac{a}{\mu}. \qquad (3.5.4)$$

The preceding result implies

$$\lim_{t \to \infty} \frac{M(t) - M(t-a)}{a} = \frac{1}{\mu}$$

and

$$\lim_{t\to\infty} \frac{dM(t)}{dt} = \frac{1}{\mu} \quad \text{and} \quad \lim_{t\to\infty} m(t) = \frac{1}{\mu}.$$

In Example 3.1.2, we saw the uses of these identities in a renewal process with an Erlang (2, 1) interarrival time distribution.

Stationary Renewal Processes

For a Poisson process, we know that

$$m(t) = \frac{1}{\mu} \qquad \text{for } \textbf{\textit{all }} t \geq 0, \tag{3.5.5}$$

where $\mu = 1/\lambda$. We may ask if there are other renewal processes that possess the property of Equation 3.5.5—known as a *stationary* property. The answer to the question is affirmative when we are dealing with what is called the *stationary* renewal process. Before we define such a process, we first introduce the notion of a *delayed* renewal process. In contrast to an ordinary renewal process, the first interarrival time X_1 of a delayed renewal process follows a different distribution K (with a density k) and all other subsequent interarrival times still follow distribution F. There are many situations in which a delayed renewal process is in effect. For example, in equipment replacements, we may start modeling the replacement process when the first unit is already in use. The interarrival time X_1 is then the excess life of the unit at the time origin whereas other interarrival times still follow F. Since the effect of X_1 becomes negligible for a large t, the asymptotic results pertaining to an ordinary renewal process still hold for a delayed process. For time-dependent behavior of a delayed renewal process, approaches for deriving various results are similar to those used earlier for an ordinary renewal process except we now replace the first F with K in such endeavors. To illustrate, we now have

$$M_D(t) = E[N_D(t)] = \sum_{n=1}^{\infty} K * F_{n-1}(t),$$

where we use the subscript D to signify a term relating to the delayed process, and an asterisk (*) denotes the convolution of the two distributions. Let $k^e(s)$ be the Laplace transform of the density k. Then the Laplace transform

$$m_D^e(s) = \int_0^\infty e^{-st} m_D(t)\,dt$$

can be found from

$$m_D^e(s) = \frac{k^e(s)}{1 - f^e(s)}. \tag{3.5.6}$$

Consider a delayed renewal process whose initial interarrival density is given by the limiting excess-life density of Equation 3.3.6, that is,

$$k(x) = \frac{1 - F(x)}{\mu}. \tag{3.5.7}$$

The Laplace transform of the preceding density is

$$k^e(s) = \frac{1}{\mu}\left[\frac{1}{s} - \frac{1}{s}f^e(s)\right] = \frac{1}{\mu s}\left[1 - f^e(s)\right].$$

Applying Equation 3.5.6, we find $m_D^e(s) = 1/s\mu$. After an inversion, we obtain $m_D(t) = 1/\mu$ for *all* t. This delayed renewal process is called the *stationary* or equilibrium renewal process.

An interesting and expected consequence of a stationary renewal process is that its excess-life distribution (not the limiting excess-life distribution) is given by the integral on the right side of Equation 3.5.7, that is,

$$\frac{\int_0^x [1 - F(y)]dy}{\mu}. \tag{3.5.8}$$

To show the preceding, for the delayed renewal process with the first interarrival time distribution K given by the limiting excess-life distribution of Equation 3.5.8 we mimic the derivation of Equation 3.3.1 and find

$$\bar{V}_t^E(x) = 1 - K(t + x) + \int_0^t \{1 - F(t + x - y)\}m_E(y)dy \tag{3.5.9}$$

$$= \frac{1}{\mu}\int_{t+x}^{\infty}[1 - F(y)]dy + \frac{1}{\mu}\int_0^t[1 - F(t + x - y)]dy$$

$$= \frac{1}{\mu}\int_{t+x}^{\infty}[1 - F(y)]dy + \frac{1}{\mu}\int_x^{t+x}[1 - F(y)]dy = \frac{1}{\mu}\int_x^{\infty}[1 - F(y)]dy,$$

where we use E to refer to the terms relating to the stationary renewal process and the fact that $m_E(y) = 1/\mu$ for all $y \geq 0$. A direct way to write Equation 3.5.9 is to condition on the time of the last renewal before t of the stationary renewal process and note that $m_E(y)\,dy$ denotes the probability that a renewal is to occur in $(y, y + dy)$.

Transient Renewal Processes

In a transient renewal process, the interarrival time distribution is *defective* in the sense that $F(\infty) < 1$. This means that there is a positive probability $1 - F(\infty)$ that the interarrival time is of infinite length. We can think of many cases in which the notion of a transient renewal process will be applicable. For example, $\{X_n\}$ may represent the times between successive purchases of a given brand of a product. However after each purchase there is a positive probability that a customer may forego the purchase of the brand indefinitely. Then the purchase behavior can be modeled as a transient renewal process.

Let $N(\infty)$ denote the total number of renewals in $(0, \infty)$. Then we see that $N(\infty)$ follows the geometric distribution

$$P\{N(\infty) = k\} = F(\infty)^k[1 - F(\infty)] \qquad k = 0, 1, \ldots$$

with its mean given by

$$E[N(\infty)] = \lim_{t \to \infty} M(t) = \frac{F(\infty)}{1 - F(\infty)}.$$

Let L denote the time of the last renewal. We call L the *lifetime* of a renewal process. We now derive the distribution for L. Let $g(t) = P\{L > t\}$. Conditioning on the length of the first interarrival time $X_1 = x < t$, we have $P\{L > t \mid X_1 = x\} = g(t - x)$. Using the law of total probability, we obtain

$$g(t) = P\{t < X_1 < \infty\} + \int_0^t g(t - x)f(x)dx = F(\infty) - F(t) + \int_0^t g(t - x)f(x)dx.$$

Let $h(t) = F(\infty) - F(t)$. The preceding expression is again in the form of Equation 3.2.1 with its solution given by Equation 3.2.2. Therefore, we find

$$g(t) = F(\infty) - F(t) + \int_0^t [F(\infty) - F(t - x)]m(x)dx$$

$$= F(\infty) - F(t) + F(\infty)M(t) - \int_0^t F(t - x)m(x)dx$$

$$= F(\infty) - F(t) + F(\infty)M(t) - [M(t) - F(t)] = F(\infty) - [1 - F(\infty)]M(t)$$

and

$$P\{L \le t\} = 1 - g(t) = [1 - F(\infty)][1 + M(t)]. \tag{3.5.10}$$

To compute the mean $E[L]$, we observe that if X_1 is infinite then $L = 0$ (since we have defined $S_0 = 0$), and if X_1 is finite then L is the sum of X_1 plus the lifetime of another identical transient renewal process originating from X_1. Taking the expectation, we obtain

$$E[L] = \int_0^\infty t f(t)dt + F(\infty)E[L] = \int_0^\infty \int_0^t dz f(t)dt + F(\infty)E[L]$$

$$= \int_0^\infty [F(\infty) - F(z)]dz + F(\infty)E[L]$$

or

$$E[L] = \frac{1}{[1 - F(\infty)]} \int_0^\infty [F(\infty) - F(z)]dz. \tag{3.5.11}$$

EXAMPLE 3.5.1 **A Street Crossing Problem** Suppose that a pedestrian standing on a corner wants to cross the street. Let $\{S_n\}$ be the successive epochs at which cars pass by the pedestrian. Assume that the interarrival times $\{X_n\}$ associated with $\{S_n\}$ are i.i.d. random variables with a common distribution G. To cross the street, the pedestrian needs s amount of time. Hence the pedestrian starts crossing the street at $L = S_n$ such that $X_1 \le s$, $X_2 \le s$, ..., $X_n \le s$, and $X_{n+1} > s$. Therefore L is the lifetime of a transient renewal process.

For this transient renewal process, the interarrival time distribution F is given by

$$F(t) = \begin{cases} G(t) & \text{if } t \le s \\ G(s) & \text{if } t > s. \end{cases}$$

Therefore F is the distribution function associated with a defective random variable. From Equations 3.5.10 and 3.5.11, we obtain

$$P\{L \le t\} = [1 - G(s)][1 + M(t)] \qquad \text{and} \qquad E[L] = \frac{1}{1 - G(s)} \int_0^s [G(s) - G(z)]dz.$$

Consider the case in which car arrivals follow a Poisson process with mean interarrival time of five seconds and the street crossing time is ten seconds. Then we have

$$E[L] = \frac{1}{e^{-2}} \int_0^{10} \left\{ [1 - e^{-2}] - [1 - e^{-2z}] \right\} dz$$

$$= e^2 \left[\int_0^{10} e^{-2z} dz - 10 e^{-2} \right] = 5(e^2 - 1) - 10.$$

The mean *waiting* time needed to cross the street is about twenty-two seconds. ■

3.6 Regenerative Processes

Consider a stochastic process $Z = \{Z(t),\ t \ge 0\}$ with state space $S = \{0, 1, \ldots\}$ having the property that the process starts afresh at $S_1, S_2, \ldots$. By "the process starts afresh at S_n," we mean the process Z that originates at S_n follows the same probability law as the process Z that originates at S_{n-1}. Such a process Z is called a regenerative process, $\{S_n\}$ the regeneration epochs, and $\{X_n\}$ the regeneration cycles, where $X_n = S_n - S_{n-1}$ and $S_0 = 0$. For such a regenerative process, we can envision that $\{S_n\}$ are the arrival epochs of a renewal process with interarrival times $\{X_n\}$. Again we assume that $\{X_n\}$ follow distribution F with a finite mean μ.

An important and useful result of regenerative processes is

$$\lim_{t \to \infty} P\{Z(t) = j\} = \frac{E[\text{amount of time in state } j \text{ in a cycle}]}{E[\text{length of a cycle}]}$$

$$= \frac{\int_0^\infty P\{Z(t) = j,\ X_1 > t\} dt}{\mu}. \tag{3.6.1}$$

To see that the numerators of the last two terms are the same, we first sum the joint distribution over all $j \ge 0$. The integral of the resulting marginal distribution is $E[X_1]$. If we integrate the joint distribution as shown in Equation 3.6.1, we will then get the expected amount of time in state j during a cycle X_1.

The derivation of Equation 3.6.1 is similar to what we have done in the last few sections. Let $g(t) = P\{Z(t) = j\}$. Conditioning on $X_1 = x < t$, we have

$g(t|X_1 = x) = g(t - x)$. Applying the law of total probability, we obtain the renewal-type equation

$$g(t) = P\left\{Z(t) = j|X_1 > t\right\}P\{X_1 > t\} + \int_0^t g(t - x)f(x)dx$$

$$= P\{X_1 > t,\ Z(t) = j\} + \int_0^t g(t - x)f(x)dx. \qquad (3.6.2)$$

Let $h(t) = P\{X_1 > t,\ Z(t) = j\}$. It can be shown that $h(t)$ is directly Riemann integrable. So Equation 3.2.3 applies and the asymptotic result of Equation 3.6.1 follows. On the other hand, if we are interested in a time-dependent solution then we use Equation 3.2.2 to find

$$g(t) = P\{X_1 > t,\ Z(t) = j\} + \int_0^t P\{X_1 > t - x,\ Z(t - x) = j\}m(x)dx, \qquad (3.6.3)$$

where $m(x)$ is the renewal density of the renewal process with interarrival times $\{X_n\}$. When the state space S of the regenerative process Z is continuous, the asymptotic result of Equation 3.6.1 extends readily.

Letting A be a subset of S, we have

$$\lim_{t \to \infty} P\{Z(t) \in A\} = \frac{E[\text{amount of time in set } A \text{ in a cycle}]}{E[\text{length of a cycle}]}$$

$$= \frac{\int_0^\infty P\{Z(t) \in A,\ X_1 > t\}dt}{\mu}. \qquad (3.6.4)$$

We now extend the notion of a reward process to a regenerative process. Assume that the reward rate $r(t)$ at time t depends on the state of the system at the time $Z(t)$. This implies that $r[Z(t)]$ is the reward rate at time t. Then a time-average result for the *regenerative reward process* similar to Equation 3.4.1 is given by

$$\lim_{t \to \infty} \frac{E\int_0^t r(t)dt}{t} = \frac{E\int_0^t r[Z(t)]dt}{t} = \frac{E[R_1]}{\mu}, \qquad (3.6.5)$$

where

$$E[R_1] = E\int_0^{X_1} r[Z(t)]dt. \qquad (3.6.6)$$

Again, the long-run expected average reward per unit time is equal to the expected reward received per regeneration cycle divided by the expected length of a regeneration cycle.

When interarrival times are discrete random variables, we call the corresponding renewal process a *discrete renewal process*. In the next section, we will study the subject in some detail. However the time-average results given so far work readily for discrete renewal processes. This includes the useful result of Equation 3.6.5 in which summation replaces integration since the time index is now discrete.

EXAMPLE
3.6.1

Assume that visitors arrive at the San Diego Zoo in accordance with a Poisson process with rate λ. They will wait for jitneys to take them for guided tours. Assume that the interarrival times $\{X_n\}$ of jitneys at the zoo entrance are i.i.d. random variables with a common density f and each jitney can accommodate all waiting visitors. Let $Z(t)$ denote the number of visitors waiting at the entrance at time t. We see that $\{Z(t), t \geq 0\}$ is a regenerative process with regeneration points defined at each jitney arrival time. This is because a Poisson process is memoryless. At each jitney arrival time, the system (waiting stand) is emptied and the whole process starts afresh. A typical sample path during a regeneration cycle is shown in Figure 3.21.

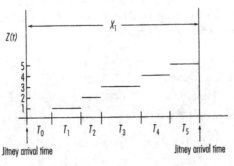

FIGURE
3.21

A sample path during a regeneration cycle.

To obtain the limiting distribution for $Z(t)$, we define T_i as the length of the interval in a cycle in which there are i visitors waiting. Applying the law of total probability, the conditional expectation for the expected length of T_i is given by

$$E\big[T_i\big|X_1 = x\big] = \sum_{j=i}^{\infty} E\big[T_i\big|X_1 = x, \ N(x) = j\big]P\{N(x) = j\},$$

where $N(x)$ is the number of visitors arrived in $(0, x]$. For a Poisson process, we know that conditioning on $N(x) = j$ the arrival times are j ordered statistics from a uniform distribution over $(0, x]$. Hence the expression simplifies to

$$E\big[T_i\big|X_1 = x\big] = \sum_{j=i}^{\infty} \frac{x}{j+1} e^{-\lambda x} \frac{(\lambda x)^j}{j!}.$$

Unconditioning on X_1, we find the mean length of the interval in a cycle with i waiting visitors

$$E[T_i] = \int_0^{\infty} f(x) \sum_{j=i}^{\infty} \frac{x}{j+1} e^{-\lambda x} \frac{(\lambda x)^j}{j!}\, dx.$$

Using Equation 3.6.2, we conclude that

$$\lim_{t \to \infty} P\{Z(t) = i\} = \frac{E[T_i]}{\mu}. \quad \blacksquare$$

EXAMPLE
3.6.2

A Regenerative Approach for Steady-State Simulation Consider a personal computer manufacturer fills orders for its Turbo 586 machines for two types of customers: its
own three retail stores and direct purchasers. Each retail store always orders in
lots of a fixed size and each direct purchaser orders one computer at a time. For
store i, we let D_i denote the constant lot size demanded at each replenishment and
T_i the number of days between successive replenishments. Assume that T_i follows
a probability mass function $q_i(t)$, $t = 1, \ldots, m_i$. For individual purchases, the probability is s_k that on a given day the number of units ordered is k and is independent of the past.

Let $Z(t)$ denote the total number of units demanded on day t from the manufacturer. Our objective is to estimate the empirical distribution $\{p(i), i = 0, 1, \ldots\}$
of $Z(t)$. Consider a regenerative reward process in which a reward of one unit is
earned when each computer is demanded. Then the long-run fraction of time the
number of units demanded is i is given by the time average of the reward rate,
that is,

$$\lim_{t \to \infty} \frac{E\left[\sum_{t=0}^{\infty} r[Z(t) = i]\right]}{t} = p(i).$$

In order to apply the asymptotic result of Equation 3.6.5 to this discrete-time
regenerative reward process, we need to find a regeneration cycle. Individual purchases are memoryless, so they will not affect the choice of a regeneration cycle.
On a given day, whether there will be an order from a retail store depends on the
conditional distribution

$$\frac{q_i(w)}{q_i(w) + q_i(w+1) + \cdots + q_i(m_i)},$$

where w is the number of days elapsed since the last replenishment for store
i—namely, it depends on the store's history. However, if on any day *all* stores
issue orders for replenishment, the future of the demand process is independent
of the past. Therefore, we define the number of days elapsed between successive
epochs at which all stores issue orders as a regeneration cycle. Let $E[X_1]$ denote
the mean length of a cycle. In Figure 3.22, we display a sample regeneration
cycle. Equation 3.6.5 implies that

$$p(i) = \frac{E[R_1]}{E[X_1]} = \frac{E\left[\sum_{t=0}^{X_1-1} r[Z(t) = i]\right]}{\mu}.$$

We note that R_1 gives the number of days within a cycle when there are i units demanded. In this simulation, we will collect sample statistics $\{(R_1, X_1), (R_2, X_2), \ldots\}$, where subscripts are used to index regeneration cycles. Each pair represents a sample independent of all other pairs. Statistically estimating the means $E[R_n]$ and $E[X_n]$ is standard. Since R_n and X_n are typically dependent, the statistical property of the ratio estimator calls for additional analysis that can be found in simulation literature. ■

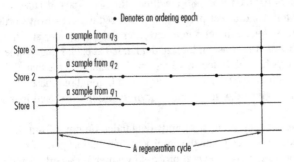

FIGURE
3.22 A sample regeneration cycle for the simulation example.

EXAMPLE
3.6.3 **The Queue Length Distribution of an M/G/1 Queue** In an $M/G/1$ queue, arrivals to a single server are Poisson with rate λ and there is no limit on the number of waiting customers. Assume that the service time distribution is G with mean $E[S]$, service is done on a first-in–first-out basis, and the queue is stable in the sense that $\lambda E[S] < 1$. Our objective is to find a way to compute the limiting distribution for the queue length—the number of customers in the system at any time. Let $Z(t)$ denote the number of customers in the system at time t. We see that the stochastic process $\mathbf{Z} = \{Z(t), t \geq 0\}$ is a regenerative process with the regeneration point defined at an arrival epoch before which the system is empty.

In Figure 3.23, we display the sample path of a regeneration cycle, say X_1. We observe that such a cycle contains a busy period with at least one customer present and an idle period with no one present. Let T_k be the amount of time in a cycle with k customers present. Then we have $X_1 = T_0 + T_1 + T_2 + \cdots$. Let p_k denote the limiting probability that there are k customers in the system at any time. From queueing theory, we know that if the queue is stable, $E[X_1]$ is finite. Consequently Equation 3.6.1 is applicable and

$$p_k = \lim_{t \to \infty} P\{Z(t) = k\} = \frac{E[T_k]}{E[X_1]}. \tag{3.6.7}$$

In finding a recursive approach to compute $\{p_k\}$, we first partition a busy period into intervals marked by service completion epochs. For example, in Figure 3.23, we have six such intervals. For $j = 1, 2, \ldots$, we let N_j denote the number of service completion epochs at which there are j remaining customers. For

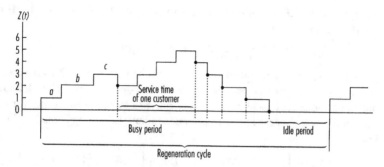

3.23   A sample path in a regeneration cycle for the $M/G/1$ queue.

example, the figure shows that $N_1 = 1$, $N_2 = 2$, $N_3 = 1$, and $N_4 = 1$. Define $E[A_{jk}]$ as the expected amount of time that there are k customers present in a service interval that starts with j customers; for example, we have $E[A_{11}] = a$, $E[A_{12}] = b$, and $E[A_{13}] = c$ in the figure. Since the sample path of $Z(t)$ can never go downward in a service interval, it is only possible to have k customers present in service intervals starting with k or fewer customers. Using Wald's equation, we have

$$E[T_k] = E[A_{1k}] + \sum_{j=1}^{k} E[N_j] E[A_{jk}] \qquad k = 1, 2, \dots . \qquad (3.6.8)$$

In Equation 3.6.8, the first term on the right side is the contribution to $E[T_k]$ made by the first arrival to an empty system, and the terms under the summation sign are contributions to $E[T_k]$ triggered by departures.

We note that $E[N_j]$ represents the expected number of service completion epochs at which there are j remaining customers. When the system is in equilibrium, we can look at the sample path of $Z(t)$ in reverse order of the time axis and, by symmetry, conclude that $E[N_j]$ also represents the expected number of arrival epochs at which there are j customers already in the system. By PASTA, specifically Equation 2.7.3, we note that the latter is given by $\lambda E[T_j]$ and hence $E[N_j] = \lambda E[T_j]$. Using this identity in Equation 3.6.8 yields

$$E[T_k] = E[A_{1k}] + \sum_{j=1}^{k} \lambda E[T_j] E[A_{jk}] \qquad k = 1, 2, \dots . \qquad (3.6.9)$$

The expected length of an idle period $E[T_0]$ is $1/\lambda$. Then Equation 3.6.7 implies that $p_0 = 1/\lambda E[X_1]$ or $E[X_1] = 1/\lambda p_0$. Dividing both sides of Equation 3.6.9 by $E[X_1]$, we find the recursive equation we are looking for:

$$p_k = \lambda p_0 E[A_{1k}] + \sum_{j=1}^{k} \lambda p_j E[A_{jk}] \qquad k = 1, 2, \dots . \qquad (3.6.10)$$

There are two possible initial values for p_0 to start the recursive computation. If we set $p_0 = 1$ (arbitrarily), we then need to normalize the resulting $\{p_k\}$ so that they sum to one. Otherwise, we can use the known result from queueing theory: $p_0 = 1 - \lambda E[S]$. To find a means to compute the input $E[A_{jk}]$ for Equation 3.6.10, we first note that

$$\int_0^\infty P\{S > t, \ N(t) = i\}dt \tag{3.6.11}$$

gives the expected amount of time in a service period S with i arrivals, where $N(t)$ is the number of arrivals in $(0, t]$ of a Poisson process. The idea underlying Equation 3.6.11 is identical to the interpretation given to the last term of Equation 3.6.1. Since S and $N(t)$ are independent, following Equation 3.6.11 we have

$$E[A_{jk}] = \int_0^\infty P\{S > t, \ N(t) = k - j\}dt = \int_0^\infty P\{S > t\}P\{N(t) = k - j\}dt$$

$$= \int_0^\infty \{1 - G(t)\}e^{-\lambda t}\frac{(\lambda t)^{k-j}}{(k-j)!}dt \qquad 1 \le j \le k. \ \blacksquare$$

EXAMPLE **Control of an M/G/1 Queue with an N Policy**   For the $M/G/1$ queue considered in the
3.6.4   previous example, we now impose a control called an N policy. Under such a policy, as soon as the facility becomes empty it is closed down. It will be reopened as soon as there are N customers present. For this facility, there is a start-up cost of K and cost of holding one customer per unit time h. Assume also that the start-up time is negligible. Our objective is find the long-run average cost per unit time for a given N.

Define the regeneration point as the time the facility is opened. The regeneration cycle X has two intervals: a server busy period B followed by a server idle period I. Let C_k, $k = B$, and I, denote the customer holding costs during the respective periods in the cycle. For the idle period, it is clear that $E[I] = N/\lambda$ and $E[C_I] = h[((N-1)/\lambda) + \cdots + (1/\lambda)] = (h/2\lambda)N(N-1)$. For the busy period, we let t_n denote the expected time until the facility becomes idle given that initially there are n customers in the system and h_n the expected holding cost incurred until the facility becomes idle given that initially there are n customers in the system. Then the long-run average cost per unit time is given by

$$\frac{E[C_I] + K + E[C_B]}{E[I] + E[B]} = \frac{\left(\dfrac{h}{2\lambda}\right)N(N-1) + K + h_N}{\dfrac{N}{\lambda} + t_N}.$$

Remaining to be determined are t_N and h_N. To this end, first we let X denote the number of arrivals in a service interval S and define $a_k = P\{X = k\}$. We see that

$$a_k = \int_0^\infty e^{-\lambda x}\frac{(\lambda x)^k}{k!}dG(x)$$

and

$$E[X] = \sum_{k=1}^\infty ka_k = E\big[E[X|S]\big] = E[\lambda S] = \lambda E[S]. \tag{3.6.12}$$

Also, we have

$$Var[X] = E\big[Var[X|S]\big] + Var\big[E[X|S]\big] = E[\lambda S] + Var[\lambda S]$$

$$= \lambda E[S] + \lambda^2 Var[S] = \lambda E[S] + \lambda^2 \{E[S^2] - E^2[S]\}$$

$$= E[X^2] - \lambda^2 E^2[S].$$

The last equality implies that

$$E[X^2] = \sum_{k=1}^{\infty} k^2 a_k = \lambda E[S] + \lambda^2 E[S^2]. \tag{3.6.13}$$

It is clear that $t_0 = 0$. For $n \geq 1$, we have

$$t_n = E[S] + \sum_{k=0}^{\infty} t_{n-1+k} a_k. \tag{3.6.14}$$

We derive the preceding expression by considering the consequences resulting from the service of the first customer. The terms under the summation signs are those contributions to t_n triggered by the arrivals during the first service time. We now argue that $t_n = nt_1$. Clearly, t_n is independent of the service discipline. Let C_k denote the kth customer of the initial n customers who are present at time 0. We choose a service discipline to serve C_1 and all arrivals during the service of C_1; we then serve C_2 and all arrivals during the service of C_2, and so on. Let W_k be the interval starting from the service of C_k and ending with the completion of the services of all arrivals during the service of C_k. We see that $E[W_k] = t_1$. This leads to the conclusion that $t_n = nt_1$. With $n = 1$ in Equation 3.6.14, we use Equation 3.6.12 to obtain

$$t_1 = E[S] + \sum_{k=0}^{\infty} t_k a_k = E[S] + \sum_{k=0}^{\infty} k a_k t_1 = E[S] + \lambda E[S] t_1$$

$$t_1 = \frac{E[S]}{1 - \lambda E[S]} \quad \text{and} \quad t_n = \frac{nE[S]}{1 - \lambda E[S]}.$$

Let H_n denote the holding cost during the service of the first customer C_1 when initially there are n customers at time 0. There are two groups of customers to consider: the initial n customers and those who arrive during the first service interval. For the former, the expected holding cost during the service of C_1 is $nhE[S]$; for the latter, it is given by (see Example 2.2.3)

$$hE\left[\int_0^S \lambda t\, dt\right] = \lambda h E\left[\frac{S^2}{2}\right] = \frac{\lambda h}{2} E[S^2].$$

Hence $E[H_n] = nhE[S] + (\lambda h/2)E[S^2]$. We mimic the derivation of Equation 3.6.14 to write

$$h_n = E[H_n] + \sum_{k=0}^{\infty} h_{n-1+k} a_k. \tag{3.6.15}$$

Following the aforementioned service discipline, during the interval W_k we observe that the expected holding cost is $h_1 + h(n - k)t_1$, where the first term is the expected holding cost associated with all customers who are served in W_k and the remaining term is the total holding cost associated with customers $C_{k+1}, \ldots, C_n$ who must wait during the interval W_k. This in turn implies that h_n is quadratic in n; specifically we have

$$h_n = \sum_{k=1}^{n} \left\{ h_1 + h(n-k)t_1 \right\} = nh_1 + \frac{1}{2} hn(n-1)t_1. \qquad (3.6.16)$$

With $n = 1$ in Equation 3.6.15, we obtain

$$h_1 = E[H_1] + \sum_{k=0}^{\infty} h_k a_k = E[H_1] + \sum_{k=0}^{\infty} \left(kh_1 + \frac{1}{2} hk(k-1)t_1 \right) a_k \qquad \text{(by Eq. 3.6.16)}$$

$$= E[H_1] + \lambda E[S] h_1 + \frac{1}{2} h t_1 \left[\sum_{k=0}^{\infty} k^2 a_k - \sum_{k=0}^{\infty} k a_k \right]$$

$$= E[H_1] + \lambda E[S] h_1 + \frac{1}{2} h t_1 \lambda^2 E[S^2] \qquad \text{(by Equations 3.6.12 and 3.6.13)}.$$

This means

$$(1 - \lambda E[S]) h_1 = hE[S] + \frac{\lambda h}{2} E[S^2] + \frac{1}{2} h \lambda^2 E[S^2] \left[\frac{E[S]}{1 - \lambda E[S]} \right]$$

$$= h \left[E[S] + \frac{\lambda E[S^2]}{2\{1 - \lambda E[S]\}} \right]$$

or

$$h_1 = \frac{h}{1 - \lambda E[S]} \left[E[S] + \frac{\lambda E[S^2]}{2\{1 - \lambda E[S]\}} \right].$$

Since h_1 and t_1 are now known, from Equation 3.6.16, we conclude that

$$h_n = \frac{h}{1 - \lambda E[S]} \left[\frac{1}{2} n(n-1) E[S] + nE[S] + \frac{\lambda n E[S^2]}{2\{1 - \lambda E[S]\}} \right].$$

Therefore the long-run expected average cost per unit time can be computed accordingly. It is interesting to note that this cost depends only on the first two moments of the service time distribution. ∎

EXAMPLE **3.6.5**   **Little's Formula in the Context of a GI/G/1 Queue**   Consider a single server service system with an infinite number of waiting spaces. Interarrival times to the system and service times of each customer are i.i.d. random variables. The two sets of random variables are mutually independent. Let $Z(t)$ denote the number of customers in the system at time t. Then a typical sample path is identical to that depicted in Figure 3.23. We define the regeneration epoch exactly as we did in Example 3.6.3—namely, the epoch of an arrival to an empty system. We let X denote the length of a regeneration cycle. Such a cycle is shown in Figure 3.23.

Our objective in this example is to derive a simple formula for the time average L of $\{Z(t), t \ge 0\}$, that is,

$$L = \lim_{t \to \infty} \frac{E\left[\int_0^t Z(u)du\right]}{t}.$$ (3.6.17)

In queueing theory, the preceding quantity is called the mean queue length.

We now use the regenerative reward approach to solve the problem. Assume the system earns one unit per unit time if one customer is in the system. Then the total reward received in a cycle is simply the area under the "curve" shown in Figure 3.23. Using Equation 3.6.5, we can rewrite it as $L = E[R]/E[X]$, where $E[R]$ is the expected total reward received in a cycle—the expected area under the curve. Let W_i denote the sojourn time of customer i in the system (the waiting time plus the service time) and N the number of arrivals in a cycle. A glance at Figure 3.23 will convince us that $E[R] = E[W_1] + \cdots + E[W_N]$ and

$$L = \frac{E\left[\sum_{i=1}^{N} W_i\right]}{E[X]}.$$ (3.6.18)

To ensure $E[X]$ is finite, we require that the queue is stable. This means that the arrival rate is less than the service rate.

Consider another regenerative reward process in which the system earns one unit per each arriving customer. Then, applying Equation 3.6.5 once more, we obtain the time average of the number of arrivals per unit time λ, that is, the arrival rate,

$$\lambda = \lim_{t \to \infty} \frac{N(t)}{t} = \frac{E[N]}{E[X]},$$ (3.6.19)

where $N(t)$ is the number of arrivals by time t. Let W denote the mean sojourn time per customer. We have

$$W = \frac{E\left[\sum_{i=1}^{N} W_i\right]}{E[N]} = \frac{E\left[\sum_{i=1}^{N} W_i\right]}{E[X]} \frac{E[X]}{E[N]} = \frac{L}{\lambda},$$

where the last equality is due to Equations 3.6.18 and 3.6.19. Hence we obtain Little's formula:

$$L = \lambda W.$$ (3.6.20)

While our derivation is for a *GI/G/*1 queue, Little's formula is applicable in more general settings. For example, Equation 3.6.20 works for a *GI/G/c* queue in which there are c (≥ 1) servers. Also, it works for a *GI/G/c/c* queue—a queue without waiting room—provided that the arrival rate is properly adjusted to reflect the *effective* arrival rate. ∎

EXAMPLE **Using Trailing Stops in a Bull Market** Consider a stock that is doing well in the sense
3.6.6 that its price is trending upward even though it fluctuates. For simplicity, we use
a simple random walk to model the price change. Let P_k denote the price at the
end of period k and Y_k denote the price change in the kth period. Assume that
$P\{Y_i = 1\} = p$, and $P\{Y_i = -1\} = 1 - p \equiv q$. With $P_0 = 0$, we have $P_n = Y_1 + \cdots + Y_n$,
$n = 1, 2, \ldots$. Since the price is trending upward, we assume $E[Y_1] = p - q > 0$.

To protect a trader from an appreciable decline in price, the trader may use a
trading strategy involving a trailing stop L. Let M_n be the maximum price reached
by the stock to date, that is, $M_n = \max\{P_k, 0 \le k \le n\}$. Following the strategy, the
stock will be sold at the end of period n when the current price P_n is equal to
$M_n - L$ for the first time. In this case, the trade is said to be "stopped out" when
the stock incurs a $\$L$ "drawdown" from its peak price. By using the trailing stop,
the trader limits the loss to $\$L$. Let $T(L)$ denote the number of periods until the
trade stops out, that is, $T(L) = \min\{n > 0 \mid P_n = M_n - L\}$. We are interested in finding the mean time to be stopped out $E[T(L)]$. Let $G(L)$ denote the total gain on
this trade. We also want to compute the first two moments of $G(L)$.

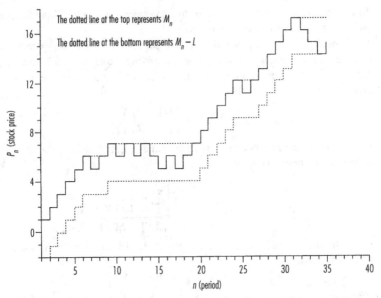

3.24 A simulated sample path of the stock price process.

In Figure 3.24, we present a simulated sample path of the price process $\{P_n\}$
with $p = 0.68$ and $L = 3$. We see in this case, $T(3) = 34$ and $G(3) = 14$. Let D_n
denote the drawdown at the end of period n. The amount $D_n \equiv M_n - P_n$ is the loss
relative to the stock's best performance to date. In the figure, $\{D_n\}$ are shown in
the form of gaps between the solid line representing the stock prices and the
dotted line at the top representing the best performances to date. We see that each
time when $D_n = 0$ the process governing $\{D_n\}$ restarts. We hence consider any

period n with $D_n = 0$ a regeneration point. Let N_k denote the kth time the price makes or ties all time high and set $N_0 = 0$. We see that $\{N_k\}$ are the regeneration points. For $k \geq 1$, define $n_k = N_k - N_{k-1}$, and $i_k = P_{N_k} - P_{N_{k-1}}$. The number of periods in regeneration cycle k is given by n_k, and the net price change over regeneration cycle k is i_k. To illustrate the use of these notations, in the figure we see for example that $N_{10} = 13$, $N_{11} = 19$, $n_{11} = 6$, and $i_{11} = 0$. In other words, the eleventh regeneration cycle lasts for six periods and the net price change over this cycle is zero. The figure suggests that when the price moves down from the best performance to date by one unit (for the first time in a regeneration cycle) the length of the cycle will always be longer than one; when the price moves up from all time high by one unit, the resulting cycle will only be of length one. We conclude that from period 0 to period $T(L)$, there are three type of regeneration cycles: B_1 cycles with $n_k = 1$ and $i_k = 1$; B_2 cycles with $n_k > 1$, $i_k = 0$, and with drawdowns in each such cycle less than L; and a final cycle with $n_k > 1$ and the last drawdown in the cycle is L. We call these cycles types 1, 2, and 3 cycles, respectively. For the sample path shown in Figure 3.24, there are seventeen type 1 cycles and five type 2 cycles. The total gain on this trade is given by $G(L) = B_1 - L$. For the example, we have $G(3) = 17 - 3 = 14$, as expected.

Consider a type 2 cycle (say, the eleventh cycle). In this type of cycle, the first step is a downward move of one unit. The probability, denoted by $\alpha(L)$, that the price process $\{P_n\}$ will move to the level that triggers stopping out (the lower boundary) before it moves back to its all time high (the upper boundary) can be linked to a well-known result for a related simple random walk. In this second random walk, each step is again $+1$ with probability p and -1 with probability q, the lower boundary is the X-axis, the upper boundary is the horizontal line with a distance of L units above the X-axis, and the initial position of the walk is one unit below the upper boundary (triggered by the first time the price makes a downward move). The probability that the second random walk will reach the lower boundary before it reaches the upper boundary is $\alpha(L)$ and given by

$$\alpha(L) = \frac{1 - (p/q)}{1 - (p/q)^L} \tag{3.6.21}$$

(see Problem 37, Chapter 4). Define $\beta(L) = 1 - \alpha(L)$. For a regeneration cycle, the probabilities that it is of type 1, 2, or 3 cycle are p, $q\beta(L)$, and $q\alpha(L)$, respectively.

Given $B_1 = n$, the total numbers of types 1 and 2 cycles that can occur before stopping out are $n, n+1, \ldots$. By independence of these regeneration cycles, for $n \geq 0$, we write

$$P(B_1 = n) = q\alpha(L) \sum_{k=n}^{\infty} \binom{k}{k-n} p^n \left(q\beta(L)\right)^{k-n} = q\alpha(L) p^n \sum_{m=0}^{\infty} \binom{m+n}{m} \left(q\beta(L)\right)^m$$

$$= q\alpha(L) p^n \left(1 - q\beta(L)\right)^{-(n+1)} = V^n U,$$

where $U \equiv q\alpha(L)/\left(1 - q\beta(L)\right)$, $V \equiv 1 - U$, and we use the transform pair Z-5 to derive the third equality. Therefore, we conclude that B_1 follows a geometric distribution with $E[B_1] = V/U = p/q\alpha(L)$ and $Var[B_1] = V/U^2 = p\left(1 - q\beta(L)\right)/\left(q\alpha(L)\right)^2$. It follows that the mean and variance of the gain $G(L)$ are: $E[G(L)] = E[B_1] - L$ and $Var[G(L)] = Var[B_1]$. For the mean of $T(L)$, we note that

$$E[G(L)] = E\left[\sum_{i=1}^{T(L)} E[Y_i]\right] = E\left[\sum_{i=1}^{T(L)} (p-q)\right] = E[T(L)](p-q),$$

where the last equality is due to Wald's equation. Hence, we find $E[T(L)] = E[G(L)]/(p-q)$. For the numerical example with $p = 0.68$ and $L = 3$, we obtain $E[G(3)] = 13.23$, $Var[G(L)] = 279.86$, and $E[T(3)] = 36.77$. ∎

3.7 Discrete Renewal Processes

In this section we consider a renewal process in which interarrival times $\{X_n\}$ are i.i.d. nonnegative integer-valued random variables. Let $N(n)$ denote the number of renewals by time n (excluding the initial renewal at time 0). The counting process $N = \{N(n), n = 0, 1, \ldots\}$ is defined on the set of nonnegative integers. It is called the discrete-time renewal process or discrete renewal process. We assume that interarrival time probabilities are given by $f_k = P\{X_1 = k\}$, $k = 0, 1, \ldots$ with distribution function $F(k) = f_0 + f_1 + \cdots + f_k$ and $f_0 = 1$. As before, we also assume that $E[X_1] < \infty$.

Let $M(n)$ denote the expected number of renewals *by* time n and $m(n)$ the expected number of renewals *at* time n. Define $m(0) = 1$. Conditioning on the time X_1 of the first renewal and applying the law of total probability, we find

$$M(n) = \sum_{k=0}^{n} f_k[1 + M(n-k)] = F(n) + \sum_{k=0}^{n} M(n-k)f_k \qquad n = 0, 1, \ldots \qquad (3.7.1)$$

$$m(n) = \delta(n) + \sum_{k=0}^{n} f_k m(n-k) \qquad n = 0, 1, \ldots, \qquad (3.7.2)$$

where $\delta(n) = 1$ if $n = 0$ and 0 otherwise. For $n \geq 1$, we note that $m(n) = M(n) - M(n-1)$. Equation 3.7.1 is a special case of a discrete-time renewal-type equation

$$g_n = h_n + \sum_{k=0}^{n} g_{n-k} f_k \qquad n = 0, 1, \ldots, \qquad (3.7.3)$$

where we assume that $\{h_n\}$ are given and bounded. The solution of Equation 3.7.3 is given by

$$g_n = \sum_{k=0}^{n} h_{n-k} m(k) \qquad n = 0, 1, \ldots. \qquad (3.7.4)$$

To establish Equation 3.7.4, we use the Z-transform approach. Let

$$f^g(z) = \sum_{k=0}^{\infty} z^k p_k, \quad m^g(z) = \sum_{k=0}^{\infty} z^k m(k), \quad h^g(z) = \sum_{k=0}^{\infty} z^k h(k), \quad g^g(z) = \sum_{k=0}^{\infty} z^k g(k).$$

Then we find from Equation 3.7.2 that $m^g(z) = 1 + f^g(z)m^g(z)$ or $m^g(z) = 1/[1 - f^g(z)]$. From Equation 3.7.3, we find $g^g(z) = h^g(z) + g^g(z)f^g(z)$ or

$g^g(z) = h^g(z)/[1 - f^g(z)] = h^g(z)m^g(z)$. Inverting the last expression, we obtain Equation 3.7.4.

For asymptotic results of $\{g_n\}$, we need to introduce the notion of *periodicity*. If $d \geq 2$ is the greatest common divisor of the set of all n such that $f_n > 0$, then X_1 (or F) is said to be periodic with *period d*. Otherwise, X_1 is said to be *aperiodic*. Note that if $f_1 > 0$ then X_1 is always aperiodic. Let $\mu = E[X_1]$. If X_1 is aperiodic, it can be shown that

$$\lim_{n \to \infty} g_n = \frac{\sum_{n=0}^{\infty} h_n}{\mu}. \tag{3.7.5}$$

This result is the discrete analog of Equation 3.2.3. If X_1 is periodic, then the time average of $\{g_n\}$ is given by

$$\lim_{n \to \infty} \frac{1}{n} \sum_{k=1}^{n} g_n = \frac{\sum_{n=0}^{\infty} h_n}{\mu}. \tag{3.7.6}$$

Let $g_n = m(n)$ and $h_n = \delta(n)$ so that Equation 3.7.2 is now in the form of Equation 3.7.3.

If X_1 is aperiodic, then Equation 3.7.5 implies that

$$\lim_{n \to \infty} m(n) = \frac{1}{\mu}, \tag{3.7.7}$$

and if X_1 is periodic, we have the time average from Equation 3.7.6 that

$$\lim_{n \to \infty} \frac{\sum_{k=1}^{n} m(k)}{n} = \lim_{n \to \infty} \frac{M(n)}{n} = \frac{1}{\mu}. \tag{3.7.8}$$

If X_1 is periodic with period d, Blackwell has shown that

$$\lim_{n \to \infty} m(nd) = \lim_{n \to \infty} E[\text{number of renewal at } nd] = \frac{d}{\mu}. \tag{3.7.9}$$

EXAMPLE **The Excess Life of a Discrete Renewal Process** The excess life of a discrete renewal
3.7.1 process is defined as $Y(n) = S_{N(n)+1} - n$, where $S_n = X_1 + \cdots + X_n$. Define $g_n(m) = P\{Y(n) > m\}$. Following the discrete analog of Figure 3.6, we write

$$P\{Y(n) > m \mid X_1 = k\} = \begin{cases} 1 & \text{if } k > n + m \\ 0 & \text{if } n < k \leq n + m \\ g_{n-k}(m) & \text{if } 0 < k \leq n. \end{cases}$$

Applying the law of total probability, we obtain

$$g_n(m) = \sum_{k>n+m} f_k + \sum_{k=0}^{n} g_{n-k}(m)f_k = [1 - F(n+m)] + \sum_{k=0}^{n} g_{n-k}(m)f_k.$$

Assuming that X_1 is aperiodic, the limiting complementary distribution is given by

$$\lim_{n\to\infty} P\{Y(n) > m\} = \frac{\sum_{k=0}^{\infty} [1 - F(k+m)]}{\mu}. \tag{3.7.10}$$

Clearly, Equation 3.7.10 is the discrete analog of Equation 3.3.4. Finally, we find that

$$\lim_{n\to\infty} P\{Y(n) = m\} = \lim_{n\to\infty} P\{Y(n) > m-1\} - \lim_{n\to\infty} P\{Y(n) > m\}$$

$$= \frac{\sum_{k=0}^{\infty}[1 - F(k+m-1)] - \sum_{k=0}^{\infty}[1 - F(k+m)]}{\mu}$$

$$= \frac{1 - F(m-1)}{\mu} = \frac{P\{X_1 \geq m\}}{\mu}. \quad\blacksquare \tag{3.7.11}$$

Problems

1 Find $P\{N(t) \geq k\}$ for a renewal process with interarrival density

$$f(x) = \begin{cases} \mu e^{-\mu(x-a)} & \text{if } x > a \\ 0 & \text{if } x \leq a, \end{cases}$$

where $a > 0$ is a fixed constant.

2 Use the Laplace transform approach to find the renewal function corresponding to the interarrival density

$$f(x) = \lambda^2 x e^{-\lambda x} \qquad x > 0.$$

3 Consider a renewal process with interarrival density given by

$$f(t) = 2(e^{-t} - e^{-2t}) \qquad t > 0.$$

(a) What is this density? (b) Find the renewal function $M(t)$. (c) Compute

$$\lim_{t\to\infty}\left\{ M(t) - \left(\frac{t}{\mu}\right) \right\}$$

and verify that this limit is given by Equation 3.1.5.

4 Consider a renewal process $N = \{N(t), t \geq 0\}$ for which the interarrival times $\{X_n\}$ are i.i.d. Poisson random variables with parameter λ. (a) What is the probability distribution for the arrival time of the nth event S_n? (b) How do you compute $P\{N(t) = k\}$? (c) Compute $P\{N(t) = n\}$ for the case with $\lambda = 4$, $t = 5:5:25$, and $n = 0:1:15$. For each t value, plot the probability mass function

of $N(t)$. Comparing these plots, what would you conjecture about the asymptotic distribution of $N(t)$? Hint: The identity $P\{N(t) = n\} = P\{S_n \leq t\} - P\{S_{n+1} \leq t\}$ also works for the case in which the interarrival times are discrete random variables.

5 A Poisson process is a renewal process in which the interarrival times are exponential random variables with parameter λ. **(a)** Find $P\{A(t) > x, Y(t) > y\}$, where $A(t)$ is the age at time t, and $Y(t)$ is the excess life at time t. (Hint: Draw a picture and see what can be said about the event implied by the two joint events.) **(b)** From Part a, obtain the marginal densities for the two random variables. **(c)** Are the two random variables independent? **(d)** Find $P\{S_{N(t)+1} > t + x\}$. **(e)** Find $P\{S_{N(t)} \leq s\}$.

6 The time-to-failure distribution of a piece of equipment follows an exponential distribution with mean $1/\mu$. The repair involves the replacement of a module; thus the repair time can be considered instantaneous. The status of the machine is only inspected at equally spaced intervals T, $2T$, ... scheduled immediately after each repair. What is the fraction of time the machine is in working condition? Use two distinct approaches to solve the problem (by defining the regeneration point differently).

*7 A Continuous-Review Inventory System with Arbitrary Interarrival Distribution between Unit Demand   Let $\{X_i\}$ be the interarrival times between successive demand occurrences. Each demand occurrence incurs a withdrawal of a single unit from the stock. We assume that $\{X_i\}$ are i.i.d. random variables with a common distribution F and density f. Let $H(t)$ be the inventory position at time t. Consider the following inventory replenishment policy: if $H(t)$ reaches the reorder level s, order $Q = S - s$ units so that the inventory position immediately reaches S again. Therefore this policy is of (s, S) variety. At time 0, a demand has just occurred resulting in an inventory position $H(0) = s + i$ $(i = 1, 2, \ldots, Q)$. A sample path of the inventory position is shown in the following graph.

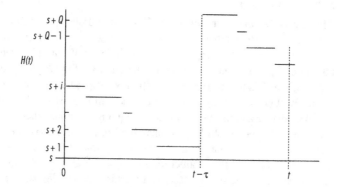

We see that the stochastic process $\{H(t), t \geq 0\}$ is a continuous-time stochastic process with a discrete state space $S = \{s + 1, s + 2, \ldots, s + Q\}$. For notational convenience, we define an equivalent process $\{\tilde{H}(t), t \geq 0\}$ with

$\tilde{H}(t) = H(t) - s$. Then the two stochastic processes are identical except that the latter is defined on the state space $\{1, 2, \ldots, Q\}$. (a) For $k \geq 1$, let $S_k = Y_1 + \cdots + Y_k$, where Y_k denote the interarrival time of the kth and $(k-1)$st reorders. Let $f_n(\cdot)$ denote the n-fold convolution of f with itself. What is the density of the random variable S_k (using the convolution notation of f)? (b) For $k \geq 1$, let $E(k, t-\tau)$ denote the following event: in the interval $[0, t-\tau]$ there are exactly k reorders and the kth reorder occurs exactly at the end of the interval. Also let event F denote that there is no reorder in the interval $(t-\tau, t]$. For $n = 1, \ldots, Q$, find $P\{\tilde{H}(t) = n | E(k, t-\tau), F\}$ in terms of the convolution notation involving F (as usual, we let F_n denote distribution function of f_n and $F_0(x) \equiv 1$ for all x). (Hint: What is demand in an interval of length τ?) (c) Let $E(0, t)$ denote the event that in the interval $[0, t]$ there are no reorders. For $n = 1, \ldots, Q$, find $P\{\tilde{H}(t) = n | E(0-t)\}$ in terms of the convolution notation involving F. (d) Conditioning on all possible values of k and S_k, find $P\{\tilde{H}(t) = n\}$ for $n = 1, \ldots, Q$. (e) Define the following Laplace transforms

$$P^e(n, v) = \int_0^\infty e^{-vt} P\{H(t) = n\} dt \quad \text{and} \quad f^e(v) = \int_0^\infty e^{-vt} f(t) dt$$

(sorry, we cannot use s as the argument of a Laplace transform here since s has already been used in a different context). For $n = 1, \ldots, Q$, find $P^e(n, v)$. (f) Use the final-value theorem of the Laplace transform to show that

$$P\{\tilde{H} = n\} = \lim_{t \to \infty} P\{\tilde{H}(t) = n\} = 1/Q,$$

for $n = 1, \ldots, Q$.

8 The Distribution of the Last Arrival Time before t of a Renewal Process
We recall that $S_{N(t)} = X_1 + \cdots + X_{N(t)}$ denotes the last arrival time before time t. (a) Show that

$$P\{S_{N(t)} \leq s\} = \bar{F}(t) + \int_0^s \bar{F}(t-y)m(y)dy \qquad 0 \leq s \leq t.$$

(b) What is the density of $S_{N(t)}$? (c) For the Poisson process with parameter λ, find $P\{S_{N(t)} \leq s\}$.

9 The Free-Replacement Product Warranty Policy Consider Example 3.3.3 again. (a) Recall that W_i represents the time between the ith purchase and the $(i-1)$st purchase made by the consumer and $\{W_i\}$ are i.i.d. random variables. Let F_W denote the distribution function of W_i. Find F_W in terms of F and $m_2(u)$, where F is the distribution function of the product lifetime and $m_2(u)$ is the renewal density of the renewal process with interarrival time distribution F. (b) Consider a consumer who has a planning horizon, called the product's life cycle, of L years. Within the planning horizon, the consumer will make purchases of the identical product if necessary so that the consumer will have a continued usage of the product for a total of L years. Assume that the product is covered by the free-replacement warranty policy described in Example 3.3.2, and the unit purchase cost is c dollars. What is the total expected purchase cost to be paid by the consumer over $(0, L]$? (c) Assume that the product under consideration is a laptop computer costing $2000. The product has a free-replacement policy covering a two-year period (that is, $t = 2$). Assume that the product lifetime follows the Erlang distribution described in Example 3.1.2. Plot $F_W(u)$ over $(0, 4]$ and $M_1(u)$ over $(0, 4]$, where $M_1(u)$ represents the expected number of purchases in $(0, 4]$. Compute the expected total purchase cost to be paid by the consumer over the four-year period $(0, 4]$. By comparing graphs of $F_W(u)$ and $M_1(u)$ over the interval $(0, 4]$, what conclusion will you draw? **(d)** Do the same for the interval $(0, 10]$, that is, with $L = 10$.

10 Statistical Process Control   Consider a continuous manufacturing process that produces a chemical compound at a rate of fifty gallons per hour. The critical measurement of the compound is its viscosity. When the process is in control, the viscosity readings follow a normal distribution with mean μ_0 and variance σ^2; when the process is out of control, the readings follow a normal distribution with mean μ_1 and the same variance σ^2, where $\mu_1 = \mu_0 + 2\sigma$. At time 0, the process has just been calibrated so that it is in control. The process, on average, will stay in control for eight hours. Once it is out of control, it will remain there until a recalibration is done so that it will be in control again. Suppose that the manufacturing process is under a process control scheme using a control chart. The manufactured product is sampled every half hour. If the sampled measurement exceeds $\mu_0 + k\sigma$, then we conclude that the process is out of control. In this case, there are two possibilities: if the process is actually in control but the sample signals that the process is out of control, then the cost per each false alarm is $500; if it is not a false alarm, then a cost of recalibration exists but we ignore it in our calculation since this corresponds to a right decision. Balancing against the costs of having to deal with false alarms, there is the cost of being out of control and without being detected. This cost is $2 per gallon when production is done while the process is out of control. Assume that sampling and recalibration times are negligible. Compute the expected average costs per hour for $k = 1 : 0.1 : 3$. What is the optimal k that minimizes the average cost per hour?

11 Consider an inventory system in which demand for an item follows a Poisson process with parameter λ. Inventory replenishments can only be done when the system is out of stock and a replenishment opportunity is available. The time between successive replenishment opportunities follows a renewal process with an interarrival time distribution F whose mean is μ, and variance is σ^2. Each replenishment order is of size Q. Unmet demands are lost. The fixed cost of replenishment is K per order. The cost of holding one unit of inventory per unit time is h. The cost of lost sales is π per each unmet demand. What is the long-run average cost per unit time?

12 The Periodic Review (s, S) Policy Revisited   We now return to Example 3.3.4 and consider the cost of holding inventory. Let the cost of holding one unit of inventory per unit time be 1. Define $h(x)$ as the expected total holding cost from the start of a week with x units on hand to the start of the week at which a replenishment is made given that $x > s$. **(a)** Conditioning on the size of demand during the week, write a renewal-type equation for $h(x)$. **(b)** Apply Equation 3.2.2 to solve the renewal-type equation so that $h(x)$ is expressed in terms of the renewal function $M(\cdot)$ whose interarrival time density is f. **(c)** What is the expected average cost of holding inventory per week?

13 Opportunity-Based Block Replacement   Consider a variant of the block replacement problem presented in Example 3.4.3. Assume now that the block replacement time is no longer a constant T. The times between successive block replacement opportunities are i.i.d. random variables with a common density g. Let $m_g(\cdot)$ denote the renewal density associated with a renewal process with interarrival time density g. Under an opportunity block replacement policy with a control limit T, an item is replaced preventively at the first opportunity that is at least T time units since the last block replacement time. Find an expression representing the long-run average cost per unit time under control limit T.

14 Consider a TV game show with one million dollars at stake. The host selects a random number X from a uniform distribution over the interval $(0, 1)$ and divides the one million into two pots: the X million dollar pot and the $(1 - X)$ million dollar pot. You will independently select a random number Y from a uniform distribution over the interval $(0, 1)$. If $Y \leq X$, you will take the X million dollar pot; otherwise you will take the $(1 - X)$ million dollar pot. **(a)** What is your expected winning? **(b)** What is the moral of the story?

15 Let $X \sim exp(\lambda)$ and Y a random variable with distribution function F and density f. Define $Z = \min\{X, Y\}$, where X and Y are assumed to be independent. **(a)** Find the probability density g of Z as a function of λ, f, and F. **(b)** Define the Laplace transforms

$$g^e(s) = \int_0^\infty e^{-st} g(t)dt \quad \text{and} \quad f^e(s) = \int_0^\infty e^{-st} f(t)dt.$$

Find $g^e(s)$ as a function of λ and $f^e(s)$. **(c)** Consider the two renewal processes with interarrival time densities g and f, respectively. Let $M_Z(t)$ and $M_Y(t)$ denote the respective renewal functions and $m_Z(t)$ and $m_Y(t)$ the respective renewal densities. Define

$$m_Z^e(s) = \int_0^\infty e^{-st} m_Z(t)dt \quad \text{and} \quad m_Y^e(s) = \int_0^\infty e^{-st} m_Y(t)dt.$$

Show that $m_Z^e(s) = \dfrac{\lambda}{s} + \left(\dfrac{\lambda}{s}\right) m_Y^e(s + \lambda) + m_Y^e(s + \lambda)$. **(d)** Invert the transform found in Part c to obtain an expression for evaluating $M_Z(t)$.

16 Consider a simple random walk with $S_n = X_1 + \cdots + X_n$, where $\{X_i\}$ are i.i.d. random variables with $P\{X_1 = 1\} = 1/2$ and $P\{X_1 = -1\} = 1/2$. Let $N = \min\{n \mid S_n = a \text{ or } S_n = b\}$, where $a < 0 < b$. We rewrite S_n as $S_n = S_{n-1} + X_n$ and define $S_0 = 0$. Let $w_i = E[N \mid S_0 = i]$. **(a)** By conditioning on the outcome of the first trial X_1, state a system of difference equations characterizing $\{w_i\}$. **(b)** Solve the system for $\{w_i\}$. **(c)** Apply Wald's equation to find $P\{S_N = a\}$ and $P\{S_N = b\}$.

17 Consider an asymmetric simple random walk with $S_n = X_1 + \cdots + X_n$, where $\{X_i\}$ are i.i.d. random variables with $P\{X_1 = 1\} = p$ and $P\{X_1 = -1\} = 1 - p$, where $p > 1/2$. For any $i > 0$, we define $N = \min\{n \mid S_n = i\}$. Use Wald's equation to find $E[N]$.

18 Consider Example 3.4.2 again. Recall that $X(t)$ denotes the inventory level at time t. **(a)** State the limiting probability $\lim_{t \to \infty} P\{X(t) = S\}$ in terms of $M_G(\cdot)$, where $M_G(\cdot)$ is the renewal function of the renewal process with interarrival distribution G. **(b)** For $s \leq x < S$, define the limiting probability $f(x)dx = \lim_{t \to \infty} P\{x < X(t) < x + dx\}$. Thus $f(x)$ is the limiting density of the inventory level at x. Express $f(x)$ in terms of $m_G(\cdot)$ and $M_G(\cdot)$, where $m_G(\cdot)$ is the renewal density of the renewal process with interarrival distribution G.

*19 Consider an inventory system involving a given commodity. Assume that intervals $\{X_i\}$ of time between successive demand arrival epochs are i.i.d. random variables, each following a distribution function $A(\cdot)$ with density $a(\cdot)$. Let $A_n(\cdot)$ denote the n-fold convolutions of $A(\cdot)$ with itself and $a_n(\cdot)$ the corresponding density. Define $N(u)$ as the number of demand occurrences in $(0, u]$ and $N(u, u + L)$ as the number of demand occurrences in $(u, u + L]$, where $u > 0$ and $L > 0$ are any given constants. We are interested in finding the joint distribution $T(u, L; n, k) \equiv P\{N(u) = n, N(u, u + L) = k\}$. **(a)** Find $T(u, L; 0, 0)$ in terms of $A(\cdot)$. **(b)** Show that, for $n \geq 1$, we have

$$T(u, L; n, 0) = \int_0^u a_n(y)[1 - A(u + L - y)]dy.$$

(c) Find $T(u, L; 0, k)$ for $k \geq 1$ in terms of $a(\cdot)$ and $A_n(\cdot)$. **(d)** Show that for $n \geq 1$ and $k \geq 1$, we have

$$T(u, L; n, k) = \int_{z=0}^u a_n(z)\int_{y=0}^L a(u + y - z)\big[A_{k-1}(L - y) - A_k(L - y)\big]dydz.$$

*20 Consider an inventory system involving a given commodity. Assume that intervals $\{X_i\}$ of time between successive demand epochs are i.i.d. random variables, each following a distribution function $A(\cdot)$. The quantities demanded $\{D_i\}$ at demand arrival epochs are i.i.d. random variables each following a distribution function $B(\cdot)$. Moreover the quantity demanded is independent of the interval between successive demand occurrences. As usual, we let $A_n(\cdot)$ and $B_n(\cdot)$ denote the n-fold convolutions of $A(\cdot)$ and $B(\cdot)$ with themselves, respectively. Assume that a demand arrives at time 0. Define $N(t)$ as the number of demand occurrences in $(0, t]$, that is,

$$N(t) = \max \left\{ n \,\Big|\, \sum_{i=1}^n X_i \leq t \right\}.$$

Then $\{N(t), t \geq 0\}$ is a renewal process with interarrival time distribution $A(\cdot)$. Let $D(t)$ denote the cumulative demand in $(0, t]$. Thus, we have

$$D(t) = \sum_{i=1}^{N(t)} D_i.$$

(a) By conditioning on $N(t)$, show that

$$P\{D(t) \leq x\} = 1 - A(t) + \sum_{n=1}^{\infty}[A_n(u) - A_{n+1}(u)]B_n(x).$$

(b) Assume that $A(\cdot)$ is hyperexponential with parameters $p_1 = 0.9$, $p_2 = 0.1$, $\mu_1 = 0.45$, $\mu_2 = 0.02$, and $E[X_1] = 7$, and $B(\cdot)$ is Erlang $(3, 0.6)$ and $E[D_1] = 5$. Compute $E[N(10)]$ and $Var[N(10)]$. **(c)** Compute $P\{D(1) \le x$ for $x = 1{:}100$. **(d)** Use discrete approximation $\sum_{x=0}^{100} P\{D(10) > x\}$ to find $E[D(10)]$.

*21 Consider the inventory problem described in Problem 20 again. Define $H(x) = \max\left\{n | \sum_{i=1}^{n} D_i \le x\right\}$. Then $H(x)$ represents the number of demand occurrences such that the cumulative demand is less than x. We see that $\{H(x), x \ge 0\}$ form a renewal process with interarrival time distribution $B(\cdot)$. We call its renewal function $R(x) = E[H(x)]$. In addition to the dynamics of the demand process, we now incorporate an (s, S) replenishment policy, where s is the reorder point and S is the order level with $s \ge 0$ and $s < S$. Let $I_p(t)$ denote the inventory position at time t (the inventory position right after all transactions occurring at t are done). $I_p(t)$ is known to the decision maker at all times and includes the inventory on hand and on order at time t. At any time t, if $I_p(t)$ is less than s, an order of size $S - I_p(t)$ is placed and arrives L time units later. The delivery lead time L is a constant. We assume that unmet demands are back ordered and filled at the next delivery. Let $\Delta = S - s$. Assume that $I_p(0) = S$. Then stochastic process $\{I_p(t), t \ge 0\}$ is a regenerative process with regeneration points at epochs for which $I_p(t) = S$. The interval between two successive regeneration points forms a regeneration cycle. Let T_n denote the length of the nth regeneration cycle. Let $K(t)$ denote the number of regenerations (replenishments) over $(0, t]$. Then $\{K(t), t \ge 0\}$ is a renewal process with interarrival times $\{T_n\}$. Let $G(t)$ and $g(t)$ be the corresponding renewal function and density. For the three processes, we summarize the notations involved in the following table.

Renewal Process	$\{N(t), t \ge 0\}$	$\{H(x), x \ge 0\}$	$\{K(t), t \ge 0\}$
Interarrival times	$\{X_i\}$	$\{D_i\}$	$\{T_i\}$
Interarrival time distribution	$A(\cdot)$	$B(\cdot)$	$C(\cdot)$
Renewal function	$M(t)$	$R(x)$	$G(t)$
Renewal density	$m(t)$	$r(x)$	$g(t)$

(a) Use the fact that $\{T_n \le t\}$ if and only if $D(t) \ge \Delta\}$ to derive the distribution function for T_n. Specifically, find $C(t) \equiv P\{T_1 \le t\}$ in terms of $A_n(\cdot)$ and $B_n(\cdot)$. **(b)** Find $E[T_1]$. **(c)** What is the asymptotic renewal density $\lim_{t \to \infty} g(t)$?

*22 Consider the inventory problem described in Problems 20 and 21 again. Recall that $g(u)\Delta u$ represents the probability that an inventory replenishment order is about to be issued in $(u, u + \Delta u)$. Assume as before that $I_p(0) = S$. **(a)** By conditioning on the time of the last replenishment before time t, prove the following:

$$P\{I_p(t) = S\} = 1 - A(t) + \int_0^t g(t - u)[1 - A(u)]du \qquad t \ge 0.$$

(b) For $s \le x < S$, we define $f_p(t, x)dx = P\{x < I_p(t) < x + dx\}$. By conditioning on the time of the last replenishment before time t, prove the following

$$f_p(t, x) = \sum_{n=1}^{\infty} [A_n(t) - A_{n+1}(t)]b_n(S - x)$$

$$+ \sum_{n=1}^{\infty} \int_0^t g(t - u)[A_n(u) - A_{n+1}(u)]b_n(S - x)du,$$

where $a_n(\cdot)$ and $b_n(\cdot)$ are the densities of $A_n(\cdot)$ and $B_n(\cdot)$, respectively.

*23 Following Problem 22, apply the key renewal theorem to show that the limiting distribution of the inventory position is given by

$$\lim_{t \to \infty} P\{I_p(t) = S\} \equiv P\{I_p = S\} = \frac{1}{1 + R(\Delta)}$$

and

$$\lim_{t \to \infty} f_p(t, x) \equiv f_p(x) = \frac{r(S - x)}{1 + R(\Delta)}.$$

We observe that the two preceding equations are identical, respectively, to the two equations for the limiting distribution of the inventory position obtained in Problem 18. There the delivery lead time is zero. Why are they the same?

*24 Consider Problems 20–23 once more. Assume that the inventory system is in steady state and we are studying its limiting behaviors. Let I denote the on-hand inventory level, I_p the inventory position, and $D_\infty(L)$ the cumulative demand during a lead time of length L. The distribution of I_p is given in Problem 23. Then we have the relation $I = I_p - D_\infty(L)$. This is true because all back orders included in I_p will be delivered in the next interval of length L and the difference becomes the inventory on hand L time units later (note that there can be more than one outstanding replenishment order at any time). **(a)** Express $E[D_\infty(L)]$ in terms of $E[X_1]$, $E[D_1]$, and L. **(b)** Express $E[I]$ in terms of $R(\cdot)$, $E[X_1]$, $E[D_1]$, and L. **(c)** For the numerical example given in Problem 20b, compute $E[I]$ and $E[I_p]$ for the case in which $L = 2$, and $(s, S) = (2, 10)$.

25 We return to Example 3.4.2. Assume that there is a fixed cost of ordering K, a holding cost of h per unit per unit time. Note that K is the fixed cost per replenishment order regardless of the order size. Let $\Delta = S - s$ and $c(s, \Delta)$ denote the long-run expected average cost per unit time. **(a)** Express $c(s, \Delta)$ as a function of s, Δ, and $M_G(\cdot)$. **(b)** Since the delivery lead time is 0, we expect that an optimal value for the reorder point s is 0. Consider a numerical example in which $K = 209.65$, $h = 1$, $E[T_1] = 7$, and the order size distribution G is hyperexponential with parameters $p_1 = 0.4$, $p_2 = 0.6$, $\mu_1 = 0.5$, and $\mu_2 = 0.2$ (hence $E[D_1] = 3.8$). Compute $c(0, \Delta)$ for order quantity $\Delta = 1{:}20$. **(c)** What is the optimal order size?

26 A pickup truck passes by a street corner every two minutes. An observer shows up at the street corner at an arbitrary point in time. Let X denote the time until the observer sees a total of three pickup trucks. Find the limiting distribution function of X.

27 Let Y be the lifetime of a car dealership. Assume that Y follows an exponential distribution with mean $1/\mu$. Let $\{X_i\}$ denote the intervals between successive car sales. Assume that $\{X_i\}$ are i.i.d. random variables. In addition, we assume Y is independent of $\{X_i\}$. Then $N(Y)$ denotes the number of car sales during the lifetime of the dealership. Find $E[N(Y)]$ in terms of $E[e^{-\mu X_1}]$.

28 A piece of equipment is subject to random failures. The time-to-failure random variable follows an exponential distribution with mean $1/\lambda$. When the equipment fails, it will be repaired. The repair time follows an exponential distribution with mean $1/\mu$. The times to failure and repair times are all mutually independent. Thus the state of the equipment alternates between working and being repaired. Assume that at time 0 the equipment is working. Let $M(t)$ denote the number of failures by time t. Find an approximation of $E[M(t)]$ for large t.

29 Consider a renewal process with interarrival time distribution F. Assume that a renewal occurs at time 0. Let W denote the time (from the time origin 0) when the interarrival time from the preceding renewal event first exceeds a fixed positive constant s. **(a)** Determine $E[W]$. **(b)** Determine an integral equation satisfied by $V(t)$, where $V(t) = \text{Prob}\{W \le t\}$.

Bibliographic Notes

Texts covering renewal theory include those of Asmussen (1987), Cox (1962), Çinlar (1975), Heyman and Sobel (1982), Karlin and Taylor (1975), Resnick (1992), Ross (1983), Taylor and Karlin (1994), Tijms (1986), and Wolff (1989). Applications of renewal theory in reliability theory are treated extensively in the two books by Barlow and Proschan (1965, 1975). A thorough analysis of inventory systems by regenerative approaches is covered in Sahin (1990). Problems 19–23 are based on Sahin (1979). The use of regenerative approach in simulation is discussed in Fishman (1978). Applying renewal theory in product warranty analysis is the focus of a book by Blischke and Murthy (1994). The books by Tijms (1986, 1994) contain many intriguing applications involving renewal theory. For statistical analyses of renewal processes, readers can consult Cox and Lewis (1978).

This text skips a mathematical elaboration of an important result—the key renewal theorem, stated as Equation 3.2.3. Readers who are interested in digging into technical details can consult Asmussen (1987) and Resnick (1992). Excellent expositions of direct Riemann integrability and the role it plays in the key renewal theorem can be found in Asmussen (1987, 118–20) and Resnick (1992, Section 3.10). Example 3.2.4 is based on Heyman and Sobel (1982, Example 5–6). The program used in Example 3.2.5 for computing the renewal function was coded in the spirit of Xie (1989). Ross (1983) spearheaded the use of renewal reward process as a tool for problem solving. A formal proof of Equation 3.4.1 can be found in Ross (1983). Example 3.4.2 is based on Ross (1983, Example 3.4(a)). Examples involving equipment replacements such as Examples 3.4.3 and 3.4.4 have their origins in Barlow and Proschan (1965). The approach used in Example 3.4.5 follows that of Wolff (1989, Example 2–3). Example 3.5.1 is based on Çinlar (1975, 291–92, Example 1.33). Example 3.6.1 is inspired by Ross (1983, Problem 3.26). Example 3.6.2 is based on an actual application by Crane and Iglehart (1975). Examples 3.6.3 and 3.6.4 are from Tijms (1986). Examples and problems on product warranty analysis can all be traced back to the references cited in Blischke and Murthy (1994). Little's formula stems from Little (1961) and is discussed in detail in Wolff (1989). Stidham (1974) proclaimed to have a last word on the subject. Example 3.6.6 is based on Glynn and Iglehart (1995). Our exposition of discrete renewal process patterns after Taylor and Karlin (1994, Section 7.6). Problems 1 and 2 are based on Problems 1 and 6 of Karlin and Taylor (1975, 230–31), respectively. Problem 5 is from Exercises 7.3.1 and 7.3.3 of Taylor and Karlin (1994, 406). Problem 7 is based on the paper by Sivazlian (1974). Problem 10 is an adaptation of the process control example given in Taylor and Karlin (1994, 425–26). Problem 11 is from Tijms (1986, Problem 1.16) and Problem 12 is also from Tijms (1986, 16). Problem 13 is based on the work of Dekker and Smeitink (1991).

References

Asmussen, S. 1987. *Applied Probability and Queues.* New York: John Wiley & Sons.

Barlow, R. E., and F. Proschan. 1965. *Mathematical Theory of Reliability.* New York: John Wiley & Sons.

Barlow, R. E., and F. Proschan. 1975. *Statistical Theory of Reliability and Life Testing: Probability Models.* New York: Holt, Rinehart and Winston.

Blischke, W. R., and D. N. P. Murthy. 1994. *Warranty Cost Analysis.* New York: Marcel Dekker.

Çinlar, E. 1975. *Introduction to Stochastic Processes.* Englewood Cliffs, NJ: Prentice-Hall.

Cox, D. R. 1962. *Renewal Theory.* London: Methuen.

Cox, D. R., and P. A. W. Lewis. 1978. *The Statistical Analysis of Series of Events.* London: Chapman and Hall.

Crane, M. A., and D. L. Iglehart. 1975. Simulating Stable Stochastic Systems: III. Regenerative Processes and Discrete-Event Simulations. *Operations Research* 23(1):33–45.

Dekker, R., and E. Smeitink. 1991. Opportunity-Base Block Replacement. *European Journal of Operational Research* 53(1):46–63.

Fishman, G. S. 1978. *Principles of Discrete Event Simulation.* New York: John Wiley & Sons.

Glynn, P. W., and D. L. Iglehart. 1995. Trading Securities Using Trailing Stops. *Management Science* 41(6):1096–1106.

Heyman, D. P., and M. J. Sobel. 1982. *Stochastic Models in Operations Research: Volume I.* New York: McGraw-Hill.

Karlin, S., and H. M. Taylor. 1975. *A First Course in Stochastic Processes.* New York: Academic Press.

Little, J. D. C. 1961. A Proof for the Queueing Formula: $L = \lambda W$. *Operations Research* 9:383–87.

Resnick, S. I. 1992. *Adventures in Stochastic Processes.* Boston: Birkhäuser.

Ross, S. M. 1983. *Stochastic Processes.* New York: John Wiley & Sons.

Sahin, I. 1979. On the Stationary Analysis for Continuous Review (s, S) Inventory Systems with Constant Lead Times. *Operations Research* 27(4):717–29.

Sahin, I. 1990. *Regenerative Inventory Systems.* New York: Springer-Verlag.

Sivazlian, B. D. 1974. A Continuous-Review (*s, S*) Inventory System with Arbitrary Interarrival Distribution between Unit Demand. *Operations Research* 22(1):65–71.

Stidham, S. 1974. A Last Word on *L* = λ*W*. *Operations Research* 22(2):417–21.

Taylor, H. M., and S. Karlin. 1994. *An Introduction to Stochastic Modeling.* Revised ed. New York: Academic Press.

Tijms, H. C. 1986. *Stochastic Modelling and Analysis: A Computational Approach.* New York: John Wiley & Sons.

Tijms, H. C. 1994. *Stochastic Models: An Algorithmic Approach.* New York: John Wiley & Sons.

Wolff, R. W. 1989. *Stochastic Modeling and the Theory of Queues.* Englewood Cliffs, NJ: Prentice-Hall.

Xie, M. 1989. On the Solution of Renewal-Type Integral Equations. *Communication in Statistics—Simulation* 18(1):281–93.

Appendix

Chapter 3: Section 2

Example 3.2.5   The MATLAB program `c3_mt_f` is for computing the renewal function based on the interarrival time distribution F. The program uses the algorithm proposed by M. Xie. Using the distribution function F as the input, the program produces an output matrix X with two columns. Each row of the matrix X gives the pair $(t, M(t))$.

```
function [X]=c3_mt_f(F,g,t)
%
%    Find the renewal function given cdf  F
%    Ref:  Xie, M. "On the Solution of Renewal-Type Integral
%    Equations"
%    Commun. Statist. -Simula., 18(1), 281-293 (1989)
%
[n,m]=size(F); g0=g(1);g(1)=[];M=F;dno=1-g0;
M(1)=F(1)/dno;
for i=2:m
   sum=F(i)-g0*M(i-1);
   for j=1:i-1
      if j==1
      sum=sum+g(i-j)*M(j);
      else
      sum=sum+g(i-j)*(M(j)-M(j-1));
      end
   end
   M(i)=sum/dno;
end
%
%    Output the results
%
d=20; x=[0]; Mt=[0];
for i=d:d:m
   y=i*t/m; x=[x y]; Mt=[Mt M(i)];
end
m=length(x); X=zeros(m,2); X(:,1)=x'; X(:,2)=Mt';

function [F,g]=c3_wibol(alpha,beta,t)
%
```

```
%    Find distribution for Weibull  F(x) = 1 - exp(-(x/alpha)^beta)
%    x in (0, t)
%    1/n  is the step length
%
n=200; n1=n*t;
tn=1/n; i=1:1:n1; x=tn*(i+0.5); x=(1/alpha)*x;
g=1-exp(-(x.^beta));
i=1:1:n1; x=tn*i; x=(1/alpha)*x;
F=1-exp(-(x.^beta));
```

Chapter 3: Section 3

Example 3.3.1   The following program is for simulating the effect of length biased sampling. For varying values of t, we sample the total lives at time t, $T_j(t)$, from a renewal process with exponential interarrival times with a mean of 1 and find their respective sample means.

```
function [T,t]=e331
%
%    Simulating mean total life
%    Ref:   Example 3.3.1
%
T=[]; t=[]; rand('seed',12345);
for i=1:10
    st=100.0*i; d=[]; t=[t st];
    for n=1:100
        at=0.0;
        while  at <= st
        x=-log(1-rand(1,1)); at=at+x;
        end
    d=[d x];
    end
    m=mean(d); T=[T m];
end
```

Example 3.3.2   The first three programs **invt_laq**, **e332b**, and **e332c** are for computing $M(t) - t/\mu$ for a given interval $[0, t]$ when the Laplace transform of the renewal function $M^e(s)$ is given. In this example, the interarrival time distribution is $U(0, 1)$. MATLAB program **invt_laq** is almost identical to **invt_lap** given in Chapter 1. The only differences are that the current program uses MATLAB function **feval**, receives a single t value, and produces a single output value $f(t)$. The program **e332b** stores the Laplace transform for which an inversion is done in **invt_laq**. In this case, it contains the Laplace transform $M^e(s)$. The third program **e332c** is the user's calling program for computing $M(t) - t/\mu$ for $t = 0:0.5:10$.

```
function [f]=invt_laq(input,t)
%
%    Numerical Inversion of Laplace transform
%    "Numerical Inversion of Laplace Transform of Probability
%    Distribution," by Abate and Whitt, ORSA J on Computing
%    Vol. 7, No. 1, 1995, pp. 36-43.  A version uses "feval"
%
```

```
m=11; c=[]; ga=8; A=ga*log(10); mm=2^m;
for k=0:m
  d=binomial(m,k); c=[c d];
end
ntr=15; u=exp(A/2)/t; x=A/(2*t); h=pi/t; su=zeros(m+2);
sm=feval(input,x,0)/2;
for k=1:ntr
  y=k*h; sm=sm+((-1)^k)*feval(input,x,y);
end
su(1)=sm;
for k=1:12
  n=ntr+k; y=n*h; su(k+1)=su(k)+((-1)^n)*feval(input,x,y);
end
av=0;
for k=1:12
  av=av+c(k)*su(k+1);
end
f=u*av2/mm;

function [z]=e332b(x,y)
s=x+y*i;
up=(1-exp(-s))/s; dm=1-up; qs=up/(s*dm); z=real(qs);

function [D]=e332c
%
%  Compute the renewal function with U(0,1) interarrival time
%
Mt=[0]; T=[0];
for t=1:0.5:10
  q=invt_laq('e332b',t; Mt=[Mt q]; T=[T t];
end
D=Mt-T/0.5;
D=[T; D]';
```

The next set of four MATLAB programs are for computing the mean excess life at time 1, $E[Y(1)]$. Program **e332g** stores the Laplace transform of the renewal density $m^e(s)$, and program **invt_laq** does the numerical inversion. Program **e332d** produces the integrands $M(y)(y-x)$ for the set of y values over the interval [ymin, ymax]. For a given x, **e332e** does the inner integration $\int_x^1 m(y)(y-x)dy$. The user's calling program **e332f** performs the outer integration of $\int_0^1 \int_x^1 M(y)(y-x)dydx$ with respect to dx. At the end of the program listing, we demonstrate such a run in which we find $E[Y(1)] = 0.3568$.

```
function out=e332d(y,ymin,ymax)
global x;
nn=length(y); out=[];
for i=1:nn
yy=y(i); z=invt_laq('e332g',yy); h=z*(yy-x); out=[out h];
end

function out=e332e(variable)
global ymin;
global ymax;
global fun;
global outvar;
eval(['global ' outvar]); out=ones(size(variable));
```

```
for i=1:length(variable)
   eval([outvar, '=variable(i);']);
   ymin=x; ymax=1;
   out(i)=quad(fun,ymin,ymax);
end

function out=e332f(fun,outvar,xmin,xmax)
global ymin;
global ymax;
global fun;
global outvar;
out=quad('e332e',xmin,xmax);

function [z]=e332g(x,w)
s=x+w*i;
up=(1-exp(-s))/s; dm=1-up; qs=up/dm; z=real(qs);

e332f('e332d','x',0.0001,1)

ans =
    0.3568
```

4

Discrete-Time Markov Chains

Tips for Chapter 4

- Examples 4.3.2 and 4.3.3 are two classic examples of single server queues. They will be of particular interest to those interested in Markov chain applications in queueing theory.
- Examples 4.4.2 (a blackjack example) and 4.4.3 (a tennis example) are for supplementary reading about applications of absorbing Markov chains.
- Examples 4.4.7 and 4.4.8 deal with the derivation of a numerically stable approach for computing the stationary probability vector of a finite-state Markov chain. When the state space is large, this approach is attractive for achieving numerical accuracy.
- Section 4.5 on Markov reward processes provides some background for the study of discrete-time Markov decision problems—a subject in stochastic optimization typically covered in a course such as dynamic programming.
- Readers who are interested in queueing networks should read Section 4.6 on reversible discrete-time Markov chains. The concepts presented there are closely related to those of Section 5.8 regarding reversible continuous-time Markov chains.

4.0 Overview

Handling temporal dependency in stochastic modeling, while often challenging, is sometimes necessary. For a discrete-time stochastic process with a discrete state space, if the future of the process depends only on the current state of the system, the process is called a Markov chain. In Section 4.1, we present many examples illustrating chains with varied underlying structures. For a Markov chain, we call the probability of making a transition from one state to another in a single step the transition probability and the matrix containing these probabilities the transition probability matrix. The structure of a transition probability matrix dictates the probabilistic behavior of a Markov chain.

In Section 4.2, we define states with distinguishing features and present ways to decompose state space into classes with differing structural properties. After the decomposition, subchains and their interactions can be analyzed independently so as to reduce the complexity involved. Of the many varieties, there are three chains that occupy our attention: the ergodic, periodic, and absorbing chains. In an ergodic chain, if we are in a state then we are able to reach any other state of the chain in finite time and the limiting probabilities in each state exist; in a periodic chain, transitions from a state back to itself can take place only in multiples of some constant. For the latter case, the limiting probabilities do not exist, but the long-run fractions of time the process is in each state do exist. Ergodic and periodic chains are covered in Section 4.3.

Some Markov chains contain closed and nonclosed classes of state. Once the process leaves a nonclosed class, it will never return to the class. States in a non-closed class are called transient states. On the other hand, the process will move around states in a closed class forever once it enters such a class. In Section 4.5, we call these chains absorbing Markov chains and study them in detail. In Section 4.6, we use time-to-absorption of an absorbing Markov chain to model a discrete probability distribution. A distribution amenable to such a representation is called a phase-type distribution. Phase-type distributions have nice closure properties and are becoming popular modeling and computation devices for problem solving. Section 4.5 deals with a Markov chain that generates rewards, and Section 4.6 covers reversible Markov chains. If a chain is reversible, in terms of probabilistic behavior of the process in steady state, we cannot distinguish between a process that moves along the time axis in a forward manner and the same process that moves along the time axis in a reversed manner. The section serves as a prelude to reversible continuous-time Markov chains, which play a prominent role in queueing networks.

4.1   Introduction

Consider a stochastic process $X = \{X_n, n = 0, 1, \ldots\}$, where X_n denotes the state of the system at "time" n. As an example, X_n can be a gambler's fortune after n plays. We assume that random variable X_n takes on values in the set S. The set S is called the *state space*. We assume that S is finite or infinitely countable. For convenience, we index the possible values of the state space so that $S = \{0, 1, \ldots\}$. If $X_n = i$, we say that the process is in state i at time n.

> The stochastic process X is a discrete-time Markov chain with state space S if
>
> $$P\{X_{n+1} = j | X_n = i, X_{n-1}, \ldots, X_1 = i_1, X_0 = i_0\} = P\{X_{n+1} = j | X_n = i\}$$
>
> holds for all $i, j, i_0, i_1, \ldots, i_{n-1}$ in S and all $n = 0, 1, \ldots$.

The preceding property is known as the Markovian property of a Markov chain. It says that the future of the process is independent of the past once the current state of the process is known. For all n, if

$$P\{X_{n+1} = j | X_n = i\} = p_{ij} \qquad (4.1.1)$$

for all i and j in S, then the Markov chain is said to be *time homogeneous*, with the transition probabilities thus staying constant over time. Markov chains that are not time homogeneous can be modified to form equivalent time–homogeneous Markov chains by enlarging their state spaces. From now on, we will only be dealing with time-homogeneous Markov chains. When the state space of a Markov chain is finite, we call it a finite Markov chain; when it is infinitely countable, we call it a denumerable Markov chain. The analysis of a denumerable Markov chain is more intricate than that of a finite Markov chain. In fact, the nature of the solution process depends on whether the chain is finite or infinite.

In Equation 4.1.1, we see that the present "time" n is a fixed constant. When the present time is a random variable T, Equation 4.1.1 still holds for the Markov chain X provided that T is a stopping time with respect to $\{X_n\}$ and $P\{T < \infty\} = 1$ (the notion of a stopping time has been defined in Section 3.2). In other words, we have

$$P\{X_{T+1} = j | X_T = i\} = p_{ij}, \qquad (4.1.2)$$

where T is a stopping time. More generally, we can say the past $\{X_m, m \le T\}$ and the future $\{X_m, m > T\}$ are conditionally independent given the present state x_T. Such a property is called the *strong Markov property*. We will find many applications of this property in the sequel.

Let $P = \{p_{ij}\}$ denote the matrix of transition probabilities, where $Pe = e$ and $P \ge 0$ and e is a column vector of ones. Define the state probability at time n by $s_j(n) = P\{X_n = j\}$ and let row vector $s(n) = \{s_j(n)\}$. The starting state probability vector $s(0)$ and the transition probability matrix P completely characterize a Markov chain. We now look at a few examples to introduce various issues to be studied in this chapter.

EXAMPLE
4.1.1

A Mouse in a Maze   A mouse starts from cell 1 in the maze shown in Figure 4.1. A cat is hiding patiently in cell 7, and there is a piece of cheese in cell 9. In the absence of learning, when the mouse is in a given cell, it will choose the next cell to visit with probability $1/k$, where k is the number of adjoining cells. Assume that once the mouse finds either the piece of cheese or the cat, it will understandably stay there forever.

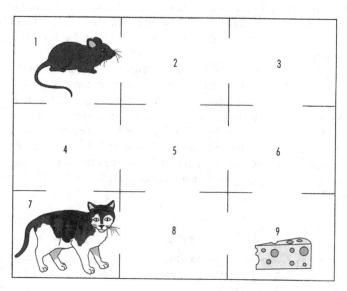

FIGURE
4.1 A mouse in a maze.

Let X_n denote the position of the mouse after n changes of cells. The stochastic process $\{X_n | n = 0, 1, \ldots\}$ is a Markov chain with state space $S = \{1, 2, \ldots, 9\}$, starting state probability vector $s(0) = (1, 0, \ldots, 0)$, and transition probability matrix

$$
P = \begin{bmatrix}
0 & 1/2 & 0 & 1/2 & 0 & 0 & 0 & 0 & 0 \\
1/3 & 0 & 1/3 & 0 & 1/3 & 0 & 0 & 0 & 0 \\
0 & 1/2 & 0 & 0 & 0 & 1/2 & 0 & 0 & 0 \\
1/3 & 0 & 0 & 0 & 1/3 & 0 & 1/3 & 0 & 0 \\
0 & 1/4 & 0 & 1/4 & 0 & 1/4 & 0 & 1/4 & 0 \\
0 & 0 & 1/3 & 0 & 1/3 & 0 & 0 & 0 & 1/3 \\
0 & 0 & 0 & 0 & 0 & 0 & 1 & 0 & 0 \\
0 & 0 & 0 & 0 & 1/3 & 0 & 1/3 & 0 & 1/3 \\
0 & 0 & 0 & 0 & 0 & 0 & 0 & 0 & 1
\end{bmatrix}.
$$

Things of interest to the mouse include (i) the probability distribution for the time—specifically, the number of cell changes—in reaching the piece of cheese before reaching the cat, (ii) the probability distribution for the time in reaching the cat before reaching the cheese, and (iii) the probability of reaching the cheese first. ∎

EXAMPLE
4.1.2 **A Single-Server Queue with a Finite Number of Waiting Rooms and a Constant Service Time**   A shoeshine boy operates in the Port Authority Bus Terminal in Manhattan. There are four chairs, three of which are for waiting customers. When all chairs are occupied, arriving customers will seek service elsewhere. Each shoeshine takes exactly ten minutes.

Let X_n denote the number of waiting customers immediately after the completion of the nth shoeshine and A_n the number of arrivals during the nth shoeshine. We assume that $P\{A_n = i\} = a_i$ for $i = 0, 1, \ldots$, and $n = 1, 2, \ldots$, and that a departure and an arrival do not occur at the same time. At the completion of the nth shoeshine, if $X_n = 0$ the following shoeshine starts when the next customer arrives and at its completion there will be $\min\{3, A_{n+1}\}$ customers waiting. At the completion of the nth shoeshine, if $X_n > 0$ then the next shoeshine starts immediately and at its completion there will be $\min\{3, X_n - 1 + A_{n+1}\}$ customers waiting. Thus we have

$$X_{n+1} = \begin{cases} \min(3, A_{n+1}) & \text{if } X_n = 0 \\ \min(3, X_n - 1 + A_{n+1}) & \text{if } X_n = 1, 2, 3. \end{cases}$$

We see that X_{n+1} is determined probabilistically if X_n is known and the stochastic process $\{X_n | n = 1, 2, \ldots\}$ is a Markov chain with state space $S = \{0, 1, 2, 3\}$ and the transition probability matrix given by

$$P = \begin{bmatrix} a_0 & a_1 & a_2 & {}^{\geq}a_3 \\ a_0 & a_1 & a_2 & {}^{\geq}a_3 \\ 0 & a_0 & a_1 & {}^{\geq}a_2 \\ 0 & 0 & a_0 & {}^{\geq}a_1 \end{bmatrix}.$$

We note that in this example the state of the Markov chain is defined at the time of a service completion to establish the Markovian property of the resulting stochastic process. This is known as the *method of embedding*. We also note that in this example the interval between successive transitions is in general not of equal length. Things of interest to the shoeshine boy include (i) the long-run expected number of lost customers per hour, (ii) the average occupancy of the shop, and (iii) the average length of an idle period—so that the shoeshine boy can take a break from his work. ∎

EXAMPLE
4.1.3 **A Periodic-Review (s, S) Inventory System**   Suppose weekly demands for a given aircraft spare part at a maintenance depot are i.i.d. random variables with

$$P\{D_n = i\} = a_i, \quad i, n = 0, 1, 2, \ldots, \quad \text{and} \quad \sum_{i=0}^{\infty} a_i = 1,$$

where D_n denotes the demand in week n. Let X_n denote the inventory position at the start of week n before the receipt of replenishment, if any. We assume that inventory replenishment is instantaneous and unfilled demands are lost. The inventory policy used at the depot is of (s, S) type: If at the beginning of a week n, the stock level X_n is s or higher, we do not order; if it is lower than s, an order is made to bring the inventory level to S. In the parlance of inventory theory, the model is known as the case of periodic review, immediate delivery, and lost sales. To see that the stochastic process $\{X_n, n = 0, 1, ...\}$ forms a Markov chain with state space $S = \{0, 1, ..., S\}$, we note that

$$
X_{n+1} = \begin{cases} \max\{X_n - D_n, \ 0\} & \text{if } X_n \geq s \\ \max\{S - D_n, \ 0\} & \text{if } X_n < s. \end{cases}
$$

Thus X_{n+1} depends on the history $\{X_n, n = 0, 1, ...\}$ only through X_n, and the process is a Markov chain with the following transition probabilities:

$$
p_{ij} = \begin{cases} r_S & \text{if } 0 \leq i < s \text{ and } j = 0 \\ a_{S-j} & \text{if } 0 \leq i < s \text{ and } 1 \leq j \leq S \\ r_i & \text{if } s \leq i \leq S \text{ and } j = 0 \\ a_{i-j} & \text{if } s \leq i \leq S \text{ and } 1 \leq j \leq i \\ 0 & \text{otherwise,} \end{cases}
$$

where $r_i = a_i + a_{i+1} + \cdots$. For the policy $(s, S) = (2, 5)$, we display the transition matrix as follows:

$$
P = \begin{array}{c|cccccc} & 0 & 1 & 2 & 3 & 4 & 5 \\ \hline 0 & r_5 & a_4 & a_3 & a_2 & a_1 & a_0 \\ 1 & r_5 & a_4 & a_3 & a_2 & a_1 & a_0 \\ 2 & r_2 & a_1 & a_0 & 0 & 0 & 0 \\ 3 & r_3 & a_2 & a_1 & a_0 & 0 & 0 \\ 4 & r_4 & a_3 & a_2 & a_1 & a_0 & 0 \\ 5 & r_5 & a_4 & a_3 & a_2 & a_1 & a_0 \end{array}.
$$

For this inventory system, things of interest include (i) the expected amount of lost sales, (ii) the average level of inventory per week, and (iii) the expected number of weeks between successive replenishments. ∎

In a Markov chain, the n-step transition probability is defined by

$$
p_{ij}^{(n)} = P\{X_{n+m} = j | X_m = i\}.
$$

This is the probability that the process goes from state i to state j in n transitions. Since we assume that the Markov chain is time homogeneous, the probability is invariant in m. We note that $p_{ij}^{(1)} = p_{ij}$ and define $P_{ij}^{(0)} = 1$ if $i = j$ and 0 otherwise. Let the n-step transition probability matrix $P^{(n)} = \{p_{ij}^{(n)}\}$. Then we have $P^{(1)} = P$ and $P^{(0)} = I$. The following equation, known as the Chapman-Kolmogorov equation, provides a means for computing $P^{(n)}$:

$$p_{ij}^{(n+m)} = \sum_{k=0}^{\infty} p_{ik}^{(n)} p_{kj}^{(m)}. \tag{4.1.3}$$

We now give an intuitive interpretation of Equation 4.1.3: Starting from state i, to be in state j in exactly $n+m$ steps, we must be in some intermediate state k in exactly n steps and state j from state k in exactly m steps. Adding the probabilities associated with these paths over all possible intermediate states k, we obtain the desired result. We can write Equation 4.1.3 in matrix form as $P^{(n+m)} = P^{(n)}P^{(m)}$. A special case of the identity is $P^{(n+1)} = P^{(n)}P^{(1)} = P^{(n)}P$. By successive substitutions, we conclude

$$P^{(n)} = P^n \qquad n = 0, 1, 2, \ldots.$$

By conditioning on the starting state probabilities, the *state probability vector* $s(n) = \{s_j(n)\}$ can now be found from

$$s(n) = s(0)P^n \qquad n = 0, 1, 2, \ldots, \tag{4.1.4}$$

where $s_j(n) = P\{X_n = j | s(0)\}$. The state probability vector reveals the time-dependent behavior of the Markov chain under the starting state probability vector $s(0)$.

EXAMPLE
4.1.4

A Mouse in a Maze Revisited The state probabilities of interest to the mouse are $s_7(n)$ and $s_9(n)$, the probabilities that the mouse will be visiting either the cat or the piece of cheese at the nth cell change. These two probabilities are plotted in Figure 4.2.

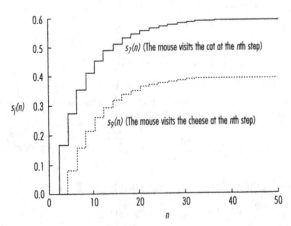

FIGURE
4.2   The state probabilities for the mouse-in-a-maze example.

The figure shows that after a large number of steps, the mouse will either be with the cat or the piece of cheese—with probabilities 0.6 and 0.4, respectively. In other words, the figure suggests that $\lim_{n\to\infty} s_7(n) = 0.6$ and $\lim_{n\to\infty} s_9(n) = 0.4$.

Suppose that the mouse had started from cell 3 instead, that is, $s(0) = (0, 0, 1, 0, ..., 0)$. By symmetry, the results shown in Figure 4.2 would have stayed the same but with the two curve labels switched and $\lim_{n \to \infty} s_7(n) = 0.4$ and $\lim_{n \to \infty} s_9(n) = 0.6$. We remark that $p_{17}^{(n)}$ and $p_{19}^{(n)}$ are zero when n is an odd number. This phenomenon is related to what will be called in the sequel the periodicity property associated with the specific transition matrix P. ▪

EXAMPLE 4.1.5   **The Periodic-Review (s, S) Inventory System Revisited**   Consider the case in which the weekly demand for the spare part follows a Poisson distribution with mean 2. In Figure 4.3, we plot the time-dependent state probabilities in state 0 under two possible starting states: 0 and 3. Unlike Example 4.1.4, we see the two sets of state probabilities quickly converge to a common limit. Depending on the type of transition probability matrices, the behaviors of Markov chains differ. We will study these issues in detail in the remainder of the chapter. ▪

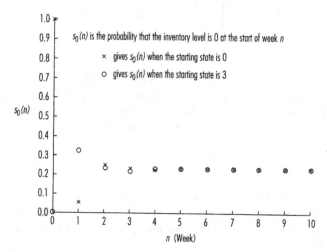

FIGURE 4.3   The state probabilities for the inventory example.

4.2   Classification of States

The behavior of a Markov chain depends on the structure of the transition matrix P. In this section, we will study various measures for characterizing states, subsets of states in a Markov chain, and Markov chains possessing different behaviors.

State j is said to be *accessible* from i if for some $n \geq 0$, $p_{ij}^{(n)} > 0$. We use $i \to j$ to denote it. If $i \to j$ and $j \to i$, we say that i and j communicate and use $i \leftrightarrow j$ to denote it. By definition, $p_{ii}^{(0)} = 1$, so we have $i \leftrightarrow i$ for all $i \in S$. Communication is an equivalence relation, that is, it is reflexive, symmetric, and transitive. When $i \leftrightarrow j$, we say i and j are in the same *equivalence class*. For brevity, we refer to an equivalence class as a class. Since communication is an equivalence relation, two classes are either disjoint or the same. We say a class is *closed* if no states outside of the class can be reached from any state within the class. A Markov chain is *irreducible* if its only closed class is the set of states in its state space *S*.

<hr>

EXAMPLE
4.2.1

Consider a Markov chain with transition probability matrix

$$
\begin{array}{c}
\quad\quad 0 \quad 1 \quad 2 \quad \cdot \quad \cdot \quad N-1 \; N \\
P = \begin{array}{c} 0 \\ 1 \\ 2 \\ \cdot \\ \\ N-1 \\ N \end{array}
\begin{bmatrix}
1 & 0 & 0 & 0 & 0 & 0 & 0 \\
q & 0 & p & 0 & 0 & 0 & 0 \\
0 & q & 0 & p & 0 & 0 & 0 \\
\cdot & \cdot & \cdot & \cdot & \cdot & \cdot & \cdot \\
\cdot & \cdot & \cdot & \cdot & \cdot & \cdot & \cdot \\
0 & 0 & 0 & 0 & q & 0 & p \\
0 & 0 & 0 & 0 & 0 & 0 & 1
\end{bmatrix},
\end{array}
$$

where $p + q = 1$ and $p > 0$ and $q > 0$. The chain is sometimes called a random walk with two absorbing barriers. We see that the chain has three classes $\{0\}$, $\{N\}$, and $\{1, \ldots, N-1\}$, and $\{0\}$ and $\{N\}$ are two closed classes. ■

<hr>

Let T_{ij} denote the number of transitions the process takes for its first entrance into state j given that $X_0 = i$. Random variable T_{ij} is known as the first passage time from i to j. We use $f_{ij}^{(n)}$ to denote the probability mass function for T_{ij}. Formally, we have

$$
f_{ij}^{(n)} = P\{T_{ij} = n\} = P\{X_n = j, X_{n-1} \neq j, \ldots, X_1 \neq j | X_0 = i\} \quad n = 1, 2, \ldots .
$$

Conditioning on the first transition out of state i, $f_{ij}^{(n)}$ is found recursively as follows:

$$
f_{ij}^{(1)} = P_{ij}
$$

$$
f_{ij}^{(n)} = \sum_{\substack{k=0 \\ k \neq j}}^{\infty} P_{ik} f_{kj}^{(n-1)} \quad\quad n = 2, 3, \ldots .
$$

We rewrite the last equation as

$$f_{ij}^{(n)} = \sum_{k=0}^{\infty} P_{ik} f_{kj}^{(n-1)} - P_{ij} f_{jj}^{(n-1)}$$

$$= \sum_{k=0}^{\infty} P_{ik} f_{kj}^{(n-1)} - \sum_{k=0}^{\infty} P_{ik} \delta_{kj} f_{kj}^{(n-1)},$$

where $\delta_{kj} = 1$ if $k = j$ and 0 otherwise. The preceding expression enables us to express first-passage-time probabilities in a matrix form. For any square matrix A, we let A_d denote a diagonal matrix formed by the diagonal elements of A and A_0 a matrix resulting from setting the diagonal elements of A to zeros. Hence we have $A = A_d + A_0$. Let $F^{(n)} = \{f_{ij}^{(n)}\}$. Then we find

$$F^{(1)} = P$$

$$F^{(n)} = PF^{(n-1)} - PF_d^{(n-1)} = PF_0^{(n-1)} \qquad n = 2, 3, \dots. \qquad (4.2.1)$$

The first-passage-time probabilities can be computed recursively from Equation 4.2.1. For any state j, a random variable of special interest is T_{jj}. It is called the *recurrence time* of state j.

Let f_{ij} denote the probability that state j is ever reached from state i. Then we have

$$f_{ij} = P\{T_{ij} < \infty\} = \sum_{n=1}^{\infty} f_{ij}^{(n)}.$$

We call f_{ij} the *reaching probability* from state i to state j. State j is called *recurrent* if $f_{jj} = 1$ and *transient* if $f_{jj} < 1$. In the former case, we have $P\{T_{jj} < \infty\} = 1$. Let $\mu_{jj} = E[T_{jj}]$ denote the mean recurrence time of state j. Even if $P\{T_{jj} < \infty\} = 1$, when the state space is infinite it is not necessary that $E[T_{jj}] < \infty$. When state j is recurrent then if $\mu_{jj} < \infty$, the state is called *positive recurrent;* if $\mu_{jj} = \infty$, the state is called *null recurrent.*

EXAMPLE
4.2.2

The Shoeshine Boy Revisited   We consider a finite-capacity single-server queue. The service time is constant and there are three waiting spaces. Assume that arrivals are Poisson with a rate of two per service period ($\lambda = 0.2$ arrivals per minute). The transition probability matrix is then given by

$$P = \begin{matrix} 0 \\ 1 \\ 2 \\ 3 \end{matrix} \begin{bmatrix} 0.1353 & 0.2707 & 0.2707 & 0.3233 \\ 0.1353 & 0.2707 & 0.2707 & 0.3233 \\ & 0.1353 & 0.2707 & 0.5940 \\ & & 0.1353 & 0.8647 \end{bmatrix}.$$

We use Equation 4.2.1 to compute the first-passage-time probability mass functions $\{f_{0j}^{(n)}\}$—since we know that the starting state is 0. For each n, the first row of $F^{(n)}$ gives $\{f_{0j}^{(n)}\}$. In Figure 4.4, the cumulative probability approaches to 1 rather rapidly for $j = 3$ and rather slowly for $j = 0$. This means that the system becomes congested quickly after it opens and takes a long time to become idle.

For example, when n is as large as five hundred, the cumulative probability $\sum_{k=0}^{n} f_{00}^{(k)}$ is still only 0.8949—even though eventually it will reach 1. The congestion shown in this example is expected because the arrival rate is twice the service rate. We note that for this problem, the reaching probabilities $\{f_{ij}\}$ are all equal to 1. This means all states communicate and the chain is irreducible. The computational details are given in the Appendix. ∎

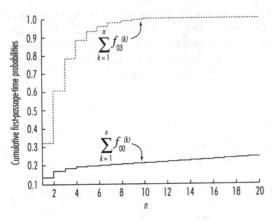

4.4 The cumulative first-passage-time probabilities.

EXAMPLE
4.2.3   **A Mouse in a Maze Revisited**   Using Equation 4.2.1 and summing the probability mass functions over all n, we find for example that $f_{13} = P\{T_{13} < \infty\} = 0.4286$, $f_{17} = P\{T_{17} < \infty\} = 0.6$, and $f_{19} = P\{T_{19} < \infty\} = 0.4$. This shows that random variables T_{13}, T_{17}, and T_{19} are all defective random variables. In other words, there are positive probabilities that first passages to these destination states may never consummate. Mathematically, these eventualities are expressed by $T_{13} = \infty$, $T_{17} = \infty$, and $T_{19} = \infty$. In the Appendix, we also show that $f_{37} = 0.4$ and $f_{39} = 0.6$. This is expected by symmetry of cells 1 and 3 with respect to cells 7 and 9 shown in Figure 4.1 (see Example 4.1.4). ∎

Given $X_0 = i$, we are now interested in counting the number of visits to state j. Define indicator random variable

$$I_{ij}(n) = \begin{cases} 1 & \text{if } X_n = j | X_0 = i \\ 0 & \text{otherwise.} \end{cases}$$

Then

$$N_{ij}(n) = \text{the number of visits to state } j \text{ by time } n = \sum_{k=1}^{n} I_{ij}(k).$$

We next show that $\{N_{ij}(n)|n \geq 1\}$ is a discrete-time delayed renewal process. The initial interarrival time is T_{ij} and the subsequent interarrival times are T_{jj}. Hence the initial interarrival time distribution G follows $\{f_{ij}^{(n)}\}$ and the subsequent inter-arrival time distribution F follows $\{f_{jj}^{(n)}\}$. If the chain is presently in a given state, the first time it will visit state j is a stopping time. By the strong Markov property, we conclude that these interarrival times are conditionally independent. Hence $\{N_{ij}(n)\}$ is a delayed renewal process. The mean state-occupancy time, $M_{ij}(n) \equiv E[N_{ij}(n)]$, can be found as follows:

$$M_{ij}(n) = E \sum_{k=1}^{n} I_{ij}(k) = \sum_{k=1}^{n} E[I_{ij}(k)] = \sum_{k=1}^{n} p_{ij}^{(k)}.$$

In matrix notation, let $M(n) = \{M_{ij}(n)\}$, then

$$M(n) = \sum_{k=1}^{n} P^{(k)}.$$

Recall that if state j is recurrent, then $f_{jj} = 1$. This means that state j will be visited infinitely often, that is, $P\{N_{jj}(\infty) = \infty\} = 1$ or $M_{jj}(\infty) = \infty$. On the other hand, if state j is transient, then $f_{jj} < 1$, and $N_{jj}(\infty)$ is a geometric random variable with probability mass function

$$P\{N_{jj}(\infty) = k\} = (f_{jj})^k (1 - f_{jj}) \qquad k = 0, 1, 2, \ldots$$

with mean

$$M_{jj}(\infty) = E[N_{jj}(\infty)] = \frac{1}{1 - f_{jj}} < \infty.$$

The preceding deliberation implies that state j is recurrent if and only if

$$\sum_{n=1}^{\infty} p_{jj}^{(n)} = \infty. \qquad (4.2.2)$$

We see that Equation 4.2.2 provides another way to characterize a recurrent state.

For state i, we let $d(i)$ denote the greatest common divisor of all integers $n \geq 1$ for which $p_{ii}^{(n)} > 0$. The integer $d(i)$ is called the *period* of state i. This definition of *periodicity* is identical to that used in characterizing the interarrival time of a discrete renewal process. A state with a period of 1 is called *aperiodic*. It can be shown that periodicity and recurrence are class properties. If one state in a class possesses a given property, all states in the class possess it. An irreducible Markov chain whose states have a period of 1 is called an *aperiodic chain*.

EXAMPLE 4.2.4

Consider a Markov chain with transition probability matrix

$$P = \begin{matrix} 1 \\ 2 \\ 3 \end{matrix} \begin{bmatrix} 0 & 1 & 0 \\ 0.4 & 0 & 0.6 \\ 0 & 0.7 & 0.3 \end{bmatrix}.$$

We see that all states communicate so the chain is irreducible. With $p_{33} > 0$, it is clear that $d(3) = 1$. Since periodicity is a class property, all states have a period of 1. ■

EXAMPLE 4.2.5

Consider a Markov chain with transition probability matrix

$$P = \begin{array}{c} 1 \\ 2 \\ 3 \\ 4 \end{array} \begin{bmatrix} 0 & 1 & 0 & 0 \\ 0 & 0 & 1 & 0 \\ 0 & 0 & 0 & 1 \\ 0.5 & 0 & 0.5 & 0 \end{bmatrix}.$$

For this chain, we see that all states communicate. Therefore it is an irreducible chain. The multistep transition probabilities $\{p_{11}^{(n)}\}$ are given as follows:

$$p_{11}^{(1)} = p_{11}^{(2)} = p_{11}^{(3)} = 0,$$

$$p_{11}^{(4)} = 0.5, \quad p_{11}^{(5)} = 0, \quad p_{11}^{(6)} = 0.25,$$

and $p_{11}^{(n)} > 0$ if $n \geq 4$ and n is an even number.

So we conclude that $d(i) = 2$ for all i. ■

Having defined the various properties characterizing a state, we are now ready to discuss the decomposition of a state space into equivalence classes so that states in the same class will have the same characterizing attributes. For state i, we let $C(i)$ denote the class containing state i. Since all states in $C(i)$ communicate with i, all states in i communicate with one another. Define $T(i)$ as the set of all states that are accessible from i (the "TO-LIST") and $F(i)$ as the set of states from which state i is accessible (the "FROM-LIST"). We note that $C(i)$ is the intersection of the two sets $T(i)$ and $F(i)$ and if $C(i) = T(i)$, then class $C(i)$ is closed. The following algorithm partitions the state space S into disjoint classes, say, $E_1, E_2, ..., E_m$.

1   For each state i in S, let $T(i) = \{i\}$ and $F(i) = \{\varnothing\}$.

2   For each state i in S, do the following:

For each state k in $T(i)$, add to $T(i)$ all states j such that $p_{kj} > 0$ (if k is not already there). Reiterate the step until no further addition is possible.

3   For each state i in S, do the following:

Add state j to $F(i)$ if state i is in $T(j)$. Reiterate the step until no further addition is possible.

4   For each state i in S, find $C(i) = F(i) \cap T(i)$.

Suppose that $E_1, ..., E_n$ ($1 \leq n \leq m$) are closed and the rest of the classes (if any) are nonclosed. We lump all nonclosed classes into one set T—the set of transient states. The transition matrix of the chain can be rewritten as

$$P = \begin{bmatrix} P_1 & 0 & 0 & 0 & 0 & 0 & 0 \\ 0 & P_2 & 0 & 0 & 0 & 0 & 0 \\ & & \cdot & & & & \\ & & & \cdot & & & \\ & & & & \cdot & & \\ 0 & 0 & 0 & 0 & 0 & P_n & 0 \\ R_1 & R_2 & \cdot & \cdot & \cdot & R_n & Q \end{bmatrix},$$

where the rows and columns are permuted, if necessary, so that they correspond to the respective transition probabilities with states listed in the order $E_1, \dots, E_n$ and the set T. A transition matrix written in the given format is considered in a *canonical form*. If we are only interested in the relations between transient states and the closed classes and not interested in transitions within closed classes, we can combine states in each closed class into one state, that is, make each closed class an *absorbing* state (for any state i, if $p_{ii} = 1$ then state i is called absorbing). Then the canonical form can be written as

$$P = \begin{bmatrix} I_n & O \\ R & Q \end{bmatrix}, \tag{4.2.3}$$

where I_n is an identity matrix of dimension n. Also, let v denote the number of states in T; then R is $v \times n$ and Q is $v \times v$.

EXAMPLE 4.2.6   Consider a Markov chain with state space $S = \{1, \dots, 10\}$ and transition matrix

$$P = \begin{array}{c} \\ 1 \\ 2 \\ 3 \\ 4 \\ 5 \\ 6 \\ 7 \\ 8 \\ 9 \\ 10 \end{array} \begin{bmatrix} 1/2 & 0 & 1/2 & 0 & 0 & 0 & 0 & 0 & 0 & 0 \\ 0 & 1/3 & 0 & 0 & 0 & 0 & 2/3 & 0 & 0 & 0 \\ 1 & 0 & 0 & 0 & 0 & 0 & 0 & 0 & 0 & 0 \\ 0 & 0 & 0 & 0 & 1 & 0 & 0 & 0 & 0 & 0 \\ 0 & 0 & 0 & 1/3 & 1/3 & 0 & 0 & 0 & 1/3 & 0 \\ 0 & 0 & 0 & 0 & 0 & 1 & 0 & 0 & 0 & 0 \\ 0 & 0 & 0 & 0 & 0 & 0 & 1/4 & 0 & 3/4 & 0 \\ 0 & 0 & 1/4 & 1/4 & 0 & 0 & 0 & 1/4 & 0 & 1/4 \\ 0 & 1 & 0 & 0 & 0 & 0 & 0 & 0 & 0 & 0 \\ 0 & 1/3 & 0 & 0 & 1/3 & 0 & 0 & 0 & 0 & 1/3 \end{bmatrix}.$$

In the Appendix, we give a MATLAB program **mc_equca** to identify the equivalent classes. We find

$T(1) = \{1, 3\}$, $T(2) = \{2, 7, 9\}$, $T(3) = \{1, 3\}$, $T(4) = \{5, 4, 9, 2, 7\}$,
$T(5) = \{4, 5, 9, 2, 7\}$, $T(6) = \{6\}$, $T(7) = \{7, 9, 2\}$,
$T(8) = \{3, 4, 8, 10, 1, 5, 2, 7, 9\}$, $T(9) = \{2, 7, 9\}$, $T(10) = \{2, 5, 10, 7, 4, 9\}$.

$F(1) = \{1, 3, 8\}$, $F(2) = \{2, 4, 5, 7, 8, 9, 10\}$, $F(3) = \{1, 3, 8\}$,
$F(4) = \{4, 5, 8, 10)$, $F(5) = \{4, 5, 8, 10\}$, $F(6) = \{6\}$,
$F(7) = \{2, 4, 5, 7, 8, 9, 10\}$, $F(8) = \{8\}$, $F(9) = \{2, 4, 5, 7, 8, 9, 10\}$,
$F(10) = \{8, 10\}$.

Consequently,

$C(1) = \{1, 3\}$, $C(2) = \{2, 7, 9\}$, $C(3) = \{1, 3\}$, $C(4) = \{4, 5\}$,
$C(5) = \{4, 5\}$, $C(6) = \{6\}$, $C(7) = \{2, 7, 9\}$, $C(8) = \{8\}$, $C(9) = \{2, 7, 9\}$,
$C(10) = \{10\}$.

Hence, we find $E_1 = \{1, 3\}$, $E_2 = \{2, 7, 9\}$, $E_3 = \{6\}$, and $T = \{4, 5, 8, 10\}$. The canonical form of the transition matrix reads

		1	3	2	7	9	6	4	5	8	10
	1	1/2	1/2								
	3	1	0								
	2			1/3	2/3	0					
	7			0	1/4	3/4					
$P =$	9			1	0	0					
	6						1				
	4	0	0	0	0	0	0	0	1	0	0
	5	0	0	0	0	1/3	0	1/3	1/3	0	0
	8	0	1/4	0	0	0	0	1/4	0	1/4	1/4
	10	0	0	1/3	0	0	0	0	1/3	0	1/3

The preceding transition matrix can also be written in the form of Equation 4.2.3:

		E_1	E_2	E_3	4	5	8	10
	E_1	1	0	0				
	E_2	0	1	0				
$P = \begin{bmatrix} I_3 & 0 \\ R & Q \end{bmatrix} =$	E_3	0	0	1				
	4	0	0	0	0	1	0	0
	5	0	1/3	0	1/3	1/3	0	0
	8	1/4	0	0	1/4	0	1/4	1/4
	10	0	1/3	0	0	1/3	0	1/3

A Markov chain whose transition probability matrix is in the form of Equation 4.2.3 is called an *absorbing Markov chain*. In Section 4.4, we present ways for computing mean times to absorption and probabilities of absorbing into specific recurrent classes. We will also introduce the discrete-time phase-type distribution—a concept useful in computational probability.

4.3 Ergodic and Periodic Markov Chains

Consider an irreducible Markov chain with state space $S = \{0, 1, \ldots\}$ constituting a single closed communication class. Let $N_{ij}(n)$ denote the number of visits to state j in n transitions given that $X_0 = i$. We can study the asymptotic behavior of $\{N_{ij}(n)\}$ using the results obtained for a discrete renewal process given in Section 3.7. Let T_{ij} denote the first passage time from state i to state j. Then $\{N_{ij}(n), n \geq 0\}$ is a delayed discrete renewal process with the first interarrival time T_{ij} and subsequent ones $\{T_{jj}\}$. Moreover, because of the strong Markov property (see Section 4.2), these interarrival times are independent. Using the results given in Section 3.7, we first note that Equation 3.7.8 implies

$$\lim_{n \to \infty} \frac{M_{ij}(n)}{n} = \lim_{n \to \infty} \frac{\sum_{k=1}^{n} p_{ij}^{(k)}}{n} = \frac{1}{\mu_{jj}}, \tag{4.3.1}$$

where $\mu_{jj} = E[T_{jj}]$ is the mean recurrence time to state j. Thus Equation 4.3.1 gives the long-run expected fraction of time the process will be visiting state j.

If state j is aperiodic, using Equation 3.7.7, we have a stronger result

$$\lim_{n \to \infty} p_{ij}^{(n)} = \frac{1}{\mu_{jj}}, \tag{4.3.2}$$

independent of the starting state i. Similarly, if the state has a period d, then Equation 3.7.9 implies

$$\lim_{n \to \infty} p_{jj}^{(nd)} = \frac{d}{\mu_{jj}}. \tag{4.3.3}$$

In Equation 4.3.3, we *require* that $X_0 = j$ for the limit to hold. Denote the *limiting state* probability by

$$\pi_j = \lim_{n \to \infty} p_{jj}^{(n)}.$$

For the aperiodic chain, we have

$$\pi_j = \frac{1}{\mu_{jj}}. \tag{4.3.4}$$

The preceding identity shows that one way to find the limiting probability is by taking the reciprocal of the mean recurrence time. A simpler way to find $\{\pi_j\}$ will be given shortly. In Section 4.2, we called state j positive recurrent if $\mu_{jj} < \infty$ and null recurrent if $\mu_{jj} = \infty$. Hence for the former case, we have $\pi_j > 0$ and for the latter case $\pi_j = 0$.

For a Markov chain with transition probability matrix P, we now introduce the idea of a stationary distribution. A probability distribution $\{P_i, i \geq 0\}$ is a stationary distribution for the Markov chain if

$$P_j = \sum_{i=0}^{\infty} P_i p_{ij} \qquad j \geq 0. \qquad (4.3.5)$$

For any Markov chain, if $P\{X_0 = j\} = P_j$ then repeated applications of Equation 4.3.5 show that $P\{X_n = j\} = P_j$ for all $n = 0, 1, \ldots$. Such a Markov chain is called a stationary Markov chain.

When an irreducible Markov chain is aperiodic and positive recurrent, the chain is called an *ergodic Markov chain*.

The limiting distribution $\{\pi_j\}$ of an ergodic chain is the unique nonnegative solution of Equation 4.3.5, that is,

$$\pi_j = \sum_{k=0}^{\infty} \pi_k p_{kj} \qquad j = 0, 1, 2, \ldots,$$

$$\sum_{j=0}^{\infty} \pi_j = 1. \qquad (4.3.6)$$

To derive Equation 4.3.6, we take the limit on both sides of Equation 4.1.3 and use Equations 4.3.2 and 4.3.4 to obtain

$$\pi_j = \lim_{n \to \infty} p_{ij}^{(n+1)} = \lim_{n \to \infty} \sum_{k=0}^{\infty} p_{ik}^{(n)} p_{kj} = \sum_{k=0}^{\infty} \lim_{n \to \infty} p_{ik}^{(n)} p_{kj} = \sum_{k=0}^{\infty} \pi_k p_{kj}.$$

When a solution of Equation 4.3.6 exists, clearly we have $\pi_j > 0$ for all j and consequently $\mu_{jj} < \infty$ for all j. This implies that the chain is positive recurrent. Hence whether Equation 4.3.6 has a solution provides one way to check the positive recurrence of an irreducible Markov chain. Unless we force a Markov chain to take the stationary distribution as the starting state distribution, the state probabilities $s(n) = \{s_j(n)\}$ will generally follow the stationary distribution only in the limit.

The ratio π_j / π_i has an interesting and useful interpretation. Consider a discrete regenerative process in which each regeneration corresponds to a return to state i. The interarrival time of this regenerative process is the recurrence time T_{ii}. Let V_j denote the number of visits to state j between two successive visits to i. Then, applying Equation 3.6.1, we find

$$\pi_j = \lim_{n \to \infty} P\{X_n = j\} = \frac{E[V_j]}{E[T_{ii}]} = E[V_j]\pi_i \qquad \text{or} \qquad E[V_j] = \frac{\pi_j}{\pi_i}. \qquad (4.3.7)$$

In other words, the ratio of the two limiting state probabilities represents the expected number of visits to state j between two successive visits to i.

When a chain is irreducible, positive recurrent, and periodic with period d, we call it a *periodic Markov chain*. Now the solution of Equation 4.3.6 can only be interpreted as the long-run fraction of time that the process will be visiting state j.

To show that this is indeed the case, we define indicator random variable $I_j(k) = 1$ if $X_k = j$ and 0 otherwise (with $X_0 = l$ for any l) and the time-average probability as

$$\pi_j = \lim_{n \to \infty} \frac{E\left[\sum_{k=1}^{n} I_j(k)\right]}{n}. \tag{4.3.8}$$

Conditioning on the possible states leading into state j in one step, we write

$$\pi_j = \lim_{n \to \infty} \frac{E\left[\sum_{k=1}^{n}\sum_{i=0}^{\infty} I_i(k-1)I_j(k)\right]}{n} = \lim_{n \to \infty} \frac{\sum_{k=1}^{n}\sum_{i=0}^{\infty} E[I_i(k-1)I_j(k)]}{n}$$

$$= \lim_{n \to \infty} \frac{\sum_{k=1}^{n}\sum_{i=0}^{\infty} E[I_i(k-1)p_{ij}]}{n} = \sum_{i=0}^{\infty} p_{ij} \lim_{n \to \infty} \frac{\sum_{k=1}^{n} E[I_i(k-1)]}{n} = \sum_{i=0}^{\infty} \pi_i p_{ij}.$$

In the previous derivations, we skip the justifications for being able to interchange expectation, limiting, and summation operations. For the periodic chain, the last equality holds because of the time-average definition of π_i used in Equation 4.3.8. Even for a periodic chain, $1/\mu_{jj}$ is still the rate of returning to state j or, equivalently, the average number of visits to state j per unit time. With the time-average interpretation for π_j, we have $\pi_j = 1/\mu_{jj}$. Therefore Equation 4.3.3 can be rewritten as

$$\lim_{n \to \infty} p_{jj}^{(nd)} = d\pi_j, \tag{4.3.9}$$

where π_j is obtained from solving Equation 4.3.6. Of course, if we solve $\pi = \pi P$ and $\pi e = d$ instead, then the right side of Equation 4.3.9 is simply π_j.

For a positive-recurrent Markov chain, Equation 4.3.6 holds a prominent position. We shall study it in some detail.

In matrix notation, Equation 4.3.6 can be stated as

$$\pi = \pi P \quad \text{and} \quad \pi e = 1, \tag{4.3.10}$$

where π is the row vector containing $\{\pi_i\}$.

The system of linear equations $\pi = \pi P$ has one redundant equation. Replacing any one of the linear equations by $\pi e = 1$, we have a system of linearly independent equations. When the state space is finite, we can write $\pi = \pi P$ as $\pi B = 0$, where $B = I - P$ and 0 is a row vector of zeros. We replace the first column of B by e and call the resulting matrix C. We replace the first element of 0 by a 1 and call the resulting row vector c. The net effect of the substitutions is to replace the first linear equation of $\pi = \pi P$ by $\pi e = 1$. Then π is the solution of $\pi C = c$ and π is simply given by the first row of C^{-1}.

We now look at the right side of Equation 4.3.6, namely, $\sum_i \pi_i p_{ij}$. Since p_{ij} is the fraction of transitions leaving state i for state j, the right side can then be interpreted as the rate of transition into j—after conditioning by the "rates" of being in each state i. The left side of Equation 4.3.6 can be stated as $\pi_j = \pi_j \sum_i p_{ji} = \sum_i \pi_j p_{ji}$. Thus the left side can be similarly interpreted as the rate of transition out of j. In steady state, the two rates are equal for each state. In the context of network theory, this says that for each node the total input should be equal to the total output. Figure 4.5 depicts the relation graphically.

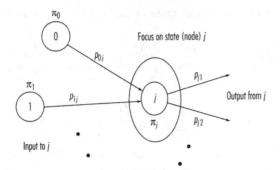

FIGURE
4.5   The transition rate balance equation at node j.

There is another way to look at the solution of Equation 4.3.6. We partition the state space S into two sets A and B such that $A \cup B = S$ and $A \cap B = \varnothing$. Let r_{AB} denote the aggregate transition rates from A to B, and we similarly define r_{BA}. In steady state, we expect that $r_{AB} = r_{BA}$. Now

$$r_{AB} = \sum_{i \in A} \pi_i \sum_{j \in B} p_{ij} = \sum_{i \in A} \sum_{j \in B} \pi_i p_{ij}$$

and

$$r_{BA} = \sum_{j \in B} \pi_j \sum_{i \in A} p_{ji} = \sum_{j \in B} \sum_{i \in A} \pi_j p_{ji}.$$

Hence we find a useful equation

$$\sum_{i \in A} \sum_{j \in B} \pi_i p_{ij} = \sum_{j \in B} \sum_{i \in A} \pi_j p_{ji}. \qquad (4.3.11)$$

In the context of network theory, states can be viewed as nodes and transitions as arcs. For a given transition probability matrix P, we can construct a *transition diagram* showing possible one-step transitions between states. A *cut* partitions the nodes into two mutually exclusive and collectively exhaustive sets. Equation 4.3.11 says that for each cut, the total transition rate traversing from one direction must be equal to that of the other direction. We show this observation graphically in Figure 4.6. In many applications, an examination of the transition diagram will reveal some special structures of a transition probability matrix. When this

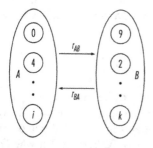

FIGURE
4.6 The transition rate balance equation at a given cut.

occurs, we can often exploit the structure using Equation 4.3.11 so that analytic results can be conveniently obtained (vis à vis Equation 4.3.6). In Examples 4.3.1 and 4.3.2, we illustrate the use of this idea.

EXAMPLE
4.3.1   **A Random Walk with a Reflecting Barrier**   Consider a Markov chain with state space $S = \{0, 1, \ldots\}$ and transition probability matrix P given by

$$\begin{bmatrix} q & p & 0 & . & . & . \\ q & 0 & p & 0 & . & . \\ 0 & q & 0 & p & 0 & . \\ . & 0 & q & 0 & p & . \\ & & . & . & . & \\ & & & . & . \end{bmatrix},$$

where $p > 0$, $q > 0$, $q > p$, and $p + q = 1$. In Figure 4.7, we display the transition diagram.

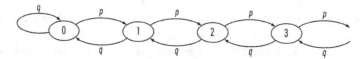

FIGURE
4.7 The transition diagram of a random walk.

The transition diagram shows that all states communicate. Hence the chain is irreducible. Since $p_{00} = q > 0$, the chain is aperiodic.

With the exception of node 0, establishing the balance equation (Equation 4.3.6) for each state involves the use of three consecutive π_i. On the other hand, with a cut between two adjacent nodes shown in Figure 4.7, the resulting balance equation (Equation 4.3.11) contains only two neighboring π_i. Thus the latter approach is slightly more convenient. This gives $p\pi_i = q\pi_{i+1}$, $i \geq 0$. Expressing all π_i in terms of π_0, we obtain

$$\pi_i = \left(\frac{p}{q}\right)^i \pi_0, \qquad i \geq 0.$$

Using the normalizing equation $\pi e = 1$, it is easy to see that $\pi_0 = (q - p)/q$. So, all $\pi_i > 0$ and the chain is positive recurrent. With Equation 4.3.7, we have $E[V_i] = \pi_i/\pi_0 = (p/q)^i$, $i \geq 1$. Thus $(p/q)^i$ represents the expected number of visits to state i between two successive visits to state 0. We remark that when $p > q$, it can be shown that the chain is transient, and when $p = q$, it is null recurrent. ∎

EXAMPLE
4.3.2

The Limiting Probabilities of an M/G/1 Queue Consider a single-server system with an infinite number of waiting rooms. Arrivals follow a Poisson process with rate λ and service times are i.i.d. random variables with a common distribution G. We assume that the service discipline is FIFO and service times are independent of interarrival times. Let $X(t)$ denote the number of customers in the system at time t. The process $\{X(t), t \geq 0\}$ is not Markovian because the service time is not memoryless. As in Example 4.1.2, we define X_n as the number of customers left behind by the nth departure. Let A denote the number of arrivals during the service of a customer. Analogous to the rationale used in Example 4.1.2, we find that $X_{n+1} = [X_n - 1]^+ + A$. Therefore X_{n+1} depends only on X_n but not $\{X_1, \ldots, X_{n-1}\}$ and $X = \{X_n | n = 1, 2, \ldots\}$ is a Markov chain embedded at service-completion epochs. Let

$$a_j = P\{A = j\} = \int_0^\infty e^{-\lambda x} \frac{(\lambda x)^j}{j!} \, dG(x) \qquad j = 0, 1, 2, \ldots.$$

Then we see that the transition probability matrix P has

$$P_{ij} = \begin{cases} a_j & i = 0, \, j \geq 0 \\ a_{j-i+1} & i > 0, \, j \geq i-1 \\ 0 & \text{otherwise.} \end{cases}$$

The transition probability matrix has a structure similar to that shown in Example 4.1.2 except that it is infinite in dimension and $\{a_j\}$ propagate to the right according to $\{p_{ij}\}$. If we use Equation 4.3.6 to write the balance equations, we obtain

$$\pi_j = \pi_0 a_j + \sum_{i=1}^{j+1} \pi_i a_{j-i+1} \qquad j = 0, 1, \ldots,$$

where $\{\pi_j\}$ gives the limiting queue length distribution at departure epochs. The preceding system can be solved by using the transform method. However, a simpler solution procedure can be devised based on Equation 4.3.11. Let A_i and B_i denote the two sets associated with the ith cut, $i = 0, 1, \ldots$. At the ith cut, we define $A_i = \{0, 1, \ldots, i\}$ and $B_i = S - A_i$ (that is, the set S with states in A_i removed). Define $b_i = a_{i+1} + a_{i+2} + \cdots = P\{A > i\}$. We see that $E[A] = \sum_{i \geq 0} P\{A > i\} = \sum_{i \geq 0} b_i = $ the expected number of arrivals in a service time. We define this quantity as ρ—the traffic intensity of the queue. In Figure 4.8, we display the transition diagram associated with P with cut 0 explicitly shown.

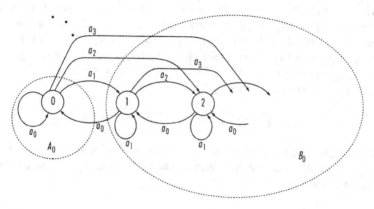

FIGURE

4.8   The transition diagram for the $M/G/1$ queue with cut 0 shown.

In the figure, we observe that the total transition rate moving from B_0 to A_0 is $\pi_1 a_0$ and the total transition rate going the other direction is $\pi_0(1 - a_0) = \pi_0 b_0$. Similarly, we write the complete system of linear equations based on cuts $i = 0, 1, \ldots$:

$$\pi_1 a_0 = \pi_0 b_0$$
$$\pi_2 a_0 = \pi_0 b_1 + \pi_1 b_1$$
$$\pi_3 a_0 = \pi_0 b_2 + \pi_1 b_2 + \pi_2 b_1$$
$$\cdots$$

Clearly, the preceding system is simpler than that implied by Equation 4.3.6. Since the system has a redundant equation, we can arbitrarily set $\pi_0 = 1$ and express the remaining π_i as multiples of π_0. We add the resulting equations and obtain

$$a_0 \sum_{i=1}^{\infty} \pi_i = \sum_{i=0}^{\infty} b_i + \sum_{i=1}^{\infty} \pi_i \sum_{k=1}^{\infty} b_k.$$

Let $\sum_{i=0}^{\infty} \pi_i = c$ and consider the case in which $c < \infty$. This implies that $\pi_0 = 1/c$. For $i \neq 0$, π_i are expressed as multiples of π_0. The last equation reduces to $a_0(c - 1) = \rho + (c - 1)(\rho - b_0)$ or $(c - 1)[a_0 + b_0 - \rho] = \rho$. Since $a_0 + b_0 = 1$, we conclude that $c = 1/(1 - \rho)$ if $\rho < 1$. This shows that for the Markov chain to be positive recurrent, we require that $\rho < 1$. In this case, we have $\pi_0 = 1 - \rho$ and other π_i can be found recursively. These are the limiting probabilities of the queue length at service-completion epochs. ∎

EXAMPLE

4.3.3   **The Limiting Probabilities of a GI/M/1 Queue**   Consider a single-server system with an infinite number of waiting rooms. Assume that successive interarrival times are i.i.d. random variables with a common distribution G with mean $1/\lambda$ and service times are i.i.d. exponential random variables with parameter μ, and service and interarrival times are mutually independent. Customers are served one at a time on a FIFO basis. Let $X(t)$ denote the number of customers in the system at time t. The process $\{X(t), t \geq 0\}$ is not Markovian because interarrival times are not memoryless. Let X_n denote the number of customers in the system immediately before the arrival of the nth customer. Then $X_{n+1} = X_n + 1 - B$, where B is the number of service completions during an interarrival interval. Since X_{n+1} depends on X_n but not on $\{X_1, \ldots, X_{n-1}\}$, $X = \{X_n | n = 1, 2, \ldots\}$ is a Markov chain embedded at arrival epochs. Define

$$q_j = \int_0^\infty e^{-\mu x} \frac{(\mu x)^j}{j!} dG(x) \qquad j = 0, 1, 2, \ldots .$$

Thus q_j gives the probability that there are j service completions in an interarrival interval. Based on the balance equation linking X_{n+1} and X_n, we find the transition matrix of the Markov chain

$$
P = \{p_{ij}\} =
\begin{array}{c}
 \\
0 \\
1 \\
2 \\
\cdot \\
\cdot
\end{array}
\begin{array}{ccccccc}
0 & 1 & 2 & \cdot & \cdot & \\
\left[\begin{array}{cccccc}
r_0 & q_0 & 0 & 0 & \cdot \\
r_1 & q_1 & q_0 & 0 & \cdot \\
r_2 & q_2 & q_1 & q_0 & \cdot \\
\cdot & \cdot & \cdot & \cdot & \cdot \\
\cdot & \cdot & \cdot & \cdot & \cdot
\end{array}\right],
\end{array}
$$

where we define $r_j = 1 - (q_0 + \cdots + q_j)$. In deriving the transition matrix P, we use the relation $j = i + 1 - B$. If $j > 0$, then, for a fixed i, $B = i + 1 - j$; if $j = 0$, then, for a fixed i, $B \geq i + 1$.

Let $\{\pi_j\}$ denote the limiting queue length distribution immediately before an arrival. Then $\{\pi_j\}$ can be found by solving Equation 4.3.6. Ignoring one redundant equation involving $j = 0$, we obtain the following system of linear equations:

$$\pi_j = \sum_{i=j-1}^\infty \pi_i q_{i+1-j} \qquad j \geq 1 \qquad \text{and} \qquad \sum_{j=0}^\infty \pi_j = 1.$$

An analysis of the structure of the preceding difference equation suggests that the solution is of the form $\pi_j = ca^j$. Then we have

$$ca^j = \sum_{i=j-1}^\infty ca^i \int_0^\infty e^{-\mu x} \frac{(\mu x)^{i+1-j}}{(i+1-j)!} dG(x) = c \int_0^\infty e^{-\mu x} a^{j-1} \sum_{i=j-1}^\infty \frac{(\mu x a)^{i+1-j}}{(i+1-j)!} dG(x)$$

$$= ca^{j-1} \int_0^\infty e^{-\mu(1-a)x} dG(x)$$

or $a = G^e(\mu(1-a))$, where $G^e(s)$ is the Laplace-Stieltjes transform of the interarrival time distribution G. Thus a is a root of the last equation. Another way to look at the last equation is by noting that the probability generating function of $\{q_j\}$ is given by

$$B(z) = \sum_{j=0}^\infty z^j q_j = \sum_{j=0}^\infty z^j \int_0^\infty e^{-\mu x} \frac{(\mu x)^j}{j!} dG(x) = \int_0^\infty e^{-\mu x} e^{\mu x z} dG(x) = G^e(\mu(1-z)).$$

Equivalently, a is a root of the equation $z = B(z)$. In other words, the solution a is given by the intersection a of $y = B(z)$ and $y = z$ on a y-z plane. We observe that $0 < B(0) = q_0 < 1$, $B(1) = 1$, and for $0 < z < 1$,

$$B^{(1)}(z) = \sum_{j=1}^{\infty} jq_j z^{j-1} > 0 \qquad \text{and} \qquad B^{(2)}(z) = \sum_{j=2}^{\infty} j(j-1)q_j z^{j-2} > 0.$$

These conditions imply that $B(z)$ is strictly convex in $(0, 1)$. The slope of $B(z)$ at $z = 1$ is the mean of the distribution $\{q_j\}$—the expected number of service completions during an interarrival interval, namely, $B^{(1)}(1) = \mu(1/\lambda) \equiv r$. We now consider the two possible scenarios shown in Figure 4.9. In the left side of the graph, we see that $r \le 1$ and there are no roots in $(0, 1)$. In the right side of the graph, we have $r > 1$ and exactly one root in $(0, 1)$. For the constant c, we see that

$$\sum_{j=0}^{\infty} \pi_j = \sum_{j=0}^{\infty} ca^j = \frac{c}{1-a} = 1.$$

Hence $c = 1 - a$ and the limiting queue length distribution is given as $\pi_j = (1-a)a^j, j = 0, 1, 2, \ldots$. ∎

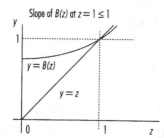
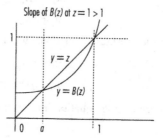

FIGURE
4.9   Plots of $y = B(z)$ and $y = z$ under the two scenarios.

EXAMPLE
4.3.4   **A Quality Control Problem**   A production line produces a product one unit at a time. Each item produced is inspected until i consecutive nondefective items are found. When this occurs, only one in r items is randomly chosen and inspected. As soon as a defective item is found, the sampling plan calls for reverting back to inspecting every item until i consecutive nondefective items are again found—and the cycle repeats. We assume that the probability that an item is found to be defective is p.

Let X_n denote the state of the system in which item n leaves the inspection station (whether or not it has been inspected). We use $X_n = 0$ to denote the event that item n is found to be defective. For $k = 1, 2, \ldots, i - 1$, we use $X_n = k$ to denote the event that there are k consecutive nondefective items found after having completed the inspection of item n. We use $X_n = i$ to denote the event that the system is in the sampling phase when item n leaves the inspection station (so item n may or may not have been inspected). The stochastic process $\{X_n | n = 1, 2, \ldots\}$ is a Markov chain with transition matrix P given by

	0	1	2	.	.	.	i
0	p	q					
1	p		q				
2	p			q			
.							
.							
$i-1$	p						q
i	$\dfrac{p}{r}$						$1 - \dfrac{p}{r},$

where $q = 1 - p$. In the previous transition matrix, we see that $P\{X_{n+1} = 0 | X_n = i\}$ $= P\{X_{n+1} = 0 | \text{the system is in the sampling phase}\} = P\{X_{n+1} = 0 | \text{item } n \text{ is selected}$ for inspection$\} \times P\{\text{item } n \text{ is selected for inspection}\} = p(1/r)$. Using Equation 4.3.6, we obtain

$$\pi_k = q\pi_{k-1}, \qquad k = 1, \ldots, i-1,$$
$$\pi_i = q\pi_{i-1} + (1 - (p/r))\,\pi_i.$$

From the preceding equations, we find

$$\pi_k = q^k\pi_0, \qquad k = 1, \ldots, i-1,$$
$$\pi_i = q^i(r/p)\pi_0.$$

The normalizing equation implies that $\pi_0\{1 + q + \cdots + q^{i-1} + q^i(r/p)\} = 1$ or

$$\pi_0 = \frac{p}{1 + (r-1)q^i}.$$

Having obtained the limiting probabilities, we compute the *average fraction inspected (AFI)* by conditioning on $\{\pi_k\}$. This gives AFI $= \pi_0 + \cdots + \pi_{i-1} + (1/r)\pi_i = 1 - \pi_i + (1/r)\pi_i$ or

$$\text{AFI} = \frac{1}{1 + (r-1)q^i}.$$

Assume that all defective items found at the inspection station are replaced by nondefective items. Then only the fraction $1 -$ AFI (of items leaving the inspection station) can contain defective items. Thus the items leaving the inspection station on the average contain the fraction of defectives $p(1 - \text{AFI})$. In the parlance of quality control, this quantity is called the *average outgoing quality (AOQ)*. Since AOQ is a function of p, we write

$$\text{AOQ}(p) = \frac{p(r-1)q^i}{1 + (r-1)q^i}.$$

In actuality, the value of p is not known. Define the average outgoing quality limit by

$$\text{AOQL} = \max_{0 \le p \le 1} \{\text{AOQ}(p)\}.$$

The AOQL places an upper bound on the defective fraction of the outgoing quality. As an example, if we use a plan in which $(i, r) = (10, 5)$, then Figure 4.10 shows that in a worst-case scenario the outgoing fraction defective will not exceed 7 percent. In the figure, we also observe that a less stringent plan $(i, r) = (5, 10)$ yields an AOQL of about 18 percent. ▪

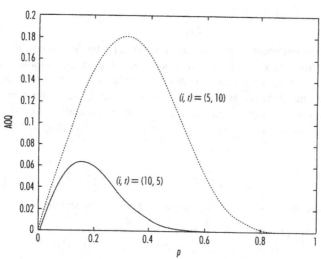

FIGURE
4.10 The average outgoing quality as a function of input fraction defective p.

We now return to a finite-state periodic Markov chain with period d. It can be shown that the state space S can be partitioned into disjoint sets $S_1, \ldots, S_d$ such that the transition probability matrix can be permuted so that in block-partitioned form it reads

$$
P = \begin{array}{c} \\ S_1 \\ S_2 \\ \cdot \\ \cdot \\ S_{d-1} \\ S_d \end{array}
\begin{array}{cccccc} S_1 & S_2 & \cdot & \cdot & S_{d-1} & S_d \end{array}
\left[\begin{array}{ccccc}
& P_1 & & & \\
& & P_2 & & \\
& & & & \\
& & & & \\
& & & & P_{d-1} \\
P_d & & & &
\end{array} \right].
\tag{4.3.12}
$$

In Equation 4.3.12, we see that $p_{ij} \geq 0$ if $i \in S_k$ and $j \in S_{k+1}$, $k = 1, \ldots, d-1$ or $i \in S_d$ and $j \in S_1$, and 0 otherwise. The transition probability matrix of a periodic Markov chain written in the form of Equation 4.3.12 is said to be in canonical form. By repeated matrix multiplications it can be shown that P^d is a block-diagonal matrix with

$$
\begin{array}{c}
\begin{array}{ccc} S_1 & \cdots & S_d \end{array} \\
P^d = \begin{array}{c} S_1 \\ \vdots \\ S_d \end{array}
\begin{bmatrix}
Q_1 & & \\
& \ddots & \\
& & Q_d
\end{bmatrix}.
\end{array}
\tag{4.3.13}
$$

For a Markov chain with a *one-step* transition probability matrix P^d, it can be shown that the sets $S_1, \ldots, S_d$ form d closed communication classes and each sub-matrix Q_k is a stochastic matrix. Thus each subchain with state space S_k is itself an ergodic Markov chain. However, each transition in S_k takes three *original* steps to consummate. Giving more precise exposition will require the introduction of additional notation. In lieu of doing so, we present the various issues at stake in the following example.

EXAMPLE
4.3.5

Consider an irreducible Markov chain with transition probability matrix

$$
\begin{array}{c}
\begin{array}{c} 1 \\ 2 \\ 3 \\ 4 \\ 5 \\ 6 \\ 7 \end{array}
\begin{bmatrix}
0 & 0 & 0 & 0.65 & 0 & 0.35 & 0 \\
0 & 0 & 0 & 0.90 & 0 & 0.10 & 0 \\
0.10 & 0.10 & 0 & 0 & 0 & 0 & 0.80 \\
0 & 0 & 0.12 & 0 & 0.88 & 0 & 0 \\
0.15 & 0.35 & 0 & 0 & 0 & 0 & 0.50 \\
0 & 0 & 0.75 & 0 & 0.25 & 0 & 0 \\
0 & 0 & 0 & 0.40 & 0 & 0.60 & 0
\end{bmatrix}.
\end{array}
$$

We use program **mc_perdo** given in the Appendix to find the period of the chain. This gives $d = 3$. We then use program **mc_canop** to find the transition matrix in the following canonical form:

$$
P = \begin{array}{c} 3 \\ 5 \\ 1 \\ 2 \\ 7 \\ 4 \\ 6 \end{array}
\begin{array}{c}
\begin{array}{ccccccc} 3 & 5 & 1 & 2 & 7 & 4 & 6 \end{array} \\
\begin{bmatrix}
& & 0.10 & 0.10 & 0.80 & & \\
& & 0.15 & 0.35 & 0.50 & & \\
& & & & & 0.65 & 0.35 \\
& & & & & 0.90 & 0.10 \\
& & & & & 0.40 & 0.60 \\
0.12 & 0.88 & & & & & \\
0.75 & 0.25 & & & & &
\end{bmatrix}.
\end{array}
$$

We use program **mc_limsr** given in the Appendix to find the long-run fractions of time the chain will be in each state: $\{0.1329, 0.2004, 0.0434, 0.0834, 0.2065, 0.1859, 0.1474\}$. Now we see that

$$
P^3 = \begin{bmatrix} P_1P_2P_3 & & \\ & P_2P_3P_1 & \\ & & P_3P_1P_2 \end{bmatrix}
$$

$$
= \begin{matrix} 3 \\ 5 \\ 1 \\ 2 \\ 7 \\ 4 \\ 6 \end{matrix}
\begin{bmatrix}
0.4508 & 0.5493 & & & & & \\
0.3641 & 0.6359 & & & & & \\
& & 0.1330 & 0.2649 & 0.6022 & & \\
& & 0.1409 & 0.3043 & 0.5549 & & \\
& & 0.1251 & 0.2255 & 0.6494 & & \\
& & & & & 0.5960 & 0.4040 \\
& & & & & 0.5094 & 0.4906
\end{bmatrix}.
$$

For the Markov chain with *one-step* transition probability matrix P^3, the chain has three closed communication classes. The state space is partitioned into three disjoint sets: $S_1 = \{3, 5\}$, $S_2 = \{1, 2, 7\}$, and $S_3 = \{4, 6\}$. The three subchains are ergodic. We let π_1, π_2, and π_3 denote their respective *limiting* probability vectors. We in turn obtain $\pi_1 = \{0.3987, 0.6013\}$, $\pi_2 = \{0.1301, 0.2503, 0.6196\}$, $\pi_3 = \{0.5577, 0.4423\}$. Since it takes exactly three *original* steps to complete one transition within each class of this new chain, the time-average probability of being in each state can also be computed from $\pi = (1/d)(\pi_1, \pi_2, \pi_3)$. This gives $\pi = \{0.1329, 0.2004, 0.0434, 0.0834, 0.2065, 0.1859, 0.1474\}$. Finally, we observe that

$$
\lim_{n \to \infty} P^{(3n)} =
\begin{matrix} & 3 & 5 & 1 & 2 & 7 & 4 & 6 \\
3 \\ 5 \\ 1 \\ 2 \\ 7 \\ 4 \\ 6 \end{matrix}
\begin{bmatrix}
0.3987 & 0.6013 & & & & & \\
0.3987 & 0.6013 & & & & & \\
& & 0.1301 & 0.2503 & 0.6022 & & \\
& & 0.1301 & 0.2503 & 0.6022 & & \\
& & 0.1301 & 0.2503 & 0.6022 & & \\
& & & & & 0.5960 & 0.4040 \\
& & & & & 0.5960 & 0.4040
\end{bmatrix}.
$$

If the process starts from one state in S_i it takes exactly one step to go to a state in S_{i+1}. Once we reach S_{i+1}, the given limiting state probabilities apply. As an example, consider the case in which $X_0 = 3$. Since state 3 is in S_1, with probability 1 it will move to a state in S_2 at the next transition. Once in a state in S_2, the process will visit a state in S_2 again every three steps. This implies that

$$\lim_{n\to\infty} P^{(3n+1)} = \begin{array}{c} \\ 3 \\ 5 \\ 1 \\ 2 \\ 7 \\ 4 \\ 6 \end{array}\begin{array}{ccccccc} 3 & 5 & 1 & 2 & 7 & 4 & 6 \\ & & 0.1301 & 0.2503 & 0.6196 & & \\ & & 0.1301 & 0.2503 & 0.6196 & & \\ & & & & & 0.5577 & 0.4423 \\ & & & & & 0.5577 & 0.4423 \\ & & & & & 0.5577 & 0.4423 \\ 0.3987 & 0.6013 & & & & & \\ 0.3987 & 0.6013 & & & & & \end{array}.$$

We can obtain $\lim_{n\to\infty} P^{(3n+2)}$ in an analogous manner. All programs used in this example are listed in the Appendix. ∎

In many random walks and queueing applications, state spaces are typically infinitely countable and Markov chains involved irreducible. The issue about whether such a Markov chain is recurrent or transient surfaces frequently. To conclude this section, we give criteria for performing the test. Let X be an irreducible Markov chain with state space $S = \{0, 1, \ldots\}$.

A necessary and sufficient condition that X is transient is that the system of equations

$$y_i = \sum_{\substack{j=0 \\ j\neq k}}^{\infty} p_{ij}y_j \qquad i \neq k \qquad (4.3.14)$$

have a bounded solution that is not identically zero.

State k in Equation 4.3.14 can be chosen arbitrarily. On the other hand, the converse to the preceding establishes that X is recurrent. A reference for a proof of this useful result can be found in the Bibliographic Notes.

4.4 Absorbing Markov Chains

Consider an absorbing Markov chain with a set T^c of absorbing states and a set T of transient states. The transition probability matrix P is given in a canonical form

	T^c	T
T^c	I	O
T	R	Q

$$(4.4.1)$$

We find the multistep transition probability matrix $P^{(n)}$ by multiplying P by itself n times. This gives

$$P^{(n)} = P^n = \begin{bmatrix} I & O \\ R_n & Q^n \end{bmatrix},$$

(4.4.2)

where $R_n = (I + Q + Q^2 + \cdots + Q^{n-1})R$. States in T are transient; eventually the process will leave T and never return. In Equation 4.4.1, we observe that Q is a *substochastic matrix* in that elements of at least one of its rows do not sum to 1. It can be shown that when Q is substochastic, Q^n converges to a zero matrix as n approaches infinity. Define the *fundamental matrix* of an absorbing Markov chain by $W = I + Q + Q^2 + \cdots$.

When the state space S is finite, we have

$$W = (I - Q)^{-1}.$$

(4.4.3)

When S is denumerable, W is the minimal nonnegative solution of $(I - Q)W = I$. The elements $\{w_{ij}\}$ in W have probabilistic interpretations. Considering $i \in T$ and $j \in T$, then

$$w_{ij} = E\left[\sum_{m=0}^{\infty} I(X_m = j \mid X_0 = i)\right] = \sum_{m=0}^{\infty} E[I(X_m = j \mid X_0 = i)] = \sum_{m=0}^{\infty} p_{ij}^{(m)},$$

where $I(A) = 1$ if A is true and 0 otherwise. With $i \in T$ and $j \in T$, the last term of the preceding expression is nothing but the (i, j)th element of $Q^{(m)}$. Therefore, $w_{ij} = E[\tau_{ij}]$, where τ_{ij} represents the number of visits (*counting* the initial state occupancy at time 0) to state $j \in T$ before absorption given $X_0 = i$, where $i \in T$. In other words, with $M(n) = \{M_{ij}(n)\}$, we have

$$W = \sum_{m=0}^{\infty} Q^{(m)} = \lim_{n \to \infty} \{I + E[M(n)]\},$$

where the identity matrix is being added on the right side of the preceding to account for the fact that $M_{ij}(n)$ defined in the last section does not include the initial occupancy at time 0. It is instructive to look at the matrix equation $W = I + QW$ in a component form:

$$w_{ij} = \delta_{ij} + \sum_{k \in T} p_{ik} w_{kj} \qquad i \in T, j \in T,$$

(4.4.4)

where $\delta_{ij} = 1$ if $i = j$ and 0 otherwise. We could have started with conditioning on the first transition out of the starting state and obtained Equation 4.4.4 by applying the law of total probability.

Let τ_i denote the number of visits to all transient states given $X_0 = i$ or, equivalently, the time to absorption given $X_0 = i$. Define $v_i = E[\tau_i]$ and let column vector $v = \{v_i\}$. Then v_i is obtained by summing over the ith row of W or, in matrix form,

$$v = We.$$

(4.4.5)

We recall that $f_{ij}^{(n)}$ denotes the probability that the first passage from i to j occurs at the nth transition and $F^{(n)} = \{f_{ij}^{(n)}\}$. We now use the same notation except that we restrict ourselves to the case in which $i \in T$ and $j \in T^c$. It is easy to see that

$$F^{(n)} = Q^{n-1}R \qquad n \ge 1. \tag{4.4.6}$$

With i and j similarly defined, the matrix $F = \{f_{ij}\}$ gives the reaching probabilities. We see that

$$F = \sum_{n=1}^{\infty} F^{(n)} = \lim_{n \to \infty} R_n = WR. \tag{4.4.7}$$

For the absorbing Markov chain with transition probability matrix of Equation 4.4.1, the limiting multistep transition probabilities are given by

$$\lim_{n \to \infty} P^{(n)} = \begin{bmatrix} I & O \\ F & O \end{bmatrix}. \tag{4.4.8}$$

EXAMPLE 4.4.1 **A Cash Balance Problem** One responsibility of the financial office of a firm is to maintain an adequate amount of cash on hand to meet the firm's needs. The firm's cash position can be adjusted upward or downward at a cost by buying or selling marketable securities such as treasury bills or notes or by increasing or decreasing short-term loans. The objective is to find an optimal cash management program so that it balances transaction costs associated with adjustment actions and interest costs associated with having idle cash available.

Consider the case in which a firm uses a three-parameter policy to control its cash balance. Let the three parameters be denoted by (α, β, γ), where $\alpha < \gamma < \beta$. Let X_n denote the cash position at the start of day n, where X_n is in the set of integers (that is, we discretize the cash position for modeling convenience). Whenever X_n is out of the interval $[\alpha, \beta]$, the firm instantaneously brings the cash position back to γ. Let c_1 (c_2) denote the variable cost of increasing (decreasing) the cash position and c_3 the interest cost for carrying a unit of cash per day. Let D_n denote the demand for cash on day n. We assume that $\{D_n\}$ are i.i.d. random variables and follow a common probability mass function $\{p_k\}$, where $p_k = P\{D_n = k\}$, $k = -b, 1-b, 2-b, \ldots, -1, 0, 1, 2, \ldots, a$, and $a > 0$ and $b > 0$. A negative k corresponds to an inflow of k units of cash.

Every time the cash position goes out of the interval (α, β) an adjustment will always bring it back to level γ and the cash fluctuation process starts afresh. So it is natural to consider an adjustment epoch a regeneration epoch and the interval between two successive regeneration epochs the regeneration cycle. In Figure 4.11, we display a sample path denoting the fluctuation of cash flow.

We now model the stochastic process $\{X_n, n = 0, 1, \ldots\}$ during a regeneration cycle as an absorbing Markov chain. There are two sets of absorbing states:

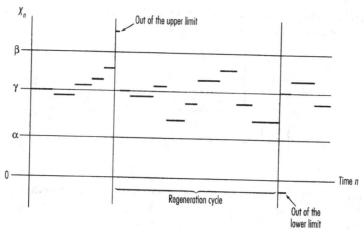

FIGURE
4.11   The cash flow.

$S_a = \{\alpha - a, \alpha - a + 1, \ldots, \alpha - 1\}$ and $S_b = \{\beta + 1, \beta + 2, \ldots, \beta + b\}$. The first set contains the states to which the process enters when it goes out of the lower limit α and the second set those leaving upper limit β. The set of transient states is given by $S_t = \{\alpha, \alpha + 1, \ldots, \beta\}$. The transition probabilities of this Markov chain are

$$p_{ij} = \begin{cases} p_{i-j} & \text{if } i \in S_t, \ i - a \le j \le i + b \\ 1 & \text{if } i \notin S_t, \ j \notin S_t, \text{ and } i = j \\ 0 & \text{otherwise}. \end{cases}$$

For $i \in S_t$, the preceding suggests that the rows of the transition probability matrix are the same (that is, $(p_a, p_{a-1}, \ldots, p_1, p_0, p_{-1}, \ldots, p_{-b})$) except each is shifted by one position to the right as we increment i (assume that the states are listed in increasing order of their labels from $\alpha - a$ to $\beta + b$). Let $w_{\gamma i}$ denote the expected number of visits to transient state i given that $X_0 = \gamma$ and $f_{\gamma i}$ the reaching probability to absorbing state i given that $X_0 = \gamma$. Using the result of Equation 3.6.5 of the regenerative reward process, the long-run expected average cost per day is given by

$$C(\alpha, \beta, \gamma) = \frac{c_1 \sum_{i \in S_a}(\gamma - i)f_{\gamma i} + c_2 \sum_{i \in S_b}(i - \gamma)f_{\gamma i} + c_3 \sum_{i \in S_t} i\, w_{\gamma i}}{\sum_{i \in S_t} w_{\gamma i}}. \tag{4.4.9}$$

To illustrate the use of the preceding approach to model the cash balance problem, we consider a specific numerical example with $\alpha = 2$, $\beta = 10$, $\gamma = 6$, $a = 3$, $b = 4$ and the one-period demand probabilities $p_3 = 0.03$, $p_2 = 0.05$, $p_1 = 0.16$, $p_0 = 0.18$, $p_{-1} = 0.25$, $p_{-2} = 0.14$, $p_{-3} = 0.12$, $p_{-4} = 0.07$. We use

MATLAB program **mc_canon** given in the Appendix to find the transition probability matrix of this chain in a canonical form:

$$
P = \begin{bmatrix} I_3 & & & \\ & I_4 & & O \\ R_1 & & & \\ & R_2 & & Q \end{bmatrix} = \begin{bmatrix} I & O \\ R & Q \end{bmatrix},
$$

where I_n is an identity of order n, and other submatrices are shown as follows:

$$
R_1 = \begin{bmatrix} .03 & .05 & .16 \\ & .03 & .05 \\ & & .03 \\ & & 0 \\ & & 0 \end{bmatrix}
\qquad
R_2 = \begin{bmatrix} .07 & & & \\ .12 & .07 & & \\ .14 & .12 & .07 & \\ .25 & .14 & .12 & .07 \end{bmatrix}
$$

$$
Q = \begin{bmatrix}
.18 & .25 & .14 & .12 & .07 & & & & \\
.16 & .18 & .25 & .14 & .12 & .07 & & & \\
.05 & .16 & .18 & .25 & .14 & .12 & .07 & & \\
.03 & .05 & .16 & .18 & .25 & .14 & .12 & .07 & \\
& .03 & .05 & .16 & .18 & .25 & .14 & .12 & .07 \\
& & .03 & .05 & .16 & .18 & .25 & .14 & .12 \\
& & & .03 & .05 & .16 & .18 & .25 & .14 \\
& & & & .03 & .05 & .16 & .18 & .25 \\
& & & & & .03 & .05 & .16 & .18
\end{bmatrix}.
$$

For the given matrix P, the states are listed in the order (-1; 0; 1; 11; 12; 13; 14; 2, 3, ..., 9, 10), where we use a semicolon to demarcate a change of equivalence class. To find $\{w_{\gamma i}\}$, we use Equation 4.4.3 to obtain W and extract its fifth row:

$$
(w_{\gamma 2}, \ldots, w_{\gamma 10}) = e_5'(I - Q)^{-1}.
$$

To find $\{f_{\gamma i}\}$, we use Equation 4.4.7 to obtain F and extract its fifth row:

$$
(f_{\gamma, -1}, \ldots, f_{\gamma 1}; f_{\gamma 11}, \ldots, f_{\gamma 14}) = e_5'(I - Q)^{-1} R.
$$

Once $\{w_{\gamma i}\}$ and $\{f_{\gamma i}\}$ are found, using Equation 4.4.9 to compute the long-run average cost $C(\alpha, \beta, \gamma)$ is straightforward. As an illustration, with $c_1 = 1$, $c_2 = 0.5$, and $c_3 = 0.05$, in the Appendix we find an expected cycle length of 6.4335 and $C(\alpha, \beta, \gamma) = 0.8126$. ∎

EXAMPLE 4.4.2

Blackjack Anyone? Consider a simplified version of the blackjack game. A player is playing against the house. The player throws a die as many times as necessary to accumulate a sum. If the sum exceeds seven, the player busts and loses a game. If the player stops before exceeding seven, the house takes over and throws the die repeatedly until the sum is four or higher. If the house's sum is more than seven, the house loses the game. Otherwise whoever has the larger sum wins. In case of a tie, the house wins. If the player adopts the strategy of throwing the die until the sum is four or higher, what is the probability that the player wins a game?

We formulate the player's problem as an absorbing Markov chain with state space $S = \{0, 1, \ldots, 9\}$, where state 0 is the starting state, states 8 and 9 are the two absorbing states (corresponding to the events that the player wins or loses, respectively), and other state labels denote the sums accumulated in the midst of the game. The transition probability matrix in a canonical form is given by

$$
P = \begin{array}{c} \\ 8 \\ 9 \\ 0 \\ 1 \\ 2 \\ 3 \\ 4 \\ 5 \\ 6 \\ 7 \end{array}
\begin{array}{cccccccccc}
8 & 9 & 0 & 1 & 2 & 3 & 4 & 5 & 6 & 7 \\
\left[\begin{array}{cccccccccc}
1 & & & & & & & & & \\
 & 1 & & & & & & & & \\
 & & & 1/6 & 1/6 & 1/6 & 1/6 & 1/6 & 1/6 & \\
 & & & & 1/6 & 1/6 & 1/6 & 1/6 & 1/6 & 1/6 \\
 & 1/6 & & & & 1/6 & 1/6 & 1/6 & 1/6 & 1/6 \\
 & 1/3 & & & & & 1/6 & 1/6 & 1/6 & 1/6 \\
140/1256 & 1116/1256 & & & & & & & & \\
483/1256 & 773/1256 & & & & & & & & \\
826/1256 & 430/1256 & & & & & & & & \\
1169/1256 & 127/1256 & & & & & & & &
\end{array}\right].
\end{array}
$$

Finding the one-step transition probabilities p_{ij} for $i = 0, 1, 2, 3, 8, 9$ is straightforward. When $i = 4, 5, 6, 7$, following the player's strategy, the player stops and the house starts throwing. Thus we trace all the possible sample paths to be traversed by the house and enumerate the corresponding probabilities. This is tedious but nevertheless can easily be done. As an example, assume that the player stops at the last throw when the sum is greater than 4 and the sum turns out to be 6. If the house's first throw is 4 or 5, the player wins with probability $1/6$; if it is 3, the player wins with probability $(1/6)^2(4)$ after one more throw; if it is 2, the player wins with probability $(1/6)^2(3)$ after one more throw and the player wins with probability $(1/6)^3(4)$ after two more throws; if it is 1, the corresponding probabilities can be similarly computed. To summarize, we have

$$p_{68} = P\{\text{the player's sum is 6 and wins the game}\}$$

$$= \left(\frac{1}{6}\right)(2) + \left[\left(\frac{1}{6}\right)^2(4)\right] + \left[\left(\frac{1}{6}\right)^2(3) + \left(\frac{1}{6}\right)^3(4)\right]$$

$$+ \left[\left(\frac{1}{6}\right)^2(2) + \left(\frac{1}{6}\right)^3(4) + \left(\frac{1}{6}\right)^3(3) + \left(\frac{1}{6}\right)^4(4)\right] = \frac{825}{1256}.$$

To show the systematic pattern associated with these computations, we also note that

$$p_{78} = P\{\text{the player's sum is 7 and wins the game}\}$$

$$= \left(\frac{1}{6}\right)(3) + \left[\left(\frac{1}{6}\right)^2 (5)\right] + \left[\left(\frac{1}{6}\right)^2 (4) + \left(\frac{1}{6}\right)^3 (5)\right]$$

$$+ \left[\left(\frac{1}{6}\right)^2 (3) + \left(\frac{1}{6}\right)^3 (5) + \left(\frac{1}{6}\right)^3 (4) + \left(\frac{1}{6}\right)^4 (5)\right] = \frac{1169}{1256}.$$

We observe that the second numbers of each pair are incremented by 1 as the first index i of p_{ij} is incremented.

We use Equation 4.4.7 to compute the reaching probability matrix F

$$F = \begin{matrix} 0 \\ 1 \\ 2 \\ 3 \\ 4 \\ 5 \\ 6 \\ 7 \end{matrix} \begin{bmatrix} .3965 & .6035 \\ .4728 & .5272 \\ .4053 & .5947 \\ .3474 & .6526 \\ .1115 & .8885 \\ .3846 & .6154 \\ .6576 & .3424 \\ .9307 & .0693 \end{bmatrix}.$$

We see that when the player stops and his sum is 4, he is in the most precarious situation. Also, we have $f_{08} = 0.3965$—the player's winning probability if the playing strategy is followed. ▪

For i, j in T, we recall that τ_{ij} denotes the number of visits to state j before absorption given that $X_0 = i$. We now consider the computation of the variance of τ_{ij} for an absorbing Markov chain with a finite state space S. Since $\text{Var}[\tau_{ij}] = E[\tau_{ij}^2] - E^2[\tau_{ij}] = E[\tau_{ij}^2] - W_{ij}^2$, we first derive an expression for the second moment of τ_{ij}. We observe that, for i and j in T,

$$\tau_{ij}^2 = \begin{cases} \delta_{ij}^2 & \text{with probability } p_{ik}, \text{ where } k \in T^c \\ [\delta_{ij} + \tau_{kj}]^2 & \text{with probability } p_{ik}, \text{ where } k \in T. \end{cases}$$

In deriving the preceding expression, we take note that if the first transition is to an absorbing state, then the initial occupancy in state j will be counted only if $i = j$ and no further occupancies of state j are possible. If the first transition is to a transient state k, then the initial occupancy in state j will again be counted only if $i = j$; once the process reaches state k the random variable τ_{kj} will keep track of the occupancy time in state j from there on. Since we are working with the square of τ_{ij}, we square the two terms on the right side accordingly. We apply the law of total probability and take expectations. This gives, for i and j in T,

$$E[\tau_{ij}^2] = \sum_{k \in T^c} p_{ik} E[\delta_{ij}^2] + \sum_{k \in T} p_{ik} E[\delta_{ij} + \tau_{kj}]^2$$

$$= \sum_{k \in S} p_{ik} \delta_{ij} + 2 \sum_{k \in T} p_{ik} \delta_{ij} E[\tau_{kj}] + \sum_{k \in T} p_{ik} E[\tau_{kj}^2]$$

$$= \delta_{ij} + 2\delta_{ij} \sum_{k \in T} p_{ik} E[\tau_{kj}] + \sum_{k \in T} p_{ik} E[\tau_{kj}^2].$$

Let square matrix $K = \{E[\tau_{ij}^2]\}$. The preceding equation can be rewritten in a matrix form as

$$K = I + 2(QW)_{dg} + QK,$$

where in the second term of the right side we only preserve the diagonal elements of QW because of the presence of the δ_{ij} term in the previous equation. We simplify the preceding as follows:

$$K = (I - Q)^{-1}[2(QW)_{dg} + I] = W[2(Q + Q^2 + \cdots)_{dg} + I]$$

$$= W[2(I + Q + Q^2 + \cdots)_{dg} - I)]$$

$$= W[2W_{dg} - I]. \tag{4.4.10}$$

Let square matrix $V = \{Var[\tau_{ij}]\}$. The variances of $\{\tau_{ij}\}$ are given by $V = K - W \square W$ (if $A = \{a_{ij}\}$, $B = \{b_{ij}\}$, and $C = \{c_{ij}\}$, then $C = A \square B$ implies $c_{ij} = a_{ij} \times b_{ij}$ for all i and j).

For transient state i, we recall that τ_i denotes the number of visits to all transient states given that $X_0 = i$ and $v_i = E[\tau_i]$. To find the second moment of τ_i, we approach it in a manner similar to the derivation of $E[\tau_{ij}^2]$. We observe that, for i and j in T,

$$\tau_i^2 = \begin{cases} 1 & \text{with probability } p_{ik}, \text{ where } k \in T^c \\ [1 + \tau_k]^2 & \text{with probability } p_{ik}, \text{ where } k \in T. \end{cases}$$

It follows that

$$E[\tau_i^2] = \sum_{k \in T^c} p_{ik} + \sum_{k \in T} p_{ik} E[1 + \tau_{kj}]^2 = \sum_{k \in S} p_{ik} + 2 \sum_{k \in T} p_{ik} E[\tau_k] + \sum_{k \in T} p_{ik} E[\tau_k^2]$$

$$= 1 + 2 \sum_{k \in T} p_{ik} E[\tau_k] + \sum_{k \in T} p_{ik} E[\tau_k^2].$$

Let column vector $\{h = E[\tau_i^2]\}$. Then the last equation can be stated in matrix notation as $h = e + 2Qv + Qh$ or $h = W(2Qv + e)$. Using the identity $W(I - Q) = I$, we obtain

$$h = 2(W - I)v + v = (2W - I)v. \tag{4.4.11}$$

Let column vector $v_\tau = \{Var[\tau_i]\}$. Then the column vector containing the variances is given by $v_\tau = h - v \square v$.

EXAMPLE
4.4.3

Serving in the Game of Tennis Suppose that a player is playing a tennis game. The player's shot is a winning shot if her opponent is unable to reach the shot or hits the shot into the net or out of bounds. Conversely, the shot is a losing shot if the player is unable to reach her opponent's shot or if she hits the shot into the net or out of bounds. If the player is able to return her opponent's shot in bounds, her shot can be considered strong (**S**) or weak (**W**) depending on its pace and depth. Let absorbing state 0 (1) denote the event that the player has a winning (losing) shot and transient state 2 (3) denote the event that the player has just hit a strong (weak) shot. The transition probability matrix describing the dynamics of a single serving is given by

$$
\begin{array}{c}
0 \\ 1 \\ 2 \\ 3
\end{array}
\begin{bmatrix}
1 & 0 & 0 & 0 \\
0 & 1 & 0 & 0 \\
p_{20} & p_{21} & p_{22} & p_{23} \\
p_{30} & p_{31} & p_{32} & p_{33}
\end{bmatrix}
=
\begin{bmatrix}
I & O \\
R & Q
\end{bmatrix}.
$$

To interpret the various transition probabilities, we use p_{23} as an example. Here, p_{23} is the probability that the player hits a strong shot that is returned by her opponent and followed by a weak shot from the player. We define the starting state probability vector $s(0) = (p_0, p_1, p_2, p_3)$, where $p_0 = P\{$the player's first serve is in the service court and is not returned by her opponent$\}$, $p_1 = P\{$the player's first serve is not in the service court$\}$, $p_2 = P\{$the player's first serve is a strong shot in the service court that is returned by her opponent$\}$, and $p_3 = P\{$the player's first service is a weak shot in the service court that is returned by her opponent$\}$. We also let $s'(0) = (p_0', p_1', p_2', p_3')$ denote the starting state probability vector associated with the second serve. The absorption probabilities $\{f_{ij}\}$ from the transient states 2 and 3 can be obtained by Equation 4.4.7:

$$
F = \begin{bmatrix} f_{20} & f_{21} \\ f_{30} & f_{31} \end{bmatrix} = (I - Q)^{-1} R.
$$

The probability that the player wins the point on her first shot is given by $p_0 + p_2 f_{20} + p_3 f_{30}$. Since the probability that the player will be forced to take a second serve is p_1, the probability that she will win on her second serve is $p_1[p_0' + p_2' f_{20} + p_3' f_{30}]$. The probability that the player will win the point is the sum of these two probabilities.

We now illustrate the use of the preceding model to evaluate the effect of different serving strategies. Consider the case in which

$$
R = \begin{bmatrix} .3 & .1 \\ .1 & .4 \end{bmatrix}
\quad \text{and} \quad
Q = \begin{bmatrix} .4 & .2 \\ .1 & .4 \end{bmatrix}.
$$

When the player uses a strong shot, the starting state probability vector is (0.4, 0.4, 0.2, 0); when the player uses a weak shot, the probability vector becomes (0.25, 0.25, 0, 0.55). We see that a strong serve, which lands in the service court 60 percent of the time, will be returned by her opponent only $0.2/(0.4 + 0.2)$ = 33.33 percent of the time if it lands in the service court. On the other hand, a weak service, which lands in the service court 80 percent of the time, will be returned by her opponent $0.55/(0.55 + 0.25) = 68.75$ percent of the time. Based on the given data, we obtain the probabilities of winning a point under the four serving strategies: **S-S**, **S-W**, **W-S**, and **W-W**, where the first (second) letter denotes the type of serving for the first (second) shot. They are, respectively, 0.725, 0.676, 0.499, and 0.475. Therefore for this particular player it is better to use the strategy of serving a strong shot first; if needed, it is to be followed by another strong shot.

We can compute the second moments of times in transient states by using Equation 4.4.10. This in turn enables us to compute in the Appendix the corresponding variances. They are

$$\begin{bmatrix} Var[\tau_{22}] & Var[\tau_{23}] \\ Var[\tau_{32}] & Var[\tau_{33}] \end{bmatrix} = \begin{bmatrix} 1.3495 & 1.1419 \\ 0.6574 & 1.3495 \end{bmatrix}.$$

We see that $Var[\tau_{22}]$, the variance for the number of shots hit by the player before winning or losing a point given that the point starts with a strong serve in the service court, is 1.3419. Similar interpretations can be given to other $Var[\tau_{ij}]$. ■

EXAMPLE 4.4.4

Discrete Phase-Type Distributions   Consider an absorbing Markov chain with state space $S = \{1, \ldots, m, m + 1\}$, where state $m + 1$ is absorbing and all other states are transient. Let the starting state probability vector $s(0)$ be denoted by (α, α_{m+1}), where $\alpha = (\alpha_1, \ldots, \alpha_m)$. The transition probability matrix is stated in a canonical form:

$$P = \begin{bmatrix} Q & r \\ 0 & 1 \end{bmatrix},$$

where r is a column vector of size m and Q is an $m \times m$ substochastic matrix. Let τ be the time until absorption into state $m + 1$.

Using Equation 4.4.2, we find the probability mass function $p_k = P\{\tau = k\}$ as follows:

$$p_0 = \alpha_{m+1}$$

$$p_k = \alpha Q^{k-1} r \qquad k = 1, 2, \ldots. \qquad (4.4.12)$$

The given probability distribution $\{p_k\}$ is called the discrete phase-type distribution with representation (α, Q) and order m.

The probability generating function of the discrete phase-type distribution is given by

$$P_\tau(z) = \sum_{k=0}^{\infty} z^k p_k = \alpha_{m+1} + \sum_{k=1}^{\infty} z^k \alpha Q^{k-1} r = \alpha_{m+1} + z\alpha \sum_{k=1}^{\infty} z^{k-1} Q^{k-1} r$$

$$= \alpha_{m+1} + z\alpha \sum_{k=0}^{\infty} (zQ)^k r = \alpha_{m+1} + z\alpha [I - zQ]^{-1} r. \qquad (4.4.13)$$

Differentiating the probability generating function with respect to z and using the identity

$$\frac{d}{dz}X^{-1} = -X^{-1}\left(\frac{d}{dz}X\right)X^{-1} \tag{4.4.14}$$

(we use X to denote $X(z)$ for notational convenience), we find

$$\frac{d}{dz}P_\tau(z) = \alpha[I - zQ]^{-1}r + z\alpha\left[-(I - zQ)^{-1}\frac{d}{dz}[I - zQ](I - zQ)^{-1}\right]r$$

$$= \alpha[I - zQ]^{-1}r + z\alpha[(I - zQ)^{-1}Q(I - zQ)^{-1}]r.$$

Hence the first (factorial) moment is given by

$$E[\tau] = \frac{d}{dz}P_\tau(z)\Big|_{z=1} = \alpha[I - Q]^{-1}r + \alpha[I - Q]^{-1}Q[I - Q]^{-1}r$$

$$= \alpha[I - Q]^{-1}[I + Q(I - Q)^{-1}]r = \alpha[I - Q]^{-1}[I + Q + Q^2 + \cdots]r$$

$$= \alpha[I - Q]^{-1}[I - Q]^{-1}r = \alpha[I - Q]^{-1}e,$$

where we use Equation 4.4.7 to obtain the last equality. Higher-order factorial moments can be found similarly. They are

$$P_\tau^{(k)} = k!\alpha Q^{k-1}[I - Q]^{-k}e \qquad k = 1, 2, \ldots. \tag{4.4.15}$$

Finding moments of a discrete random variable typically involves summations of infinite numbers of terms. When the random variable is of phase type, we see that the operations reduce to simple matrix manipulations. ∎

EXAMPLE 4.4.5

The Generalized Negative Binomial Distribution   For $i = 1, \ldots, m$, we let X_i be a geometric random variable with the following probability mass function

$$P\{X_i = n\} = q_i^{n-1}p_i \qquad n = 1, 2, \ldots,$$

where $q_i = 1 - p_i$. Let $Y = X_1 + \cdots + X_m$. We can give an interpretation to random variable Y. Suppose that we are to play m rounds of a game, each round consisting of repeated plays, to obtain a success for the first time. At the ith round, the probability of being successful in a single play is p_i. As soon as we obtain a success in round i, we move to round $i + 1$. Hence X_i denotes the number of plays needed to win the ith round and Y the total number of plays needed to conclude the game.

   We now construct an absorbing Markov chain with an absorbing state $m + 1$ and the transition diagram found in Figure 4.12.

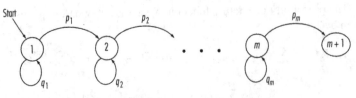

FIGURE 4.12   The transition diagram for the generalized negative binomial random variable.

From the figure, we see that the first passage time to state $m + 1$ corresponds to the value of Y, the number of trials needed to conclude the game. This implies that Y follows a discrete phase-type distribution with representation (α, Q) and order m, where the starting state probability vector $\alpha = (1, 0, \ldots, 0)$ and

$$Q = \begin{bmatrix} q_1 & p_1 & & & \\ & q_2 & p_2 & & \\ & & \cdot & \cdot & \\ & & & q_{m-1} & p_{m-1} \\ & & & & q_m \end{bmatrix}.$$

A formal way to check the validity of the previous representation is by first noting that the probability generating function for random variable X_i is $p_i z/(1 - q_i z)$. Since $\{X_i\}$ are independent random variables, the probability generating function of Y is given by

$$P_Y(z) = \prod_{i=1}^{m} \left(\frac{p_i z}{1 - q_i z} \right).$$

Using Equation 4.4.13, we should obtain a probability generating function identical to the preceding. ∎

EXAMPLE
4.4.6

The Coupon Collection Problem Consider a football team that has fifty-two players on its roster. The team offers the following promotional campaign. Every box of cereal sold in the city contains a coupon bearing the picture of one of its players. A person who collects a complete set of coupons (which contains at least one of each type) receives an award enabling him or her to attend all of the team's home games for free in the forthcoming season. Assume that each time one opens a box of cereal it is equally likely to be one of the fifty-two types. Let T denote the number of coupons needed until one obtains a complete set. We are interested in $E[T]$, $Var[T]$, and the probability distribution of T.

Let X_i denote the number of additional coupons needed to obtain the ith distinct type given that the player has already amassed $i - 1$ distinct types. For generality, we let m denote the number of distinct types of coupons (here we have $m = 52$). For $i = 1, 2, \ldots, m$, we see that $P\{X_i = n\} = (q_i)^{n-1} p_i$, $n = 1, 2, \ldots$, where

$$p_i = \frac{m - (i - 1)}{m}.$$

It is now obvious that T follows a generalized negative binomial distribution and has the phase-type representation described in Example 4.4.5. Using Equation 4.4.15, we compute the first two factorial moments of T. They in turn give $E[T] = 235.98$ and a standard deviation of 64.5. With Equation 4.4.12, we compute the probability mass function for T in the Appendix. The function is shown in Figure 4.13. ∎

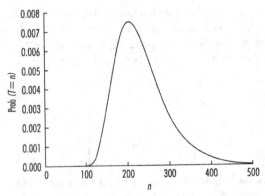

4.13   The probability mass function of the number of coupons needed to amass a complete set.

The next example presents an interesting use of ideas about absorbing chains in working with ergodic chains. The results obtained from the example have applications in theory and computation. Before we proceed, we recall the notion of the *strong Markov property* introduced in Section 4.1. If stochastic process $X = \{X_n, n \geq 0\}$ is a Markov chain with transition probability matrix $P = \{p_{ij}\}$ and state space S, and T is a stopping time with respect to X with $P\{T < \infty\} = 1$, then for all i and j in S, $P\{X_{T+1} = j | X_T = i\} = p_{ij}$. In other words, the Markov property is preserved at a random time T (vis à vis the deterministic time index n).

EXAMPLE   **A Censored Markov Chain** Let $X = \{X_n, n \geq 0\}$ be an ergodic Markov chain with a
4.4.7*   finite state space S and transition probability matrix P. Let $S = A \cup B$ and $A \cap B = \emptyset$ so that A and B form a partition of A. We now form a stochastic process $Y = \{Y_n, n \geq 0\}$ by defining

$$T_0 = \min\{m \geq 0 \,|\, X_m \in A\} \quad \text{and} \quad T_{n+1} = \min\{m > T_n \,|\, X_m \in A\} \quad n \geq 0$$

and $Y_n = X_{T_n}$, $n \geq 0$. Since $\{T_n, n \geq 0\}$ are stopping times and the ergodicity of X implies that $P\{T_n < \infty\} = 1$, the strong Markov property implies that $\{Y_n, n \geq 0\}$ form a Markov chain. We call stochastic process Y a *censored* Markov chain (imagining a censor hides all visits to states in B from an observer).

To find the transition probability matrix $S = \{s_{ij}\}$ of Y, we partition the transition probability matrix P as follows

$$P = \begin{matrix} & A & B \\ A & \begin{bmatrix} T & U \\ B & R & Q \end{bmatrix} \end{matrix}.$$

Define $T = \{t_{ij}\}$, $U = \{u_{ij}\}$, $R = \{r_{ij}\}$, and $Q = \{q_{ij}\}$. Conditioning on the first transition out of a state $i \in A$, we construct one-step transition probabilities between states in A. This gives, for $i, j \in A$,

$$s_{ij} = t_{ij} + \sum_{k \in B} \sum_{l \in B} [u_{ik} q_{kl}^{(0)} r_{lj} + u_{ik} q_{kl}^{(1)} r_{lj} + u_{ik} q_{kl}^{(2)} r_{lj} + \cdots],$$

where $\{q_{kl}^{(n)}\}$ are the n-step transition probabilities of the chain X when states are restricted to the set B. We see that $Q^n = \{q_{ij}^{(n)}\}$. We write the preceding equation in matrix notation and obtain

$$S = T + U(I + Q + Q^2 + \cdots)R = T + U(I - Q)^{-1}R, \qquad (4.4.16)$$

where we use the fact that Q is a substochastic matrix to justify the existence of its inverse (if Q were a stochastic matrix then each row of Q would have summed to 1 and X could not possibly be ergodic). To show that S is a stochastic matrix, we postmultiply Equation 4.3.12 by a column vector e_A (e_A is one vector of dimension $|A|$, the number of elements in the set A—and e_B is defined similarly) and obtain

$$Se_A = Te_A + U(I - Q)^{-1} Re_A = Te_A + UFe_A$$

$$= Te_A + UFe_A = Te_A + Ue_B = [T \; U]\begin{bmatrix} e_A \\ e_B \end{bmatrix} = [T \; U]e = e_A,$$

where $Fe_A = e_B$ since $F = \{f_{ij}\}$ gives the reaching probabilities from $i \in B$ to $j \in A$ and adding all j yielding 1 guarantees the chain is ergodic. The last equality is due to $Pe = e$.

Let π be the limiting probability vector of X and define

$$\eta_i = \frac{\pi_i}{\sum_{k \in A} \pi_k} \qquad i \in A. \qquad (4.4.17)$$

Then $\eta = \{\eta_i\}$ is the limiting probability vector of the censored chain. One way to see this is by noting

$$\lim_{n \to \infty} P\{Y_n = i\} = \lim_{n \to \infty} P\{X_n = i \,|\, X_n \in A\} = \frac{\pi_i}{\sum_{k \in A} \pi_k} \qquad i \in A.$$

An algebraic way to establish Equation 4.4.17 is by showing $\eta = \eta S$, where η is defined by Equation 4.4.17 and S is given by Equation 4.4.16. We leave the proof as an exercise (see Problem 12). ∎

EXAMPLE 4.4.8*   **State Reduction for Computing the Limiting Probabilities of an Ergodic Markov Chain**   We now use the results obtained in the previous example to develop a numerically stable method for computing the limiting probability vector. In real-world applications of Markov chains, typically state spaces are rather large. Algorithms for computing the vector are susceptible to numerical errors because computers can only handle numbers of finite word lengths. The procedure to be presented in this example is known as the most stable means available for computing the limiting probability vector and has received ample attention in recent literature. Without any modifications, the procedure is also applicable for solving for the limiting probabilities of an ergodic continuous-time Markov chain to be introduced in Chapter 5.

Consider an ergodic Markov chain with state space $\{1, \ldots, N\}$ and transition probability matrix $P = \{p_{ij}\}$. We are interested in computing the limiting probability vector $\{\pi_1, \ldots, \pi_N\}$. We devise a computation procedure that has two passes: a backward pass and a forward pass. We start with the backward pass consisting of steps $N-1, N-2, \ldots, 1$. At step n, we want to construct a censored Markov chain with state space $\{1, \ldots, n\}$ and transition probability matrix $P(n) = \{p_{ij}(n)\}$ based on our knowledge of $P(n+1) = \{p_{ij}(n+1)\}$. Clearly, we have $P(N) = P$. We partition $P(n+1)$ as follows:

$$P(n+1) = \begin{bmatrix} T(n+1) & u(n+1) \\ r(n+1) & P_{n+1,n+1}(n+1) \end{bmatrix},$$

where $T(n+1)$ is $n \times n$, $u(n \times 1)$ is $1 \times n$, $r(n+1)$ is $n \times 1$, and $P_{n+1,n+1}(n+1)$ is scalar. Following Equation 4.4.16, the transition probability matrix of the censored chain $P(n)$ can be constructed from

$$P(n) = T(n+1) + u(n+1)\left(1 - P_{n+1,n+1}(n+1)\right)^{-1} r(n+1)$$

$$= T(n+1) + a(n+1)r(n+1),$$

where the column vector (of size n) $a(n+1) \equiv u(n+1)/(p_{n+1,1} + \cdots + p_{n+1,n})$. The preceding construction can be stated in component form as

$$p_{ij}(n) = p_{ij}(n+1) + a_i(n+1)r_j(n+1), \tag{4.4.18}$$

where $a_i(n+1)$ and $r_j(n+1)$ are the ith and jth elements of the respective vectors. In the course of computation in the backward pass we save the column vectors $a_2(\cdot), \ldots, a_N(\cdot)$.

What roles are to be played by the vectors $a_2(\cdot), \ldots, a_N(\cdot)$? Stating the last linear equation of the balance equation (Equation 4.3.10) $\pi(n) = \pi(n)P(n)$ in component form, we obtain

$$\pi_1(n)p_{n,1}(n) + \cdots + \pi_{n-1}(n)p_{n,n-1}(n) + \pi_n(n)p_{n,n}(n) = \pi_n(n)$$

or

$$\pi_n(n) = \left(\pi_1(n)p_{n,1}(n) + \cdots + \pi_{n-1}(n)p_{n,n-1}(n)\right) / \left(1 - p_{n,n}(n)\right)$$

$$= \pi_1(n)a_{n,1}(n) + \cdots + \pi_{n-1}(n)a_{n,n-1}(n). \tag{4.4.19}$$

Hence, we see that $a_n(\cdot)$ gives the multipliers for expressing $\pi_n(n)$ as linear combinations of $\pi_1(n), \ldots, \pi_{n-1}(n)$.

The key to finding the limiting probability vector π of the chain X lies with Equation 4.4.17. The equation implies that for a censored chain with transition probability matrix $P(n)$, the limiting probability vectors $\pi(n)$ have the relative weights, shown as the numerator of Equation 4.4.17, identical to those of the uncensored chain X. The actual values of $\{\pi_i(n)\}$ differ from those of $\{\pi_i\}$ by only the normalizing constants used in the denominator of Equation 4.4.17. Thus the forward pass entails: (i) setting $\pi_1(1) = 1$ (arbitrarily), (ii) computing $\pi_2(2), \ldots, \pi_N(N)$ successively using Equation 4.4.19, and (iii) normalizing these $\{\pi_n(n)\}$ to obtain π. The complete solution procedure is sometimes called the method of state reduction for finding the limiting probability vector. The complete procedure does not include operations involving negative numbers; hence it is numerically stable. The program **mc_limsr** shown in the Appendix (listed as a part of Example 4.3.5) is a MATLAB implementation of the procedure. ∎

4.5 Markov Reward Processes

In this section, we restrict our attention to an irreducible positive-recurrent Markov chain in which each occupancy of state j generates a reward r_j. Such a chain is called a *Markov reward process*. We assume that $0 \le r_j < \infty$. Our interests in this class of problems are in ways of evaluating reward generation. The total rewards received in an infinite number of transitions are, in general, infinite. In this case, the time average of the expected reward received per transition is a measure of rate associated with the reward generation process. For brevity, we call it the *gain rate*. In cost models, we call it the *cost rate*.

We see that the gain rate g of a Markov reward process can be computed from

$$g = \sum_{j=0}^{\infty} r_j \pi_j, \tag{4.5.1}$$

where π_j is the long-run fraction of time the process is in state j.

Suppose now that rewards received in future periods are discounted by a discount factor $0 < \alpha < 1$. Specifically, the present value of one unit of reward received one transition from now is worth amount α. Let g_i denote the present value of all rewards received in an infinite number of transitions given that $X_0 = i$ and define column vector $g = \{g_i\}$. Then we see that

$$g = r + \alpha Pr + (\alpha P)^2 r + \cdots = \sum_{k=0}^{\infty} (\alpha P)^k r = \sum_{k=0}^{\infty} Q^k r,$$

where $Q = \alpha P$. Since $0 < \alpha < 1$, Q is a substochastic matrix.

When the state space is finite, we recall that $W = I + Q + Q^2 + \cdots$ and hence

$$g = (I - Q)^{-1} r. \tag{4.5.2}$$

Of course, if each state occupancy of i yields a reward of one unit, then Equation 4.5.2 reduces to Equation 4.4.5. This is expected if we envision that when the process is in any given state there is a probability $1 - \alpha$ that the next transition will be to an absorbing state not in S and the process will receive no further rewards when absorption occurs.

EXAMPLE **The Periodic-Review (s, S) Inventory System Revisited** We now return to Example 4.1.3.
4.5.1 In the inventory model considered there, we assume that there is a holding cost of h per unit of item left at the end of a week, a penalty cost of p per unit of unmet demand, a fixed cost of ordering K and a variable cost of ordering of c per unit ordered. Let L_i denote the expected amount of unmet demand in a week if the week has i units available to meet the demand in the week. Then

$$L_i = \sum_{k=i}^{\infty} (k-i) a_k.$$

The expected cost per transition out of state i is then given by

$$r_i = \begin{cases} K + c(S-i) + ih + pL_S & i = 0, 1, \dots, s-1 \\ ih + pL_i & i = s, s+1, \dots, S. \end{cases}$$

Once the cost rates are found, we use Equation 4.5.1 to compute the long-run average cost per week g. For a numerical example with $K = 15$, $c = 4$, $h = 1$, and $p = 2$, we do the computation in the Appendix and present our results in Figure 4.14 for the cases in which $s = 2$ and 4, $5 \le S \le 20$, and the weekly demand follows a Poisson distribution with mean 5. ∎

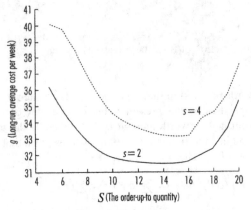

FIGURE
4.14 Average cost per week for the inventory model.

EXAMPLE **A Fleet-Car Insurance Problem** A utility company owns a fleet of minivans to meet its
4.5.2 transportation needs. The company has a contract with an insurance company to insure the fleet. There are five premium classes denoted by $1, \dots, 5$. For class i, the premium is h_i and the deductible is d_i. The premium structure is such that $h_i > h_{i+1}$ and $d_i > d_{i+1}$ for all i. Hence we see that class 1 is most expensive and others are indexed in increasing order of their relative desirability. In a year, if the

company does not file a claim then its premium class will be moved to the next higher class. On the other hand, if it files a claim then its premium class will be moved to the next lower class. When the premium class reaches class 1, it stays there the next year if a claim is filed for the year; when the premium class reaches class 5, it stays there if no claim is filed for the year. The contract stipulates that a lump-sum claim can be made at the end of each year for all the damages incurred during the year. We assume that damages incurred in successive years are i.i.d. random variables with distribution function F and density f.

To minimize the expected total discounted cost in an infinite planning horizon, the company decides to establish a claim limit s_j if its current premium class is j, $j = 1, \ldots, 5$. Under the claim-limit policy $(s_1, \ldots, s_5)$, the company will not file a claim for the year if the total damage incurred in the year is less than s_i and its premium class is i for the year. Assume that $s_i > d_i$ for all i. We are now ready to compute the expected total discounted cost.

Let X_n denote the premium class for year n. Then $\{X_n | n = 1, 2, \ldots\}$ is a Markov chain with transition probability matrix

$$P = \begin{matrix} 1 \\ 2 \\ 3 \\ 4 \\ 5 \end{matrix} \begin{bmatrix} 1 - F(s_1) & F(s_1) & & & \\ 1 - F(s_2) & 0 & F(s_2) & & \\ & 1 - F(s_3) & 0 & F(s_3) & \\ & & 1 - F(s_4) & 0 & F(s_4) \\ & & & 1 - F(s_5) & F(s_5) \end{bmatrix}.$$

The expected cost per year associated with premium class j is given by

$$r_j = h_j + \alpha \int_0^{s_j} xf(x)dx + \alpha d_j \int_{s_j}^\infty f(x)dx = h_j + \alpha \left[\int_0^{s_j} xf(x)dx + d_j \left[1 - F(s_j) \right] \right].$$

In the previous expression, we see that when in premium class j the company has to pay the yearly premium, the damages absorbed by the utility company by not filing the claim, and the expected deductibles due to damages exceeding the claim limit established for the year. The discount factor α appears in the preceding expression to reflect the timing of the respective payment. For any specified claim-limit policy $(s_1, \ldots, s_5)$, the expected total discounted cost can be computed from Equation 4.5.2.

We now look at a numerical example. Suppose that a preliminary analysis shows that the past yearly damages can be approximated by a Weibull distribution with mean \$46,000 and standard deviation \$12,500. Since Weibull distributions are used extensively in stochastic modeling, we digress from our subject and briefly review a few pertinent aspects about the distribution.

Let W be a random variable whose distribution function is given by

$$F_W(w) = 1 - e^{-\left(\frac{w}{v}\right)^\beta} \qquad w > 0. \tag{4.5.3}$$

The distribution of Equation 4.5.3 is known as the Weibull distribution with *scale* parameter v and *shape* parameter β. The distribution is one of the most frequently used distributions in reliability theory and quality control. Its density is given by

$$f_W(w) = \left(\frac{\beta}{v}\right)\left(\frac{w}{v}\right)^{\beta-1} e^{-\left(\frac{w}{v}\right)^{\beta}} \qquad w > 0. \qquad (4.5.4)$$

The random variable W is related to an exponential random variable X through the identity $W = vX^{1/\beta}$, where $F_x(x) = 1 - \exp(-x)$, $x > 0$. The identity enables us to write the first two moments of W as

$$E[W] = v\Gamma\left(\frac{1}{\beta}+1\right) \qquad \text{and} \qquad E[W^2] = v^2\Gamma\left(\frac{2}{\beta}+1\right), \qquad (4.5.5)$$

where Γ denotes the complete gamma function

$$\Gamma(a) = \int_0^\infty e^{-x}x^{a-1}dx.$$

Returning to the insurance problem, given $E[W] = 46{,}000$ and $Var[W] = (12{,}500)^2$, we need to estimate the Weibull parameters v and β. One approach is to find the two parameters so that they match the given two moments (this is known as *the method of moments* in statistics). From Equation 4.5.5, simple algebra shows that

$$\frac{\Gamma\left(\frac{2}{\beta}+1\right)}{\Gamma^2\left(\frac{1}{\beta}+1\right)} = \frac{Var(W)+E^2(W)}{E^2(W)}. \qquad (4.5.6)$$

The right side of Equation 4.5.6 is given. Tracing the left side over the range of β to find its matching value is easy to do on a computer. Once β is known, we find v from Equation 4.5.5. For this example, we find $v = 50{,}647.61$ and $\beta = 4.145$. The Weibull density with the given parameters is shown in Figure 4.15.

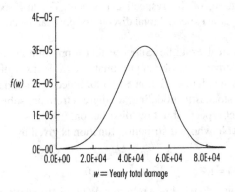

FIGURE
4.15 The Weibull density with $v = 50{,}647.61$ and $\beta = 4.145$.

We now consider the case in which the company starts from premium class 5 (a clean record) and the yearly interest rate is 10 percent. This implies that the discount rate is $1/(1 + 0.1) = 0.9091$. If the company does not have an insurance contract, the expected total discounted cost over an infinite horizon is $46{,}000(\alpha/(1 - \alpha)) = 460{,}051$. Suppose that the premiums $h_1, \ldots, h_5$ are 40,000, 30,000, 20,000, 10,000, and 8,000, respectively; the deductibles $d_1, \ldots, d_5$ are 8,000, 7,000, 6,000, 5,000, and 4,000, respectively. One claim-limit policy is $s_i = d_i$ for all i. This policy implies the company will always claim damages exceeding the deductibles. Under this policy, we find in the Appendix $g(5) = 428{,}327$ (we use the last element of the vector g because $X_0 = 5$). To see whether a more conservative claim-limit policy will yield a lower expected total discounted cost, we first try a simple policy of setting $s_i = 30{,}000$ for all i. This yields $g(5) = 420{,}506$—only a slight improvement over the "always-claim" policy. We now consider a claim-limit policy with $s = (70{,}000, 60{,}000, 50{,}000, 40{,}000, 30{,}000)$. This policy will reduce the likelihood of moving to the more expensive premium classes. Under this policy, we find $g(5) = 363{,}667$—about 15 percent reduction in expected discounted cost over the "always-claim" policy. ∎

4.6 Reversible Discrete-Time Markov Chains

If we look at a stochastic process with its time axis reversed, some processes retain all the properties of the forward process. Such processes, called reversible processes, have many interesting features and play an important role in queueing networks.

To make the notion of reversibility more precise, we define a stochastic process $X = \{X(t), t \geq 0\}$ as *reversible* if $\{X(t_1), \ldots, X(t_n)\}$ follows the same distribution as $\{X(s - t_1), \ldots, X(s - t_n)\}$ for all $t_1, \ldots, t_n$ and s.

If X is reversible, with $s \leftarrow 0$ and $s \leftarrow -s$ in $\{X(s - t_1), \ldots, X(s - t_n)\}$, respectively. it is easy to see that both $\{X(t_1), \ldots, X(t_n)\}$ and $\{X(t_1 + s), \ldots, X(t_n + s)\}$ have the same distribution as $\{X(-t_1), \ldots, X(-t_n)\}$. Therefore a reversible process is a stationary process.

We now restrict our attention to discrete-time Markov chains. Let $X = \{X_n, n \geq 0\}$ be an irreducible stationary Markov chain with state space $S = \{0, 1, \ldots\}$ and transition probability matrix $P = \{p_{ij}\}$. If we construct a related stochastic process X^* by looking at the sample path of X in reversed order of the time axis, we see that

$$P\{X_m = j | X_{m+1} = i, X_{m+2} = i_2, \ldots, X_{m+k} = i_k\}$$

$$= \frac{P\{X_m = j, X_{m+1} = i, X_{m+2} = i_2, \ldots, X_{m+k} = i_k\}}{P\{X_{m+1} = i, X_{m+2} = i_2, \ldots, X_{m+k} = i_k\}}$$

$$= \frac{P\{X_{m+2} = i_2, \ldots, X_{m+k} = i_k | X_m = j, X_{m+1} = i\} P\{X_{m+1} = i | X_m = j\} P\{X_m = j\}}{P\{X_{m+2} = i_2, \ldots, X_{m+k} = i_k | X_{m+1} = i\} P\{X_{m+1} = i\}}$$

$$= \frac{P\{X_{m+2} = i_2, \ldots, X_{m+k} = i_k | X_{m+1} = i\} P\{X_{m+1} = i | X_m = j\} P\{X_m = j\}}{P\{X_{m+2} = i_2, \ldots, X_{m+k} = i_k | X_{m+1} = i\} P\{X_{m+1} = i\}}$$

$$= \frac{P\{X_{m+1} = i | X_m = j\} P\{X_m = j\}}{P\{X_{m+1} = i\}} = \frac{p_{ji} \pi_j}{\pi_i},$$

where $\{\pi_j\}$ are the stationary probability vectors of the Markov chain **X**. Let H denote the history of the process **X*** "before" (since for **X***, the time axis has been reversed) time $m + 1$, that is, $H = \{X_{m+1}, \ldots, X_{m+k}\}$. The preceding derivations show that if **X** is stationary then

$$P\{X_m = j | X_{m+1} = i, H\} = P\{X_m = j | X_{m+1} = i\} = \frac{p_{ji} \pi_j}{\pi_i}.$$

In other words, the reversed stochastic process **X*** is a Markov chain whose transition probability $\{p_{ij}^*\}$ is given by the last expression, namely,

$$p_{ij}^* = \frac{p_{ji} \pi_j}{\pi_i}. \tag{4.6.1}$$

If the transition probability matrix $P^* = \{p_{ij}^*\} = P$ (or equivalently, $p_{ij}^* = p_{ij}$ for all i and j), we now argue that the Markov chain is reversible. Consider a sample path involving the observation of a sequence of states $\{j_0, j_1, \ldots, j_k\}$ at epochs $\{m, m + 1, \ldots, m + k\}$. The probability for observing this sample path is given by

$$P\{X_m = j_0, X_{m+1} = j_1, \ldots, X_{m+k} = j_k\} = \pi_{j_0} p_{j_0, j_1} p_{j_1, j_2} \cdots p_{j_{k-1}, j_k}. \tag{4.6.2}$$

Consider another sample path of **X** involving the observation of the sequence $\{j_k, j_{k-1}, \ldots, j_0\}$. Then the corresponding probability is given by

$$P\{X_{m'} = j_k, X_{m'+1} = j_{k-1}, \ldots, X_{m'+k} = j_0\} = \pi_{j_k} p_{j_k, j_{k-1}} p_{j_{k-1}, j_{k-2}} \cdots p_{j_1, j_0}. \tag{4.6.3}$$

Using Equation 4.6.1 with $p_{ij}^* = p_{ij}$ for all i and j, by successive substitutions it can be shown that the right sides of Equations 4.6.2 and 4.6.3 are the same. Define $\tau = m + m' + k$. We rewrite Equation 4.6.3 as

$$P\{X_{\tau-m-k} = j_k, X_{\tau-m-(k-1)} = j_{k-1}, \ldots, X_{\tau-m} = j_0\}$$

$$= P\{X_{\tau-m} = j_0, X_{\tau-m-1} = j_1, \ldots, X_{\tau-m-k} = j_k\}.$$

Hence we have shown that $\{X_m, \ldots, X_{m+k}\}$ and $\{X_{\tau-m}, \ldots, X_{\tau-(m+k)}\}$ follow the same distribution and consequently the chain X is reversible.

> Thus a necessary and sufficient condition for a stationary Markov chain to be reversible is
> $$\pi_i p_{ij} = \pi_j p_{ji} \qquad \text{for all } i, j \in S. \tag{4.6.4}$$

The preceding equations are called the *detailed balance equations* (see the balance equations $\pi = \pi P$). Computationally, if we can solve the following system of linear equations:

$$x_i p_{ij} = x_j p_{ji} \qquad \text{for all } i, j \qquad \text{and} \qquad \sum_{i \in S} x_i = 1, \tag{4.6.5}$$

then the Markov chain is reversible and the $\{x_i\}$ so obtained are the stationary distributions of the chain. To establish the latter, we add the first part of Equation 4.6.5 over i. This yields the balance equations $\pi = \pi P$ for a stationary distribution.

EXAMPLE
4.6.1

Consider a two-state Markov chain with transition probability matrix

$$P = \begin{bmatrix} 0 & 1 \\ 1/2 & 1/2 \end{bmatrix}.$$

Is this chain reversible? We find that the stationary probability vector $\pi = [1/3, 2/3]$. We see that $\pi_1 p_{12} = (1/3)(1) = \pi_2 p_{21} = (2/3)(1/2)$. Hence the chain is reversible. ▪

EXAMPLE
4.6.2

Consider a Markov chain with state space $S = \{1, 2, 3\}$ and transition probability matrix

$$P = \begin{bmatrix} 0 & 0.6 & 0.4 \\ 0.1 & 0.8 & 0.1 \\ 0.5 & 0 & 0.5 \end{bmatrix}.$$

We find the stationary probability vector $\pi = [5/27, 15/27, 7/27]$. The chain is not reversible; for example, we have $\pi_1 p_{12} = (5/27)(6/10) = 1/9 \neq \pi_2 p_{21} = 1/18$. ▪

Using Equation 4.6.4 or 4.6.5 to check reversibility of a chain requires the solution of the stationary probability vector π. Another way to accomplish the same without using π is by the Kolmogorov criteria: a stationary Markov chain is reversible if and only if the product of all one-step transition probabilities associated with the arcs traversing from each node back to itself along each one of the possible paths is equal to the product of their counterparts when the directions of the arcs are reversed. Specifically, for any state i and any path $(i, i_1, \ldots, i_k, i)$, the criteria call for

$$p_{i,i_1} p_{i_1,i_2} \cdots p_{i_k,i} = p_{i,i_k} p_{i_k,i_{k-1}} \cdots p_{i_1,i}. \tag{4.6.6}$$

The necessity of the criteria can be established by using an approach similar to the establishment of the equality of Equations 4.6.2 and 4.6.3 (with $j_0 = j_k = i$). To show the sufficiency, we choose any two states i and j and a path $(i, i_1, \ldots, i_k, j, i)$. Then Equation 4.6.6 implies that

$$p_{i,i_1} p_{i_1,i_2} \cdots p_{i_k,j} p_{j,i} = p_{i,j} p_{j,i_k} p_{i_k,i_{k-1}} \cdots p_{i_1,i}.$$

Adding the given equality over all states $i_1, \ldots, i_k$, gives $p_{ij}^{(k+1)} p_{ji} = p_{ij} p_{ji}^{(k+1)}$. Summing the last equality over $k = 1, \ldots, n$, dividing the resulting sum by n, and letting $n \to \infty$, by Equation 4.3.8 we conclude that $\pi_j p_{ji} = \pi_i p_{ij}$ and the reversibility condition of Equation 4.6.4 is met.

EXAMPLE
4.6.3

Consider any irreducible stationary two-state Markov chain with state space $S = \{1, 2\}$. There are only two paths for which each node is only traversed once; they are $1 \to 2 \to 1$ and $2 \to 1 \to 2$. The products of one-step transition probabilities along these two paths are $p_{12} p_{21}$ and $p_{21} p_{12}$. When the directions of these two paths are reversed, the two corresponding products stay the same. Hence they satisfy Kolmogorov criteria. Along each path, if a state is visited more than once, the added multiplier shows up in both products (the ones associated with the forward and reversed paths). Hence we conclude that the two-state Markov chain is reversible. ∎

EXAMPLE
4.6.4

Returning to Example 4.6.2, we see that for the path $1 \to 2 \to 3 \to 1$ the product of the associated transition probabilities is $p_{12} p_{23} p_{31} = (6/10)(1/10)(5/10) = 3/100$. For the reversed path, the product is $p_{13} p_{32} p_{21} = (4/10)(0)(1/10) = 0$. Thus it fails to meet the Kolmogorov criteria and the chain is not reversible. ∎

For a Markov chain in equilibrium, the reversed chain always exists and can be constructed by Equation 4.6.1. When the chain is not reversible (this occurs when $P \neq P^*$), the notion of a reversed chain can be useful in the following context. For the Markov chain with transition probability matrix P, if we can find $\{\pi_i\}$ and $P^* = \{p_{ij}^*\}$ such that Equation 4.6.1 is satisfied. Then $\{\pi_i\}$ are the limiting probabilities of the Markov whose transition probability matrix is P and the limiting probabilities of the reversed Markov chain whose transition probability matrix is P^*. To establish the last two assertions, we first rewrite Equation 4.6.1 as $\pi_i p_{ij}^* = \pi_j p_{ji}$. Adding the last expression over all i, we obtain $\pi P^* = \pi$; summing it over all j yields $\pi = \pi P$. The two assertions follow accordingly. The following example illustrates an application of this useful idea.

<hr>

EXAMPLE 4.6.5

The Limiting Excess Life of a Discrete Renewal Process   In Example 3.7.1, we derived the limiting excess life of a discrete renewal process with interarrival time mass function $\{f_k\}$, where $f_k = P\{X = k\}$ and X denotes the interarrival time. Let Y_n denote the excess life of the renewal process at time n. Then $Y = \{Y_n, n \geq 0\}$ is a discrete-time Markov chain with transition probability matrix

$$P = \begin{array}{c} \\ 1 \\ 2 \\ 3 \\ \cdot \\ \cdot \end{array} \begin{array}{cccccc} 1 & 2 & \cdot & \cdot & \cdot & \\ \left[\begin{array}{ccccc} f_1 & f_2 & f_3 & \cdot & \cdot \\ 1 & & & & \\ & 1 & & & \\ & & \cdot & & \\ & & & \cdot & \end{array}\right] \end{array}.$$

Using $\pi = \pi P$ and $\pi e = 1$, it is easy to show that for this chain, the limiting probabilities are given by $\pi_1 = 1/\mu$, $\pi_i = P\{X \geq i\}/\mu$, $i > 1$, where μ is the mean interarrival time (see Problem 21). This result was derived originally as Equation 3.7.11.

Let X_n denote the age of the renewal process at time n. Define $g_i = f_i / \sum_{k=i}^{\infty} f_k$. We see that g_i represents the conditional probability that a renewal occurs at the end of age i given that the interarrival time is at least of length i. Then $X = \{X_n, n \geq 1\}$ is a Markov chain with transition probability matrix

$$P^* = \begin{array}{c} \\ 1 \\ 2 \\ 3 \\ \cdot \\ \cdot \end{array} \begin{array}{cccccc} 1 & 2 & 3 & \cdot & \cdot & \\ \left[\begin{array}{ccccc} g_1 & 1-g_1 & & & \\ g_2 & & 1-g_2 & & \\ g_3 & & & 1-g_3 & \\ \cdot & & & & \\ \cdot & & & & \end{array}\right] \end{array}.$$

We will next show that $\pi_i p_{ij}^* = \pi_j p_{ji}$ for all i and j and hence establish that P^* is the transition probability matrix of the reversed chain. This in turn implies that $\{\pi_i\}$ are also the limiting probabilities of the reversed chain and equivalently the limiting age distribution. To establish the identities, there are two cases to consider:

(i) For $j = 1$, we have

$$\pi_i p_{ij}^* = \frac{P\{X \geq i\}}{\mu} g_j = \frac{P\{X \geq i\}}{\mu} \frac{f_i}{P\{X \geq i\}} = \frac{f_i}{\mu}.$$

Since $\pi_1 p_{1i} = (1/\mu) f_i$, the two sides are the same.

(ii) For $j > 1$, the only nontrivial case is with $j = i + 1$ for any i; hence

$$\pi_i p^*_{i,i+1} = \frac{P\{X \geq i\}}{\mu}(1 - g_i) = \frac{P\{X \geq i\}}{\mu}\frac{P\{X \geq i+1\}}{P\{X \geq i\}} = \frac{P\{X \geq i+1\}}{\mu}.$$

Since $\pi_{i+1}p_{i+1,i} = (P\{X \geq i+1\}/\mu)(1)$, we are done. It is important to observe that for this example $P \neq P^*$, and the chain is not reversible. ∎

EXAMPLE
4.6.6 **If an Ergodic Markov Chain Is Reversible, So Is the Censored Chain** Let X and Y be the two Markov chains defined in Example 4.4.7. From Equation 4.4.18, we see that the limiting probability vectors differ only by the constant term used in the denominator of Equation 4.4.18. Hence the reversibility conditions of Equation 4.6.4 are also satisfied by the chain Y. Thus the censored chain Y is also reversible. ∎

Problems

1 Determine the communicating classes and period for each state of the Markov chain whose transition probability matrix is

	0	1	2	3	4	5
0	1/2				1/2	
1			1			
$P = $ 2				1		
3					1	
4						1
5			1/3	1/3		1/3

2 Which states are transient and which are recurrent in the Markov chain whose transition probability matrix is

	0	1	2	3	4	5
0	1/3		1/3			1/3
1	1/2	1/4	1/4			
$P = $ 2					1	
3	1/4	1/4	1/4			1/4
4			1			0
5						1

3 Recall that $f_{ij}^{(n)}$ denote the first passage time probability from state i to state j, that is, $f_{ij}^{(n)} = P\{X_n = j, X_{n-1} \neq j, \ldots, X_1 \neq j \mid X_0 = i\}$, for $n = 1, 2, \ldots$. For the Markov chain with transition matrix P, determine $f_{00}^{(n)}$ and $f_{03}^{(n)}$ for $n = 1, 2, 3, 4, 5$.

$$
P = \begin{array}{c|cccc}
 & 0 & 1 & 2 & 3 \\
\hline
0 & & 1/2 & & 1/2 \\
1 & & & 1 & \\
2 & & & & 1 \\
3 & 1/2 & & & 1/2
\end{array}
$$

4 Consider a Markov chain with the following transition probability matrix

$$
P = \begin{array}{c}
1 \\ 2 \\ 3 \\ 4 \\ 5 \\ 6 \\ 7 \\ 8 \\ 9 \\ 10
\end{array}
\left[\begin{array}{cccccccccc}
 & & & & & & & & & 1 \\
.3 & .3 & .1 & .3 & & & & & & \\
 & .6 & & & & & .4 & & & \\
 & & & 1 & & & & & & \\
.4 & & & & & .3 & & .3 & & \\
 & & .9 & & & .1 & & & & \\
 & & & & & & & & 1 & \\
.8 & & & & & & .2 & & & \\
 & & & & & & & & & 1 \\
1 & & & & & & & & &
\end{array}\right]
$$

(a) Decompose the state space into equivalence classes. (b) Identify the closed equivalence classes and find their respective periods. (c) State the transition probability matrix in a canonical form.

5 Hospital Admissions   The surgical ward of a small hospital has three beds. Arriving patients requiring admission to the ward consist of two types: emergency and scheduled arrivals. Let A_t denote the number of emergency arrivals during day t. $\{A_t \mid t = 0, 1, \ldots,\}$ is a sequence of i.i.d. random variables with probability distribution

$$P\{A_t = j\} = a_j \qquad t = 0, 1, 2, \ldots; j = 0, 1, 2, \ldots,$$

and

$$\sum_{j=0}^{\infty} a_j = 1.$$

Let X_t denote the number of beds occupied at the start of day t and D_t the number of patients that leaves the ward during day t. Thus X_t is a discrete variable with possible values $0, 1, \ldots, 3$. We assume that $\{D_t \mid t = 0, 1, \ldots\}$ is a sequence of random variables that is dependent on the occupancy of the ward, specifically

$$P\{D_t = j \mid X_t = i\} = d_{ij} \qquad i = 0, \ldots, 3; \, j = 0, \ldots, i; \, t = 0, 1, 2, \ldots,$$

and

$$\sum_{j=0}^{i} d_{ij} = 1 \qquad i = 0, \ldots, 3.$$

To facilitate the formulation, we assume that discharges (determined probabilistically through $\{d_{ij}\}$) can be made when new patients arrive on any given day. (a) Consider the policy of only admitting emergency arrivals to the ward. The stochastic process $\{X_t \mid t = 0, 1, 2, \ldots\}$ resulting from the policy forms a Markov chain with transition probability matrix $P = \{p_{ij}\}$, where $p_{ij} = P\{X_{t+1} = j \mid X_t = i\}$. State the transition probability matrix P in terms of a_i and d_{ij}. (b) Consider the policy of admitting two scheduled arrivals when $X_t = 0$; one scheduled arrival when $X_t = 1$ and no scheduled arrivals when $X_t = 2$ or 3. Here, we assume that scheduled arrivals are always available and will not be preempted by emergency arrivals, and emergency arrivals will be admitted to the ward when space is available. The stochastic process $\{X_t \mid t = 0, 1, 2, \ldots\}$ resulting from the policy forms a Markov chain with transition probability matrix $P = \{p_{ij}\}$, where $p_{ij} = P\{X_{t+1} = j \mid X_t = i\}$. State the transition probability matrix P in terms of a_i and d_{ij}.

6 The Cash Management Problem   Consider the cash balance problem given in Example 4.4.1. Another commonly used policy has two critical parameters: α and β. When the cash position exceeds β, it is decreased to level β; when the cash position drops below α, it is increased to α. We assume that all other factors remain unchanged. Following the regenerative approach used in Example 4.4.1, (a) find the long-run average cost per day in terms of known parameters; (b) compute the long-run average cost per day based on the data given in the example.

7 The Cash Management Problem   Use the Markov reward process approach to do Problem 6.

8 A fugitive moves about three cities, Houston, Galveston, and Sugarland, to escape a bounty hunter. Initially, the fugitive is in Houston and the bounty hunter is in Galveston. The fugitive and bounty hunter move independently of each other and each follows the respective transition probability matrix of a Markov chain:

$$P_{\text{fugitive}} = \begin{array}{c} \text{Houston} \\ \text{Galveston} \\ \text{Sugarland} \end{array} \begin{bmatrix} .4 & .6 \\ .7 & .3 \\ .1 & .9 \end{bmatrix} \qquad P_{\text{bounty hunter}} = \begin{array}{c} \text{Houston} \\ \text{Galveston} \\ \text{Sugarland} \end{array} \begin{bmatrix} .3 & .2 & .5 \\ .8 & .1 & .1 \\ .5 & .2 & .3 \end{bmatrix}.$$

We assume that each party makes the move (transition) at the end of each day. When the two are in the same city on a given day there is a 50 percent chance the fugitive will be captured. What is the expected time for the bounty hunter to catch the fugitive?

9 Tax Auditing   The auditor's manual for the Internal Revenue Service dictates the following review procedures for tax returns for private individuals.

(i) All submitted returns are subject to a computer check for arithmetic errors and "unusual" deviations from typical returns. Seventy percent pass the test; 10 percent must be returned to the taxpayer for correction of errors; the remainder are sent to trained auditors for review. In addition, a spot check is made on those returns that passed the test by selecting 5 percent at random to be audited.

(ii) Of those returns selected for audit, 40 percent are judged satisfactory without involving the taxpayer. The others require that the taxpayer be contacted to provide additional information or substantiation. After hearing the taxpayer's case, the department may assess additional taxes, which happens in seven out of ten cases, or it may accept the return as submitted.

(iii) If a taxpayer is assessed additional taxes and wants to appeal, he or she may do so by applying for review by the IRS District Appeal Board. The chance that the appeal board will overturn the auditor's decision is only one in fifty. Nevertheless a third of such taxpayers do appeal.

(iv) As a last resort, a taxpayer may go to court to obtain a favorable ruling, but his or her chances are only one in one hundred of emerging without any additional tax assessment. Of those who go to court, half do.

(a) Develop a Markov chain to model the dynamics of the flow of auditing so that it can be used in answering the subsequent questions. Define the states, give the corresponding transition probability matrix, and draw the transition diagram. (b) Of a million submitted returns, how many (on the average) go to court? (c) If a return is audited, what is the probability that the taxpayer must pay additional taxes? (d) Including the computer check, what is the expected number of times that a return will be reviewed? (e) What is the expected number of times an audited return will be reviewed?

10 Consider a Markov chain with the following transition probability matrix:

$$
P = \begin{array}{c} 1 \\ 2 \\ 3 \\ 4 \\ 5 \\ 6 \\ 7 \end{array}
\left[\begin{array}{ccccccc}
.4 & .6 & & & & & \\
.9 & .1 & & & & & \\
& & & .7 & .3 & & \\
& & 1 & & & & \\
& & 1 & & & & \\
.2 & & & .2 & .3 & .1 & .2 \\
& & & & & & 1
\end{array} \right].
$$

(a) Find the limiting probabilities $\lim_{n \to \infty} p_{ij}^{(n)}$ for the following (i, j) pairs: $(1, 1)$, $(2, 1)$, $(3, 1)$, $(3, 3)$, $(6, 1)$, $(6, 7)$, $(6, 3)$, $(6, 4)$—when they exist. (b) For those cases in which the limit does not exist for all n, find the limiting probabilities as a function of n (for example, for n even or odd).

11 Consider Problem 5 again. Assume that emergency arrivals follow a Poisson process with a rate of one arrival per day and the number of discharges per day, given that there are i patients at the start of the day, follows a binomial distribution with parameters (i, p), where $p = 0.5$. For the two patient admission policies, find **(a)** the one-step transition probability matrices; **(b)** the long-run expected occupancies per day; **(c)** the long-run expected number of emergency patients who are turned away per day. Under the current assumption of binomial discharges, what type of length-of-stay distributions are we assuming?

12 Establish Equation 4.4.17 algebraically as suggested at the end of Example 4.4.7.

13 Consider an automobile painting shop with a waiting area that can hold a maximum of M cars. Cars arrive at the waiting area in accordance with a Poisson process with rate λ. When the waiting area is full, arriving cars simply leave without receiving any service. At the start of each hour, the painting shop opens and takes in a maximum of N ($N < M$) waiting cars. All cars in the painting shop will receive a one-hour paint job done simultaneously by robots. The scenario is depicted in the following figure.

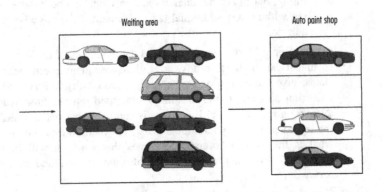

We are interested in the long-run fraction of cars that find upon arrival that the waiting area is full and thus are lost as customers. Let X_n denote the number of cars in the waiting area at the start of hour n just before entering the painting shop. **(a)** Show that the stochastic process $\{X_n, n \geq 0\}$ can be formulated as a Markov chain and derive the transition probability matrix. **(b)** Derive an expression for computing the long-run fraction of lost cars. **(c)** Assuming $M = 8$, $\lambda = 2$, and $N = 3$, compute the stationary probability vector of the Markov chain and the long-run fraction of lost cars. **(d)** If the long-run fraction of lost cars is very small (for example, the size of the waiting area M is very large), how would you estimate the average number of cars in the waiting area?

14 **The Game of Squash**   In a game of squash, there are two players called A and B. A score of $i - j$ gives the points made by A and B, respectively, at a given time. Under the international scoring system, the first player to score nine points wins the game. If the game is tied at eight, the player who reaches eight points has two options: play to nine points (called "set one") or ten points (called "set two"). Set one means that the next player to score wins the game; set two means the first player to score two points wins the game. Under this scoring system, points are only scored by the player who is serving and a player who wins a rally must serve the next rally. Consequently, if the game is tied at eight, the person who makes the decision must be receiving in the rally that causes the tie and hence will not serve the next rally.

When player A is serving, we let p denote the probability that player A wins a rally; when player B is serving, we let q denote the probability that player A wins a rally. Consider the case in which a game is tied at eight and player A must choose the set. Use an analysis based on the theory of Markov chains to determine the best strategy for player A **(a)** if $p = 0.5$ and $q = 0.5$, **(b)** if $p = 0.7$ and $q = 0.6$, and **(c)** $p = 0.25$ and $q = 0.4$.

15 Cab driver Jimmy plies his trade in three neighboring cities: Houston, Galveston, and Sugarland, indexed by 1, 2, and 3, respectively. The transition probability matrix governing his movement is given as follows:

$$P = \begin{array}{c} \\ 1 \\ 2 \\ 3 \end{array} \begin{array}{ccc} 1 & 2 & 3 \\ \left[\begin{array}{ccc} .6 & .1 & .3 \\ .7 & .2 & .1 \\ 0 & .5 & .5 \end{array}\right] \end{array}$$

For example, if the last trip ended in Houston then with probability 0.6, 0.1, and 0.3, his next trip will end in Houston, Galveston, and Sugarland, respectively. Jimmy lives in Houston and he starts his day there. When a trip involves a change of cities, Jimmy has an inclination to drive above the speed limit. A trip involving a change of cities will include a 50 percent probability of Jimmy getting a speeding ticket. **(a)** What are the mean and variance of the number of trips made before receiving a ticket? **(b)** What is the probability that Jimmy will visit Sugarland before receiving a ticket?

16 An input conveyor passes by a work station every twelve minutes with an item to be processed by the operator at the work station. The amount of work for each item differs depending on the instructions stated on the tag of each item. There are three types of items: 2/7 of them require six-minutes of work, 4/7 of them require twelve-minutes of work, and 1/7 of them require eighteen-minutes of work. The worker at the work station works on one item at a time. The station has a storage space for holding one unprocessed item. If an item is completed in the middle of the twelve-minute interval, it is immediately loaded on an output conveyor and the item at the storage space, if any, is moved to the work station for immediate processing. At the end of a twelve-minute interval, the worker loads the finished item, if any, on the output conveyor and moves

the waiting item at the storage space, if any, to the work station for immediate processing. In addition, the arriving item from the input conveyor is moved to the work station or the storage space, depending on the space available at either place. Assume that loading and unloading times are negligible. When an item passes by the work station and the work station does not have a place to store it, the item leaves the system unprocessed. The figure above depicts the complete system. Define the work load at the work station as the amount of work remaining (in minutes of work yet to be completed) right before the conveyor passes by the work station. **(a)** Compute the average amount of work load at the work station. **(b)** Compute the fraction of items left in the work station unprocessed.

17 Consider the $M/G/1$ queue described in Example 4.3.2 again. Assume that the queue is stable, that is, $\rho = \lambda/\mu < 1$, where λ is the arrival rate and μ is the service rate. Recall that the balance equations for the limiting queue length distribution $\{\pi_j\}$ defined at departure epochs are given by

$$\pi_j = \pi_0 a_j + \sum_{i=1}^{j+1} \pi_i a_{j-i+1} \quad j \geq 0, \quad \text{where} \quad a_j = \int_0^\infty pos(j;\ \lambda,\ t)dG(t)$$

and G is the service time distribution. Define the probability generating functions

$$\pi^g(z) = \sum_{j=0}^\infty \pi_j z^j \quad \text{and} \quad a^g(z) = \sum_{j=0}^\infty a_j z^j.$$

(a) Show that $\pi^g(z) = \dfrac{\pi_0(1-z)a^g(z)}{a^g(z) - z}$. **(b)** Why do we have $\dfrac{d}{dz} a^g(z)\big|_{z=1}$

$= \rho$? **(c)** Find π_0 using $\lim_{z \to 1} \pi^g(z) = 1$. **(d)** Show that $a^g(z) = G^e(\lambda - \lambda z)$,

where $G^e(s) = \int_0^\infty e^{-sx}dG(x)$. (When the results of Parts c and d are used in Part a, the expression is known as the Pollaczek-Khintchine formula for the $M/G/1$ queue.)

18 Use the Pollaczek-Khintchine formula to find the queue length distribution at the departure epochs of an $M/M/1$ queue in which the service time distribution is exponential with parameter μ.

19 Consider an $M/G/1$ queue in which the arrival rate is five per hour and the service time distribution is a two-point distribution: with probability 2/3 it is eight minutes long and with probability 1/3 it is seventeen minutes long.

(a) Use the notations introduced in Problem 17 to find $a^g(z)$. (b) Invert the generating function $\pi^g(z)$ numerically to find $\{\pi_1, \ldots, \pi_5\}$.

20 Consider an $M/D/1$ queue in which the service time takes a constant of $1/\mu$ time units. The arrival process is Poisson with rate λ. Use the Pollaczek-Khintchine formula to find the probability generating function for the queue length distribution at departure epochs.

21 Consider a Markov chain with state space $S = \{1, 2, 3, \ldots\}$ and transition probability matrix

$$P = \begin{matrix} 1 \\ 2 \\ 3 \\ \cdot \\ \cdot \end{matrix} \begin{bmatrix} f_1 & f_2 & f_3 & \cdot & \cdot \\ 1 & & & & \\ & 1 & & & \\ & & \cdot & & \\ & & & \cdot & \end{bmatrix},$$

where $f_k = P\{X = k\}$ and $E[X] = \mu$. Let $\{\pi_i\}$ denote the limiting probabilities of this Markov chain. Show that $\pi_1 = 1/\mu$ and $\pi_i = P\{X \geq i\}/\mu$, $i > 1$.

22 A database contains n records ordered in a linear list. Exactly one record of the database is accessed at each one of the distinct epochs 1, 2, 3, …. With probability α_i record i is accessed at a given epoch. After each access, the record accessed changes its position in the linear list. There are two reordering rules to consider. Under Rule 1, the accessed record is placed at the head of the linear list and the positions of other records shift accordingly. Under Rule 2, the accessed record is moved one position closer to the head of the list (if it is not already so) and the positions of other records shift accordingly. To model the system showing the locations of the records in the database by a Markov chain, the states of the system will be the $n!$ permutations of the integers 1, …, n representing the locations of the records in the linear list. For simplicity, we consider a numerical example with $n = 3$ and $\alpha_1 = 0.5$, $\alpha_2 = 0.3$, and $\alpha_3 = 2$. Under each reordering rule, display the transition probability matrix of the Markov chain, compute the stationary probability vector of the chain, and check the reversibility of the chain.

23 Daily demand D for a given product follows a probability mass function $\{p_d\}$ where $p_d = P\{D = d\}$, $d = 0, 1, 2, 3, 4$, and $\{p_0, \ldots, p_4\} = \{0.7, 0.15,$

less than

0.1, 0.04, 0.01$\}$. The inventory on hand is reviewed in the morning every four days. Thus we consider the four-day interval between successive reviews a period. If the inventory at the start of a period is i, where $i < s$, it is ordered to S ($s < S$) and hence the order quantity is $S - s$. Assume that the delivery lead time is 0 so items ordered will be available immediately. Demand that cannot be satisfied by inventory on hand is lost. The cost of holding inventory is \$0.01 per unit per day and the cost of unmet demand is \$0.50 per unit. The fixed cost of restocking is \$0.20. (a) Let X_n denote the inventory on hand at the start of period n (after replenishment if it is so requested). Define $p_{ij} = P\{X_{n+1} = j | X_n = i\}$. Under a given (s, S) policy, find the transition probability matrix $P = \{p_{ij}\}$. (b) Under a given (s, S) policy, give an expression for computing the expected costs per period $\{R_i\}$, where $X_n = i$.

24 We return to Problem 23. **(a)** Find the average total costs per day under the policy $(s, S) = (3, 7)$. **(b)** Do the same for policies $(3, 16)$ and $(15, 16)$.

25 Consider a variant of Problem 23. Assume that there is a two-day delivery lead time. Thus if k units are ordered at the start of a period, they will be added to the inventory on hand at the start of the third day of the same period. Do the two parts of Problem 23 for this variant.

26 We return to Problem 25. Find the average total costs per day under the (s, S) policies with $s = 3:6$ and $S = s:10$. What is the minimal cost (s, S) policy among those being considered?

27 At the beginning of each day, a piece of equipment is inspected to determine its working condition. The equipment can be found in one of four working conditions indexed by $1, \ldots, 4$. Working condition i is better than working condition $i + 1$. The equipment deteriorates over time. If the present working condition is i and no repair is done, then at the beginning of the next day the equipment will be in working condition j with probability q_{ij}. Working condition 5 represents a malfunction that requires a repair that takes three days. Such a repair restores the equipment to working condition 1. If the equipment is in working state 4, a preventive maintenance can be considered. The preventive maintenance takes only one day and also restores the equipment to working condition 1. The $\{q_{ij}\}$ are given in the following table:

$i\backslash j$	1	2	3	4	5
1	0.75	0.2	0.05	0	0
2	0	0.5	0.2	0.2	0.1
3	0	0	0.5	0.25	0.25
4	0	0	0	0.3	0.7

(a) Compute the long-run fraction of time the equipment is in repair in the absence of preventive maintenance. **(b)** Compute the long-run fraction of time the equipment is in repair when preventive maintenance is done.

28 Consider Example 4.3.3 again. Assume that the customer interarrival time is a discrete random variable with possible values $1, 2, \ldots, 6$ and respective probabilities $0.1, 0.2, 0.3, 0.2, 0.1, 0.1$. Hence the mean interarrival time is 3.3. Assume that the mean service time is 3.0. **(a)** Solve $B(z) = z$ for $0 \le z \le 1$. **(b)** Find the queue length distribution $\{\pi_j\}$ before an arrival. **(c)** Assume now that the mean service time is 4.0. Verify that the solution to $B(z) = z$ for $0 \le z \le 1$ is 1.

29 Consider a random walk on the set of nonnegative integers with the following transition probability matrix

$$
\begin{array}{c c c c c}
 & 0 & 1 & 2 & 3 & \cdots \\
\end{array}
$$

$$
P = \begin{array}{c}
0 \\
1 \\
2 \\
\cdot \\
\cdot \\
\cdot \\
\cdot
\end{array}
\left[
\begin{array}{c c c c c}
0 & 1 & 0 & 0 & \cdots \\
q_1 & 0 & p_1 & 0 & \cdots \\
0 & q_2 & 0 & p_2 & \cdots \\
 & & \cdot & & \cdots \\
 & & & & \cdots \\
 & & & & \cdots \\
 & & & &
\end{array}
\right],
$$

where $p_i > 0$, $q_i > 0$, and $p_i + q_i = 1$, for all i. (a) What is the period of this Markov chain? (b) What is the condition for positive recurrence of this chain?

*30 Consider a random walk on the set of nonnegative integers with the following transition probability matrix

$$
\begin{array}{c c c c c}
 & 0 & 1 & 2 & 3 & \cdots \\
\end{array}
$$

$$
P = \begin{array}{c}
0 \\
1 \\
2 \\
\cdot \\
\cdot \\
\cdot \\
\cdot
\end{array}
\left[
\begin{array}{c c c c c}
r_0 & p_0 & 0 & 0 & \cdots \\
q_1 & r_1 & p_1 & 0 & \cdots \\
0 & q_2 & r_2 & p_2 & \cdots \\
 & & \cdot & & \cdots \\
 & & \cdot & & \cdots \\
 & & \cdot & & \cdots \\
 & & & & \cdot
\end{array}
\right],
$$

where $p_i > 0$, $q_i > 0$, $r_i > 0$, $p_i + r_i + q_i = 1$, for all i except $q_0 = 0$. Define

$$
\pi_0 = 1, \qquad \pi_n = \frac{p_0 p_1 \cdots p_{n-1}}{q_1 q_2 \cdots q_n}.
$$

(a) Use Equation 4.3.14 with $k = 0$ to verify that

$$
y_0 = 0, \qquad y_n = \sum_{i=0}^{n-1} \frac{1}{p_i \pi_i}
$$

is a solution to Equation 4.3.14. If $\sum_{i=0}^{\infty} (1/p_i \pi_i) < \infty$, then $\{z_n\}$ is a bounded nondegenerate solution and hence the random walk is transient. On the other hand, if $\sum_{i=0}^{\infty} (1/p_i \pi_i) = \infty$, the random walk is recurrent. (b) If the random walk is recurrent, give the conditions under which it will be, respectively, positive and null recurrent.

*31 A Markov Chain of QBD Type We consider an irreducible, positive recurrent, and aperiodic Markov chain X whose transition probability matrix is of block tridiagonal form:

$$
P = \left[
\begin{array}{c c c c}
B_1 & A_0 & & \\
A_2 & A_1 & A_0 & \\
 & A_2 & A_1 & A_0 \\
 & & \cdot & \cdot & \cdot
\end{array}
\right],
$$

where all submatrices are of size $m \times m$. Let $S = \{(i, j), i \geq 0, 1 \leq j \leq m\}$ denote the state space of the chain X. We call the subset $\{(n, j), 1 \leq j \leq m\}$ of the state space *level* $l(n)$. Thus the state space is partitioned into levels $l(0), l(1), \ldots$. The structure of P implies that the changes of state are restricted to states in the same or adjacent levels. Such a process is called a *quasi birth and death (QBD) process*. With the exception of the first m rows and columns, the transition matrix has repeating rows and columns. When $m = 1$, the process reduces to a birth and death process.

A *taboo* set is a level that X is not allowed to visit. We define the following:

$G_{ij} = P\{$the first state visited by X in $l(n)$ is $(n, j)|X$ starts from $(n + 1, i)\}$,

$U_{ij} = P\{$the first return to $l(n + 1)$ is via $(n + 1, j)|X$ starts from $(n + 1, i)$ and the taboo set is $l(n)\}$,

$R_{ij} = E[$number of visits to $(n + 1, j)$ before ever visiting $l(n)|X$ starts from $(n, i)]$.

Let square matrices $G = \{G_{ij}\}$, $U = \{U_{ij}\}$, and $R = \{R_{ij}\}$. If we remove states associated with levels $\{l(0), l(1), \ldots, l(n)\}$ from the chain by removing the corresponding rows and columns of P, then the remaining transition matrices are identical for any $n \geq 0$. Such a property is called the spatial homogeneity of P. The property suggests that the three given terms are invariant in n. **(a)** By conditioning on the first transition out of state (n, i), show that $U = A_1 + A_0 G$. **(b)** Argue that we can construct a censored Markov chain with state space $\{l(0), l(1), \ldots, l(n)\}$ and transition probability matrix

$$
\begin{bmatrix}
B_1 & A_0 & & & & \\
A_2 & A_1 & A_0 & & & \\
 & A_2 & A_1 & A_0 & & \\
 & & \cdot & \cdot & \cdot & \\
 & & & A_2 & A_1 & A_0 \\
 & & & & A_2 & U
\end{bmatrix}
$$

Also, argue that U is substochastic and hence the inverse of $I - U$ exists. **(c)** What is the meaning of the (i, j)th element of U^k, $k \geq 1$? What is the meaning of the (i, j)th element of $(I - U)^{-1}$? **(d)** Show that $R = A_0 (I - U)^{-1}$, $G = (I - U)^{-1} A_2$, and $U = A_1 + RA_2$. **(e)** Show that $R = A_0 + RA_1 + R^2 A_2$ and $G = A_2 + A_1 G + A_0 G^2$. **(f)** Let row vector $\pi_n = \{\pi(n, 1), \pi(n, 2), \ldots, \pi(n, m)\}$ denote the stationary probabilities associated with states in $l(n)$. Show that

$$ \pi_n = \pi_0 R^n, \qquad n \geq 0, $$

where π_0 can be found by solving the following system of linear equations:

$$ \pi_0 (I - R)^{-1} e = 1 \quad \text{and} \quad \pi_0 = \pi_0 (B_1 + RA_2). $$

We remark that a simple but inefficient way to compute the matrix R is by successive approximation.

32 Consider the $M/G/1$ problem described in Problem 17 once more. Let W denote the sojourn time (time in the system) of a given customer. **(a)** Conditioning on the length of W, show that the limiting probability π_i that there are i waiting customers at a departure epoch of this customer is given by

$$\pi_i = \int_0^\infty pos(i;\, \lambda,\, t)\,dW(t).$$

(b) Define the Laplace-Stieltjes transform of the sojourn time distribution, that is,

$$W^e(s) = \int_0^\infty e^{-st}\,dW(t).$$

Show that $\pi^g(z) = W^e[\lambda(1-z)]$, where $\pi^g(z)$ was defined in Problem 17. **(c)** Show that we can write $W^e(s) = W_q^e(s)G^e(s)$, where $G^e(s)$ is the Laplace-Stieltjes transform of the service time distribution G and

$$W_q^e(s) \equiv \int_0^\infty e^{-st}\,dW_q(t) = \frac{(1-\rho)s}{s - \lambda[1 - G^e(s)]}.$$

(d) What is the role played by the distribution $W_q(t)$? (Hint: What is the implication of the relation $W^e(s) = W_q^e(s)G^e(s)$?) **(e)** Show that $W_q^e(s) = \sum_{n=0}^\infty (1-\rho)\rho^n (G_E^e(s))^n$, where $G_E^e = (\mu/s)[1 - G^e(s)]$, where $G_E^e(s)$ is the Laplace-Stieltjes transform of the limiting excess life (see Equation 3.5.7).

*33 **A Markov Chain of $GI/M/1$ Type**   We consider an irreducible, positive-recurrent, and aperiodic Markov chain X whose transition probability matrix is of lower block tridiagonal form:

$$P = \begin{bmatrix} B_0 & A_0 & & & & \\ B_1 & A_1 & A_0 & & & \\ B_2 & A_2 & A_1 & A_0 & & \\ B_3 & A_3 & A_2 & A_1 & A_0 & \cdot \\ & \cdot & & & \cdot & \\ & & \cdot & & & \cdot \end{bmatrix},$$

where all submatrices are of size $m \times m$. Let $S = \{(i,\, j),\ i \geq 0,\ 1 \leq j \leq m\}$ denote the state space of the chain X. We call the subset $\{(n,\, j),\ 1 \leq j \leq m\}$ of the state space *level* $l(n)$. Thus the state space is partitioned into levels $l(0)$, $l(1)$, …. The structure of P implies that moving to a state in a higher level must traverse through at least one state in each intermediate level. This structural characteristic is known as the skip-free-to-the-right property. Moreover, the form of P is identical to that of a $GI/M/1$ queue except now each entry is actually a submatrix of size $m \times m$. The stochastic process is hence called a Markov chain of $GI/M/1$ type. For level $l(n)$, we let $\pi_n = (\pi(n,\, 1), \ldots, \pi(n,\, m))$ denote the stationary probability level for states in the level. **(a)** Use an approach similar to that given in Problem 31 to show that $\pi_i = \pi_0 R^i$, $i \geq 0$. **(b)** Argue that the matrix R must satisfy the relation $R = \sum_{n=0}^\infty R^n A_n$. **(c)** Define $B(R) = \sum_{n=0}^\infty R^n B_n$. How will you solve for the stationary probability vector π_0?

*34 A Markov Chain of QBD Type—A Variant Consider a variant of the Markov chain described in Problem 31. Suppose that now the transition probability matrix P has the following form:

$$P = \begin{bmatrix} B_1 & B_0 & & & \\ B_2 & A_1 & A_0 & & \\ & A_2 & A_1 & A_0 & \\ & & \cdot & \cdot & \cdot \\ & & & \cdot & \cdot \end{bmatrix},$$

where submatrices $\{A_i\}$ remain unchanged but B_0 is $m_1 \times m$, B_1 is $m_1 \times m_1$, and B_2 is $m \times m_1$, and m_1 may not be the same as m. For this Markov chain we define the stationary probability vector $\pi = \{\pi_n\}$ as in Problem 31. However, for the subvector associated with level $l(0)$, we have $\pi_0 = \{\pi(0, 1), \ldots, \pi(0, m_1)\}$. (a) For this Markov chain, show that $\pi_n = \pi_1 R^{n-1}$, $n \ge 1$. (b) Express π_1 in terms of π_0. (c) How do you find π_0?

35 A Discrete-Time $M/PH/1$ Queue Consider a discrete-time version of an $M/G/1$ queue (see Example 4.3.2) in which individual arrivals may occur only at distinct time points $0, 1, \ldots$, and the service time follows a discrete-time phase-type distribution with representation (α, T) and is of order m, that is, $G \sim PH(\alpha, T)$. At a given time the probability that there will be an arrival is p and the probability that there will be no arrivals is $q = 1 - p$. At a service completion epoch, if there is at least one customer remaining, then the next service phase starts from state i with probability α_i (here $\alpha_{m+1} = 0$, that is, service time is strictly positive). Let X_n denote the number of customers in the system at the start of period n and Y_n the service phase at the start of period n if $X_n > 0$. Then $\{(X_n, Y_n), n \ge 1\}$ is a Markov chain of QBD type with state space $\{0\} \cup \{(i, j), i \ge 1, 1 \le j \le m\}$ and with a transition probability matrix P of the form shown in Problem 34. (a) What is the interarrival time distribution? (b) Find the submatrices $\{B_i\}$ and $\{A_i\}$ for this chain.

36 A Discrete-Time $PH/M/1$ Queue Consider a discrete-time version of $GI/M/1$ queue (see Example 4.3.3) in which the interarrival time follows a discrete-time phase-type distribution with representation (α, T) and is of order m, that is, $G \sim PH(\alpha, T)$, and if there is at least one customer in the system then the probability that a service completion will occur in a period is p and the probability that no service completions will occur in the period is $q = 1 - p$. A customer arrival corresponds to the event that in the absorbing Markov chain governing the phase-type distribution, an absorption to state $m + 1$ has just occurred. When this event takes place, an arrival phase is restarted instantaneously from state i with probability α_i (here $\alpha_{m+1} = 0$, that is, an arrival will take at least one period). Let X_n denote the number of customers in the system at the start of period n and Y_n the arrival phase at the start of period n if $X_n > 0$. Then $\{(X_n, Y_n), n \ge 1\}$ is a Markov chain of QBD type with state space $\{(i, j), i \ge 0, 1 \le j \le m\}$ and with a transition probability matrix P of the form shown in Problem 34. **(a)** What is the service time distribution? **(b)** Find the submatrices $\{B_i\}$ and $\{A_i\}$ for this chain.

37 Consider the random walk model described in Example 4.2.1. Assume that $p \neq q$. Let P_i denote the probability of reaching state N before reaching state 0 given that a walk starts from state i, where $0 \leq i \leq N$. **(a)** Derive a closed-form expression for P_i. **(b)** Use the result obtained in Part a to verify (3.6.21). **(c)** Let v_i denote the mean time to reach either state 0 or state N for the first time given that a walk starts from state i, where $0 \leq i \leq N$. Derive a closed-form expression for v_i.

38 A student goes to a Lake Tahoe casino during a spring break to play at a roulette table. The student begins with $5 and wants to simply win $15 dollars (or go broke) and go home. The student bets on red at each play. In this case, the probability of winning her bet (and hence doubling her ante) at each play is 18/38. To achieve her goal of reaching a total of $20 of wealth quickly, the student will use the strategy of betting her entire fortune or enough money so that reaching her goal is imminent. **(a)** What is the probability that the student will go home as a winner? **(b)** How many plays will be needed, on average, to reach a closure at the roulette table? **(c)** Will the probability of going home as a winner change if the student makes $1 bets at each play? What is that probability? What about the expected number of plays until the student goes home in this case?

Bibliographic Notes

Howard's (1971) book contains a comprehensive and lively treatment of discrete-time Markov chains by the geometric transform approach. Çinlar (1975) presents a formal and thorough exposition of the subject. The two sources complement each other well. The algorithm for partitioning the state space given in Section 4.2 is based on Gaver and Thompson (1973, 446–48). The simpler version of the approach for finding the limiting probabilities in Example 4.3.2 follows that of Heyman and Sobel (1982, 248–49). Examples 4.3.2 and 4.3.3 are staples of Markov chains and are covered in many texts such as Çinlar (1975), Gross and Harris (1985), Heyman and Sobel (1982), Ross (1983), and Wolff (1989). The quality control problem given as Example 4.3.4 is based on Taylor and Karlin (1994). The MATLAB program **mc_perdo** for finding periods of a Markov chain was coded à la Denardo (1977). A proof of Equation 4.3.14 can be found in Asmussen (1987, 18), or Çinlar (1975, 135). Many of the results on absorbing Markov chains can be found in the seminal book by Kemeny and Snell (1960). Example 4.4.3 about serving in tennis is based on Hannan (1976). More on discrete-time phase-type distributions can be found in Neuts (1975, 1981). Example 4.4.7 deals with a censored Markov chain and is based on a paper by Sheskin (1985). Subsequent extensions to Markov chains with denumerable state space are considered in Grassmann, Taksar, and Heyman (1985). The approach for applying the idea of a censored Markov chain for solving for limiting probabilities of an ergodic chain is known as the method of state reduction; in terms of numerical stability, it is the best method available to date. When a chain has a denumerable state space, invariably the transition matrix will have a structure—with repeating rows and/or columns. The three most prominent types of structure are QBD, $GI/M/1$, and $M/G/1$. For further reading about work in this direction, there are the books by Neuts (1981, 1989, 1995) and papers by Grassmann and Heyman (1990) and Neuts (1994). In Kao (1991), an implementation of state reduction is given for dealing with transition matrices of quasi birth and death type. Example 4.5.2 is inspired by Tijms (1986, Example 2.3). Feller's Volume 1 (1968) contains a good introduction to reversed Markov chains. A rich source of reference regarding reversible Markov chains is Kelly (1979). Example 4.6.5 is related to Ross (1983, Example 4.3(c)). Problem 5 is based on Kolesar (1970). Problem 10 is similar to Taylor and Karlin (1994, Exercise 4.5.2). Problem 14 is based on a paper on squash by Broadie and Joneja (1993). Problem 16 was developed in the spirit of Clarke and Disney (1985, 254, Problem 32). Problem 30 is related to Karlin and Taylor (1975, 106–8). Problem 31 is related to the materials presented in Latouche and Ramaswami (1993); there they give an efficient method for finding the matrix R iteratively for a QBD queue. A thorough treatment of issues relating to those considered in Problems 33 and 34 can be found in Neuts (1981). Problems 35 and 36 are based on Ramswami and Latouche (1986).

References

Asmussen, S. 1987. *Applied Probability and Queues.* New York: John Wiley & Sons.

Broadie, M., and D. Joneja. 1993. An Application of Markov Chain Analysis to the Game of Squash. *Decision Sciences* 24(5):1023–35.

Çinlar, E. 1975. *Introduction to Stochastic Processes.* Englewood Cliffs, NJ: Prentice-Hall.

Clarke, A. B., and R. L. Disney. 1985. *Probability and Random Processes: A First Course with Applications.* 2nd ed. New York: John Wiley & Sons.

Denardo, E. 1977. Periods of Connected Networks and Powers of Nonnegative Matrices. *Mathematics of Operations Research* 2(1):23–24.

Feller, W. 1968. *An Introduction to Probability Theory and Its Applications, Vol. 1.* 3rd ed. New York: John Wiley & Sons.

Gaver, D. P., and G. L. Thompson. 1973. *Programming and Probability Models in Operations Research.* Monterey, CA: Brooks/Cole.

Grassmann, W. K., M. I. Taksar, and D. P. Heyman. 1985. Regenerative Analysis and Steady State Distributions for Markov Chains. *Operations Research* 33:1107–16.

Grassmann, W. K., and D. P. Heyman. 1990. Equilibrium Distribution of Block-Structured Markov Chains with Repeating Rows. *Journal of Applied Probability* 27:557–76.

Gross, D., and C. M. Harris. 1985. *Fundamentals of Queueing Theory.* 2nd ed. New York: John Wiley & Sons.

Hannan, E. L. 1976. An Analysis of Different Serving Strategies in Tennis. In *Management Science in Sports. TIMS Studies in Management Sciences.* Edited by R. E. Machol, S. P. Ladany, and D. G. Morrison. 4:125–44.

Heyman, D. P., and M. J. Sobel. 1982. *Stochastic Models in Operations Research, Vol. I.* New York: McGraw-Hill.

Howard, R. A. 1971. *Dynamic Probabilistic Systems, Vol. I.* New York: John Wiley & Sons.

Kao, E. P. C. 1991. Computing the R Matrices in Matrix-Geometric Solutions for a Class of QBD Queues: A Phase Substitution Approach. *Communication in Statistics—Stochastic Models* 7(4):629–43.

Karlin, S., and H. M. Taylor. 1975. *A First Course in Stochastic Processes.* New York: Academic Press.

Kelly, F. P. 1979. *Reversibility and Stochastic Networks.* New York: John Wiley & Sons.

Kemeny, J. G., and J. L. Snell. 1960. *Finite Markov Chains.* Princeton, NJ: Van Nostrand.

Kolesar, P. 1970. A Markovian Model for Hospital Admission Scheduling. *Management Science* 16(6):B384–96.

Latouche, G., and V. Ramaswami. 1993. A Logarithmic Reduction Algorithm for Quasi-Birth-Death Processes. *Journal of Applied Probability* 30:650–74.

Neuts, M. F. 1975. Computational Uses of the Method of Phases in the Theory of Queues. *Computers and Mathematics with Applications* 1:151–66.

Neuts, M. F. 1981. *Matrix-Geometric Solutions in Stochastic Models.* Baltimore: The Johns Hopkins University Press.

Neuts, M. F. 1989. *Structured Stochastic Matrices of* M/G/1 *Type and Their Applications.* New York: Marcel Dekker.

Neuts, M. F. 1995. *Algorithmic Probability.* London: Chapman and Hall.

Neuts, M. F. 1995. Matrix-Analytic Methods in Queueing Theory. From *Advances in Queueing.* Edited by Jewgeni Dshalalow. Boca Raton, Florida: CRC Press.

Ramswami, V., and G. Latouche. 1986. A General Class of Markov Processes with Explicit Matrix-Geometric Solution. *OR Spektrum* 8:209–18.

Ross, S. M. 1983. *Stochastic Processes.* New York: John Wiley & Sons.

Sheskin, T. J. 1985. A Markov Chain Partitioning Algorithm for Computing Steady State Probabilities. *Operations Research* 33(1):228–35.

Taylor, H. M., and S. Karlin. 1994. *An Introduction to Stochastic Modeling.* Revised edition. New York: Academic Press.

Tijms, H. C. 1986. *Stochastic Modelling and Analysis: A Computational Approach.* New York: John Wiley & Sons.

Wolff, R. W. 1989. *Stochastic Modeling and the Theory of Queues.* Englewood Cliffs, NJ: Prentice-Hall.

Appendix

Chapter 4: Section 2

Example 4.2.2   The program **fpss_pro** solves Equation 4.2.1. The first F contains the first-passage time probabilities for $n = 1:500$, where each row of F gives $\{f_{00}^{(n)}, \ldots, f_{03}^{(n)}\}$. Summing the columns of F shows that the first element of the row gives $\sum_{k=0}^{500} f_{00}^{(k)} = 0.8949$. The second F gives the same probabilities for $n = 1:2000$. Now we see that $\sum_{k=0}^{2000} f_{00}^{(k)} = 0.9998$.

```
» P=e422_dat(2);
» P
P =

   0.1353    0.2707    0.2707    0.3233
   0.1353    0.2707    0.2707    0.3233
        0    0.1353    0.2707    0.5940
        0         0    0.1353    0.8647

» [F]=fpss_pro(P,1,500);
» sum(F)

   0.8949    1.0000    1.0000    1.0000

» [F]=fpss_pro(P,1,2000);
» sum(F)

   0.9998    1.0000    1.0000    1.0000

» function [P]=e422_dat(m)
%
```

```
%  Transition matrix for Example 4.2.2
%
p=c4_pos(m); s=3; S=s+1; P=zeros(S,S);
P(1,1:s)=p(1:s); P(1,S)=c4_cpos(m,s); P(2,:)=P(1,:); P(3,2:3)=p(1:2);
P(3,4)=c4_cpos(m,2); P(4,3)=p(1); P(4,4)=c4_cpos(m,1);

function [p]=c4_pos(m)
%
%  It produces a Poisson mass function with mean  m
%  It stops at the term when  CDF > .9999
%
x=exp(-m); s=x; p=[x]; n=1;
while  s < .9999
x=x*m/n; p=[p x]; s=s+x; n=n+1;
end

function [z]=c4_cpos(m,x)
%
%  It finds the complementary Poisson with mean m at x
%
y=c4_pos(m); m=length(y);
if x <= m
z=1-sum(y(1:x));
else
z=1;
end

>> function [F]=fpss_pro(P,m,n)
%
%  For the Markov chain with transition matrix  P,
%  we compute the first passage probabilities from state  m
%  to all states in 1:n  steps.  The output is the matrix F
%  the (k,j)th element of  F  is  fm,j(k),  where  k = 1:n
%
G=P; F=[P(m,:)]; Eo=1-eye(size(P));
for i=2:n
  G=P*(G.*Eo); F=[F; G(m,:)];
end
```

Example 4.2.3 This example shows that $\{T_{1j}\}$ and $\{T_{3j}\}$ are defective random variables for all j. It also shows that $P\{T_{17} < \infty\} = P\{T_{39} < \infty\} = 0.6$ and $P\{T_{19} < \infty\} = P\{T_{37} < \infty\} = 0.4$ because of symmetry.

```
>> [P]=e411_dat;
>> [F]=fpss_pro(P,1,100);
>> sum(F)
ans =

  Columns 1 through 7

    0.5238    0.6977    0.4286    0.7972    0.6667    0.4615    0.6000
  Columns 8 through 9

    0.3158    0.4000

>> [F]=fpss_pro(P,3,100);
>> sum(F)
ans =
```

```
    Columns 1 through 7

      0.4286      0.6977      0.5238      0.4615      0.6667      0.7972      0.4000

    Columns 8 through 9

      0.3158      0.6000
» function [P]=e411_dat
%
%   Data for Example 4.1.1
%
P=zeros(9,9);
P(1,2)=.5; P(1,4)=.5;
P(2,1)=1/3; P(2,3)=1/3; P(2,5)=1/3;
P(3,2)=.5; P(3,6)=.5;
P(4,1)=1/3; P(4,5)=1/3; P(4,7)=1/3;
P(5,2)=1/4; P(5,4)=1/4; P(5,6)=1/4; P(5,8)=1/4;
P(6,3)=1/3; P(6,5)=1/3; P(6,9)=1/3;
P(7,7)=1;
P(8,5)=1/3; P(8,7)=1/3; P(8,9)=1/3;
P(9,9)=1;
```

Example 4.2.6 The program **mc_equca** does the decomposition of state space into communication classes. The program **mc_canon** organizes the transition matrix into a canonical form using the output vector g obtained from **mc_equca**, where g is the list of indices of all closed classes. Example 4.2.6 illustrates the uses of these programs.

```
» [P]=e426_dat;
» function [P]=e426_dat
%
%   Example 4.2.6
%
P=zeros(10,10); P(1,1)=1/2; P(1,3)=1/2; P(2,2)=1/3; P(2,7)=2/3;
P(3,1)=1; P(4,5)=1; P(5,4)=1/3; P(5,5)=1/3; P(5,9)=1/3;
P(6,6)=1; P(7,7)=1/4; P(7,9)=3/4;
P(8,3)=1/4; P(8,4)=1/4; P(8,8)=1/4; P(8,10)=1/4;
P(9,2)=1; P(10,2)=1/3; P(10,5)=1/3; P(10,10)=1/3;

» [g]=mc_equca(P,1);
 This class is closed:   C(1) =        1      3
 This class is closed:   C(2) =        2      7      9
 This class is closed:   C(3) =        1      3

   C(4) =         4      5
   C(5) =         4      5

 This class is closed:   C(6) =        6
 This class is closed:   C(7) =        2      7      9
   C(8) =         8

 This class is closed:   C(9) =        2      7      9
   C(10) =        10

» function [g]=mc_equca(P,po)
%
```

```
%  Find the equivalent classes of a Markov Chain
%  po is the print option switch; po = 1  means output is needed
%  Output  g  is the list of indices of all closed classes.
%
[m,m]=size(P); T=zeros(m,m); i=1;
while  i <= m
       a=[i]; b=zeros(1,m); b(1,i)=1; old=1; new=0;
       while  old ~= new
              old=sum(find(b>0)); [k,n]=size(a);
              if n == 1
                     c=P(a,:);
              else
                     c=sum(P(a,:));
              end
              d=find(c>0);
              [k,n]=size(d); b(1,d)=ones(1,n); new=sum(find(b>0)); a=b;
       end
       T(i,:)=b; i=i+1;
end;
F=T'; C=T&F; C=reshape(C,m,m);
%  Output the result
i=1; g=[];
while  i <= m
       a=find(C(i,:)>0); b=sum(T(i,:)>0); c=sum(C(i,:)>0);
       if  b==c
              x=sum(g==a(1,1));
              if  x ~= 1
              g=[g a];
              end
              if  po == 1
              fprintf(' This class is closed:');
              end
       end
       if  po == 1
       fprintf('  C(%g) = ',i); disp(a);
       end
       i=i+1;
end
end
```

```
» [PC]=mc_canon(P,g);
 list in original state labels:
    1     3     2     7     9     6     4     5     8    10
» PC
PC =
Columns 1 through 7
    0.5000    0.5000         0         0         0         0         0
    1.0000         0         0         0         0         0         0
         0         0    0.3333    0.6667         0         0         0
         0         0         0    0.2500    0.7500         0         0
         0         0    1.0000         0         0         0         0
         0         0         0         0         0    1.0000         0
         0         0         0         0         0         0         0
         0         0         0         0    0.3333         0    0.3333
         0    0.2500         0         0         0         0    0.2500
         0         0    0.3333         0         0         0         0
Columns 8 through 10
         0         0         0
         0         0         0
         0         0         0
         0         0         0
         0         0         0
         0         0         0
         0         0         0
         0         0         0
         0         0         0
    1.0000         0         0
```

```
     0.3333        0         0
          0    0.2500    0.2500
     0.3333        0    0.3333
```

```
» function [PC]=mc_canon(P,g)
%  It converts  P  into its canonical form
%  First do:  [g]=mc_equca(P,0)  to get the  g  vector
[m,m]=size(P); a=1:m; h=[g]; i=1;
while  i <= m
     x=sum(g==i);
     if  x == 0
     h=[h i];
     end
     i=i+1;
end
PC=P(h,h);
fprintf(' list in original state labels: \n');
disp(h);
```

Chapter 4: Section 3

Example 4.3.5 The MATLAB program **mc_perio** finds the period of a Markov chain. The program **mc_canop** states the transition matrix in a canonical form for a periodic Markov chain. The program **mc_limsr** finds the limiting probability vector by state reduction. The program **e435** is the calling program for the example.

```
» e435;
  The original  P  matrix:
P =
```

0	0	0	0.6500	0	0.3500	0
0	0	0	0.9000	0	0.1000	0
0.1000	0.1000	0	0	0	0	0.8000
0	0	0.1200	0	0.8800	0	0
0.1500	0.3500	0	0	0	0	0.5000
0	0	0.7500	0	0.2500	0	0
0	0	0	0.4000	0	0.6000	0

```
period =   3
list of state labels:

f =
```

3	5	1	2	7	4	6

```
  The transition matrix in canonical form
PC =
```

0	0	0.1000	0.1000	0.8000	0	0
0	0	0.1500	0.3500	0.5000	0	0
0	0	0	0	0	0.6500	0.3500
0	0	0	0	0	0.9000	0.1000
0	0	0	0	0	0.4000	0.6000
0.1200	0.8800	0	0	0	0	0
0.7500	0.2500	0	0	0	0	0

```
  the time average  pi
p =
```

0.1329	0.2004	0.0434	0.0834	0.2065	0.1859	0.1474

```
   the  P3  matrix

P3 =
    0.4508    0.5492         0         0         0         0         0
    0.3641    0.6359         0         0         0         0         0
         0         0    0.1330    0.2649    0.6022         0         0
         0         0    0.1408    0.3042    0.5549         0         0
         0         0    0.1251    0.2255    0.6494         0         0
         0         0         0         0         0    0.5960    0.4040
         0         0         0         0         0    0.5094    0.4906

   p1:
p1 =

    0.3987    0.6013

   p2:
p2 =

    0.1301    0.2503    0.6196

   p3:
p3 =

    0.5577    0.4423


pp =

    0.1329    0.2004    0.0434    0.0834    0.2065    0.1859    0.1474

»
```

```
function e435
%
%   Example 4.3.5
%
P=zeros(7,7);
P(1,:)=[0 0 0 .65 0 .35 0];
P(2,:)=[0 0 0 .9 0 .1 0];
P(3,:)=[.1 .1 0 0 0 0 .8];
P(4,:)=[0 0 .12 0 .88 0 0];
P(5,:)=[.15 .35 0 0 0 0 .5];
P(6,:)=[0 0 .75 0 .25 0 0];
P(7,:)=[0 0 0 .4 0 .6 0];
fprintf(' The original  P  matrix: \n'); P
[PC,f]=mc_canop(P);
fprintf('   list of state labels: \n',f); f
fprintf(' The transition matrix in canonical form \n');
PC
[p]=mc_limsr(PC);
fprintf(' the time average  pi  \n'); p
%
% Look at the three closed communication classes
%
P3=PC^3;
fprintf('  the  P3  matrix \n'); P3
Q1=P3(1:2,1:2); Q2=P3(3:5,3:5); Q3=P3(6:7,6:7);
[p1]=mc_limsr(Q1); [p2]=mc_limsr(Q2); [p3]=mc_limsr(Q3);
fprintf('  p1: '); p1
fprintf('  p2: '); p2
fprintf('  p3: '); p3
pp=[p1 p2 p3]; pp=pp/3
```

```
function [q,v]=mc_perdo(P)
%
%       The algorithm is from "Periods of Connected Networks and Powers of
%       Nongegative Matrices," by Eric V. Denardo, Mathematics of Operations
%       Research, Vol. 2, No. 1, 1977, pp. 23-24.
%       Input:  P = Transition matrix of an irreducible Markov chain
%       output: q = the period of the chain
%       *** It uses subroutine MC_GCD ***
%
[n,n]=size(P); v=zeros(1,n); v(1,1)=1; PS=[]; q=0; T=[1]; [m1,m2]=size(T);
while (m2>0)&(q~=1)
        i=T(1,1); T(:,[1])=[]; PS=[PS i]; j=1;
        while j <= n
                if (P(i,j) > 0)
                PUT=[PS T]; k=sum(j==PUT);
                        if (k > 0), b=v(1,i)+1-v(1,j); [q]=mc_gcd(q,b);
                        else T=[T j]; v(1,j)=v(1,i)+1;
                        end
                end
                j=j+1;
        end
        [m1,m2]=size(T);
end
fprintf('\n'); fprintf(' period = %3.0f \n',q); fprintf('\n');

function [gcd]=mc_gcd(a,b)
%
%     Finding the greatest common divisor of "a" and "b"
%     Euclid's algorithm
%
m=min(abs(a),abs(b)); M=max(abs(a),abs(b));
if m == 0, gcd=M; return, end; z = 1;
while z ~= 0
        z=rem(M,m); if z == 0, gcd = m; return, end; M = m; m = z;
end

function [Q,f]=mc_canop(P)
%
% For a periodic irreducible Markov chain with transition probability matrix
% P, this problem put  P  in a canonical form  Q
% *** It uses subroutine PERIOD ***
%
[q,v]=mc_perdo(P); f=[]; k=rem(v,q); h=1;
while  h <= q
        g=find(k==h-1); f=[f,g]; h=h+1;
end
Q=P(f,f);


function [p]=mc_limsr(Q)
%
% Find the stationary probability vector  pi  of an irreducible,
% recurrent discrete-time Markov chain by state reduction
%
% *** this program also works for the continuous-time Markov chain ***
%
% input: the transition probability matrix or the generator  Q
%
[ns,ms]=size(Q);
n=ns;
while n > 1
        n1=n-1;
        s=sum(Q(n,1:n1));
```

```
        Q(1:n1,n)=Q(1:n1,n)/s;
        n2=n1;
        while n2 > 0
             Q(1:n1,n2)=Q(1:n1,n2)+Q(1:n1,n)*Q(n,n2);
             n2=n2-1;
        end
  n=n-1;
end
%
%  backtracking
%
p(1)=1;
j=2;
while j <= ns
        j1=j-1;
        p(j)=sum(p(1:j1).*(Q(1:j1,j))');
        j=j+1;
end
p=p/(sum(p));
```

Chapter 4: Section 4

Example 4.4.1 The program **e441_dat** provides the transition probability matrix for **e441_run** to compute the long-run expected average cost.

```
» e441_run
  list in original state labels:
  Columns 1 through 12
      1     2     3    13    14    15    16     4     5     6     7     8

  Columns 13 through 16
      9    10    11    12

    expected cycle length       =    6.4335
    average cost per unit time  =    0.8126

» function [P]=e441_dat
%
%    Data set for Example 4.4.1 - the cash balance problem
%
alpha=2; beta=10; gamma=6; a= 3; b=4; start=alpha-a; endo=beta+b;
nstate=endo-start+1; P=eye(nstate);
p=[.03 .05 .16 .18 .25 .14 .12 .07];
for i=4:12
  P(i,i-3:i-3+7)=p;
end
» function e441_run
%
%   Do Example 4.4.1 - the Cash Balance Problem
%   PC  is the transition matrix of P in canonical form
%
[P]=e441_dat; [g]=mc_equca(P,0); [PC]=mc_canon(P,g);
[m,n]=size(g); [mp,np]=size(PC); ns=mp-n; i=1:ns;
R=PC(n+i,1:n); Q=PC(n+i,n+i); I=eye(ns);
W=inv(I-Q); w=W(5,:); F=W*R; f=F(5,:);
h1=-1:1; h2=11:14; h3=2:10;
f1=f(1,1:3); f2=f(1,4:7); f3=w; ga=6;
h1=ga-h1; h2=h2-ga; c1=1; c2=0.5; c3=0.05; deno=sum(w);
```

```
t1=c1*sum(h1.*f1); t2=c2*sum(h2.*f2); t3=c3*sum(h3.*f3);
avg=(t1+t2+t3)/deno;
fprintf('  expected cycle length      = %8.4f \n',deno);
fprintf('  average cost per unit time = %8.4f \n',avg);
```

Example 4.4.3 This example solves the problem about serving strategy in tennis.

```
» e443
p1 =
      0.4000

  SS:  0.518    0.207    0.725
  SW:  0.518    0.158    0.676
  WS:  0.396    0.104    0.499
  WS:  0.396    0.079    0.475
  the variance of times in transient states
V =

    1.3495    1.1419
    0.6574    1.3495

» function e443
%
%   Example 4.4.3
%
Q=[.4 .2; .1 .4]; R=[.3 .1; .1 .4]; I=eye(2);
W=inv(I-Q); F=W*R;
s1=[.2 0]; s2=[0 .55];
%
%  strong-strong strategy
%
p0=.4; p1=.4
w1=p0+s1*F(:,1); w2=p1*(p0+s1*F(:,1)); w=w1+w2;
fprintf(' SS: %6.3f    %6.3f    %6.3f \n', w1, w2, w);
%
%  strong-weak strategy
%
p0=.4; p1=.4; p0p=.25;
w1=p0+s1*F(:,1); w2=p1*(p0p+s2*F(:,1)); w=w1+w2;
fprintf(' SW: %6.3f    %6.3f    %6.3f \n', w1, w2, w);
%
%  weak-strong strategy
%
p0p=.25; p1p=.2; p0=.4;
w1=p0p+s2*F(:,1); w2=p1p*(p0+s1*F(:,1)); w=w1+w2;
fprintf(' WS: %6.3f    %6.3f    %6.3f \n', w1, w2, w);
%
%  weak-weak strategy
%
p0p=.25; p1p=.2;
w1=p0p+s2*F(:,1); w2=p1p*(p0p+s2*F(:,1)); w=w1+w2;
fprintf(' WS: %6.3f    %6.3f    %6.3f \n', w1, w2, w);
%
%  Find the variance
%
Wdg=W.*I; K=W*(2*Wdg-I); V=K-W.*W;
fprintf('  the variance of times in transient states '); V
```

Example 4.4.6 The output matrix X from **e446** gives the data for constructing Figure 4.1.3. The first column of X is the vector 1:500 and the second column is the vector containing the corresponding probabilities.

```
» [X]=e446;
  E[T] =      235.9783
Var[T] =       64.5013
» function [X]=e446
%
%  Example 4.4.6 - the coupon collection example
%
m=52;
x=ones(1,52); x=cumsum(x); pi=(m-(x-1))/m;
qi=1-pi; Q=diag(qi); p1=pi; pi(52)=[];   A=diag(pi,1);
Q=Q+A; alpha=zeros(1,52); alpha(1,1)=1;
r=zeros(52,1); r(m,1)=p1(m); I=eye(m);
Q1=inv(I-Q); E1=sum(alpha*Q1);      % 1st moment
EF2=2*sum(alpha*Q*Q1*Q1);           % 2nd factorial moment
E2=EF2+E1;                          % by (1.2.4)
Var=E2-E1*E1; Sdv=sqrt(Var);
fprintf('  E[T] = %12.4f \n',E1);
fprintf('Var[T] = %12.4f  \n',Sdv);
pk=[]; S=alpha*I;
for k=1:500
   x=S*r; S=S*Q; pk=[pk x];
end
X=zeros(500,2); x=ones(1,500); x=cumsum(x); X(:,1)=x';
X(:,2)=pk';
```

Chapter 4: Section 5

Example 4.5.1 The output matrix z contains the data for producing Figure 4.14. The program **e451_run** calls the program **e413_dat** to obtain the transition matrix for the specified values of (s, S) and the mean demand. The program **c4_pos** computes the Poisson mass function for the given mean.

```
» [z]=e451_run;
» function [z]=e451_run
%
%  Example 4.5.1 The (s, S) inventory model with costs
%
m=5; s=4; z=[];
for S=5:20
[P]=e413_dat(m,s,S); [a]=c4_pos(m); S1=S+1; s1=s+1;
[n1,n2]=size(a); n=n2-1; L=zeros(1,S1);
if n2 < S
   a1=zeros(1,S-n2); a=[a a1];
end
%
%  Find  Li
%
for i=0:S
   k=i:n; d=k-i;
   L(1,i+1)=sum(d.*a(1,i+1:n2));
end
%
K=15; c=4; p=2; h=1; g=zeros(1,S1);
```

```
    ia=0:s-1; g(1:s)=K+c*(S-ia)+h*ia+p*L(1,1:s);
    ia=s:S; g(1,s1:S1)=h*ia+p*L(1,s1:S1);
    [pi]=mc_limsr(P); cost=g*pi';
    za=[S cost]; z=[z; za];
    end


» function [P]=e413_dat(m,s,S)
    %
    %  Find the transition matrix for the (s,S) inventory example
    %  m = mean of the Poisson demand distribution
    %
    S1=S+1; P=zeros(S1,S1); [p]=c4_pos(m);
    [n1,n2]=size(p);
    if n2 < S
      p1=zeros(1,S-n2); p=[p p1];
    end
    %  Part 1
    P(1:s,1)=c4_cpos(m,S)*ones(s,1);
    %  Part 2
    j=2;
    for j=2:S1
        P(1:s,j)=p(S1-j+1)*ones(s,1);
    end
    %  Part 3
    i=s;
    for i=s:S
        P(i+1,1)=c4_cpos(m,i);
    end
    %  Part 4
    i=s;
    for i=s:S
        P(i+1,2:i+1)=p(i-(1:i)+1);
    end


» function [p]=c4_pos(m)
    %
    %  It produces a Poisson mass function with mean  m
    %  It stops at the term when  CDF > .9999
    %
    x=exp(-m); s=x; p=[x]; n=1;
    while  s < .9999
    x=x*m/n; p=[p x]; s=s+x; n=n+1;
    end


function [z]=c4_cpos(m,x)
    %
    %  Find the complementary Poisson with mean  m  at  x
    %
    y=c4_pos(m); [n,m]=size(y);
    if x <= m
    z=1-sum(y(1:x));
    else
    z=1;
    end;
```

5

Continuous-Time Markov Chains

Tips for Chapter 5

- Section 5.8 can be covered without covering Sections 5.5–5.7. Section 5.8 introduces many important queueing networks used in modeling manufacturing, computer, and telecommunication systems.

- The following examples can be used as supplementary reading: Examples 5.2.4, 5.4.2, 5.5.1, 5.5.5, 5.5.6, 5.6.2, 5.6.3, and 5.6.4.

- Phase-type distributions are widely used to facilitate stochastic modeling. This is seen in the large amount of related work appearing in recent literature. Section 5.5 provides a short introduction to the subject.

- The idea of uniformization (Section 5.6) is conceptually important and computationally useful. The introduction of a continuous-time Markov jump process in this section sets the stage for treating continuous-time Markov reward processes involving chains that allow virtual transitions in the next section.

- Section 5.7 on continuous-time Markov reward processes provides some background for the study of continuous-time Markov decision problems—a subject in stochastic optimization typically covered in a course such as dynamic programming.

- The problems in this chapter are comparatively easy (relative to those given in the earlier chapters, particularly Chapter 3). However, some problems involving computer solutions require a little clever use of programming skills.

5.0 Overview

Continuous-time Markov chains are frequently used in modeling manufacturing systems, computer networks, and communication systems. If we skim through the performance evaluation literature in computer science or automated manufacturing systems, we will find applications of continuous-time Markov chains omnipresent. Continuous-time Markov chains also have applications in physical, social, and biological sciences. The basic idea of a continuous-time Markov chain is quite simple. If we are in a state i, our sojourn in the state will follow an exponential distribution with parameter v_i. At the end of each sojourn, the process makes a transition to a different state. The probability that a transition will be made to state j is the ratio of the transition rate from i to j and the total rate of transition out of state i, v_i. While the continuous-time Markov chain paradigm allows modeling opportunities with a high level of sophistication, the pivotal idea centers around the notion of competing exponential distributions and a clever definition of state space. Many times, extending the state definition to multiple dimensions enables the Markov property to prevail and renders a problem solvable.

In modeling a problem by a continuous-time Markov chain, we start with a state definition and then construct the infinitesimal generator Q of the chain containing the transition rates. The time-dependent probabilities that the process is in each state at time t are characterized by the Kolmogorov equation presented in Section 5.2 whose solution is actually a matrix exponential of Qt. The related limiting probabilities are discussed in Section 5.3. For state i, if $v_i = 0$, once the process enters the state it will stay there forever. Such a state is called an absorbing state. When a continuous-time Markov chain has only one absorbing state, the time-to-absorption distribution is called a phase-type distribution by Neuts (1981). Phase-type distributions have many attractive properties and are used extensively in computational probability. A brief survey of the subject is given in Section 5.5. The idea of uniformization presented in Section 5.6 provides an alternative for computing time-dependent state probabilities. A continuous-time Markov reward process is straightforward if state-dependent reward rates are the only rewards present; however, complications arise when reward structures assume more complex forms. In Section 5.7, we study the related reward accounting. The chapter is concluded with reversible continuous-time Markov chains. This eventually leads to open Jackson queueing networks and Gordon-Newell's closed queueing networks presented in Section 5.8.

5.1 Introduction

Consider a continuous-time stochastic process $X = \{X(t),\ t \geq 0\}$ with state space $S = \{0, 1, 2, \ldots\}$.

> The process X is a continuous-time Markov chain if, for $i, j \in S$,
>
> $$P\{X(t+s) = j | X(s) = i,\ X(u) = x(u),\ 0 \leq u < s\} = P\{X(t+s) = j | X(s) = i\},$$
>
> for all $s \geq 0,\ t \geq 0$, and $x(u),\ 0 \leq u \leq s$.

X is a time-homogeneous continuous-time Markov chain if

$$P\{X(t+s) = j | X(s) = i\} = P_{ij}(t) \quad \text{independent of } s.$$

We note that $P_{ij}(t)$ is called a transition function. It plays a role analogous to that of $P_{ij}^{(n)}$ in a discrete-time Markov chain. In this chapter, we will restrict our attention to time-homogeneous continuous-time Markov chains. In Example 2.3.8 we have already looked at a continuous-time Markov chain that is not time homogeneous.

The transition functions $\{P_{ij}(t)\}$ have the following properties: for all $i, j \in S$, $t \geq 0$, and $s \geq 0$,

$$P_{ij}(0) = \delta_{ij}, \qquad P_{ij}(t) \geq 0, \qquad \sum_{j \in S} P_{ij}(t) = 1,$$

$$P_{ij}(t+s) = \sum_{k \in S} P_{ik}(t) P_{kj}(s), \qquad (5.1.1)$$

where $\delta_{ij} = 1$ if $i = j$ and 0 otherwise. We note that Equation 5.1.1 is the continuous analog of the Chapman-Kolmogorov equation (Equation 4.1.3) for the discrete-time Markov chain. In matrix notation, it is simply

$$P(t+s) = P(t)P(s) \qquad t \geq 0, \ s \geq 0,$$

where the matrix function $P(t) = \{P_{ij}(t)\}$. Let row vector $s(t) = \{s_j(t)\}$, where $s_j(t) = P\{X(t) = j\}$. Then $s(0)$ is the starting state probability vector of the continuous-time Markov chain. Given $s(0)$ and $\{P(t)\}$, the state probability vector can be found from

$$s(t) = s(0)P(t). \qquad (5.1.2)$$

The previous equation is the continuous analog of Equation 4.1.4. In the next section, we will discuss the computation of $\{P(t)\}$.

In Figure 5.1, we depict a sample path of a four-state continuous-time Markov chain. There we see that the sample path is drawn to show that it is defined to be right continuous, that is,

$$X(t) = \lim_{s \to t^+} X(s)$$

(the solid dot indicates the value of $X(t)$ at the epoch of state change). In the figure, we show that the process starts from state 1 at time 0, stays there for τ_1 amount of time, makes an instantaneous state change to state 3 at time T_1, stays

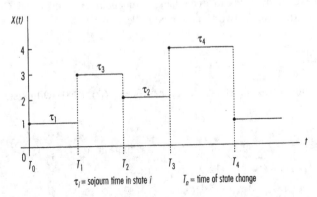

τ_i = sojourn time in state i \qquad T_n = time of state change

FIGURE
5.1   A sample path of a continuous-time Markov chain.

there for τ_3 amount of time, makes an instantaneous state change to state 2 at time $T_2 = \tau_1 + \tau_3$, and so on. We now examine a few important properties about such a sample path.

Since τ_i denotes the sojourn time of X in state i before making a transition to a *different* state, the Markovian property of the process implies that, for all $s \geq 0$ and $t \geq 0$,

$$P\{\tau_i > s + t | \tau_i > s\} = P\{\tau_i > t\}.$$

In other words, the random variable τ_i possesses the memoryless property. Letting $f(t) = P\{\tau_i > t\}$, the preceding can be written as $f(t + s) = f(t)f(s)$. This implies that τ_i follows the exponential distribution, say with parameter v_i. Let $Y(T_n)$ denote the state of the process immediately after the nth state change. Then we have

$$P\{\tau_i > t | Y(T_n) = i\} = e^{-v_i t} \qquad t \geq 0; \; n = 0, 1, \dots. \qquad (5.1.3)$$

By the memoryless property of the exponential distribution, we also observe that when $X(t) = i$, the rate of leaving the state is the constant v_i. This observation enables us to write

$$P_{ii}(h) = P\{X(t + h) = i | X(t) = i\} = 1 - v_i h + o(h).$$

We write the preceding as $v_i h = 1 - P_{ii}(h) + o(h)$. Dividing both sides by h and letting $h \to 0$ yields

$$v_i = \lim_{h \to 0} \frac{1 - P_{ii}(h)}{h}.$$

We call state i *absorbing* if $v_i = 0$, *stable* if $0 < v_i < \infty$, and *instantaneous* if $v_i = \infty$. In this text we restrict our attention to continuous-time Markov chains whose states are of the first two types. This eliminates the possibility of allowing a process to make an infinite number of transitions in a finite interval and alleviates many concomitant mathematical complications. When the process enters an absorbing state i, it stays there forever. When it enters a stable state i, with probability 1 it stays there for a finite time τ_i, where τ_i follows an exponential distribution with parameter v_i. We call v_i the *transition rate* associated with state i.

Let $H = \{X(t), 0 \leq t \leq T_n\}$ denote the history of the process up to time T_n. By invoking the strong Markov property, it can be shown that

$$P\{Y(T_{n+1}) = j, T_{n+1} - T_n > t | Y(T_n) = i, H\}$$

$$= P\{Y(T_{n+1}) = j, \tau_i > t | Y(T_n) = i\} = p_{ij} e^{-v_i t}, \qquad (5.1.4)$$

where $\{p_{ij}\}$ are elements of a stochastic matrix. We see that Equation 5.1.4 gives the joint distribution of random variables $Y(T_{n+1})$ and $(T_{n+1} - T_n)$ given $Y(T_n)$. It says that the two are conditionally independent given $Y(T_n)$. This is an important structural property of a continuous-time Markov chain. Also, if we sum Equation 5.1.4 over all j, then it reduces to Equation 5.1.3. Letting $t = 0$ in Equation 5.1.4, we see that p_{ij} gives the probability that given the process is in state i, the next state to which it will move is j. Moreover, we note that $p_{ii} = 0$ for each i. Let $P = \{p_{ij}\}$ and $Y_n = Y(T_n)$. Then P is the transition probability matrix of the Markov chain embedded at the state-change epochs. For $i \neq j$, we define q_{ij} as the *transition rate* of moving from state i to state j or, formally,

$$q_{ij} = \lim_{h \to 0} \frac{P_{ij}(h)}{h}.$$

Since $P_{ij}(h) = (v_i p_{ij})h + o(h)$ if $i \neq j$, we have

$$q_{ij} = v_i p_{ij} \qquad i \neq j \qquad (5.1.5)$$

and $p_{ij} = q_{ij}/v_i$, $i \neq j$. Define $q_{ii} = -v_i$ for all i and form a matrix $Q = \{q_{ij}\}$.

The matrix

$$Q = \begin{matrix} 0 \\ 1 \\ 2 \\ \\ \end{matrix} \begin{bmatrix} -v_0 & q_{01} & q_{02} & \cdot & \cdot \\ q_{10} & -v_1 & q_{12} & \cdot & \cdot \\ q_{20} & q_{21} & -v_2 & \cdot & \cdot \\ \cdot & \cdot & & \cdot & \cdot \\ \cdot & \cdot & \cdot & \cdot & \cdot \end{bmatrix}$$

is called the *infinitesimal generator* of a continuous-time Markov chain.

Clearly we have $Qe = 0$; the row sums are zeros. The generator Q and the starting-state probability vector $s(0)$ completely characterize a continuous-time Markov chain. As in the case of a discrete-time Markov chain, sometimes it is useful to construct a transition diagram for a continuous-time Markov chain. In such a diagram, nodes correspond to states, and arcs again represent transitions—except now with their respective transition rates shown. When a state has only outgoing arcs, it means that once the process leaves the state it will not return back to it. Such a state is called a transient state. Transient states in a continuous-time Markov chain can also be identified by analyzing the embedded discrete-time Markov chain.

EXAMPLE 5.1.1 **Birth and Death Processes** A continuous-time Markov chain is called a birth and death process if state space $S = \{0, 1, \dots\}$ and $q_{ij} = 0$ if $|i - j| > 1$. When the process is in state i, the transition rates are

$$q_{i,i+1} = \lambda_i \qquad i = 0, 1, \dots$$
$$q_{i,i-1} = \mu_i \qquad i = 1, 2, \dots$$

and $q_{ij} = 0$, otherwise. The parameters $\{\lambda_i\}$ are called the birth rates and $\{\mu_i\}$ the death rates. Note that $v_i = \lambda_i + \mu_i$ and

$$p_{i,i+1} = \frac{\lambda_i}{v_i} \qquad \text{and} \qquad p_{i,i-1} = \frac{\mu_i}{v_i},$$

and $p_{ij} = 0$ otherwise. When we are in state i, we let X denote the next birth time, Y the next death time, and $Z = \min\{X, Y\}$ the next event time. Then the use of a continuous-time Markov chain model implies that X, Y, and Z are exponential random variables with respective parameters λ_i, μ_i, and $v_i = \lambda_i + \mu_i$ (see the competing exponential random variables in Example 1.3.2). The transition diagram of the birth and death process is shown in Figure 5.2. ∎

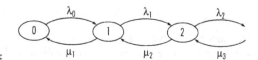

FIGURE
5.2 The transition diagram of a birth and death process.

EXAMPLE **A Barber Shop** Consider a barber shop with two barbers and two waiting chairs.
5.1.2 Customers arrive at a rate of five per hour. Each barber serves customers at a rate
of two per hour. Customers arriving to a fully occupied shop leave without being
served. When the shop opens at 8 A.M., there are already two waiting customers.
We assume that arrivals are Poisson and service times are exponential and the
arrival process is independent of service times. Hence the arrival rate is five per
hour and the service rate is two per hour for each barber.

Let $X(t)$ represent the state of the system at time t. Since arrivals or departures
can only occur one at a time, the state change can only be made to a neighboring
state. In other words, if $X(t) = i$, then the next state will either be $i + 1$ or $i - 1$. The
system $\{X(t), t \geq 0\}$ can then be modeled as a birth and death process with state
space $S = \{0, 1, 2, 3, 4\}$. The arrival rate is a constant for all $t \geq 0$. This implies
that the birth rate is 5 for all i. When there is only one customer in the shop, the
service rate is 2; when two or more are in the shop, the service rate is 4. Another
way to see the latter is by letting X_i be the service completion time of the customer served by barber i—noting that X_i follows an exponential distribution with
parameter 2 for each i and the next service completion time $Z = \min\{X_1, X_2\}$ follows an exponential distribution with parameter 4. The transition diagram is given
in Figure 5.3.

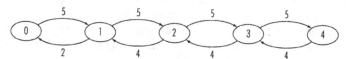

FIGURE
5.3 The transition diagram for the barber shop example.

The corresponding infinitesimal generator is given by

$$
Q = \begin{bmatrix}
-5 & 5 & & & \\
2 & -7 & 5 & & \\
& 4 & -9 & 5 & \\
& & 4 & -9 & 5 \\
& & & 4 & -4
\end{bmatrix}.
$$

The initial condition implies that the starting state vector $s(0) = [0\ 0\ 1\ 0\ 0]$. ■

EXAMPLE **An M/M/s Queue** In an $M/M/s$ queue, we have s identical servers in a service sys-
5.1.3 tem with an unlimited number of waiting spaces. Arrivals follow a Poisson
process with rate λ and each service time follows an exponential distribution with
rate μ. The arrival process and service times are independent. When an arrival
occurs and a server is idle, the arriving customer receives service immediately;
otherwise the customer stays in the waiting line. At a service completion, the customer at the head of the waiting line, if any, enters service. Let $X(t)$ denote the
number of customers in the system at time t. Following the argument similar to
that used in Example 5.1.2, we find that $\{X(t), t \ge 0\}$ is a birth and death process
with state space $S = \{0, 1, \ldots\}$ and transition rates

$$\lambda_n = \lambda \qquad n = 0, 1, \ldots$$

$$\mu_n = \begin{cases} n\mu & 1 \le n \le s \\ s\mu & n > s. \end{cases} \blacksquare$$

EXAMPLE **The Poisson Process** Consider a Poisson process with arrival rate λ. Let $X(t)$ denote
5.1.4 the number of arrivals in $(0, t]$. Then $\{X(t), t \ge 0\}$ is a birth and death process with
state space $S = \{0, 1, 2, \ldots\}$. For this process, all death rates are zeros and the
birth rate is the constant λ for all states. $\blacksquare$

EXAMPLE **Response Areas for Two Emergency Units** Consider two urban emergency service units
5.1.5 (such as fire engines or ambulances) that cooperate in responding to alarms or
calls from the public in a city. Suppose that the city is divided into two service
regions A and B. Service unit 1 is designated to serve region A and service unit 2
is designated to serve region B. A call arriving from either region when exactly
one service unit is available is served by the available unit. When both service
units are busy, an arriving call will be lost.

Assume that arrivals of emergency calls from region j ($j = A$ or B) follow a
Poisson process with rate λ_j. Service times of these calls are independent exponential random variables with rates μ_{ij}, where μ_{ij} denotes the service rate when a
region j call is served by service unit i ($i = 1, 2$).

Let $X_i(t)$ denote the status of service unit i at time t, where $X_i(t) = 0$ if the unit
is idle, 1 if it is serving a call from region A, and 2 if it is serving a call from
region B. If we define the state of the system at time t by a bivariate stochastic
process $X(t) = (X_1(t), X_2(t))$, then the process $\{X(t)|t \ge 0\}$ is a continuous-time
Markov chain. It has state space $S = \{(0, 0), (0, 1), (0, 2), (1, 0), (1, 1), (1, 2),
(2, 0), (2, 1), (2, 2)\}$ and infinitesimal generator

$$Q = \begin{array}{c} \\ (0,0) \\ (0,1) \\ (0,2) \\ (1,0) \\ (1,1) \\ (1,2) \\ (2,0) \\ (2,1) \\ (2,2) \end{array} \begin{array}{c} \begin{matrix} (0,0) & (0,1) & (0,2) & (1,0) & (1,1) & (1,2) & (2,0) & (2,1) & (2,2) \end{matrix} \\ \left[\begin{matrix} * & & & \lambda_2 & \lambda_1 & & & & \\ \mu_{21} & * & & & \lambda_1 & & & \lambda_2 & \\ \mu_{22} & & * & & & \lambda_1 & & & \lambda_2 \\ \mu_{11} & & & * & \lambda_1 & \lambda_2 & & & \\ & \mu_{11} & & \mu_{21} & * & & & & \\ & & \mu_{11} & \mu_{22} & & * & & & \\ \mu_{12} & & & & & & * & \lambda_1 & \lambda_2 \\ & \mu_{12} & & & & & \mu_{21} & * & \\ & & \mu_{12} & & & & \mu_{22} & & * \end{matrix} \right] \end{array},$$

where the diagonal elements are not shown but known to assume values so that the row sums are zeros. To see how the entries in Q are found, we use state $(0, 2)$ as an example. There are three possible events to consider: a call from Region A with rate λ_1, a call from Region B with rate λ_2, and a service completion at Region B using service unit 2 at a rate μ_{22}. The first occurrence of these three possible events will result in a transition to the destination state—$(1, 2)$, $(2, 2)$, or $(0, 0)$, respectively. Other entries can be obtained similarly. We note that when both service units are busy, all arriving calls are lost. Hence for states (i, j) with both $i > 0$ and $j > 0$, new arrivals will not be served. Consequently, λ_1 and λ_2 are absent from the corresponding rows. ■

5.2   The Kolmogorov Differential Equations

In studying the time-dependent behaviors of a continuous-time Markov chain, we need to know about the transition functions $\{P_{ij}(t)\}$ for any finite t of interest. To accomplish this, we first derive the Kolmogorov differential equations characterizing the transition functions and then examine the solution of these equations.

Using the Chapman-Kolmogorov equation (Equation 5.1.1), we write

$$P_{ij}(t+h) = \sum_{k \in S} P_{ik}(h)P_{kj}(t) = \sum_{k \in S, k \neq i} P_{ik}(h)P_{kj}(t) + P_{ii}(h)P_{ij}(t).$$

Subtracting $P_{ij}(t)$ from both sides yields

$$P_{ij}(t+h) - P_{ij}(t) = \sum_{k \in S, k \neq i} P_{ik}(h)P_{kj}(t) - (1 - P_{ii}(h))P_{ij}(t).$$

Dividing the preceding expression by h and letting $h \to 0$, we obtain

$$\lim_{h \to 0} \frac{P_{ij}(t+h) - P_{ij}(t)}{h} = \lim_{h \to 0} \frac{\sum_{k \in S, k \neq i} P_{ik}(h)P_{kj}(t)}{h} - \lim_{h \to 0}\left(\frac{1 - P_{ii}(h)}{h}\right)P_{ij}(t)$$

$$= \sum_{k \in S, k \neq i} \lim_{h \to 0} \frac{P_{ik}(h)}{h} P_{kj}(t) - \lim_{h \to 0}\left(\frac{1 - P_{ii}(h)}{h}\right)P_{ij}(t).$$

Using the defining relations of q_{ij} and v_i given in the last section, we find the following systems of the Kolmogorov differential equations

$$P'_{ij}(t) = \sum_{k \in S, k \neq i} q_{ik} P_{kj}(t) - v_i P_{ij}(t) \qquad t \geq 0, \ i, j \in S. \qquad (5.2.1)$$

The initial condition is given by $P_{ij}(0) = \delta_{ij}$, for all i, j. Traditionally, Equation 5.2.1 is called the *backward* Kolmogorov equation (perhaps because the segment involving h appears at the initial part of $(0, t)$).

In matrix notations, the backward Kolmogorov equations can be stated succinctly as

$$\frac{d}{dt} P(t) = Q P(t) \qquad t \geq 0$$

$$P(0) = I. \qquad (5.2.2)$$

Similarly we can derive the forward Kolmogorov equations as follows:

$$P'_{ij}(t) = \sum_{k \in S, k \neq j} P_{ik}(t) q_{kj} - P_{ij}(t) v_j \qquad t \geq 0, \ i, j \in S,$$

$$P_{ij}(0) = \delta_{ij} \qquad\qquad i, j \in S. \qquad (5.2.3)$$

In matrix notations, the forward Kolmogorov equations read

$$\frac{d}{dt} P(t) = P(t) Q \qquad t \geq 0$$

$$P(0) = I. \qquad (5.2.4)$$

In applications, the forward version occupies a more prominent position.

When the state space is finite, we can use several approaches to solve the preceding system of differential equations to find the transition function $P_{ij}(t)$. One way is by numerical methods for solution of differential equations. Another is by inverting the corresponding Laplace transforms either algebraically or numerically. The third approach is the use of the uniformization method described in Example 5.6.2. When the state space is infinite, exploiting the structure of the infinitesimal generator may enable us to derive closed-form solutions of transition functions. We will demonstrate the use of all these approaches in the subsequent sections.

Consider a continuous-time Markov chain with a finite state space S. Define the Laplace transform of the transition function $P_{ij}(t)$ as

$$P^e_{ij}(s) = \int_0^\infty e^{-st} P_{ij}(t) dt.$$

Define the matrix function $P^e(s) = \{P_{ij}^e(s)\}$. We take the Laplace transform of Equation 5.2.2 and obtain

$$sP^e(s) - P(0) = QP^e(s), \quad sP^e(s) - I = QP^e(s), \quad \text{or} \quad [sI - Q]P^e(s) = I.$$

It can be shown that the inverse of $[sI - Q]$ exists if $\text{Re}(s) \geq 0$. Hence we have $P^e(s) = \{sI - Q\}^{-1}$. Inverting this Laplace transform, we find

$$P(t) = e^{Qt} = \sum_{n=0}^{\infty} \frac{(Qt)^n}{n!} \qquad t \geq 0. \tag{5.2.5}$$

The term e^{Qt} is called a matrix exponential. MATLAB has a routine **expm** for computing the matrix for a finite t.

<hr/>

EXAMPLE
5.2.1

A Two-State Continuous-Time Markov Chain   Consider a two-state continuous-time Markov chain with state space $S = \{0, 1\}$ and infinitesimal generator

$$Q = \begin{bmatrix} -\lambda & \lambda \\ \mu & -\mu \end{bmatrix}.$$

Using Equation 5.1.2, we state the matrix Laplace transform

$$P^e(s) = [sI - Q]^{-1} = \begin{bmatrix} s+\lambda & -\lambda \\ -\mu & s+u \end{bmatrix}^{-1} = \frac{1}{s(s+\lambda+\mu)} \begin{bmatrix} s+\mu & \lambda \\ \mu & s+\lambda \end{bmatrix}.$$

Partial-fraction expansion gives

$$P^e(s) = \frac{1}{s} \begin{bmatrix} \dfrac{\mu}{\lambda+\mu} & \dfrac{\lambda}{\lambda+\mu} \\[2ex] \dfrac{\mu}{\lambda+\mu} & \dfrac{\lambda}{\lambda+\mu} \end{bmatrix} + \frac{1}{s+\lambda+\mu} \begin{bmatrix} \dfrac{\lambda}{\lambda+\mu} & -\dfrac{\lambda}{\lambda+\mu} \\[2ex] -\dfrac{\mu}{\lambda+\mu} & \dfrac{\mu}{\lambda+\mu} \end{bmatrix}.$$

Inverting the preceding transform, we find the matrix exponential e^{Qt} in a closed form:

$$P(t) = \begin{bmatrix} \dfrac{\mu}{\lambda+\mu} & \dfrac{\lambda}{\lambda+\mu} \\[2ex] \dfrac{\mu}{\lambda+\mu} & \dfrac{\lambda}{\lambda+\mu} \end{bmatrix} + e^{-(\lambda+\mu)t} \begin{bmatrix} \dfrac{\lambda}{\lambda+\mu} & -\dfrac{\lambda}{\lambda+\mu} \\[2ex] -\dfrac{\mu}{\lambda+\mu} & \dfrac{\mu}{\lambda+\mu} \end{bmatrix}.$$

The previous equation also shows that

$$\lim_{t \to \infty} P(t) = \begin{bmatrix} \dfrac{\mu}{\lambda+\mu} & \dfrac{\lambda}{\lambda+\mu} \\[2ex] \dfrac{\mu}{\lambda+\mu} & \dfrac{\lambda}{\lambda+\mu} \end{bmatrix}.$$

The limiting behavior of a continuous-time Markov chain will be discussed in detail in the next section. ∎

EXAMPLE **The Barber Shop Revisited** Suppose that we are interested in the state of the barber
5.2.2 shop in the first two hours of operation given that there are two waiting customers
when the shop opens at 8 A.M. In the Appendix, we compute the state probability
vector $s(t)$ for $0 \leq t \leq 2$. In the MATLAB program, we use the routine **expm** to
compute the matrix exponential e^{Qt}. The results are shown in Figure 5.4. The figure indicates that the system's state probabilities reach a steady state after one
hour of operation. The figure also shows that the probability that the shop is full
at 10 A.M. is about 0.32. Other useful time-dependent information can be obtained
similarly. ∎

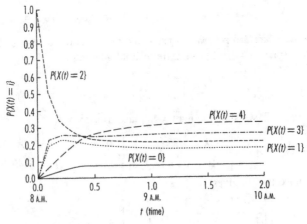

FIGURE
5.4 The time-dependent behaviors of the barber shop.

EXAMPLE **Response Areas for Two Emergency Units Revisited** Suppose that we are interested in the
5.2.3 use of service unit 1 and the percent of incoming calls lost because of capacity
limitation under the assumption that $\lambda_1 = 0.1$, $\lambda_2 = 0.125$, $\mu_{11} = 0.5$, $\mu_{12} = 0.25$,
$\mu_{21} = 0.2$, and $\mu_{22} = 0.4$ (all in rate per hour). The data imply that serving a
region by a nondesignated unit takes a longer time. For example, it takes an
average of $1/\mu_{11} = 2$ hours to serve a call from Region A by unit 1 whereas it
takes an average of $1/\mu_{21} = 5$ hours to serve a call from Region B by the same
unit. Let the two-dimensional state probability $s_{i,j}(t) = P\{X_1(t) = i, X_2(t) = j\}$.
The use of unit 1 can be expressed in terms of the following time-dependent
probabilities: $h_i(t) = \sum_j s_{i,j}(t)$, where $i = 0$ denotes that unit 1 is idle at time
t, 1 denotes that the unit is serving Region A at time t, and 2 denotes that it is serving Region B at time t. Similarly, the probability $g(t)$ that at time t an incoming
call will be lost is given by

$$g(t) = \sum_{i=1}^{2} \sum_{j=1}^{2} s_{i,j}(t).$$

Assume that at 8 A.M. a given day all units are idle. We present the results found in the Appendix for a twenty-four-hour period in Figure 5.5. We see that in about nine hours, the time-dependent probabilities reach their respective steady state values. The probability that an arriving call will be lost because of the capacity constraint approaches to a steady state value of about 10 percent. ∎

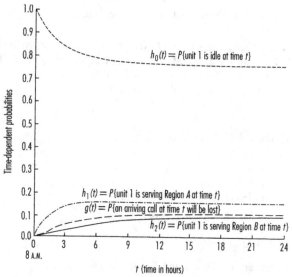

FIGURE
5.5 A numerical example for the two emergency units.

When the state space is not finite yet the infinitesimal generator has a special structure, we can often find a closed-form expression for the transition function by an algebraic method or transform approach. This is illustrated in Example 5.2.4.

EXAMPLE
5.2.4* **The Yule Process** In a birth and death process, if all death rates are zero then the corresponding process is called a *pure birth* process. One version of a Yule process is a pure birth process with state space $S = \{1, 2, \ldots\}$, birth rates $\lambda_j = j\lambda$, $j = 1, 2, \ldots$, and $X(0) = 1$, where $X(t)$ denotes the population size at time t. Another version is one with state space $S = \{N, N+1, \ldots\}$, birth rates $\lambda_j = j\lambda$, $j = N, N+1, \ldots$, and $X(0) = N$. Both versions assume that each member in a population has a probability $\lambda h + o(h)$ of giving birth to a new member in an interval of length h and there are no interactions among members of the population. To gain some insight about the identity $\lambda_j = j\lambda$, we consider the case in which $X(t) = j$. Let T_k denote the next birth time of individual k, $k = 1, \ldots, j$. Then $T = \min\{T_1, \ldots, T_j\}$ is the next birth time (so the process will move to state $j+1$

in the next transition). Clearly, T is exponential with parameter $j\lambda$. Another way to look at this is by noting

$$P[X(t+h) - X(t) = 1|X(t) = j] = \binom{j}{1} [\lambda h + o(h)] [1 - \lambda h + o(h)]^{j-1}$$

$$= j\lambda h + o(h).$$

Hence, $\lambda_j = j\lambda$. To derive the transition function, we start with the version $X(0) = 1$.

Since $X(0) = 1$, we are interested in $\{P_{1j}(t)\}, j = 1, 2, \ldots$, namely, the first row of the matrix $P(t)$. Using the forward Kolmogorov equations $P'(t) = P(t)Q$, we write

$$[P'_{11}(t), P'_{12}(t), \ldots] = [P_{11}(t), P_{12}(t), \ldots] \begin{bmatrix} -\lambda & \lambda & & & \\ & -2\lambda & 2\lambda & & \\ & & -3\lambda & 3\lambda & \\ & & & \ddots & \ddots \\ & & & & \ddots \end{bmatrix}.$$

Writing the preceding in component form, we have

$$P'_{1j}(t) = (j-1)\lambda P_{1,j-1}(t) - j\lambda P_{1j}(t) \qquad j = 2, 3, \ldots$$

$$P'_{11}(t) = -\lambda P_{11}(t).$$

The initial conditions are $P_{11}(0) = 1$ and $P_{1j}(0) = 0, j = 2, 3, \ldots$.

Define Laplace transform $P^e_{1j}(s) = \int_0^\infty e^{-st} P_{1j}(t)dt$. Then

$$sP^e_{11}(s) - 1 = -\lambda P^e_{11}(s)$$

$$P^e_{11}(s) = \frac{1}{s+\lambda}.$$

Inverting the above yields $P_{11}(t) = e^{-\lambda t}, t \geq 0$. Now for $j = 2, 3, \ldots$, we find

$$sP^e_{1j}(s) + j\lambda P^e_{1j}(s) = \lambda(j-1)P^e_{1,j-1}(s)$$

$$(s+j\lambda)P^e_{1j}(s) = \lambda(j-1)P^e_{1,j-1}(s)$$

$$P^e_{1j}(s) = \frac{(j-1)\lambda}{(s+j\lambda)} P^e_{1,j-1}(s) \qquad j = 2, 3, \ldots$$

For $j = 2$, a partial fraction expansion gives

$$P^e_{12}(s) = \frac{\lambda}{s+2\lambda} P^e_{11}(s) = \frac{\lambda}{(s+2\lambda)} \frac{1}{(s+\lambda)} = \frac{-1}{(s+2\lambda)} + \frac{1}{(s+\lambda)}$$

$$P_{12}(t) = e^{-\lambda t} - e^{-2\lambda t} = e^{-\lambda t}(1 - e^{-\lambda t}) = e^{-\lambda t}(1 - e^{-\lambda t})^{2-1}.$$

For $j = 3$, we similarly obtain

$$P^e_{13}(s) = \frac{2\lambda}{s+3\lambda} P^e_{12}(s) = \left(\frac{2\lambda}{s+3\lambda}\right)\left(\frac{\lambda}{s+2\lambda}\right)\left(\frac{1}{s+\lambda}\right)$$

$$P_{13}(t) = e^{-\lambda t}\left[1 - e^{-\lambda t}\right]^{3-1}.$$

In general, we have

$$P_{1j}(t) = e^{-\lambda t}\left[1 - e^{-\lambda t}\right]^{j-1} \qquad j = 1, 2, \dots \qquad (5.2.6)$$

We now consider the more general case with $X(0) = N$. Let $X_i(t)$ denote the population size at time t of a Yule process started with individual i. We see that $P\{X_i(t) = j\}$ is given by Equation 5.2.6 for each i. Moreover, the random variable $X_i(t)$ follows a geometric distribution with parameter $p = e^{-\lambda t}$. For the Yule process with $X(0) = N$, we have $X(t) = X_1(t) + \dots + X_N(t)$, where $\{X_i(t)\}$ are i.i.d. random variables. Hence $X(t)$ follows a negative binomial distribution with parameters (N, p). This implies that

$$P\{X(t) = j\} = \binom{j-1}{N-1}p^N(1-p)^{j-N} \qquad j = N, N+1, \dots$$

or, equivalently,

$$P_{N,j}(t) = P\{X(t) = j \mid X(0) = N\} = \binom{j-1}{N-1}e^{-\lambda t N}(1 - e^{-\lambda t})^{j-N}$$

$$j = N, N+1, \dots \quad \blacksquare$$

EXAMPLE
5.2.5

A Car Repair Problem A car is sent to a garage for a major overhaul. There are three operations to be carried out sequentially: engine tune-up, air-conditioning overhaul, and braking system replacement. The mean times for these three operations are 1.2, 1.5, and 2.5 hours, respectively. Assume that the times are mutually independent and there are no delays between operations. If the time to perform each operation follows an exponential distribution, what is the probability that four hours later the car is in the braking system replacement stage?

We use 1, 2, 3 to index the three operations and 4 to denote state of repair completion. Then $\{X(t), t \geq 0\}$ is a continuous-time Markov chain with state space $S = \{1, 2, 3, 4\}$ and infinitesimal generator

$$Q = \begin{array}{c} 1 \\ 2 \\ 3 \\ 4 \end{array}\begin{bmatrix} -1/1.2 & 1/1.2 & & \\ & -1/1.5 & 1/1.5 & \\ & & -1/2.5 & 1/2.5 \\ & & & \end{bmatrix}.$$

State 4 is an absorbing state with transition rate $v_4 = 0$. Once the process enters state 4, it stays there forever. We see that an absorbing state is actually a modeling device. The subject of absorbing continuous-time Markov chains will be examined in Section 5.4. For this problem, the starting state probability vector is $s = (1, 0, 0, 0)$. Let column vector $e_3 = [0, 0, 1, 0]$. We are looking for $P\{X(4) = 3\}$, the probability that at time 4 the car is still in state 3. We have $P\{X(4) = 3\} = s(0)e^{(4*Q)}e_3 = 0.3765$. In other words, the probability we want is the $(1, 3)$th element of the matrix exponential Qt, where $t = 4$. $\blacksquare$

5.3 The Limiting Probabilities

In Examples 5.2.2 and 5.2.3, we see that the state probabilities approach their respective limiting values for a large t. In this section, we will study this limiting behavior in detail. Earlier we saw that for each continuous-time Markov chain, there is an embedded discrete-time Markov chain defined at the transition epochs. This embedded chain enables us to keep track of the transition counts of the continuous-time Markov chain but not how long it takes to complete these transitions. Assume that this embedded chain is irreducible and recurrent. Using a renewal-theoretic type of argument, it can be shown that the transition function $P_{ij}(t)$ of the corresponding continuous-time Markov chain possesses a limit. Moreover, this limit is independent of the starting state i. When the embedded chain is null recurrent, we have $\lim_{t \to \infty} P_{ij}(t) = 0$. When the embedded chain is positive recurrent, we have

$$\pi_j \equiv \lim_{t \to \infty} P_{ij}(t) = \lim_{t \to \infty} P\{X(t) = j \mid X(0) = i\} = c > 0.$$

We differentiate the previous with respect to t and interchange the order of the differentiation and limiting operations. This gives

$$\frac{d}{dt} \lim_{t \to \infty} P_{ij}(t) = \lim_{t \to \infty} \frac{d}{dt} P_{ij}(t) = 0.$$

In matrix notation, the previous equation can be stated as $\lim_{t \to \infty} P'(t) = 0$, where the right side is a matrix of zeros. Using the forward Kolmogorov equation (Equation 5.2.4), we conclude that $\lim_{t \to \infty} P(t)Q = 0$. In component form, the last identity is given by

$$
\begin{bmatrix}
\pi_0 & \pi_1 & \pi_2 & \cdot & \cdot \\
\pi_0 & \pi_1 & \pi_2 & \cdot & \cdot \\
\cdot & & & & \\
\cdot & & & & \\
\cdot & & & &
\end{bmatrix}
\begin{bmatrix}
-v_0 & q_{01} & q_{02} & \cdot & \cdot \\
q_{10} & -v_1 & q_{12} & \cdot & \cdot \\
\cdot & & & & \\
\cdot & & & &
\end{bmatrix}
=
\begin{bmatrix}
0 & 0 & \cdot & \cdot & \cdot \\
0 & 0 & \cdot & \cdot & \cdot \\
\cdot & & & & \\
\cdot & & & & \\
\cdot & & & &
\end{bmatrix}.
$$

Let row vector $\pi = (\pi_0, \pi_1, \ldots)$. Then the previous matrix equation implies that $\pi Q = 0$ or

$$
[\pi_0 \ \pi_1 \ \pi_2 \cdots]
\begin{bmatrix}
-v_0 & q_{01} & q_{02} & \cdot & \cdot \\
q_{10} & -v_1 & q_{12} & \cdot & \cdot \\
\cdot & & & & \\
\cdot & & & &
\end{bmatrix}
= [0 \ 0 \ 0 \cdots].
$$

The jth equation of the preceding identity is

$$\sum_{k \ne j} \pi_k q_{kj} - \pi_j v_j = 0 \qquad j \in S. \tag{5.3.1}$$

Since π_j is the limiting probability of being in state j and v_j is the rate of leaving state j, we can view the second term of Equation 5.3.1, $\pi_j v_j$, as the steady state output rate from j. In the same vein, the first term of Equation 5.3.1 is the sum of all transition rates to j from all other states weighted by the limiting probabilities of their being in the various states. We can view the latter as the steady state input rate to j.

> Thus for each state j in steady state, we have
>
> $$\text{Input rate to } j = \text{output rate from } j.$$

Equation 5.3.1 and the normalizing equation $\pi e = 1$ will enable us to find the limiting distribution $\{\pi_j\}$ uniquely.

We now give another insightful way to derive Equation 5.3.1. For the continuous-time Markov chain, we consider each return to state 0 a regeneration epoch and the interval between two successive returns to state 0 a regeneration cycle. The irreducibility and positive recurrence assumptions of the embedded Markov chain imply that the expected length of a regeneration cycle is finite. Let p denote the limiting-state probability vector of the embedded Markov chain (that is, p satisfies Equation 4.3.6); then the mean cycle time is $\sum_i p_i / p_0 v_i$ (recall that $1/v_j$ is the mean sojourn time in state j, and from Equation 4.3.7, p_i/p_0 gives the expected number of visits to state i between successive visits to state 0). Using Equation 3.6.1 pertaining to a regenerative process, we obtain

$$\pi_j = \lim_{t \to \infty} P\{X(t) = j | X(0) = i\} = \frac{\dfrac{p_j}{p_0 v_j}}{\displaystyle\sum_{i \in S} \dfrac{p_i}{p_0 v_i}} = \frac{\dfrac{p_j}{v_j}}{\displaystyle\sum_{i \in S} \dfrac{p_i}{v_i}}. \tag{5.3.2}$$

From Equation 5.3.2, we have $c\pi_j = p_j/v_j$, where c is the denominator of Equation 5.3.2. Let $\{p_{ij}\}$ denote the transition probabilities of the embedded chain. Equation 4.3.6 implies that

$$p_j = \sum_{k \in S} p_k p_{kj} = \sum_{k \neq j} p_k p_{kj} \qquad \text{(since } p_{jj} = 0 \text{ in the embedded chain)}$$

$$= \sum_{k \neq j} \frac{p_k q_{kj}}{v_k} \qquad \text{(by Equation 5.1.5).}$$

Since $p_j = c\pi_j v_j$ and $p_k/v_k = c\pi_k$, upon substitution the preceding reduces to

$$c\pi_j v_j = \sum_{k \neq j} c\pi_k q_{kj}.$$

Dividing both sides by c, we obtain Equation 5.3.1.

In Equation 5.3.2, we have a transition rate balance equation for each state (equivalently, for each node of the transition diagram). Similar to the result of Equation 4.3.11 given for a discrete-time Markov chain, we have a transition rate balance equation for each *cut* of the transition diagram. Following this approach, we partition the state space S into two sets, A and B. Let r_{AB} denote the aggregate transition rate from A to B conditioned by the limiting probability of being in set A; we similarly define r_{BA}. In steady state, we expect that $r_{AB} = r_{BA}$ or equivalently,

$$\sum_{i \in A} \pi_i \sum_{j \in B} q_{ij} = \sum_{j \in B} \pi_j \sum_{i \in A} q_{ji}. \tag{5.3.3}$$

We can view Equation 5.3.3 as the continuous analog of Equation 4.3.11.

As in the case of a discrete-time Markov chain, there is the notion of a *stationary* distribution for a continuous-time Markov chain.

> The stationary distribution for a continuous-time Markov chain is the row vector $p = \{p_j\}$ that satisfies
>
> $$pQ = 0 \quad \text{and} \quad pe = 1. \tag{5.3.4}$$

To show that such a vector p indeed induces stationarity, we need to prove that if $s(0) = p$ then $s(t) = p$ for all $t \geq 0$. This is done as follows: for any $t \geq 0$,

$$s(t) = s(0)P(t) = s(0)\left[\int_{0^+}^{t} P'(\tau)d\tau + I\right] = s(0)\left[\int_{0^+}^{t} QP(\tau)d\tau + I\right]$$

$$(\text{since } P'(t) = QP(t))$$

$$= pQ\int_{0^+}^{t} P(\tau)d\tau + s(0) = s(0) = p \qquad (\text{since } pQ = 0).$$

Hence we see that the limiting distribution of a continuous-time Markov chain is also the stationary distribution.

When the state space is finite, a simple way to compute the limiting distribution is by replacing the first linear equation of $\pi Q = 0$ by $\pi e = 1$ and solving the resulting system of linear equations. Specifically, we form matrix Q_1 by replacing its first column by e and form a row vector $b = [1\ 0\ 0\ ...]$. So the system of equations becomes $\pi Q_1 = b$. We see that the vector π is given by the first row of Q_1^{-1}.

EXAMPLE
5.3.1

The Barber Shop Example Revisited To find the limiting probabilities for the barber shop example, we replace the first column of the generator Q by a column of ones and invert the resulting matrix Q_1. The first row of the inverse of the matrix Q_1 gives the limiting probability vector $\pi = [0.0649\ 0.1622\ 0.2027\ 0.2534\ 0.3168]$. These are the limiting values shown in Figure 5.4. ∎

EXAMPLE
5.3.2

The Birth and Death Process Revisited Returning to Example 5.1.1, we now want to find its limiting probabilities. Figure 5.2 suggests that a natural *cut* is between two adjacent nodes. For example, the cut between nodes j and $j+1$ is given in Figure 5.6.

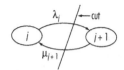

FIGURE
5.6   A cut of the state space.

We note that the cut shown in Figure 5.6 does not reveal the full detail. The set A actually contains states $\{0, 1, ..., j\}$. However, the only positive transition rate passing through the cut from left to right emanates from state j; hence only node j is shown. A similar remark can be made for the set B. Also, transition rates that do not traverse through the cut in either direction are not shown in the figure. The transition rate balancing equation Equation 5.3.3 for each cut implies that $\lambda_j \pi_j = \mu_{j+1} \pi_{j+1}$, $j = 0, 1, ...$ Define $\rho_j = \lambda_j / \mu_{j+1}$, $j = 0, 1, ...$ Then we have $\pi_{j+1} = \rho_j \pi_j$, $j = 0, 1, ...$ Repeated substitutions yield $\pi_j = \pi_0 \rho_{j-1} \cdots \rho_0$, $j = 1, 2, ...$ Using the normalizing equation, we obtain

$$\pi_0 \left[1 + \sum_{j=1}^{\infty} \rho_{j-1} \cdots \rho_0 \right] = 1. \tag{5.3.5}$$

If the summation in the previous equation is finite, then $\pi_0 > 0$ and consequently $\pi_j > 0$ for all $j \geq 0$. Hence the condition for positive recurrence for the birth and death process is

$$\sum_{j=1}^{\infty} \rho_{j-1} \cdots \rho_0 = \sum_{j=1}^{\infty} \frac{\lambda_{j-1} \cdots \lambda_0}{\mu_j \cdots \mu_1} < \infty. \tag{5.3.6}$$

Once π_0 is found, all other π_j can be obtained using $\pi_j = \rho_{j-1} \pi_j$. ∎

EXAMPLE
5.3.3
The M/M/s Queue Revisited   For the $M/M/s$ queue introduced in Example 5.1.3, it is easy to verify that $\rho \equiv \lambda/s\mu < 1$ implies Equation 5.3.6. This says that the arrival rate must be less than the maximum service rate to ensure the stability of the queue. Based on the birth and death rates given in Example 5.1.3, we see that $\rho_j = \lambda/(j+1)\mu$, $j = 0, 1, ..., s - 1$ and $\rho_j = \rho$, $j \geq s$. Equation 5.3.5 becomes

$$\pi_0 \left[1 + \left(\sum_{j=1}^{s-1} \rho_{j-1} \cdots \rho_0 \right) + \left(\sum_{j=s}^{\infty} \rho_{j-1} \cdots \rho_s \rho_{s-1} \cdots \rho_0 \right) \right]$$

$$= \pi_0 \left[\sum_{j=0}^{s-1} \frac{\lambda^j}{j! \mu^j} + \sum_{j=s}^{\infty} \rho^{j-s} \frac{\lambda^s}{s! \mu^s} \right] = \pi_0 \left[\sum_{j=0}^{s-1} \frac{\lambda^j}{j! \mu^j} + \sum_{j=s}^{\infty} \left(\frac{\lambda}{s\mu} \right)^{j-s} \frac{\lambda^s}{s! \mu^s} \right]$$

$$= \pi_0 \left[\sum_{j=0}^{s-1} \frac{\lambda^j}{j! \mu^j} + \frac{s^s}{s!} \sum_{j=s}^{\infty} \frac{\lambda^j}{(s\mu)^j} \right] = \pi_0 \left[\sum_{j=0}^{s-1} \frac{\lambda^j}{j! \mu^j} + \frac{(\rho s)^s}{s!} \frac{1}{1-\rho} \right] = 1.$$

Hence, we have

$$\pi_0 = \left[\sum_{j=0}^{s-1} \frac{1}{j!} \left(\frac{\lambda}{\mu} \right)^j + \frac{1}{s!} \left(\frac{\lambda}{\mu} \right)^s \left[\frac{s\mu}{s\mu - \lambda} \right] \right]^{-1}. \tag{5.3.7}$$

Once π_0 is known, we use $\pi_j = \pi_0 p_{j-1} \cdots p_0$ to obtain the rest of the limiting probabilities.

When $s = 1$, we have an $M/M/1$ queue. In this case Equation 5.3.7 reduces to

$$\pi_0 = 1 - \rho \qquad \text{and} \qquad \pi_j = (1 - \rho)\rho^j \qquad j \geq 1, \tag{5.3.8}$$

where $\rho = \lambda/\mu$. ∎

5.4 Absorbing Continuous-Time Markov Chains

In a continuous-time Markov chain, an absorbing state i is one whose transition rate v_i is zero. This means that once the process enters state i, it will stay there forever. A process with at least one absorbing state is called an absorbing continuous-time Markov chain. When the chain contains several absorbing states, one thing of interest is the probability that an eventual absorption occurs in a particular state (before it reaches other absorbing states). This probability can be found by working with the embedded discrete-time Markov chain and using the approach introduced in Section 4.4. A direct approach is introduced in this section.

Sojourn time related statistics are of general interest in the study of absorbing chains. To derive relevant results for chains with a finite state space, we first state the infinitesimal generator Q in a canonical form

$$Q = T^c \begin{array}{c|cc} & T^c & T \\ \hline T^c & O & O, \\ T & R & V \end{array}$$

where we partition the state space S into two sets—T^c and T, the set of absorbing states and transient states, respectively. The matrix of transition functions is also partitioned similarly

$$P(t) = T^c \begin{array}{c|cc} & T^c & T \\ \hline T^c & I & O \\ T & S(t) & T(t) \end{array} .$$

We define the matrix of the expected durations of stay in the partitioned form as

$$H(t) = T^c \begin{array}{c|cc} & T^c & T \\ \hline T^c & tI & O \\ T & M(t) & N(t) \end{array} = \int_0^t P(x)dx.$$

The (i, j)th element of $H(t)$ gives the expected time the process spent in state j by time t given the process starts from state i at time 0. Using Equation 5.2.5, we find that

$$H(t) = \int_0^t P(x)dx = \int_0^t \sum_{k=0}^\infty \frac{x^k Q^k}{k!}dx = \sum_{k=0}^\infty \frac{Q^k}{k!}\int_0^t x^k dx = \sum_{k=0}^\infty \frac{Q^k t^{k+1}}{(k+1)!}.$$

To derive a simple expression for computing $N(t)$, we first observe that

$$QH(t) = \sum_{k=0}^\infty \frac{(Qt)^{k+1}}{(k+1)!} = P(t) - I.$$

Stating in a block partitioned form as before, the previous identity reads

$$\begin{bmatrix} O & O \\ R & V \end{bmatrix}\begin{bmatrix} tI & O \\ M(t) & N(t) \end{bmatrix} = \begin{bmatrix} O & O \\ tR + VM(t) & VN(t) \end{bmatrix}$$

$$= \begin{bmatrix} I & O \\ S(t) & T(t) \end{bmatrix} - \begin{bmatrix} I & O \\ O & I \end{bmatrix} = \begin{bmatrix} O & O \\ S(t) & T(t) - I \end{bmatrix}.$$

Hence we have $VN(t) = T(t) - I$. It can be shown that the inverse of V exists. So we find that

$$N(t) = V^{-1}[T(t) - I]. \tag{5.4.1}$$

The (i, j)th element of $N(t)$ gives the expected time the process spends in transient state j by time t given that it starts from transient state i at time 0. The matrix $N(t)$ can readily be computed since $T(t)$ is a submatrix of $P(t) = exp(Qt)$. It is clear that $T(t) \to 0$ as $t \to \infty$.

This gives

$$N(\infty) = -V^{-1}. \tag{5.4.2}$$

The (i, j)th element of $N(\infty)$ gives the expected time the process spends in transient state j given that it starts from transient state i at time 0. We note that $-V^{-1}$ is the continuous analog of the *fundamental matrix* $W = (I - Q)^{-1}$ for Markov chains.

Conditioning on the state and time of the process immediately before absorption, the absorbing block $S(t)$ can be written as

$$S(t) = \int_0^t T(x)Rdx = \int_0^t T(x)dxR = N(t)R.$$

Using Equation 5.4.1, we obtain

$$S(t) = V^{-1}[T(t) - I]R. \tag{5.4.3}$$

The previous expression is the continuous analog of Equation 4.4.6 for Markov chains. Using the fact that $T(t) \to 0$ as $t \to \infty$, we obtain

$$S(\infty) = -V^{-1}R. \tag{5.4.4}$$

We see that Equation 5.4.4 is the continuous analog of Equation 4.4.7 for computing reaching probabilities of a discrete-time Markov chain.

The (i, j)th element of $M(t)$ gives the expected time the process spends in absorbing state j by time t given that it starts from transient state i at time 0. Since $S(t) = tR + VM(t)$, we have $VM(t) = S(t) - tR$. This gives

$$M(t) = V^{-1}[S(t) - tR]. \tag{5.4.5}$$

Let $D_{ij}(t)$ be the random variable denoting the duration of stay in transient state j during the interval $(0, t)$ given that $X(0) = i$, where i is also a transient state. The first moments of $\{D_{ij}(t)\}$ have already been found earlier in Equation 5.4.1. To motivate the derivation of the second moments of $\{D_{ij}(t)\}$, we define the indicator random variable

$$I_{ij}(t) = \begin{cases} 1 & \text{if } X(0) = i \text{ and } X(t) = j \\ 0 & \text{otherwise} \end{cases}$$

and observe that

$$N_{ij}(t) = E[D_{ij}(t)] = E\int_0^t I_{ij}(x)dx = \int_0^t E[I_{ij}(x)]dx = \int_0^t P\{X(x) = j | X(0) = i\}dx$$

$$= \int_0^t T_{ij}(x)dx.$$

Extending the previous approach to second moments, we find

$$E[D_{ij}^2(t)] = E\left[\int_0^t \int_0^t I_{ij}(x)I_{ij}(y)dydx\right] = 2 \iint_{0<y<x<t} E\left[I_{ij}(x)I_{ij}(y)\right]dydx$$

$$= 2\int_0^t \int_y^t P\{X(y) = j,\ X(x) = j |\ X(0) = i\}dxdy$$

$$= 2\int_0^t \int_y^t P\{X(x) = j | X(y) = j\}P\{X(y) = j | X(0) = i\}dxdy$$

$$= 2\int_0^t \left[\int_y^t T_{jj}(x - y)dx\right]T_{ij}(y)dy = 2\int_0^t T_{ij}(y)N_{jj}(t - y)dy.$$

In matrix notation, the variances of $\{D_{ij}(t)\}$ are given by

$$Var[D(t)] = 2\int_0^t T(x)[N(t - x)\square I]dx - (N(t)\square N(t)). \tag{5.4.6}$$

For a finite t, we can evaluate $Var[D(t)]$ numerically by noting $T(x)$ is a submatrix of $exp[Qt]$ and $N(t)$ is given by Equation 5.4.1. When $t \to \infty$, we see that the asymptotic variances are given in matrix notation by

$$Var[D(\infty)] = 2V^{-1}[V^{-1}\square I] - [V^{-1}\square V^{-1}]. \tag{5.4.7}$$

We observe that the previous variances can also be derived by an approach similar to that of Equation 4.4.10.

EXAMPLE
5.4.1

A Trauma Center Consider a trauma center specialized in treating victims of violent crimes. The center has four operating beds and three beds for holding waiting patients. Arrival episodes of ambulances carrying patients follow a Poisson process with a rate of one arrival per two hours (each arrival episode could include the arrivals of several ambulances, each carrying a patient). Let p_i denote the probability that a given arrival episode carries a total of i patients. Assume that $p_1 = 0.7$, $p_2 = 0.2$, and $p_3 = 0.1$. The patient's length of stay on an operating bed follows an exponential distribution with a mean of 2.5 hours. We also assume that service times are mutually independent and independent of the arrival process. The trauma center has a policy of not admitting new arrivals as soon as at least one of its waiting beds is filled. The center is interested in studying statistics relating to center closures caused by capacity limitation starting from an epoch at which all beds are empty.

Let $X(t)$ denote the number of patients in the center at time t. Assume for the moment that the center remains open at all times as long as there are beds available. Then the stochastic process $X = \{X(t), t \geq 0\}$ is a continuous-time Markov chain with state space $S = \{0, 1, ..., 7\}$. Let λ denote the rate of arrival episodes per hour and μ the rate of service completion per hour. Then $\lambda = 0.5$ and $\mu = 0.4$. Let λ_i denote the rate of arrival with i patients. Then $\lambda_i = \lambda p_i$. The infinitesimal generator of X is given by

$$
\begin{array}{c}
0 \\ 1 \\ 2 \\ 3 \\ 4 \\ 5 \\ 6 \\ 7
\end{array}
\left[
\begin{array}{cccccccc}
-\lambda & \lambda_1 & \lambda_2 & \lambda_3 & & & & \\
\mu & -(\lambda+\mu) & \lambda_1 & \lambda_2 & \lambda_3 & & & \\
 & 2\mu & -(\lambda+2\mu) & \lambda_1 & \lambda_2 & \lambda_3 & & \\
 & & 3\mu & -(\lambda+3\mu) & \lambda_1 & \lambda_2 & \lambda_3 & \\
 & & & 4\mu & -(\lambda+4\mu) & \lambda_1 & \lambda_2 & \lambda_3 \\
 & & & & 4\mu & -(\lambda+4\mu) & \lambda_1 & \lambda_2+\lambda_3 \\
 & & & & & 4\mu & -(\lambda+4\mu) & \lambda \\
 & & & & & & 4\mu & -4\mu
\end{array}
\right]
$$

Since center closures occur when $X(t)$ reaches states 5, 6, 7, we construct an absorbing chain with the set of absorbing states $T^c = \{5, 6, 7\}$. The generator written in a canonical form corresponding to this absorbing chain is given by

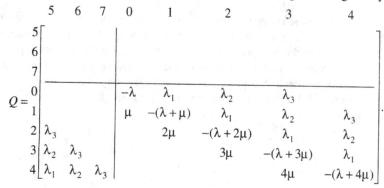

$$
Q =
\begin{array}{c}
5 \\ 6 \\ 7 \\ 0 \\ 1 \\ 2 \\ 3 \\ 4
\end{array}
\left[
\begin{array}{ccc|ccccc}
 & & & & & & & \\
 & & & & & & & \\
 & & & & & & & \\
 \hline
 & & & -\lambda & \lambda_1 & \lambda_2 & \lambda_3 & \\
 & & & \mu & -(\lambda+\mu) & \lambda_1 & \lambda_2 & \lambda_3 \\
 \lambda_3 & & & & 2\mu & -(\lambda+2\mu) & \lambda_1 & \lambda_2 \\
 \lambda_2 & \lambda_3 & & & & 3\mu & -(\lambda+3\mu) & \lambda_1 \\
 \lambda_1 & \lambda_2 & \lambda_3 & & & & 4\mu & -(\lambda+4\mu)
\end{array}
\right].
$$

From the data given, we find that

$$
R = \begin{bmatrix} 0 & 0 & 0 \\ 0 & 0 & 0 \\ 0.05 & 0 & 0 \\ 0.10 & 0.05 & 0 \\ 0.35 & 0.10 & 0.05 \end{bmatrix}
\qquad
V = \begin{bmatrix} -0.5 & 0.35 & 0.1 & 0.05 & 0 \\ 0.4 & -0.9 & 0.35 & 0.1 & 0.05 \\ 0 & 0.8 & -1.3 & 0.35 & 0.10 \\ 0 & 0 & 1.2 & -1.7 & 0.35 \\ 0 & 0 & 0 & 1.6 & -2.1 \end{bmatrix}.
$$

Using Equation 5.4.2, we obtain in the Appendix the expected times the center spends in transient states $0, \ldots, 4$. Since $X(0) = 0$, the first row of $N(\infty)$ gives these expected times $\{E[D_{0j}(\infty)]\}$. They are

$$[7.8419 \quad 7.3024 \quad 4.7844 \quad 2.3997 \quad 0.8016].$$

Summing this row, we get 23.13 hours. This says that on average it will take about twenty-three hours before the center refuses additional admissions because of the capacity consideration. We remark here that had the process started with $X(0) = 4$, the comparable figure would have been 13.90 hours.

Using Equation 5.4.7, the variances of duration of stay in transient states are found in the Appendix. Again, with $X(0) = 0$, we find $\{Var[D_{0j}(\infty)]\}$ from the first row of $Var[D(\infty)]$. These are

$$[61.4961 \quad 60.0549 \quad 25.9400 \quad 6.9742 \quad 1.0698].$$

The reaching probabilities are found from Equation 5.4.4. Since $X(0) = 0$, the first row of $S(\infty)$ gives $[0.76 \ 0.20 \ 0.04]$. For example, this says that with probability 0.76, refusing admission occurs when the total occupancy reaches five. ∎

When the state space is not finite, the matrix-based results shown previously will no longer be useful. The next example demonstrates a general approach to deal with the situation. The approach hinges on conditioning on the first transitions out of the given state and solving the resulting system of difference equations.

EXAMPLE 5.4.2*   **The Birth and Death Process Revisited**   Returning to Example 5.1.1, we consider the case in which state 0 is absorbing. This implies that $\lambda_0 = 0$. If $X(0) = i$, where $i \geq 1$, it is unclear whether the process will ever return to state 0 or if the expected time to reach state 0 will be finite. We now address these two issues in this example.

Let u_i denote the probability of ever reaching state 0 given $X(t) = i$. From Example 5.1.1, we know that for the embedded discrete-time Markov chain we have

$$
p_{i,i+1} = \frac{\lambda_i}{v_i} \equiv r_i
\qquad
p_{i,i-1} = \frac{\mu_i}{v_i} \equiv s_i,
$$

where $v_i = \lambda_i + \mu_i$. Then the transition probability matrix of the embedded chain is given by

$$P = \begin{bmatrix} 1 & & & & \\ s_1 & & r_1 & & \\ & s_2 & & r_2 & \\ & & & \ddots & \\ & & & & \ddots \\ & & & & & \ddots \end{bmatrix}.$$

For this embedded Markov chain, we condition on the first transition out of state i and obtain

$$u_i = r_i u_{i+1} + s_i u_{i-1} \qquad i = 1, 2, \ldots \qquad (5.4.8)$$
$$u_0 = 1.$$

We rewrite Equation 5.4.8 as

$$(r_i + s_i)u_i = r_i u_{i+1} + s_i u_{i-1}.$$

Rearranging terms yields $u_{i+1} - u_i = h_i(u_i - u_{i-1})$, where $h_i = \mu_i/\lambda_i$. Define $V_i = u_{i+1} - u_i$. Then $V_i = h_i V_{i-1}$, $i \geq 1$. Repeated applications of the identity yields $V_i = h_i \cdots h_1 V_0$. For $i \geq 1$, let $g_i = h_1 \cdots h_i$. Define $g_0 = 1$ and $A(n) = g_1 + \cdots + g_n$. Then $V_i = g_i V_0$ for all $i \geq 0$.

For $m \geq 1$, we have

$$\sum_{i=1}^{m-1} V_i = u_m - u_1 = V_0 \sum_{i=1}^{m-1} g_i = (u_1 - u_0)A(m-1) = (u_1 - 1)A(m-1). \qquad (5.4.9)$$

Clearly, $u_i \leq 1$ for all i. If

$$A(\infty) = \lim_{n \to \infty} \sum_{i=1}^{n} \frac{\mu_1 \cdots \mu_i}{\lambda_1 \cdots \lambda_i} = \infty, \qquad (5.4.10)$$

then we must have $u_1 = 1$ (since the left side of Equation 5.4.9 $u_m - u_1$ is finite). This implies that when Equation 5.4.10 holds, $u_m = 1$ for all $m \geq 1$. If $0 < u_1 < 1$, then we must have $A(\infty) < \infty$. To move from state m to state 0, the process must pass through each intermediate state $m-1, \ldots, 1$. The last property is called the skip-free-to-the-left property of a birth and death process. This property implies that u_m is decreasing in m. Since u_m is bounded from below by 0, we have $u_m \to 0$ as $m \to \infty$. Letting $m \to \infty$ in Equation 5.4.9, we obtain

$$-u_1 = (u_1 - 1)A(\infty) \qquad \text{or} \qquad u_1 = \frac{A(\infty)}{1 + A(\infty)}. \qquad (5.4.11)$$

From Equation 5.4.9, we find that

$$u_m = u_1 + (u_1 - 1)A(m-1) = u_1(1 + A(m-1)) - A(m-1)$$

$$= \frac{A(\infty)}{1 + A(\infty)}(1 + A(m-1)) - A(m-1) = \frac{\displaystyle\sum_{i=m}^{\infty} g_i}{\displaystyle\sum_{i=0}^{\infty} g_i}. \qquad (5.4.12)$$

Assume that $A(\infty) = \infty$, so absorption into state 0 from any state is certain. Let w_i denote the mean absorption time into state 0 given $X(0) = i$. Conditioning on the first transition out of state i enables us to write the following recursion

$$w_i = (1/v_i) + r_i w_{i+1} + s_i w_{i-1} \qquad i = 1, 2, \ldots$$

$$w_0 = 0. \tag{5.4.13}$$

We rewrite Equation 5.4.13 as $(r_i + s_i)w_i = (1/v_i) + r_i w_{i+1} + s_i w_{i-1}$. Define $Z_i = w_i - w_{i+1}$. Then the last equation can be stated as $r_i Z_i = (1/v_i) + s_i Z_{i-1}$, or $\lambda_i Z_i = 1 + \mu_i Z_{i-1}$, or

$$Z_i = (1/\lambda_i) + h_i Z_{i-1} \qquad i = 1, 2, \ldots \tag{5.4.14}$$

From Equation 5.4.14, we see that

$$Z_1 = \frac{1}{\lambda_1} + h_1 Z_0 = \frac{1}{\lambda_1}\frac{g_1}{g_1} + g_1 Z_0$$

$$Z_2 = \frac{1}{\lambda_2} + h_2 Z_1 = \frac{1}{\lambda_2} + h_2\left[\frac{1}{\lambda_1}\frac{g_1}{g_1} + g_1 Z_0\right] = \frac{1}{\lambda_2}\frac{g_2}{g_2} + \frac{1}{\lambda_1}\frac{g_2}{g_1} + g_2 Z_0$$

$$Z_3 = \frac{1}{\lambda_3} + h_3 Z_2 = \frac{1}{\lambda_3} + h_3\left[\frac{1}{\lambda_2}\frac{g_2}{g_2} + \frac{1}{\lambda_1}\frac{g_2}{g_1} + g_2 Z_0\right]$$

$$= \frac{1}{\lambda_3}\frac{g_3}{g_3} + \frac{1}{\lambda_2}\frac{g_3}{g_2} + \frac{1}{\lambda_1}\frac{g_3}{g_1} + g_3 Z_0$$

and in general

$$Z_m = \sum_{i=1}^{m}\frac{1}{\lambda_i}\frac{g_m}{g_i} + g_m Z_0 = g_m\left[\sum_{i=1}^{m}\frac{1}{\lambda_i g_i} + Z_0\right]. \tag{5.4.15}$$

Since $Z_0 = w_0 - w_1 = -w_1$, we rewrite Equation 5.4.15 as

$$\frac{1}{g_m}\left[w_m - w_{m+1}\right] = B(m) - w_1, \tag{5.4.16}$$

where we define

$$B(m) = \sum_{i=1}^{m}\frac{1}{\lambda_i g_i}.$$

The skip-free-to-the-left property of the process implies that $w_{m+1} > w_m$ for all m. Hence the left side of Equation 5.4.16 is negative. When $B(\infty) = \infty$ and $w_1 < \infty$, then the left-side negativity of Equation 5.4.16 is violated. This shows that if $B(\infty) = \infty$, we have $w_1 = \infty$ and consequently $w_m = \infty$ for all $m \geq 1$. When $B(\infty) < \infty$, Equation 5.4.16 implies that

$$w_1 = B(m) - (1/g_m)[w_m - w_{m+1}].$$

It can be shown that the second term on the right of the preceding equality goes to 0 as $m \to \infty$. Consequently, $w_1 = B(\infty)$. To find other w_m, we first define

$$^{>}B(m) = \sum_{i=m+1}^{\infty}\frac{1}{\lambda_i g_i}.$$

Using the previous notation, Equation 5.4.16 becomes $w_m - w_{m+1} = -g_m^> B(m)$ or $w_{m+1} = w_m + g_m^> B(m)$. Iterating the last expression yields

$$w_2 = w_1 + g_1^> B(1)$$

$$w_3 = w_2 + g_2^> B(2) = w_1 + g_1^> B(1) + g_2^> B(2)$$

and in general

$$w_m = w_1 + \sum_{i=1}^{m-1} g_i^> B(i) \qquad m \geq 2. \quad \blacksquare \qquad (5.4.17)$$

EXAMPLE
5.4.3

An M/M/1 Queue The M/M/1 queue is a special case of the M/M/s queue, with $s = 1$, described in Example 5.1.3. The underlying birth and death process has the simple structure $\lambda_i = \lambda$ for $i \geq 0$ and $\mu_i = \mu$ for $i \geq 1$. Hence $g_i = (\mu/\lambda)^i$ and

$$A(\infty) = \sum_{i=1}^{\infty} \left(\frac{\mu}{\lambda}\right)^i \qquad B(\infty) = \sum_{i=1}^{\infty} \frac{1}{\lambda}\left(\frac{\lambda}{\mu}\right)^i.$$

When $\lambda = \mu$, we have $A(\infty) = \infty$ and $B(\infty) = \infty$. This implies $u_m = 1$ for all $m \geq 1$ and $w_m = \infty$. Hence when the service rate and arrival rate are the same, while the probability of returning to state 0 from any other state is 1 the mean absorption time is infinite. This is a case in which the underlying continuous-time Markov chain is null recurrent. $\blacksquare$

EXAMPLE
5.4.4

The M/M/s Queue Revisited Consider the M/M/s queue introduced in Example 5.1.3 again. Assume that $\rho = \lambda/s\mu < 1$, so that the queue is stable. Suppose we are interested in finding the expected time for the system to become empty given there are i customers in the system at a certain epoch. Thus we let state 0 be the absorbing state. Using the results derived in Example 5.4.2, we now compute w_i. The underlying birth and death process has $\lambda_i = \lambda$ for all $i \geq 0$, $\mu_i = i\mu$ if $1 \leq i \leq s$, and $\mu_i = s\mu$ if $i > s$. Therefore $h_i = (i\mu/\lambda)$ if $1 \leq i \leq s$, $(s\mu/\lambda)$ if $i > s$, and

$$g_i = \begin{cases} i!\left(\dfrac{\mu}{\lambda}\right)^i & i = 1, \ldots, s-1 \\[2ex] s! s^{i-s}\left(\dfrac{\mu}{\lambda}\right)^i & i = s, s+1, \ldots. \end{cases}$$

We have

$$w_1 = B(\infty) = \frac{1}{\lambda}\left[\sum_{i=1}^{s-1}\frac{1}{i!}\left(\frac{\lambda}{\mu}\right)^i + \sum_{i=s}^{\infty}\frac{s^s}{s!}\left(\frac{\lambda}{s\mu}\right)^i\right] = \frac{1}{\lambda}\left[\sum_{i=1}^{s-1}\frac{1}{i!}\left(\frac{\lambda}{\mu}\right)^i + \frac{1}{s!}\left(\frac{\lambda}{\mu}\right)^s\left[\frac{s\mu}{s\mu-\lambda}\right]\right].$$

$$(5.4.18)$$

We rewrite Equation 5.4.17 as

$$w_m = w_1 + \sum_{i=1}^{m-1} g_i [B(\infty) - B(i)]. \tag{5.4.19}$$

We can use Equations 5.4.18 and 5.4.19 to find the mean times for absorption into state 0.

Let T_{00} denote the interval between two successive returns to state 0. We call T_{00} a busy cycle. In a busy cycle, there are exactly two subintervals—one with at least one customer present and another with no customers present. The expected lengths of the two subintervals are w_1 and $1/\lambda$, respectively. The expected cycle length is given by $w_1 + (1/\lambda)$. Using Equation 5.3.7, we rewrite Equation 5.4.19 as

$$w_1 = \frac{1}{\lambda}\left[\frac{1}{\pi_0} - 1\right] \quad \text{or} \quad w_1 + \frac{1}{\lambda} = \frac{1}{\lambda \pi_0}. \tag{5.4.20}$$

So we see that the expected length of a busy cycle is also given by $1/\lambda\pi_0$. To take another look at Equation 5.4.20, we consider a busy cycle a regeneration cycle. An application of the regenerative theory gives

$$\pi_0 = \frac{\text{expected length of a period during which there are no arrivals}}{\text{expected length of a cycle}}$$

$$= \frac{\dfrac{1}{\lambda}}{w_1 + \dfrac{1}{\lambda}}.$$

The preceding is just Equation 5.4.20.

To illustrate the previous exposition numerically, we consider the problem with $s = 5$, $\lambda = 5$, and $\mu = 1.2$. Since $\rho = 5/6 < 1$, the queue is stable. Using Equation 5.4.18, we obtain $w_1 = 20.0511$; hence the expected length of a busy cycle is 20.2511. From Equation 5.4.19, we find the other w_i. As an example, we have $w_2 = 24.6634$, $w_3 = 26.6773$, $w_4 = 27.9273$, and so on. It is easy to verify that these $\{w_i\}$ satisfy the recursion of Equation 5.4.13. ∎

5.5 Phase-Type Distributions

Consider a finite-state absorbing continuous-time Markov chain with a single absorbing state. The time-to-absorption distribution of such a Markov chain has received extensive attention in the literature of stochastic modeling and computational probability. Such a distribution has many attractive properties from modeling and computation perspectives. In this section, we will give an expository review.

Let state 0 be the single absorbing state and states $1, \ldots, m$ the m transient states. We display the infinitesimal generator of such a chain as follows:

$$Q = \begin{bmatrix} 0 & O \\ T^0 & T \end{bmatrix} = \begin{array}{c|ccccc} & 0 & 1 & 2 & \cdot \cdot & m \\ \hline 0 & 0 & 0 & 0 & \cdot \cdot & 0 \\ 1 & T_{10} & T_{11} & T_{12} & \cdot \cdot & T_{1m} \\ 2 & T_{20} & T_{21} & T_{22} & \cdot \cdot & T_{2m} \\ \cdot & \cdot & \cdot & & \cdot & \\ \cdot & \cdot & \cdot & & \cdot & \\ m & T_{m0} & T_{m1} & T_{m2} & \cdot \cdot & T_{mm} \end{array},$$

where T is $m \times m$ and T^0 is $m \times 1$. We assume that the matrix Q is irreducible. For this process, absorption from each transient state is certain. It can be shown that the matrix T is nonsingular. Let $\{\alpha_i\}$ denote the elements of the starting probability vector and row vector $\alpha = \{\alpha_1, ..., \alpha_m\}$. Define τ as the time to absorption given the starting-state probability vector and let F be its distribution function.

Let $x_j(t) = P\{X(t) = j\}$ and row vector $x(t) = \{x_1(t), ..., x_m(t)\}$. The part of the forward Kolmogorov equation relating to transient states $1, ..., m$ is given in matrix notation as $x'(t) = x(t)T$ and $x(0) = \alpha$. Define the Laplace transform (in row-vector notation):

$$x^e(s) = \int_0^\infty e^{-st} x(t)dt.$$

Taking the Laplace transform, we obtain $sx^e(s) - x(0) = x^e(s)T$ or $x^e(s)[sI - T] = \alpha$. It can be shown that the inverse of $[sI - T]$ exists for $\text{Re}(s) \geq 0$. Hence $x^e(s) = \alpha[sI - T]^{-1}$. Inverting the last identity gives $x(t) = \alpha \exp(Tt)$, $t \geq 0$. Now, we see that

$$F(t) = P\{\tau \leq t\} = 1 - x(t)e = 1 - \alpha \exp(Tt)e \qquad t \geq 0. \qquad (5.5.1)$$

The probability distribution characterized by Equation 5.5.1 is called the *phase-type* distribution with representation (α, T) with *order m*. Sometimes, we use $PH(\alpha, T)$ to denote it. When $t = 0$, Equation 5.5.1 gives $F(0) = 1 - \alpha e = \alpha_0$. Thus α_0 gives the probability mass at 0. For simplicity in exposition, we assume that $\alpha_0 = 0$. For $t > 0$, the density portion of F is given by

$$f(t) = \frac{d}{dt}F(t) = \frac{d}{dt}[1 - \alpha \exp(Tt)e] = -\alpha \frac{d}{dt}\exp(Tt)e = -\alpha \exp(Tt)Te. \qquad (5.5.2)$$

The last equality is due to $\frac{d}{dt}\exp(At) = \exp(At)A$, since $Qe = 0$, or $T^0 + Te = 0$.
Thus the right side of Equation 5.5.2 can be stated as

$$f(t) = \alpha \exp(Tt)T^0 = x(t)T^0 \qquad t > 0. \qquad (5.5.3)$$

The last expression $x(t)T^0$ has an interesting probabilistic interpretation. It says that the rate of absorption into state 0 in the small interval around t is equal to the sum of the instantaneous transition rates of moving to state 0 from transient states weighted by the probabilities that the process is in these transient states at time t. Define the Laplace transform

$$f^e(s) = \int_0^\infty e^{-st} f(t)dt.$$

From Equation 5.5.3, we obtain

$$f^e(s) = \alpha[sI - T]^{-1}T^0.$$ (5.5.4)

To find the mean of $PH(\alpha, T)$, we see that

$$\frac{d}{ds}f^e(s) = \alpha\left(\frac{d}{ds}[sI - T]^{-1}\right)T^0 = \alpha\left(-[sI - T]^{-1}\frac{d}{ds}[sI - T]^{-1}[sI - T]^{-1}\right)T^0$$

$$= \alpha\left(-[sI - T]^{-1}[sI - T]^{-1}\right)T^0.$$ (5.5.5)

In establishing the preceding, we use the result

$$\frac{d}{dt}X^{-1} = -X^{-1}\frac{dX}{dt}X^{-1},$$

where X is a matrix function whose argument is t. Using Equation 5.5.5, we obtain

$$E[\tau] = -\frac{d}{ds}f^e(s)\Big|_{s=0} = \alpha[(-T)^{-1}(-T)^{-1}]T^0.$$

Recall that $Te = -T^0$ and hence $e = -T^{-1}T^0$. Using the last identity in the previous equation gives

$$E[\tau] = -\alpha T^{-1}e.$$ (5.5.6)

Iterating the given procedure, we can obtain higher-order moments

$$E[\tau^i] = (-1)^i i!(\alpha T^{-i}e) \qquad \text{for } i \geq 1.$$ (5.5.7)

Phase-type distributions possess many useful closure properties. One such property says that the convolution of two phase-type distributions is again phase type. The details concerning this property are given in Example 5.5.1.

EXAMPLE **The Convolution of Two Phase-Type Distributions**   Suppose we want to find the convolu-
5.5.1*   tion of two independent distributions F and G, where F is $PH(\alpha, T)$ and of order m and G is $PH(\beta, S)$ and of order n. For simplicity, we again assume that $\alpha_0 = 0$ and $\beta_0 = 0$. Let $H = F*G$. We now show that H is $PH(\gamma, L)$ and of order $m + n$, where

$$\gamma = [\alpha, \; \mathbf{0}]$$

$$L = \begin{bmatrix} T & T^0\beta \\ & S \end{bmatrix},$$

where γ is a row vector of size $m + n$ and L is $(m + n) \times (m + n)$. Let $h^e(s), f^e(s)$, and $g^e(s)$ be the Laplace transforms of the respective densities h, f, and g. If we can show that $h^e(s) = f^e(s)g^e(s)$, then H is of phase type. Since

$$\begin{bmatrix} T & T^0\beta & L_1^0 \\ & S & L_2^0 \\ \mathbf{0} & \mathbf{0} & 0 \end{bmatrix}\begin{bmatrix} e \\ e \\ 1 \end{bmatrix} = \begin{bmatrix} \mathbf{0} \\ \mathbf{0} \\ 0 \end{bmatrix},$$

we find that $Te + T^0\beta e + L_1^0 = 0$ or $L_1^0 = 0$ (since $\beta e = 1$ and $Te + T^0 = 0$), and $Se + L_2^0 = 0$ or $L_2^0 = -Se = S^0$. For any invertible block-partitioned matrix

$$E = \begin{bmatrix} A & B \\ & C \end{bmatrix},$$

its inverse is given by

$$E^{-1} = \begin{bmatrix} A^{-1} & -A^{-1}BC^{-1} \\ & C^{-1} \end{bmatrix}. \tag{5.5.8}$$

Analogous to Equation 5.5.4, the Laplace transform $h^e(s)$ is given by

$$[\alpha, 0][sI - L]^{-1}\begin{bmatrix} 0 \\ S^0 \end{bmatrix} = [\alpha, 0]\left[\begin{bmatrix} sI & \\ & sI \end{bmatrix} - \begin{bmatrix} T & T^0\beta \\ & S \end{bmatrix}\right]^{-1}\begin{bmatrix} 0 \\ S^0 \end{bmatrix}$$

$$= [\alpha, 0]\begin{bmatrix} sI - T & -T^0\beta \\ & sI - S \end{bmatrix}^{-1}\begin{bmatrix} 0 \\ S^0 \end{bmatrix}.$$

Using Equation 5.5.8, we reduce the preceding to

$$h^e(s) = [\alpha, 0]\begin{bmatrix} (sI - T)^{-1} & (sI - T)^{-1}T^0\beta(sI - S)^{-1} \\ & (sI - S)^{-1} \end{bmatrix}\begin{bmatrix} 0 \\ S^0 \end{bmatrix}$$

$$= [\alpha, 0]\begin{bmatrix} (sI - T)^{-1}T^0\beta(sI - S)^{-1}S^0 \\ (sI - S)^{-1}S^0 \end{bmatrix}$$

$$= [\alpha(sI - T)^{-1}T^0][\beta(sI - S)^{-1}S^0] = f^e(s)g^e(s).$$

That finishes the proof. For continuous random variables, convolution typically involves numerical integration. When the corresponding densities are of phase type, the operation can be carried out by simple matrix operations. This is an important advantage of working with phase-type distributions. ∎

The next few examples show that many well-known continuous distributions are of phase type.

EXAMPLE
5.5.2 **The Generalized Erlang Distribution**   Let $X_1, \ldots, X_m$ be m independent random variables. For each i, we assume X_i follows an exponential distribution with parameter λ_i. Let $Y = X_1 + \cdots + X_m$. Then random variable Y follows a generalized Erlang distribution with parameters $\{\lambda_i\}$. In the special case in which $\lambda_i = \lambda$ for all i, we have the Erlang density with parameters (m, λ). The phase representation of the distribution has $\alpha = (1, 0, \ldots, 0)$ and

$$T = \begin{bmatrix} -\lambda_1 & \lambda_1 & & & \\ & -\lambda_2 & \lambda_2 & & \\ & & \cdot & \cdot & \\ & & & -\lambda_{m-1} & \lambda_{m-1} \\ & & & & -\lambda_m \end{bmatrix}.$$

The order of this representation is clearly m. The Laplace transform of the density is given by

$$\prod_{i=1}^{m} \frac{\lambda_i}{s + \lambda_i}. \quad \blacksquare$$

EXAMPLE
5.5.3   **The Hyperexponential Distribution**   The hyperexponential distribution F is a probabilistic mixture of exponential distributions with parameters $\{\lambda_i\}$, where

$$F(x) = \sum_{i=1}^{m} \alpha_i (1 - e^{-\lambda_i x}) \qquad\qquad x \geq 0, \qquad\qquad (5.5.9)$$

where $0 < \alpha_i < 1$ and $\alpha_1 + \cdots + \alpha_m = 1$. Figure 5.7 gives hints about the phase representation of F.

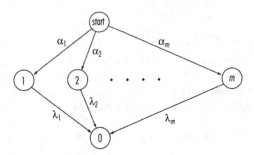

FIGURE
5.7   A transition diagram for the hyperexponential distribution.

The figure suggests that the phase representation of the hyperexponential density is given by

$$\alpha = (\alpha_1, \dots, \alpha_m) \text{ and}$$

$$T = \begin{bmatrix} -\lambda_1 & & \\ & \cdot & \\ & & \cdot \\ & & -\lambda_m \end{bmatrix}.$$

The order of the representation is again m. Since T is diagonal, powers of T are simply powers of individual diagonal elements. It is easy to verify that

$$\exp(Tx) = \sum_{i=0}^{\infty} \frac{(Tx)^i}{i!} = \begin{bmatrix} e^{-\lambda_1 x} & & \\ & \ddots & \\ & & e^{-\lambda_m x} \end{bmatrix}.$$

Hence

$$\alpha \exp(Tx)e = \sum_{i=1}^{m} \alpha_i e^{-\lambda_i x}.$$

This implies that $F(x)$ has the form of Equation 5.5.9. The Laplace transform of the density is given by

$$\sum_{i=1}^{m} \frac{\alpha_i \lambda_i}{s + \lambda_i}. \quad \blacksquare$$

EXAMPLE **A Mixture of Generalized Erlang Distributions**   The Laplace transform of a probabilistic
5.5.4   mixture of generalized Erlang distribution is given by

$$\sum_{i=1}^{m} \beta_1 \cdots \beta_{i-1} \eta_i \prod_{k=1}^{i} \frac{\lambda_k}{s + \lambda_k}, \qquad (5.5.10)$$

where $\beta_0 = 1$, $\beta_m = 0$, and $\eta_i + \beta_i = 1$, $i = 1, \ldots, m$. Let $q_i = \beta_1 \beta_2 \cdots \beta_{i-1} \eta_i$. Then the probabilistic mixture of generalized Erlang distribution can be stated as

$$F(x) = \sum_{i=1}^{m} q_i G_i(x),$$

where $0 < q_i < 1$ and $q_1 + \cdots + q_m = 1$, and G_i is the generalized Erlang distribution of order i defined in Example 5.5.2. We construct the transition diagram implied by Equation 5.5.10 in Figure 5.8.

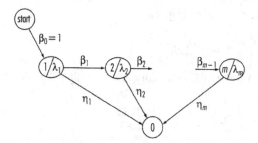

FIGURE
5.8   The transition diagram for modeling the mixture of generalized Erlangs.

In Figure 5.8, node i represents an exponential sojourn phase with rate λ_i. A path traversing i of these nodes before absorption corresponds to a generalized Erlang distribution G_j. We see that the mixture of generalized Erlangs is phase type with representation: $\alpha = [1, 0, 0, ..., 0]$ and

$$T = \begin{bmatrix} -\lambda_1 & \beta_1\lambda_1 & & & \\ & -\lambda_2 & \beta_2\lambda_2 & & \\ & & \cdot & \cdot & \\ & & & \cdot & \cdot \\ & & & & -\lambda_m \end{bmatrix}. \quad \blacksquare$$

While the aforementioned distributions are special cases of phase-type distributions, other distributions frequently can be approximated by phase-type distributions. This is illustrated in the following example.

EXAMPLE
5.5.5* **Approximating a Weibull Density by a Phase-Type Distribution**   Consider a Weibull distribution with shape parameter 2 and scale parameter 1. Its density is given by

$$f(x) = 2xe^{-x^2} \qquad x > 0.$$

The first two moments are $\mu_1 = 0.8862$ and $\mu_2 = 1$. Suppose we are interested in obtaining a phase-type approximation that matches the first two moments and the distribution function at n selected points to within an error tolerance of $\pm\varepsilon$.

There are many ways to select a set of parameters for the phase-type representation (α, T). Moreover, such a representation is not unique. For simplicity, we choose to use a probabilistic mixture of generalized Erlangs with $\lambda_i = \lambda$ for all i. In other words, we choose to use a probabilistic mixture of Erlang distributions with different orders but the same intensity λ as the phase representation of the Weibull density. Specializing Equation 5.5.10, the Laplace transform $g^e(s)$ for the density of such a mixture, called $g(x)$, is given by

$$g^e(s) = \sum_{i=1}^{m} q_i \left(\frac{\lambda}{s+\lambda} \right)^i. \tag{5.5.11}$$

Let Z denote the random variable whose density is $g(x)$. Using Equation 5.5.11, we find that

$$E[Z] = -\frac{d}{ds} g^e(s)\Big|_{s=0} = \sum_{i=1}^{m} \frac{iq_i}{\lambda}$$

$$E[Z^2] = \frac{d^2}{ds^2} g^e(s)\Big|_{s=0} = \sum_{i=1}^{m} \frac{i(i+1)q_i}{\lambda^2}.$$

Equating the previous two expressions with μ_1 and μ_2, we obtain the following two constraints

$$\sum_{i=1}^{m} i q_i = \lambda \mu_1 \quad \text{and} \quad \sum_{i=1}^{m} i(i+1) q_i = \lambda^2 \mu_2. \tag{5.5.12}$$

To match the hypothesized distribution with that of the Weibull distribution at the selected points, we require

$$\left| \sum_{i=1}^{m} q_i G_i(t_k) - F(t_k) \right| \le \varepsilon \qquad k = 1, \ldots, n, \tag{5.5.13}$$

where G_i can be computed from Equation 5.5.1. Any set of parameters m, λ, and $\{q_i\}$ that satisfy Equations 5.5.12 and 5.5.13, $q_1 + \cdots + q_m = 1$, $0 < q_i < 1$ for $i = 1, \ldots, m$, will be satisfactory.

To illustrate the given procedure of fitting, we select $\varepsilon = 0.01$ and require distributional matchings at $t = 0:0.05:2$. Using a nonlinear programming solution, a feasible solution is found with $m = 6$, $\lambda = 5.5997$, and $q = (0, 0, 0.2559, 0.1348, 0, 0.6093)$. This gives a phase-type representation with $\alpha = (1, 0, 0, 0, 0, 0)$ and

$$T = \begin{bmatrix} -5.5997 & 5.5997 & & & & \\ & -5.5997 & 5.5997 & & & \\ & & -5.5997 & 4.1667 & & \\ & & & -5.5997 & 4.5853 & \\ & & & & -5.5997 & 5.5997 \\ & & & & & -5.5997 \end{bmatrix}. \quad \blacksquare$$

EXAMPLE
5.5.6* **Phase-Type Renewal Processes**   Consider a renewal process $\{N(t), t \ge 0\}$ whose inter-arrival time distribution is $PH(\alpha, T)$ and of order m. Again we assume $\alpha_0 = 0$. Such a renewal process is called a *phase-type renewal process*. Whenever the continuous-time Markov chain reaches the absorbing state 0, we can assume that a renewal has just occurred. Immediately after such a renewal, the Markov chain is restarted with initial probability vector α. With the exception of these instantaneous sojourns in state 0, the Markov chain is perpetually moving about states $\{1, \ldots, m\}$. The dynamics of these movements are governed by an m-state irreducible recurrent continuous-time Markov chain with state space $\{1, \ldots, m\}$, generator

$$Q^* = T + T^0 \alpha,$$

and starting state probability vector α. We see that the transitions among states $1, \ldots, m$ are governed by the transition rate submatrix T plus those induced by instantaneous sojourns to the absorbing state 0.

Let $J(t)$ denote the state of this recurrent Markov chain at time t. We define matrices $P(n, t) = \{P_{ij}(n, t)\}$, where

$$P_{ij}(n, t) = P\{N(t) = n, J(t) = j | N(0) = 0, J(0) = i\},$$

for $t \geq 0$, $n \geq 0$, $1 \leq i \leq m$, and $1 \leq j \leq m$. The matrices $\{P(n, t)\}$ give the joint distribution on $N(t)$ and $J(t)$. They provide an important key to the study of phase-type renewal processes. For example, we see that the marginal distribution of $J(t)$ is given by

$$P\{J(t) = j | J(0) = i\} = \sum_{n=0}^{\infty} P(n, t) = \exp(Q * t), \qquad (5.5.14)$$

and the renewal function is given by

$$E[N(t)] = \alpha \sum_{j=1}^{m} nP(n, t)e.$$

Conditioning on the time and the various starting states at the end of the first renewal after time 0, we can write the following recursion

$$
\begin{aligned}
&P(n, t) = \int_0^t \exp(Tu)T^0 \alpha P(n-1, t-u)du && n \geq 1, \ t \geq 0 \\
&P(0, t) = \exp(Tt) && t > 0 \\
&P(n, 0) = \delta_{0n}I && n \geq 0, \qquad (5.5.15)
\end{aligned}
$$

where $\delta_{ij} = 1$ if $i = j$ and 0 otherwise. Define the following transforms

$$P^{ge}(z, s) = \sum_{n=0}^{\infty} z^n \int_0^{\infty} e^{-st} P(n, t)dt \qquad\qquad P^e(n, s) = \int_0^{\infty} e^{-st} P(n, t)dt.$$

For $t > 0$, we take the Laplace transform with respect to the second argument of Equation 5.5.15. This gives

$$
\begin{aligned}
&P^e(n, s) = (sI - T)^{-1} T^0 \alpha P^e(n-1, s) && n \geq 1, \\
&P^e(0, s) = (sI - T)^{-1}.
\end{aligned}
$$

Multiplying the nth equation of the preceding system by z^n and adding the resulting equations, we obtain

$$P^{ge}(z, s) = (sI - T)^{-1} zT^0 \alpha P^{ge}(z, s) + (sI - T)^{-1} \quad \text{or}$$

$$\left[I - (sI - T)^{-1} zT^0 \alpha \right] P^{ge}(z, s) = (sI - T)^{-1}.$$

It can be shown that the inverse of the matrix in the brackets is nonsingular. Hence we find that

$$
\begin{aligned}
P^{ge}(z, s) &= [I - (sI - T)^{-1} zT^0 \alpha]^{-1} [sI - T]^{-1} \\
&= \left[[sI - T][I - (sI - T)^{-1} zT^0 \alpha] \right]^{-1} = \left[sI - (T + zT^0 \alpha) \right]^{-1}.
\end{aligned}
$$

Inverting the previous double transform with respect to the second argument, we obtain the matrix generating function

$$P^g(z, t) = \sum_{n=0}^{\infty} z^n P(n, t) = \exp\left[(T + zT^0 \alpha)t \right].$$

We see that

$$P^g(1, t) = \sum_{n=0}^{\infty} P(n, t) = \exp\left[(T + T^0 \alpha)t \right] = \exp(Q * t).$$

This result has been shown in Equation 5.5.14. To find the renewal function $M(t) = E[N(t)]$, we use

$$M(t) = \alpha \left[\frac{d}{dz} P^g(z, t) \Big|_{z=1} \right] e.$$

After some algebra, we see that the renewal function for the phase-type renewal process is given by

$$M(t) = \frac{t}{\mu} - \frac{t}{\mu} \alpha \exp(Q * t) T^{-1} e - 1 \qquad t \geq 0,$$

where μ is the mean interarrival time, that is, $\mu = -\alpha T^{-1} e$. The phase-type renewal process also has helpful properties. For example, it can be shown that the excess-life distribution at time t is also of phase type—now with representation (α', T), where $\alpha' = \alpha \exp(Q * t)$. ∎

5.6 Uniformization

Uniformization is a useful approach for modeling and computation. It establishes a linkage between a continuous-time Markov chain and a related discrete-time Markov chain in the sense that we construct an equivalent stochastic process called a Markov chain subordinated to a Poisson process.

Consider a stochastic process with state space $S = \{1, \ldots, n\}$ and let $X(t)$ denote the state of the system at time t. Assume that state changes are governed by a discrete-time Markov chain with transition probability matrix P and state space S. Here we allow *virtual transitions* (transitions that return to the same state in a single transition). Assume also that the epochs of state change are governed by an independent Poisson process $\{N(t), t \geq 0\}$ with rate v. At a state-change epoch, if the current state is i the process selects its next destination j with probability p_{ij}; the process stays in state i until it reaches the next state change epoch at which time it enters state j, and the scenario is repeated *ad infinitum*. The stochastic process $\{X(t), t \geq 0\}$ so constructed is called a *Markov chain subordinated to a Poisson process*. Its transition function is given by

$$P_{ij}(t) = P\{X(t) = j | X(0) = i\}$$

$$= \sum_{n=0}^{\infty} P\{X(t) = j, N(t) = n | X(0) = i\}$$

$$= \sum_{n=0}^{\infty} P\{X(t) = j | X(0) = i, N(t) = n\} P\{N(t) = n | X(0) = i\}$$

$$= \sum_{n=0}^{\infty} p_{ij}^{(n)} e^{-vt} \frac{(vt)^n}{n!}. \tag{5.6.1}$$

The process $\{X(t), t \geq 0\}$ is a Markov process because

$$P\{X(s+t) = j | X(u), u \leq s\} = \sum_{n=0}^{\infty} p_{X(s),j}^{(n)} e^{-vt} \frac{(vt)^n}{n!}.$$

For any continuous-time Markov chain with a constant transition rate v, that is, $v_i = v$ for all i, a derivation identical to that of Equation 5.6.1 will show that such a continuous-time Markov chain is equivalent to a special case of a Markov chain subordinated to a Poisson process in which virtual transitions are not permitted and $\{p_{ij}\}$ are the transition probabilities of the embedded Markov chain.

We now consider a continuous-time Markov chain in which $v_i \leq v$ for all i (the case in which transition rates may differ among states but are bounded from above). When in state i, the process leaves the state at a rate v_i. Consider a related process. In this related process, transitions out of each state occur at a constant rate of v. However, when the process is in state i only a fraction v_i/v are transitions out of i to state $j \neq i$ and the rest are transitions back to state i again (the virtual transitions).

For this related process, the transition probabilities are then given by

$$\tilde{P}_{ij} = \begin{cases} \dfrac{v_i}{v} p_{ij} & \text{if } i \neq j \\[2mm] 1 - \dfrac{v_i}{v} & \text{if } i = j, \end{cases} \tag{5.6.2}$$

where $\{p_{ij}\}$ are the transition probabilities of the embedded Markov chain underlying the continuous-time Markov chain defined at the onset of Section 5.3.

This related process has been constructed in the paradigm of a Markov chain subordinated to a Poisson process. Its transition function is thus given by

$$\tilde{P}_{ij}(t) = \sum_{n=0}^{\infty} \tilde{p}_{ij}^{(n)} e^{-vt} \frac{(vt)^n}{n!}. \tag{5.6.3}$$

The question is whether this *related* process is equivalent to the original continuous-time Markov chain in which $v_i \leq v$ for all i. In other words, is $P_{ij}(t) = \tilde{P}_{ij}(t)$ for all i, j, and $t \geq 0$? For the original chain, we recall that $P(t) = \exp(Qt)$. We write Equation 5.6.2 in matrix notation

$$\tilde{P} = I + \frac{1}{v} Q \tag{5.6.4}$$

or $Q = v\tilde{P} - vI$. Using the last identity, we find

$$P(t) = e^{Qt} = e^{(v\tilde{P} - vI)t} = e^{-vtI} e^{vt\tilde{P}} = \sum_{n=0}^{\infty} \frac{(-vtI)^n}{n!} \sum_{n=0}^{\infty} \frac{(vt\tilde{P})^n}{n!}$$

$$= \sum_{n=0}^{\infty} \frac{(-vt)^n}{n!} I \sum_{n=0}^{\infty} \frac{(vt)^n}{n!} \tilde{P}^n = \sum_{n=0}^{\infty} e^{-vt} \frac{(vt)^n}{n!} \tilde{P}^{(n)} = \tilde{P}(t).$$

Hence the two processes are probabilistically equivalent.

EXAMPLE
5.6.1

A Numerical Example   Consider a continuous-time Markov chain with state space $\{1, 2, 3\}$ and infinitesimal generator

$$Q = \begin{bmatrix} -1.8 & 1.2 & 0.6 \\ 0.8 & -2.8 & 2 \\ 3.6 & 1.2 & -4.8 \end{bmatrix}.$$

The uniformized process now has $v = \max_{i}\{v_i\} = 4.8$ and the transition probability matrix

$$\tilde{P} = I + \frac{1}{4.8}Q = \begin{bmatrix} 5/8 & 2/8 & 1/8 \\ 2/12 & 5/12 & 5/12 \\ 3/4 & 1/4 & 0 \end{bmatrix}.$$

As an alternative to using $\exp(Qt)$ to compute $P(t)$, we may instead use

$$P(t) = \sum_{n=0}^{\infty} e^{-4.8t}\frac{(4.8t)^n}{n!}\tilde{P}^{(n)}. \quad \blacksquare$$

EXAMPLE
5.6.2*

Computing the State Probability Vector   Assume we have a continuous-time Markov chain with state space $\{1, \ldots, n\}$, infinitesimal generator Q, and a starting state probability vector $x(0)$. Suppose that we are interested in computing the state probability vector $x(t) = \{x_1(t), \ldots, x_n(t)\}$ over a finite interval $(0, T)$. From Equation 5.6.3 we have

$$x(t) = x(0)P(t) = \sum_{k=0}^{\infty} x(0)e^{-vt}\frac{(vt)^k}{k!}\tilde{P}^{(k)}, \tag{5.6.5}$$

where v is the maximal transition rate and $\tilde{P}$ is defined by Equation 5.6.4. Let row vector $x(k) = x(0)\tilde{P}^k$ and $g(k; v, t)$ denote the Poisson mass function

$$e^{-vt}\frac{(vt)^k}{k!}.$$

We find $x(n)$ recursively from $x(k) = x(k-1)\tilde{P}$, $n \geq 1$. Then Equation 5.6.5 can be stated as

$$x(t) = \sum_{k=0}^{\infty} x(k)g(k; v, t).$$

For each i, we have

$$x_i(t) = \sum_{k=0}^{\infty} x_i(k)g(k; \lambda, t) = \sum_{k=0}^{E} x_i(k)g(k; \lambda, t) + \sum_{k=E+1}^{\infty} x_i(k)g(k; \lambda, t)$$

$$= \sum_{k=0}^{E} x_i(k)g(k; \lambda, t) + R_E,$$

where R_E is the second summation on the left of the last equality. It is clear that

$$0 \le R_E \le \sum_{k=E+1}^{\infty} g(k; \lambda, t).$$

We can choose a truncation point E so that the last summation is less than ε. Therefore the errors for all i are bounded by ε. This procedure of finding state probability vector is used frequently in applications. The procedure is numerically stable because it uses only additions of nonnegative numbers so that the roundoff error problem is mitigated. Since Q has negative entries on the diagonal, the same cannot be said about the computing of matrix exponential $\exp(Qt)$. ∎

EXAMPLE 5.6.3*
A Continuous-Time Markov Jump Process   In a continuous-time Markov chain, we do not allow for virtual transitions. A generalization of the process is one in which virtual transitions are permitted. In this case, state changes are governed by a discrete-time Markov chain with transition probability matrix P, where diagonal elements of P are not necessarily zero. For each i, the sojourn time in state i is exponential with rate v_i. With the exception of the possibility that $\{p_{ii}\}$ may not be zero, the process is identical to that of a continuous-time Markov chain. We call such a process a continuous-time Markov jump process. The process is equivalent to a continuous-time Markov chain with the same state space and infinitesimal generator

$$Q = \begin{bmatrix} -v_1 q_1 & v_1 p_{12} & v_1 p_{13} & \cdot \\ v_2 p_{21} & -v_2 q_2 & v_2 p_{23} & \cdot \\ \cdot & \cdot & \cdot & \cdot \\ \cdot & \cdot & \cdot & \cdot \end{bmatrix} \qquad (5.6.6)$$

where $q_i = 1 - p_{ii}$. In the equivalent continuous-time Markov chain, we see that the sojourn time in state i is exponential but with rate $v_i q_i$. Because of virtual transitions, the net rate of leaving state i is reduced by an amount $v_i p_{ii}$ from its original value v_i. ∎

EXAMPLE 5.6.4*
A Continuous-Time Markov Jump Process: A Numerical Example   Consider a stochastic process for which the state changes are governed by a discrete-time Markov chain with state space $S = \{1, 2, 3\}$ and transition probability matrix

$$P = \begin{bmatrix} 0.2 & 0.7 & 0.1 \\ 0.8 & 0 & 0.2 \\ 0.3 & 0.4 & 0.3 \end{bmatrix}.$$

For $i = 1, 2, 3$, the sojourn time in state i before making a transition follows an exponential distribution with parameter v_i, where $\{v_i\} = \{2, 10, 4\}$. Based on Example

5.6.3, this continuous-time Markov jump process is equivalent to a continuous-time Markov chain with the same state space and infinitesimal generator

$$Q = \begin{bmatrix} -1.6 & 1.4 & 0.2 \\ 8 & -10 & 2 \\ 1.2 & 1.6 & -2.8 \end{bmatrix}. \quad \blacksquare$$

5.7 Continuous-Time Markov Reward Processes

Consider the continuous-time Markov chain with state space S and infinitesimal generator Q. For each state $j \in S$, we let d_j denote the reward rate in state j. We assume that the reward is discounted continuously at a rate of β. Given that the process starts from state i at time 0, we let W_i denote the discounted total reward received in an infinite horizon. Then we see that

$$W_i = \sum_j \int_0^\infty e^{-\beta t} P_{ij}(t) d_j \, dt = \sum_j P_{ij}^e(\beta) d_j,$$

where $P_{ij}^e(\cdot)$ is the Laplace transform of $P_{ij}(t)$ discussed in Section 5.2. Let $W = \{W_i\}$ denote the discounted reward vector given the respective starting states.

Then, from the result given in Section 5.2, we find

$$W = P^e(\beta)d = [\beta I - Q]^{-1}d, \tag{5.7.1}$$

where column vector $d = \{d_j\}$. If we use the long-run average reward per unit as a criterion to do the accounting, then

$$g = \sum_{j \in S} \pi_j d_j, \tag{5.7.2}$$

where we call g the *gain rate* of the process and the limiting probabilities are defined by Equation 5.3.1.

We see that Equation 5.7.1 is the continuous analog of Equation 4.5.2. When there are lump-sum costs associated with each transition into or out of a state, then the cost rate vector d must be modified accordingly. These scenarios are covered in a more general setting where we also allow virtual transitions.

A Markov Jump Process with Rewards: The Discounted Reward Case*

Consider the continuous-time Markov jump process described in Example 5.6.3 in which we allow for virtual transitions. Denote this Markov process by $X = \{X(t), t \geq 0\}$. Let $\{T_n\}$ be the sequence of jump times (state-change epochs; each virtual transition counts as one state change) and let $Y_n = X(T_n^+)$, the state of the process immediately after the nth jump. Let r_i be the expected discounted reward received per each sojourn in state i. We again assume that the reward is discounted continuously at a rate of β. If the process is in state i and its next destination is state j, the reward to be received consists of three components: (i) $\rho_0(i, j)$, the lump-sum reward received at the start of the sojourn, (ii) $\rho_1(i, j) = d_i$, the reward *rate* during its sojourn in i, and (iii) $\rho_2(i, j)$, the lump-sum reward received at the end of the sojourn. Assume that at time 0 the process is in state i ($Y_0 = i$) and the next transition will occur at random time T_1 and will move to state Y_1 as a result of the transition. The present value of the expected reward received from this first sojourn is given by

$$r_i = \underset{T_1, Y_1}{E}\left[\rho_0(i, Y_1) + \int_0^{T_1} e^{-\beta t}\rho_1(i, Y_1)dt + e^{-\beta T_1}\rho_2(i, Y_1)\Big|Y_0 = i\right]. \qquad (5.7.3)$$

For the second term in the preceding brackets, we note that T_1 follows an exponential distribution with parameter ν_i and

$$\underset{T_1}{E}\int_0^{T_1} e^{-\beta t}dt = \int_0^\infty \int_0^\tau e^{-\beta t}dt\,\nu_i e^{-\nu_i\tau}d\tau = \int_0^\infty \int_t^\infty \nu_i e^{-\nu_i\tau}d\tau\,dt$$

$$= \int_0^\infty e^{-\nu_i t}e^{-\beta t}dt = \frac{1}{\beta + \nu_i}. \qquad (5.7.4)$$

For the term relating to ρ_2, we see that

$$\underset{T_1}{E}\left[e^{-\beta T_1}\right] = \int_0^\infty e^{-\beta\tau}\nu_i e^{-\nu_i\tau}d\tau = \frac{\nu_i}{\beta + \nu_i}. \qquad (5.7.5)$$

Using Equations 5.7.4 and 5.7.5 and conditioning on the destination states, we obtain an explicit expression for Equation 5.7.3:

$$r_i = \sum_j P_{ij}\left[\rho_0(i, j) + [\rho_1(i, j) + \rho_2(i, j)\nu_i]\left(\frac{1}{\beta + \nu_i}\right)\right]. \qquad (5.7.6)$$

Define the expected total discounted reward for an infinite horizon given $Y_0 = i$ as

$$W_i = \underset{\{T_n\},\{Y_n\}}{E}\left[\sum_{n=0}^\infty e^{-\beta T_n}r_{Y_n}\Big|Y_0 = i\right]. \qquad (5.7.7)$$

As indicated in Example 5.6.3, Markov process X can be converted to an equivalent continuous-time Markov chain with the infinitesimal generator of Equation 5.6.6. Under this latter representation, sojourns in a given state caused by successive virtual transitions are considered a single sojourn in the state. Assume that $\nu = \max_i\{\nu_i q_i\}$, where $q_i = 1 - p_{ii}$ as defined in Example 5.6.3. The continuous-time Markov chain is now uniformized to form a Markov chain subordinated to a Poisson process with rate ν and transition probability matrix $\tilde{P} = \{\tilde{p}_{ij}\}$, where

$$\tilde{p}_{ij} = \begin{cases} \dfrac{\nu_i p_{ij}}{\nu} & \text{if } i \neq j \\[2ex] 1 - \dfrac{\nu_i q_i}{\nu} & \text{if } i = j. \end{cases} \qquad (5.7.8)$$

We use $\tilde{X}$ to denote this uniformized Markov process and $\tilde{W}_i$ the corresponding expected discounted reward for an infinite horizon under starting state i. In

Figure 5.9, we depict the relations of these three equivalent stochastic processes. Since processes X and $\tilde{X}$ are two representations of the identical system, the expected total discounted rewards W_i and $\tilde{W}_i$ will be the same if the expected discounted rewards associated with each *actual* sojourn of state i are the same for each i. By an actual sojourn, we mean the sojourn that may contain multiple visits to the same state caused by contiguous virtual transitions. (The term *sojourn* without the adjective *actual* refers to one involving only a single visit.) Let w_i and $\tilde{w}_i$ represent the discounted expected rewards associated with an actual sojourn in state i under processes X and $\tilde{X}$, respectively, and $\tilde{r}_i$ denote the expected discounted reward associated with a sojourn in state i under process $\tilde{X}$. We want to find an expression for $\tilde{r}_i$ so that $w_i = \tilde{w}_i$ and consequently $W_i = \tilde{W}_i$.

For the original Markov process X, we consider an actual sojourn in state i. Let N be the number of virtual transitions to state i. We see that

$$P\{N = n\} = (p_{ii})^n (1 - p_{ii}) \qquad n \geq 0. \qquad (5.7.9)$$

Let τ_k be the time of the kth virtual transition to state i. Clearly, τ_k is the sum of k i.i.d. exponential random variables, each with rate v_i. The expected discounted reward associated with an actual sojourn in state i when $N = n$ is given by

$$\underset{\{\tau_k\}}{E}\left[\sum_{k=0}^{n} e^{-\beta \tau_k} r_i\right] = r_i \underset{\{\tau_k\}}{E}\left[\sum_{k=0}^{n} e^{-\beta \tau_k}\right] = r_i \sum_{k=0}^{n} E[e^{-\beta \tau_k}] = r_i \sum_{k=0}^{n}\left(\frac{v_i}{\beta + v_i}\right)^k = r_i \sum_{k=0}^{n} \gamma_i^{\,k},$$

where we use Equation 5.7.5 to establish the third equality and define $\gamma_i = v_i/(\beta + v_i)$ in the last term. Using Equation 5.7.7, we have

$$w_i = r_i \underset{N}{E}\left[\sum_{k=0}^{N} \gamma_i^{\,N}\right] = r_i \sum_{n=0}^{\infty} (p_{ii})^n (1 - p_{ii}) \sum_{k=0}^{n} \gamma_i^{\,k} = r_i \sum_{k=0}^{\infty} \gamma_i^{\,k} \sum_{n=k}^{\infty} (p_{ii})^n (1 - p_{ii})$$

$$= r_i \sum_{k=0}^{\infty} (\gamma_i\, p_{ii})^k \sum_{n=k}^{\infty} (p_{ii})^{n-k} (1 - p_{ii}) = \frac{r_i}{1 - \gamma_i\, p_{ii}}. \qquad (5.7.10)$$

Continuous-time jump process $X = \{X(t), t \geq 0\}$

r_i = discounted reward per sojourn in i
w_i = discounted reward per actual sojourn in i
W_i = discounted reward in $[0, \infty)$ given $Y_0 = i$

via Equation 5.6.6

An equivalent continuous-time Markov chain

via Equation 5.7.8

An equivalent discrete-time Markov chain subordinated to a Poisson process $\tilde{X} = \{\tilde{X}(t), t \geq 0\}$

$\tilde{r}_i$ = discounted reward per sojourn in i
$\tilde{w}_i$ = discounted reward per actual sojourn in i
$\tilde{W}_i$ = discounted reward in $[0, \infty)$ given $Y_0 = i$

FIGURE
5.9 The three equivalent processes.

Under process $\tilde{X}$, the transition probabilities are $\{\tilde{p}_{ij}\}$, the constant transition rate is v, and the expected discounted reward associated with a sojourn in state i is $\tilde{r}_i$. A derivation identical to that of Equation 5.7.10 will show that

$$\tilde{w}_i = \frac{\tilde{r}_i}{1 - \tilde{\gamma}\, \tilde{p}_{ii}}, \tag{5.7.11}$$

where $\tilde{\gamma} = v/(\beta + v)$. If we set

$$\tilde{r}_i = \frac{r_i(\beta + v_i)}{\beta + v}, \tag{5.7.12}$$

then the following shows that $w_i = \tilde{w}_i$ and hence $W_i = \tilde{W}_i$:

$$\tilde{w}_i = \frac{\dfrac{r_i(\beta + v_i)}{\beta + v}}{1 - \left(\dfrac{v}{\beta + v}\right)\left[1 - \dfrac{v_i q_i}{v}\right]} = \frac{r_i(\beta + v_i)}{(\beta + v) - [v - v_i q_i]}$$

$$= \frac{r_i(\beta + v_i)}{\beta + v_i - v_i p_{ii}} = \frac{r_i}{1 - \left(\dfrac{v_i}{\beta + v_i}\right)p_{ii}} = \frac{r_i}{1 - \gamma_i p_{ii}} = w_i.$$

We see that Equations 5.7.8 and 5.7.12 enable us to convert a continuous-time Markov reward process to an equivalent discrete-time Markov reward chain subordinated to a Poisson process with rate v. For the latter process, we let $\tilde{Y}_n$ denote the state of the system right after the nth Poisson event. Then the total expected discounted reward is given by

$$\tilde{W}_i = E\left[\sum_{n=0}^{\infty} \alpha^n \tilde{r}_n \,\middle|\, \tilde{Y}_0 = i\right],$$

where the single-transition discounted factor $\alpha = E[e^{-\beta \tau_1}] = \tilde{\gamma} = v/(\beta + v)$ since τ_1 is an exponential random variable with parameter v. The results pertaining to the discounted Markov reward process presented in Section 4.5 are readily applicable.

A Markov Jump Process with Rewards: The Average Reward Case*

Under Markov reward process X, we define the expected time-average reward per unit time

$$G_i = \lim_{t \to \infty} E\left[\frac{1}{t}\sum_{n=0}^{N_t} r_{Y_n} \,\middle|\, Y_0\right],$$

where $N_t = \max_{t \to \infty}\{n | T_n \leq t\}$ and in Equation 5.7.6, we let $\beta = 0$ so that r_i represents the expected reward associated with a transition out of state i. For Markov reward process $\tilde{X}$, we use $\tilde{G}_i$ to denote the corresponding reward similarly. Applying Abelian theorem, we obtain

$$G_i = \lim_{\beta \to 0} \beta W_i = \lim_{\beta \to 0} \beta \, \tilde{W}_i = \tilde{G}_i \qquad (5.7.13)$$

The preceding result suggests that if we set β equal to zero in Equation 5.7.12, the two expected time-average rewards will be identical. For the uniformized process $\tilde{X}$, another measure of time-average reward will be the expected average reward per transition defined by

$$g_i = \lim_{n \to \infty} \left[\frac{1}{n+1} \sum_{k=0}^{n} \tilde{r}_i \Big| \tilde{Y}_0 = i \right].$$

Rewriting $\alpha = v/(\beta + v)$ as $\beta = v(1 - \alpha)/\alpha$, we have

$$\tilde{G}_i = \lim_{\beta \to 0} \beta \, \tilde{W}_i = \lim_{\alpha \to 1} v \left(\frac{1-\alpha}{\alpha} \right) E \left[\sum_{n=0}^{\infty} \alpha^n \, \tilde{r}_n \mid \tilde{Y}_0 = i \right]$$

$$= v \lim_{\alpha \to 1} \left(\frac{1}{\alpha} \right) \lim_{\alpha \to 1} E \left[\sum_{n=0}^{\infty} \alpha^n \, \tilde{r}_n \mid \tilde{Y}_0 = i \right] = v g_i,$$

where the last equality is obtained by an application of Abelian theorem. The previous result is anticipated because the identity simply says that the expected average reward per transition is equal to the average reward rate multiplied by the expected transition time. Since we restrict our attention to an ergodic Markov chain, we remark that the given quantities relating to gain rates are actually independent of the starting states.

EXAMPLE 5.7.1* **A Markov Jump Process with Rewards** Consider the Markov jump process of Example 5.6.4. Assume that the process generates rewards in accordance with the parameters

$$\{\rho_0(i, j)\} = \begin{bmatrix} 5 & 2 & 1 \\ 2 & 0 & 4 \\ 7 & 3 & 11 \end{bmatrix} \qquad \{\rho_2(i, j)\} = \begin{bmatrix} 12 & 4 & 7 \\ 3 & 0 & 9 \\ 5 & 3 & 15 \end{bmatrix}$$

and $\{d\} = \{1, 4, 2\}$ and the discount factor $\beta = 0.1$. Using Equation 5.7.8, we construct a discrete-time Markov chain subordinated to a Poisson process with rate $v = 10$. The transition probability matrix of the chain is given by

$$\tilde{P} = \begin{bmatrix} .84 & .14 & .02 \\ .80 & 0 & .20 \\ .12 & .16 & .72 \end{bmatrix}$$

Following Equation 5.7.6, we find in the Appendix the discounted rewards per sojourn in each state of the original Markov jump process $\{r_i\} = \{8.6, 6.95, 14.11\}$. For the Markov chain subordinated to the Poisson process, Equation 5.7.12 gives the corresponding $\{\tilde{r}_i\} = \{1.79, 6.95, 5.73\}$. Coincidentally, since $p_{22} = 0$ and $\max_i \{v_i q_i\} = v_2 = 10$, the single sojourn reward in state 2 is unaffected by the two transformations shown in Figure 5.9 and consequently $r_2 = \tilde{r}_2 = 6.95$ as expected.

The single-transition discounted factor for the Markov chain subordinated to the Poisson process is $\alpha = 10/10.1 = 0.9901$. Using Equation 4.5.2, we obtain the discounted total returns for the three possible starting states

$$\tilde{W} = [I - \alpha \tilde{P}]^{-1} \tilde{r} = \begin{bmatrix} 299.65 \\ 306.26 \\ 312.92 \end{bmatrix}. \quad \blacksquare$$

EXAMPLE
5.7.2*

A Markovian System Failure Model   Consider a system whose condition can be represented by $M + 1$ states indexed by $0, \ldots, M$. State 0 denotes a "like-new" state and state M a complete system failure. Other states are ordered and indexed in increasing degree of deterioration. The system's deterioration process is described by a Markov process with transition rates $\{\mu_{ij}\}$, where $\mu_{ij} = 0$ if $i > j$. When the system reaches state M, a restoration effort will bring the system back to state 0. The time needed for restoration follows an exponential distribution with rate η. When in state i, where $0 \le i \le M$, there is an occupancy cost of A_i incurred per unit time in the state. In addition, there is a fixed cost of restoration C payable at the end of the restoration effort. Let β be the continuous discount factor. We would like to compute the expected discounted cost for an infinite planning horizon.

Since $\mu_{ij} = 0$ if $i > j$, transitions cannot be made to a "better" state. Define $\lambda_i = \sum_{j \le i} \mu_{ij}$ and $p_{ij} = \mu_{ij}/\lambda_i$. Then the process is a continuous-time Markov jump process with transition probability matrix $P = \{p_{ij}\}$ and transition rates $\{v_i\} = \{\lambda_0, \ldots, \lambda_{M-1}, \eta\}$, where P is upper triangular. Let $v = \max\{\lambda_0, \ldots, \lambda_{M-1}, \eta\}$. The costs per sojourn of this jump process are $r_i = A_i/(\beta + \lambda_i)$, $i = 0, 1, \ldots, M - 1$, and $r_M = C\eta/(\beta + \eta)$. Using Equations 5.7.8 and 5.7.12, we can construct the corresponding Markov chain subordinated by the Poisson process with rate v. The discounted total costs for each one of the starting states can be computed accordingly.   ∎

EXAMPLE
5.7.3

A Cash Balance Problem   In Example 4.4.1, we used a Markov chain to model a discrete-time cash balance problem. In this example, we consider a continuous-time version of the problem. We assume that the time between successive cash withdrawals follows an exponential distribution with parameter λ and the time between successive cash deposits follows an exponential distribution with parameter μ. The size of each cash withdrawal follows a discrete distribution $w(i)$, $i = 1, 2, \ldots$ and the size of each cash deposit follows a discrete distribution $d(i)$, $i = 1, 2, \ldots$. Thus the cash withdrawal and deposit processes are two independent compound Poisson processes.

Assume that the firm uses a two-parameter policy to control its cash balance. Let the two parameters be denoted by (α, β), where $0 < \alpha < \beta$. Let $X(t)$ denote the cash position at time t. If $X(t) < \alpha$, the cash position is increased to α at a variable cost of c_1 per dollar of cash increased; if $X(t) > \beta$, it is decreased to β at a variable cost of c_2 per dollar of cash decreased. The transaction times are assumed to be negligible. Also, c_3 is the cost of carrying a dollar of cash per unit time (the cash holding cost rate). While the cost structure is identical to that of Example 4.4.1, the control policy used here is different from the earlier example.

The stochastic process $X = \{X(t),\ t \geq 0\}$ under the control policy (α, β) is a continuous-time Markov chain with state space $S = \{\alpha, \alpha + 1, \ldots, \beta - 1, \beta\}$ with transition diagram shown in Figure 5.10.

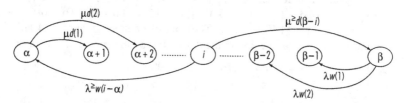

FIGURE
5.10   The transition diagram for the cash balance problem.

In the figure, we use the notations $^2d(i) = d(i) + d(i+1) + \cdots$ and $^2w(i) = w(i) + w(i+1) + \cdots$. The infinitesimal generator Q of the chain is given by

$$
Q = \begin{array}{c@{\quad}c}
 & \begin{array}{ccccccc} \alpha & \alpha+1 & \alpha+2 & \cdot & & \beta-1 & \beta \end{array} \\
\begin{array}{c} \alpha \\ \alpha+1 \\ \alpha+2 \\ \\ i \\ \\ \beta-2 \\ \beta-1 \\ \beta \end{array} &
\left[\begin{array}{ccccccc}
* & \mu d(1) & \mu d(2) & \cdot & & \cdot & \mu^2 d(\beta-\alpha) \\
\lambda^2 w(1) & * & \mu d(1) & \cdot & & \cdot & \mu^2 d((\beta-(\alpha+1)) \\
\lambda^2 w(2) & \lambda w(1) & * & & & & \mu^2 d((\beta-(\alpha+2)) \\
 & & & \cdot & & & \\
\lambda^2 w(i-\alpha) & \cdot & \cdot & & & & \mu^2 d(\beta-i) \\
 & & & & & & \\
\lambda^2 w((\beta-2)-\alpha) & \cdot & \cdot & & * & \mu d(1) & \mu^2 d(2) \\
\lambda^2 w((\beta-1)-\alpha) & \cdot & \cdot & & \lambda w(1) & * & \mu^2 d(1) \\
\lambda^2 w(\beta-\alpha) & \cdot & \cdot & & \lambda w(2) & \lambda w(1) & *
\end{array}\right]
\end{array}
$$

where the diagonal elements are not shown explicitly (they have values so that each row sums to zero).

Let q_i denote the cost rate in state i. If $X(t) = i$, then $\lambda w(i-j)$ is the transition rate of moving to state j because of a withdrawal in the next infinitesimal interval. If $j \leq \alpha - 1$, we need to move the cash position to α. Let h denote the amount of cash replenishment. Then we require $j + h = \alpha$ or $h = \alpha - j$. So a part of q_i is $\sum_{j \leq \alpha-1} \lambda w(i-j)(\alpha - j)c_1$. Similarly, if $X(t) = i$ then $\mu d(j-i)$ is the transition rate of moving to state j because of a deposit in the next infinitesimal interval.

If $j \geq \beta + 1$, we must reduce the cash position to β. Let k denote the amount of cash disposal. Then we require $j - k = \beta$ or $k = j - \beta$. Another part of q_i is then $\sum_{j \geq \beta + 1} \mu d(j - i)(j - \beta) c_2$. Combining these two parts with the cost rate of carrying cash, the cost rate in state i is given by

$$q_i = c_1 \lambda \sum_{j \leq \alpha - 1} w(i - j)(\alpha - j) + c_2 \mu \sum_{j \geq \beta + 1} d(j - i)(j - \beta) + c_3 i.$$

Using Equation 5.7.2, we find the long-run expected average cost per unit time

$$g = \sum_{j=\alpha}^{\beta} \pi_j q_j .$$

Consider a numerical example with $\{d(1), \ldots, d(4)\} = \{0.4, 0.3, 0.2, 0.1\}$, $\{w(1), \ldots, w(4)\} = \{0.2, 0.3, 0.3, 0.2\}$, $\lambda = 0.5$, and $\mu = 0.4$. The cost parameters are $\{c_1, c_2, c_3\} = \{1, 0.5, 0.05\}$. The control parameters are $\alpha = 4$ and $\beta = 10$. Based on these data, we find in the Appendix the infinitesimal generator

$$Q = \begin{array}{r} 4 \\ 5 \\ 6 \\ 7 \\ 8 \\ 9 \\ 10 \end{array} \left[\begin{array}{rrrrrrr} -.4 & .16 & .12 & .08 & .04 & & \\ .5 & -.9 & .16 & .12 & .08 & .04 & \\ .4 & .1 & -.9 & .16 & .12 & .08 & .04 \\ .25 & .15 & .1 & -.9 & .16 & .12 & .12 \\ .1 & .15 & .15 & .1 & -.9 & .16 & .24 \\ & .1 & .15 & .15 & .1 & -.9 & .4 \\ & .1 & .15 & .15 & .1 & -.5 \end{array} \right].$$

The limiting probability vector is given by $\pi = \{0.3637, 0.119, 0.1221, 0.112, 0.0915, 0.0618, 0.13\}$. The cost rates are $\{q_i\} = \{1.45, 1, 0.65, 0.47, 0.48, 0.65, 0.9\}$. Together, they yield the long-run expected average cost per unit time of 0.9794. ∎

5.8 Reversible Continuous-Time Markov Chains

In Section 4.6, we introduced the notion of reversed and reversible discrete-time Markov chains. We now examine these issues in the context of a continuous-time Markov chain. Consider a continuous-time Markov chain $X = \{X(t), t \geq 0\}$ in steady state and let $\{\pi_j\}$ be the steady state probabilities. We look at the process in reversed order of time axis as shown in Figure 5.11.

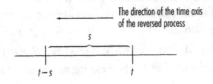

FIGURE
5.11   A reversed continuous-time Markov chain.

We see that

$P\{X$ in state i during $[t-s,\, t]\,|\, X(t)=i\}$

$$= \frac{P\{X \text{ in state } i \text{ during } [t-s,\, t]\}}{P\{X(t)=i\}} = \frac{P\{X(t-s)=i\}P\{\tau_i > s\}}{P\{X(t)=i\}} = \frac{\pi_i e^{-v_i s}}{\pi_i} = e^{-v_i s}.$$

This shows that in the reversed process the holding time in state i is again exponential with rate v_i. For the forward process, we let $P = \{p_{ij}\}$ denote the transition probability matrix of the embedded discrete-time Markov chain whose stationary probabilities are denoted by $\{\eta_j\}$. Corresponding to this discrete-time process, there exists a reversed discrete-time Markov chain. For this reversed chain to be reversible, Equation 4.6.4 requires that

$$\eta_i p_{ij} = \eta_j p_{ji} \qquad \text{for all } i \neq j. \tag{5.8.1}$$

From Equation 5.3.2, we have $\pi_i = (\eta_i/v_i)/C$, where C is a normalizing constant, or $\eta_i = C\pi_i v_i$.

> Using the last identity in Equation 5.8.1, we obtain the necessary and sufficient condition for an irreducible continuous-time Markov chain to be reversible:
>
> $$\pi_i v_i p_{ij} = \pi_j v_j p_{ji} \qquad \text{or} \qquad \pi_i q_{ij} = \pi_j q_{ji} \qquad \text{for all } i \neq j. \tag{5.8.2}$$

Equation 5.8.2 is the continuous analog of Equation 4.6.1. The Kolmogorov criterion for reversibility of a continuous-time Markov chain is identical to that of the discrete-time Markov chain except with transition rates replacing transition probabilities in its stipulation. The criterion says that for any state, the product of the transition rates along any path going back to the same state remains unchanged if the direction of state transition is reversed.

EXAMPLE
5.8.1
Birth and Death Processes Revisited   For the birth and death process defined in Example 5.1.1, the balance equation obtained at each cut in Example 5.3.2 is none other than that of Equation 5.8.2. Hence any ergodic birth and death process is reversible. ∎

EXAMPLE
5.8.2
The Departure Process from an M/M/s Queue   The $M/M/s$ queue has been defined in Example 5.1.3. There we let $X(t)$ denote the number of customers at time t. Since the continuous-time Markov process $\{X(t),\, t \geq 0\}$ is a birth and death process, we conclude that the process is reversible. If we construct a sample path as shown in Figure 5.12, we see that for the process with its time axis reversed, each point of increase (epoch at which a jump occurs) is an arrival epoch of a Poisson process with rate λ because the forward and reversed processes follow the same

probabilistic law. These points of increase of the reverse process correspond to the departure epochs of the forward process. Hence we see that the departure process of the queue is a Poisson process with rate λ. ∎

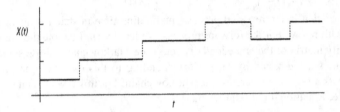

FIGURE
5.12   A sample path of an $M/M/s$ queue.

EXAMPLE
5.8.3   **A Markovian Queue with Two Heterogeneous Servers**   Consider a service system with two exponential servers with different service rates. For server i, $i = 1, 2$, the service rate is μ_i. Define $\mu = \mu_1 + \mu_2$. The arrival process is Poisson with rate λ. Assume that the queue is stable so we require $\mu > \lambda$. When a customer arrives and both servers are idle, the customer will be served by server i with probability p_i, where $p_1 + p_2 = 1$; if only one server is idle, the arriving customer will be served by the idle server; if all servers are busy the arriving customer will stay in a waiting line. Assume, as usual, that the arrival process and server times are independent.

Let $X(t)$ denote the state of the system at time t. Then $X = \{X(t), t \geq 0\}$ is a continuous-time Markov chain with state space $S = \{0, (1A), (1B), 2, 3, \ldots\}$ and transition rates shown in the transition diagram given in Figure 5.13. State $(1A)$ indicates that Server 1 is busy while Server 2 is idle and state $(1B)$ is interpreted similarly. Other states are defined as in the case of an $M/M/2$ queue.

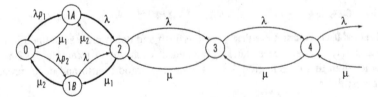

FIGURE
5.13   The transition diagram for the two-server queue.

The structure of the transition diagram shown in Figure 5.13 suggests that the Kolmogorov criterion is satisfied if the product of the transition rates along the path marked by the four darkened arcs is equal to its counterpart when the direction of the path is reversed, namely, if

$$(\lambda p_1)\lambda\mu_1\mu_2 = (\lambda p_2)\lambda\mu_1\mu_2.$$

Thus we conclude that, in equilibrium, the chain is reversible if $p_1 = p_2 = 1/2$ and not reversible otherwise. ∎

If $X = \{X(t),\ t \geq 0\}$ and $Y = \{Y(t),\ t \geq 0\}$ are two independent reversible continuous-time Markov chains with respective generators $R = \{r_{ij}\}$ and $S = \{s_{ij}\}$ and limiting probabilities $\{v_i\}$ and $\{\eta_i\}$, then the continuous-time Markov chain $(X,\ Y) = \{(X(t),\ Y(t)),\ t \geq 0\}$ is also reversible. To establish this result, we let $Q = \{q_{(i,j),(m,n)}\}$ denote the generator of the process $(X,\ Y)$ and $\{\pi_{(i,j)}\}$ its limiting probabilities. With the above definitions, we have $q_{(i,j),(i,n)} = s_{jn}$, $q_{(i,j),(m,j)} = r_{im}$, and $\pi_{(i,j)} = v_i \eta_j$. Applying Equation 5.8.2, the reversibility conditions for $(X,\ Y)$ read $\pi_{(i,j)} q_{(i,j),(m,n)} = \pi_{(m,n)} q_{(m,n),(i,j)}$. We consider two cases: $n = j$, and $m = i$. for the first case, we have

$$\begin{aligned}
\pi_{(i,j)} q_{(i,j),(m,j)} &= v_i \eta_j r_{im} \\
&= v_m r_{mi} \eta_j \qquad \text{(by reversibility of } X) \\
&= \pi_{(m,j)} q_{(m,j),(i,j)}.
\end{aligned}$$

For the second case, we have

$$\begin{aligned}
\pi_{(i,j)} q_{(i,j),(i,n)} &= v_i \eta_j s_{jn} \\
&= v_i \eta_n s_{nj} \qquad \text{(by reversibility of } Y) \\
&= \pi_{(i,n)} q_{(i,n),(i,j)}.
\end{aligned}$$

This shows that the process $(X,\ Y)$ is reversible.

We now consider another useful result of reversible continuous-time Markov chains. Let X be such a chain with state space S and limiting probabilities $\{\pi_i\}$. We call process Z a *truncated* continuous-time Markov chain with state space A, where $A \subset S$, if we set $q_{ij} = 0$ for all $i \in A$ and $j \notin A$. The chain so truncated is also reversible with limiting probabilities given by

$$\pi_j^* = \frac{\pi_j}{\displaystyle\sum_{k \in A} \pi_k} \qquad j \in A. \tag{5.8.3}$$

To verify that the truncated chain is also reversible, we need to check that the conditions of Equation 5.8.2 are again met by the truncated process. Since the original chain is reversible, this implies that $\pi_i q_{ij} = \pi_j q_{ji}$ for all $i,\ j \in A$. Dividing the last equality by the denominator of Equation 5.8.3, we obtain $\pi_i^* q_{ij} = \pi_j^* q_{ji}$ for all $i,\ j \in A$ and the result follows.

EXAMPLE
5.8.4
An M/M/s/s Queue This is a variant of an *M/M/s* queue in which the system does not have a waiting room. For an *M/M/s* queue, the continuous-time Markov chain associated with it has a state space $S = \{0, 1, 2, \ldots\}$ and in Example 5.8.2 we concluded that the chain is reversible. The continuous-time Markov chain associated

with an $M/M/s/s$ queue is a truncation of the continuous-time Markov chain under an $M/M/s$ queue with $A = \{0, 1, ..., s\}$. Using the results obtained in Examples 5.3.2 and 5.3.3 and following Equation 5.8.3, we find that the limiting probabilities of queue length are given by

$$\pi_j = \frac{\dfrac{\rho^j}{j!}}{\displaystyle\sum_{k=0}^{s} \dfrac{\rho^k}{k!}} \qquad 0 \le j \le s, \qquad\qquad (5.8.4)$$

where $\rho = \lambda/\mu$. Equation 5.8.4 is known as *Erlang's loss formula.* ∎

EXAMPLE
5.8.5

Two Queues with a Common Waiting Room   Consider two independent and stable $M/M/1$ queues, called Queue 1 and Queue 2, with traffic intensities $\rho_i = \lambda_i/\mu_i < 1$ for $i = 1$ and 2, respectively. For $i = 1, 2$, we let $X_i(t)$ denote the number of customers in Queue i at time t. Following our earlier discussions, the two-dimensional continuous-time Markov chain $\{(X_1(t), X_2(t)), t \ge 0\}$ has a state space $S = \{(i, j), i \ge 0, j \ge 0\}$ and is reversible. Moreover, the limiting queue length probabilities are

$$\pi_{(i,j)} = [(1-\rho_1)\rho_1^i][(1-\rho_2)\rho_2^j] \qquad i \ge 0, \ j \ge 0.$$

Assume now that the two queues share a common waiting room of size C. The structure of the transition diagram of the resulting two-dimensional continuous-time Markov chain remains unchanged except now the state space is truncated with $A = \{(i, j), i \ge 0, j \ge 0, i + j \le C\}$. Following Equation 5.8.3), we obtain the limiting queue length probabilities below:

$$\pi_{(i,j)}^* = \pi_{(0,0)}^* \rho_1^i \rho_2^j \qquad (i, j) \in A,$$

where $\pi_{(0,0)}^*$ can be found by the normalizing equation. ∎

EXAMPLE
5.8.6

Two Single-Server Exponential Queues in Tandem   Consider two $M/M/1$ queues arranged in series with Queue 1 followed by Queue 2 as shown in Figure 5.14. A customer completing service by the first server will be served by the second server immediately if the second server is idle and will join the queue for the second server otherwise. The arrival process to Queue 1 is Poisson with rate λ and the service rates of the two servers are, respectively, μ_1 and μ_2. Define $\rho_i = \lambda/\mu_i$, for $i = 1$ and 2. From Example 5.8.2, we conclude that in steady state, departures from Queue 1 are Poisson with parameter λ. Hence arrivals to Queue 2 are again Poisson with rate λ and Queue 2 *viewed in isolation* is an $M/M/1$ queue with traffic intensity ρ_2. For $i = 1, 2$, using Equation 5.3.8 we have

$$\pi_n^i = (1-\rho_i)\rho_i^n \qquad n \ge 0,$$

where $\pi_n^i = \lim_{t \to \infty} P\{X_i(t) = n\}$ is the limiting probability that Queue i has n customers and $X_i(t)$ denotes that the number of customers in Queue i at time t.

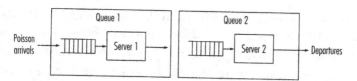

5.14   Two single-server queues in tandem.

We will next find the joint limiting queue length distribution of $X_1(t)$ and $X_2(t)$. Let $\{a_i | a_i < t\}$ denote the sequence of departure epochs from Queue 1 before time t. If we can show that $X_1(t)$ and $a_i | a_i < t\}$ are independent, these departures are arrivals to Queue 2 before time t. Consequently, $X_1(t)$ and $X_2(t)$ are independent. Reversing the time axis of the sample path of $\{X_1(t), t \geq 0\}$, in the reversed sample path, $\{a_i | a_i < t\}$ become arrival epochs to Queue 1 after time t. Since arrivals follow the Poisson process, clearly $\{a_i | a_i < t\}$ and $X_1(t)$ are independent. From this, we conclude that $\lim_{t \to \infty} P\{X_1(t) = n, X_2(t) = m\} = \pi_n^1 \pi_m^2$ for $n \geq 0$ and $m \geq 0$. ∎

If $X = \{X(t), t \geq 0\}$ is a stationary continuous-time Markov chain with state space S, generator $Q = \{q_{ij}\}$, and limiting probabilities $\{\pi_j\}$, then we can construct a reversed continuous-time Markov chain with the same state space, generator $Q^* = \{q_{ij}^*\}$, and the same limiting probabilities, where

$$q_{ij}^* = \frac{\pi_j q_{ji}}{\pi_i} \qquad i \neq j \text{ and } i, j \in S. \qquad (5.8.5)$$

It is possible that $Q \neq Q^*$ and the chain is not reversible. However, the idea of the reversed chain can be rather useful in a way similar to the stationary discrete-time Markov chain discussed in Section 4.6.

The idea (known as Kelly's lemma) boils down to this: If we can find a collection of nonnegative numbers $\{q_{ij}^*, i \neq j\}$ and a nonnegative vector $\{\pi_i\}$ whose elements sum to 1 such that

$$v_i^* = v_i \qquad i \in S \qquad (5.8.6)$$

and

$$\pi_i q_{ij}^* = \pi_j q_{ji} \qquad i \neq j, \qquad (5.8.7)$$

where as before we have $v_i = \sum_{j \neq i} q_{ij}$ and $v_i^* = \sum_{j \neq i} q_{ij}^*$, then $Q^* = \{q_{ij}^*\}$ is the generator of the reversed chain and $\{\pi_i\}$ is the limiting probabilities for both chains.

The first result follows directly from comparing Equations 5.8.5 and 5.8.7. The second result can be established by adding both sides of Equation 5.8.7, first with respect to all $i \ne j$ yielding $\sum_{i \ne j} \pi_i q^*_{ij} = \pi_j \sum_{i \ne j} q_{ij} = \pi_j v_j = \pi_j v^*_j$, or in matrix notation $\pi Q^* = 0$, and then (starting afresh) with respect to all $j \ne i$ yielding $\pi Q = 0$. The second result follows accordingly. In applications, typically we will guess the forms of $\{q^*_{ij}\}$ and $\{\pi_i\}$ and assess them by using Equations 5.8.6 and 5.8.7. This approach has been employed fruitfully in solving problems in queueing networks. We will now look at some examples.

EXAMPLE **5.8.7**   **Open Jackson Networks**   Consider a system of s servers. For server i, service times follow an exponential distribution with rate μ_i, arrivals from *outside* of the system follow a Poisson process with rate r_i, and there is no limit on the queue size. When a customer completes service at server i, the customer joins the queue in front of server j with probability P_{ij} and leaves the system with probability P_{i0}. Hence we have $\sum_{j=0}^{s} P_{ij} = 1$ for each i. This type of routing is called *Markov routing*. The system is depicted in Figure 5.15.

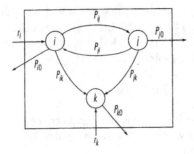

FIGURE **5.15**   A queueing network.

Let λ_j denote the total arrival rate to server j. The arrival rates can be found by solving the following system of linear equations:

$$\lambda_j = r_j + \sum_{i=1}^{s} \lambda_i P_{ij} \qquad j = 1, \dots, s. \qquad (5.8.8)$$

Equation 5.8.8 is called the *traffic equation*. In matrix notation, we have $\lambda = r + \lambda \tilde{P}$, where $\lambda = \{\lambda_i\}$ and $r = \{r_i\}$ are row vectors, and $\tilde{P} = \{P_{ij}\}$ is a substochastic matrix of size $s \times s$ (it is substochastic because in an open network, for $1 \le i \le s$, we must have at least one $P_{i0} > 0$). Thus a unique solution $\lambda = r(I - \tilde{P})^{-1}$ exists (see Section 4.4). Define $\rho_i = \lambda_i / \mu_i$ and assume $\rho_i < 1$ for all i. For $i = 1, \dots, s$, we let $X_i(t)$ denote the queue length at server i at time t. The stochastic process $\{(X_1(t), \dots, X_s(t)), t \ge 0\}$ forms a continuous-time Markov chain with generator $\{q(\cdot, \cdot)\}$. Define the limiting queue length distribution $\pi(n)$ as

$$\pi(n) = \lim_{t \to \infty} P\{X_1(t) = n_1, \ldots, X_s(t) = n_s\},$$

where row vector $n = \{n_1, \ldots, n_s\}$. Our goal is to show that

$$\pi(n) = \prod_{i=1}^{s} (1 - \rho_i)\rho_i^{n_i}. \tag{5.8.9}$$

The preceding equation is known as the product-form solution of a Jackson network. It gives the impression that the joint queue length distribution is given by the product of s independent marginal queue length distributions, each of an $M/M/1$ type. This is not true. In fact the arrival process to a server is generally not Poisson. To see this, we look at a single-server system with feedback, that is, with $P_{11} > 0$ and $P_{10} > 0$. While arrivals from outside of the system follow a Poisson process, the arrival process to the server does not; the arrival rate to the server at service completion epochs differs from that during an interval in which service is in progress.

Embarking on the derivation of the limiting queue length distribution, we first *conjecture* that the solution is of product form with $\pi(n) = \prod_{i=1}^{s} \pi_i(n_i)$. For each server in steady state, the input rate is equal to the output rate. For the system in steady state, the rate of moving through an arc from i to j should equal the rate of moving through the arc from j to i. The rate from i to j is $\lambda_i P_{ij}$. We let $\{P_{ij}^*\}$ denote the routing probabilities of the reversed process. For the reversed process, the output rate from j is λ_j and the rate of moving from j to i is $\lambda_j P_{ji}^*$. Since the rates passing through i and j must be the same in either direction, we have

$$P_{ji}^* = \frac{\lambda_i P_{ij}}{\lambda_j}. \tag{5.8.10}$$

For a given n, we let $A_i(n)$ denote the vector n with its ith element incremented by 1 denoting an arrival from outside to server i; we let $D_i(n)$ denote the vector n with its ith element decremented by 1 denoting a departure from the system via server i; we let $T_{ij}(n)$ denote the vector n with its ith element decremented by 1 and its jth element incremented by 1 denoting a transfer from i to j. Applying Equation 5.8.7, we require

$$\pi(n)q(n, A_i(n)) = \pi(A_i(n))q*(A_i(n), n) \tag{5.8.11}$$

$$\pi(n)q(n, D_i(n)) = \pi(D_i(n))q*(D_i(n), n) \tag{5.8.12}$$

$$\pi(n)q(n, T_{ij}(n)) = \pi((T_{ij}(n))q*(T_{ij}(n), n) \tag{5.8.13}$$

where $\{q*(\cdot,\cdot)\}$ denotes the generator of the reversed chain.

To establish Equation 5.8.11, we see that in the reversed process a transition from $A_i(n)$ to n corresponds to a departure from the system after a service completion from server i. This implies

$$q*(A_i(n), n) = \mu_i \left(1 - \sum_{k=1}^{s} P_{ik}^*\right) = \mu_i \left(1 - \sum_{k=1}^{s} \frac{\lambda_k P_{ki}}{\lambda_i}\right) \qquad \text{by Equation 5.8.10}$$

$$= \frac{\mu_i}{\lambda_i}\left(\lambda_i - \sum_{k=1}^{s} \lambda_k P_{ki}\right) = \frac{r_i}{\rho_i} \qquad \text{by Equation 5.8.8.}$$

Rewriting Equation 5.8.11, we obtain

$$\left(\prod_{k=1}^{s} \pi_k(n_k)\right) r_i = \pi_i(n_i + 1) \left(\prod_{\substack{k=1 \\ k \neq i}}^{s} \pi_k(n_k)\right) \frac{r_i}{\rho_i}.$$

After cancellations, we obtain $\pi_i(n_i + 1) = \rho_i \pi_i(n_i)$, $n_i \geq 0$. This implies that for each i

$$\pi_i(n) = \rho_i^n (1 - \rho_i), \quad n \geq 0, \tag{5.8.14}$$

and Equation 5.8.11 is satisfied by the solution.

We now consider Equation 5.8.12. It is clear that $q(\mathbf{n}, D_i(\mathbf{n})) = \mu_i P_{i0}$. We see that

$$q^*(D_i(\mathbf{n}), \mathbf{n}) = r_i = \lambda_i - \sum_{k=1}^{s} \lambda_k P_{ki}^* \qquad \text{by Equation 5.8.8}$$

$$= \lambda_i - \sum_{k=1}^{s} \lambda_k \frac{\lambda_i P_{ik}}{\lambda_k} \qquad \text{by Equation 5.8.10}$$

$$= \lambda_i \left(1 - \sum_{k=1}^{s} P_{ik}\right) = \lambda_i P_{i0}. \tag{5.8.15}$$

Using these observations and Equation 5.8.14, we rewrite Equation 5.8.12 as

$$\left((1 - \rho_i) \rho_i^{n_i} \prod_{\substack{k=1 \\ k \neq i}}^{s} \pi_k(n_k)\right) \mu_i P_{i0} = \left((1 - \rho_i) \rho_i^{n_i - 1} \prod_{\substack{k=1 \\ k \neq i}}^{s} \pi_k(n_k)\right) \lambda_i P_{i0}.$$

The equality of both sides of the preceding expression is clear. To evaluate Equation 5.8.13, we use arguments similar to those presented earlier to write

$$\left((1 - \rho_i) \rho_i^{n_i} (1 - \rho_j) \rho_j^{n_j} \prod_{\substack{k=1 \\ k \neq i,j}}^{s} \pi_k(n_k)\right) \mu_i P_{ij}$$

$$= \left((1 - \rho_i) \rho_i^{n_i - 1} (1 - \rho_j) \rho_j^{n_j + 1} \prod_{\substack{k=1 \\ k \neq i,j}}^{s} \pi_k(n_k)\right) \mu_j P_{ji}^*.$$

Making the substitution of the last term using Equation 5.8.10, equality of both sides is now obvious. We skip the verification of Equation 5.8.6 because it is tedious and not insightful. In conclusion, we show that the limiting queue length distribution of the Jackson network assumes the product form given by Equation 5.8.9. Consequently, the limiting queue length distributions at individual queues are independent at any given moment in time.

Arrivals to server i from outside of the system of the reversed process correspond to departures from the system via server i. From Equation 5.8.15, the former event has a rate $\lambda_i P_{i0}$ of occurrence. Moreover, these arrivals follow independent Poisson processes by a reversibility argument. Therefore we conclude that departures from the system via different servers form independent Poisson processes with rate $\lambda_i P_{i0}$ for server i.

We derive various performance measures of the network. The mean queue length at server i is given by $L_i = \sum_{n=0}^{\infty} n(1-\rho_i)\rho_i^n = \rho_i/(1-\rho_i)$. The mean queue length of the network L is then the sum of these L_i over $1, \ldots, s$. Let W denote the average time a customer spends in the network. Using Little's law, we find $W = L/\lambda$, where $\lambda \equiv r_1 + \cdots + r_s$, the sum of all external arrival rates. Let W_i denote the expected sojourn time of a customer at server i for each visit to the server. Then $W_i = L_i/\lambda_i = 1/(\mu_i(1 - \rho_i))$. For a customer arriving at server i, we let S_i denote the expected sojourn time in the network. Conditioning on the first visit to server i, we write the following system of linear equations for finding $\{S_i\}$:

$$S_i = W_i + \sum_{j=1}^{s} P_{ij}S_j \qquad i = 1, \ldots, s.$$

We mention in passing that when there are c_i (>1) identical servers at service station i, the product-form solution prevails, except individual queue length distributions $\{\pi_i(\cdot)\}$ are now defined by results relating to Equation 5.3.7. Moreover, various performance measures can be determined similarly. ∎

EXAMPLE
5.8.8

Closed Queueing Networks: Gordon-Newell Networks   Consider a situation in which there are N customers who circulate perpetually among the s servers in the queueing network described in Example 5.8.7. This model differs from the open Jackson network in two specific ways: (i) $P_{i0} = 0$ for $i = 1, \ldots, s$ (no transitions out of the system) and (ii) $r_i = 0$ for all $i = 1, \ldots, s$ (no external arrivals). Let $\boldsymbol{n} = (n_1, \ldots, n_s)$ and define the set $S_n = \{\boldsymbol{n}|n_1 \geq 0, \ldots, n_s \geq 0, n_1 + \cdots + n_s = n\}$. Then S_n is the set of states containing a total of exactly n customers in the system. The state space of this closed network is S_N. Because of (i) and (ii),

$$\lambda_j = \sum_{i=1}^{s} \lambda_i P_{ij} \qquad j = 1, \ldots, s. \tag{5.8.16}$$

Using the notations introduced in Example 5.8.7 for open Jackson networks, the previous traffic equations can be stated in matrix form as $\lambda = \lambda\tilde{P}$. The routing $s \times s$ matrix $\tilde{P}$ is stochastic. We assume it is irreducible; the equations do not have a unique solution. We can set any one λ_i equal to an arbitrary constant $a > 0$ and solve for the remaining unknowns. The term λ_i is still considered the arrival rate to server i.

In a manner similar to the derivation of Equation 5.8.9 for the open queueing network, we can derive the limiting probability vector of the closed network. It is given by

$$\pi_N(\boldsymbol{n}) = \frac{\pi(\boldsymbol{n})}{\sum_{\boldsymbol{n} \in S_N} \pi(\boldsymbol{n})} = \frac{1}{G(N)} \prod_{i=1}^{s} \rho_i^{n_i} \qquad \boldsymbol{n} \in S_N, \tag{5.8.17}$$

where the normalizing constant is defined by

$$G(N) = \sum_{n \in S_N} \left(\prod_{i=1}^{s} \rho_i^{n_i} \right) \qquad (5.8.18)$$

and $\pi(n)$ is the limiting queue length distribution of the open Jackson network. Let M_i denote the queue length at server i. The marginal queue length distribution is given by

$$P\{M_i = k\} = \sum_{n \in S_N \text{ and } n_i = k} \pi_N(n).$$

A simple way to find the distribution is by first noting

$$P\{M_i \geq k\} = \sum_{n \in S_N \text{ and } n_i \geq k} \pi_N(n) = \sum_{n \in S_N \text{ and } n_i \geq k} \frac{1}{G(N)} \prod_{l=1}^{s} \rho_l^{n_i}$$

$$= \frac{\rho_i^k}{G(N)} \left(\sum_{n \in S_{N-k}} \prod_{l=1}^{s} \rho_l^{n_i} \right) = \frac{\rho_i^k}{G(N)} G(N-k),$$

where the last equality is obtained by applying Equation 5.8.18 to the closed system with $N - k$ customers. Then we have

$$P\{M_i = k\} = P\{M_i \geq k\} - P\{M_i \geq k+1\} = \frac{\rho_i^k}{G(N)} \left[G(N-k) - \rho_i G(N-k-1) \right]. \quad (5.8.19)$$

For server i, the mean queue length L_i and its utilization U_i are, respectively,

$$L_i = E[M_i] = \sum_{k=1}^{N} P\{M_i \geq k\} = \sum_{k=1}^{N} \rho_i^k \frac{G(N-k)}{G(N)} \qquad (5.8.20)$$

and

$$U_i = P\{M_i \geq 1\} = \frac{\rho_i G(N-1)}{G(N)}. \qquad (5.8.21)$$

Buzen's Algorithm for Finding the Normalizing Constant $G(N)$

For $i = 1, \ldots, s$, we define the generating function

$$C_i^g(z) = \sum_{l=0}^{\infty} C_i(l) z^l = \prod_{k=1}^{i} (1 - \rho_k z)^{-1} = \prod_{k=1}^{i} (1 + \rho_k z + \rho_k^2 z^2 + \cdots).$$

Applying the preceding definition to the closed queueing network yields

$$C_s(n) = \sum_{n \in S_n} \prod_{k=1}^{s} \rho_k^{n_k} \qquad \text{and} \qquad C_s(N) = G(N).$$

Now we see that $C_i^g(z) = (1 - \rho_i z)^{-1} C_{i-1}^g(z)$, $i = 2, \ldots, s$ or

$$C_i^g(z) - \rho_i z C_i^g(z) = C_{i-1}^g(z) \qquad \text{or} \qquad C_i^g(z) = C_{i-1}^g(z) + \rho_i z C_i^g(z).$$

Equating the coefficients of z^k on both sides of the last expression, we obtain the recursion for $k = 1, 2, \ldots, N$ and $i = 2, \ldots, s$,

$$C_i(k) = C_{i-1}(k) + \rho_i C_i(k-1) \qquad (5.8.22)$$

with initial conditions $C_1(k) = \rho_1^k$, $k = 0, 1, \ldots, N$ and $C_i(0) = 1$, $i = 2, \ldots, s$.

Waiting Times at Each Server

To obtain the waiting time distributions at a given server during each visit to the server, we need to know the queue length distribution of the server immediately before a "jump" to the server. In other words, if a customer is about to jump from server i to server j, we would like to know the queue length distribution as seen by this customer. Let T_m denote the epoch of the mth such jump and $P\{X_{T_m^-} = n\}$ the probability that the network is in state n immediately before T_m. As before, we assume that m is large and the system is in steady state. Define $v(n)$ as the rate of jumps from i to j when the state of the network is n. Then the total rate of jumps from i and j is given by

$$v = \sum_{n \in S_N} v(n).$$

Conditioning on the event that a jump from i to j is about to occur, the probability that this jump emanates from state n can be found by an argument based on competing exponential distributions. This yields the useful identity

$$P\{X_{T_m^-} = n\} = \frac{v(n)}{v}.$$

Since the customer is currently in i, we know $n_i > 0$ and

$$v(n) = \pi_N(n)\mu_i P_{ij} = \frac{1}{A}\pi(n)\mu_i P_{ij} = \frac{1}{A}\lambda_i P_{ij}\pi(D_i n),$$

where we let A denote the denominator of the second term of Equation 5.8.17, and the last equality is found by using the results of Example 5.8.7. The preceding in turn gives

$$v = \sum_{n \in S_N} v(n) = \frac{1}{A}\lambda_i P_{ij}\sum_{n \in S_N}\pi\big(D_i(n)\big) = \frac{1}{A}\lambda_i P_{ij}\sum_{n \in S_{N-1}}\pi\big(D_i(n)\big).$$

Combining these results, we obtain the following joint queue length distribution immediately before a jump

$$P\{X_{T_m^-} = n\} = \frac{v(n)}{v} = \frac{\pi\big(D_i(n)\big)}{\displaystyle\sum_{n \in S_{N-1}}\pi\big(D_i(n)\big)} = \pi_{N-1}\big(D_i(n)\big). \tag{5.8.23}$$

The preceding implies that the joint queue length distribution as seen by the customer who is about to enter a queue is given by the joint queue distribution of a closed network with $N-1$ customers. This important result is called the *arrival theorem* for closed networks. (By a similar argument, it can be shown that the arrival theorem also works for open networks.) Conditioning on the marginal queue length distribution obtained from Equation 5.8.23, the waiting time distribution can easily be constructed.

For a closed network with N customers, we let $W_j(N)$ denote the mean waiting time per each visit to server j and $L_j(N)$ the mean queue length of server j. The average number of customers found in queue j by an arriving customer is

$L_j(N-1)$. Hence the average time spent in the queue by the customer is $L_j(N-1) + 1$ times the average service time $1/\mu_j$. We find

$$W_j(N) = \frac{1}{\mu_j}\left[1 + L_j(N-1)\right]. \qquad (5.8.24)$$

For Gordon-Newell networks, no restrictions of the type $\rho_i < 1$ are needed since the queue will be stable in any event because of the capacity limitations. However, we need the irreducibility of the routing matrix P so that all nodes communicate with one another. When service stations have capacity limitations, closed queueing networks generally do not have product-form solutions. Finding their limiting distributions typically require solutions of multidimensional continuous-time Markov chains. References about closed queueing networks with blocking are given in the Bibliographic Notes. ∎

EXAMPLE **A Flexible Manufacturing System**   Consider a flexible manufacturing system in which
5.8.9   each part requires two operations. Operation 1 is done at Machine 1, with probability p Operation 2 is done at Machine 2 and with probability $q = 1 - p$ Operation 2 is done at machine 3. The service time at Machine i is exponential with parameter μ_i. The system uses a pallet that holds N fixtures, each holding one work piece. A fixtured work piece on a pallet moves about the system. As soon as a work piece finishes the two operations, a new work piece replaces it in the fixture. Hence the system can be considered a closed queueing network with N customers circulating perpetually among the three servers. Let $X_i(t)$ denote the number of work pieces at Machine i. Then $\{(X_1(t), X_2(t), X_3(t)), t \geq 0\}$ can be modeled as a continuous-time Markov chain. The system is depicted in Figure 5.16.

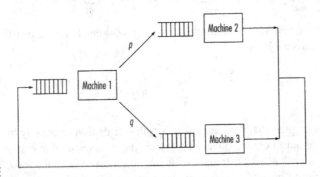

FIGURE
5.16   A closed queueing network.

Consider the case in which $p = 0.75$, $N = 10$, and $\{\mu_1, \mu_2, \mu_3\} = \{0.25, 0.48, 0.08\}$. We are interested in finding the marginal queue length distributions at each one of the three machines in the center. For this closed queueing network, the Markov routing matrix is given by

$$P = \begin{array}{c} \\ 1 \\ 2 \\ 3 \end{array} \begin{array}{c} 1 \quad 2 \quad 3 \\ \left[\begin{array}{ccc} & p & q \\ 1 & & \\ 1 & & \end{array} \right]. \end{array}$$

Solving Equation 5.8.16 with $\lambda_1 = 1$, we obtain $\rho_1, \rho_2, \rho_3 = (1/\mu_1, p/\mu_2, q/\mu_3)$. The number of states in the state space is given by the solution of the occupancy problem about the number of ways to distribute N indistinguishable balls to s distinguishable urns:

$$\binom{N-s+1}{N} = \binom{10-3+1}{10} = 66.$$

Using Buzen's algorithm, we find in the Appendix that $G(N) = C_3(10) = 7{,}231{,}882.912$. Using the normalizing coefficient in Equation 5.8.17, we can compute the three limiting marginal distributions. They are plotted in Figure 5.17. From the figure, we observe that the bottleneck is at Machine 1. Based on the joint and marginal queue length distributions, many performance characteristics of the production system can be computed. ∎

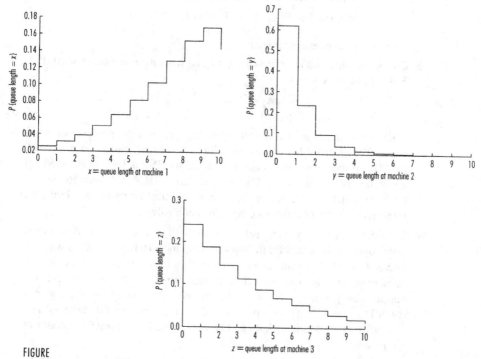

FIGURE
5.17   The marginal queue length distributions at the three machine centers.

Problems

1 We consider the car repair problem given as Example 5.2.5 again. If the time to perform each operation follows a two-stage Erlang distribution with the corresponding mean, what is the probability that four hours later the car is in the braking system replacement stage?

2 Consider a production system consisting of a machine center followed by an inspection station. Arrivals from outside of the system follow a Poisson process with rate λ. The machine center and inspection station each can only process one job at a time. When a job is complete at the machine center, it moves to the inspection queue for inspection. A job passes the inspection p percent of the time and fails the inspection $q = 1 - p$ percent of the time. When a job passes the inspection, it leaves the system; otherwise it moves to the machining queue for rework. For simplicity, we call the machining queue and inspection queue servers 1 and 2, respectively. The service time at server i follows an exponential distribution with rate μ_i. Moreover, the service times at the two servers are mutually independent. Let $X_i(t)$ denote the number of jobs at server i. Then $\{(X_1(t), X_2(t)), t \geq 0\}$ can be modeled as a continuous-time Markov chain. Find the limiting probabilities

$$\pi(i,j) = \lim_{t \to \infty} P\{X_1(t) = i, \ X_2(t) = j\} \qquad i \geq 0, j \geq 0$$

in terms of the given parameters.

3 Consider a birth and death process $\{X(t), t \geq 0\}$ with state space $S = \{0, 1, 2, 3, 4, 5\}$ and parameters

$$\lambda_0 = 0, \ \lambda_1 = 1, \ \lambda_2 = 2, \ \lambda_3 = 3, \ \lambda_4 = 4, \ \lambda_5 = 0 \qquad \text{and}$$
$$\mu_0 = 0, \ \mu_1 = 4, \ \mu_2 = 3, \ \mu_3 = 2, \ \mu_4 = 1, \ \mu_5 = 0.$$

Note that 0 and 5 are the absorbing states. Suppose the process begins in state $X(0) = 2$. **(a)** What is the probability of eventual absorption in state 0? **(b)** What is the mean time to absorption? (Hints: Part a, let $U_i = Pr\{\text{eventual absorption occurs in state } 0 | X(0) = i\}$; in Part b, let $W_i = E[\text{time to absorption} | X(0) = i]$. Conditioning on the outcomes of the first transition, derive the respective system of difference equations and solve.)

4 Consider an infinitely many server queue with an exponential service time distribution with parameter μ. Suppose customers arrive in batches with the interarrival time following an exponential distribution with parameter λ. The number of arrivals in each batch is assumed to follow the geometric distribution with parameter $\rho(0 < \rho < 1)$, namely, $Pr\{\text{number of arrivals in a batch has size } k\} = \rho^{k-1}(1 - \rho) \ (k = 1, 2, \ldots)$. Formulate this process as a continuous-time Markov chain and determine the infinitesimal generator of the process.

5 We consider a lost sales inventory problem with probabilistic lead times. Demands occur for single units in accordance with a Poisson process with rate λ. When the on-hand inventory reaches the reorder point R, an order is placed with a supplier for a lot of size Q units. The lead time, defined as the time from placing the order until its receipt, is a random variable whose

distribution is a two-stage Erlang with mean $2/\mu$. If a demand occurs and the system is out of stock, the sale is lost. Consider the case in which $Q = 3$, $R = 2$, $\lambda = 2$, and $\mu = 3$. **(a)** Model the inventory process as a continuous-time Markov chain by defining the states, drawing the transition diagram, and displaying the appropriate transition rates along the arcs of the transition diagram. **(b)** Let $P(j)$ denote the limiting probability that there are j units on hand. Find $P(j)$, $j = 0, 1, \ldots, 5$.

6 Consider the inventory problem with probabilistic lead times and lost sales described in Problem 5 again. Assume now that the lead time distribution follows the hyperexponential density

$$f(t) = 0.4(2)e^{-2t} + 0.6(4)e^{-4t} \qquad t \geq 0$$

and $R = 1$, $Q = 2$, and $\lambda = 5$. **(a)** Model the inventory process as a continuous-time Markov chain by defining the states, drawing the transition diagram, and displaying the appropriate transition rates along the arcs of the transition diagram. **(b)** Display the infinitesimal generator of the process. **(c)** Let $P(j)$ denote the limiting probability that there are j units on hand. Find $P(j)$, $j = 0, 1, 2, 3$. **(d)** What is the long-run rate of lost sales?

7 Consider an $M/M/1$ queue with arrival rate λ and service rate μ. Assume that customers are impatient. A customer who finds the server busy will wait a time Z for service to begin and will leave without service if service has not begun by then. Assume that Z follows the exponential distribution with mean $1/\gamma$. Define $X(t)$ as the number of customers in the system at time t. Model the stochastic process $\{X(t), t \geq 0\}$ as a continuous-time Markov chain by stating the corresponding infinitesimal generator of the chain.

8 A Two-Machine Transfer Line   Consider a manufacturing system containing two machines, Machines 1 and 2, arranged in tandem. A work piece is processed by the two machines in sequence and then leaves the system. There is a buffer of finite size N between the two machines to store work in progress. The system is depicted in the following diagram:

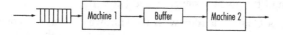

We see that in front of Machine 1 there is an unlimited number of waiting rooms. In the parlance of automated manufacturing systems, Machine 1 is said to be never *starved*. Once a work piece leaves Machine 2, it is out of the system. Hence Machine 2 is never *blocked*. However, when the buffer is full, at the service completion epoch Machine 1 becomes blocked since there is no place to store additional work in progress. The work piece on a blocked machine stays there until the buffer has room to store the additional piece. For Machine i, we assume that processing, machine failure, and machine repair times are independent exponential random variables with parameters μ_i, τ_i, and η_i, respectively; these times are mutually independent between machines. We also assume that neither a starved machine nor a blocked machine can fail. One way to model the system is by constructing

a continuous-time Markov chain using a state definition (i, j, k), where i = the queue size at Machine 2 (the number of work pieces in the buffer plus the number of work pieces being processed by Machine 2), j = the state of Machine 1 (1 = working and 0 = in repair), and k = the state of Machine 2 (1 = working and 0 = in repair). For simplicity, we look at the case in which $N = 1$. In this case, the state space has the following states: (000), (001), (010), (011), (100), (101), (110), (111), (200), (201), (210), and (211).
(a) Draw the transition diagram for this continuous-time Markov chain.
(b) Give the infinitesimal generator of the continuous-time Markov chain.
(c) Identify the transient states and recurrent states for the chain.

9 Consider Problem 8 again. Suppose that we have $\{\mu_1, \mu_2\} = (1.4, 0.9\}$, $\{\tau_1, \tau_2\} = \{0.15, 0.12\}$, $\{\eta_1, \eta_2\} = \{12, 8\}$. (a) Define F_i as the long-run fraction of time Machine i is producing. Find F_1 and F_2. (b) Find the expected size of work in progress in the buffer, I. (c) What is the output rate from the transfer line?

10 Consider Problem 8 again. At time 0, both machines are operational and Machine 2 is starved. In other words, initially the state of the system is 011. (a) What is the mean time to reach a situation in which both machines fail for the first time? (b) What is the mean time to reach a situation in which Machine 1 is blocked for the first time?

11 Consider a single-server loss system that oscillates between two levels denoted by 1 and 2. When the system is at level i, the arrival process is Poisson with rate λ_i. The time interval during which the system functions at level i is an exponential random variable with rate η_i. The service times under both levels are exponential with rate μ. (a) Find the average arrival rate of the system. (b) The system can be modeled as a continuous-time Markov chain $\{(X_1(t), X_2(t)), t \geq 0\}$, where $X_1(t)$ denotes the number of customers in the system at time t and $X_2(t)$ the level in effect at time t. The state space of the process is given by $S = \{(m, i), m = 0, 1$ and $i = 1, 2\}$. Find the infinitesimal generator of the process. (c) Let $\pi(m, i) = \lim_{t \to \infty} P\{(X_1(t), X_2(t)) = (m, i)\}$. Find the fraction of lost customers, F, as a function of the given parameters and $\{\pi(m, i)\}$.

12 **Modeling a Coal Unloader with Breakdowns** A utility company owns and operates a coal-fired power plant. Coal is brought by train to the plant from mines in nearby states and is unloaded by a single unloader system. There are four time components of the system's operations: the unloading time U (the time to unload a train), the repair time R (the time to repair a broken unloader), the cycle time C (the time it takes for an unloaded train to go to the coal fields and return with a load of coal), and the failure time F (the time between breakdowns when the unloader is operating continuously). Assume that the four random variables follow independent exponential distributions with respective parameters μ_1, μ_2, λ_1, and λ_2. Let K denote the total number of trains in the system. We see that λ_1 represents the arrival rate of an individual train when not in queue for unloading at the unloader.
(a) Let $X_1(t)$ denote the number of trains queued at the unloader for unloading at time t (for example, if $X_1(t) = 2$, it means that there is one train being

unloaded and another waiting to be unloaded) and $X_2(t)$ the number of broken unloaders at time t. The process $\{(X_1(t), X_2(t)), t \geq 0\}$ can be modeled as a continuous-time Markov chain with state space $\{(i, j), 0 \leq i \leq K$ and $j = 0, 1\}$. Construct a transition diagram for the process. What can you say about state $(0, 1)$? **(b)** Let L denote the average number of trains queued at the unloader for unloading, T the average arrival rate at the unloader (in units of train loads), W the average time a train has to wait until unloading is done (average total time at the unloader), and V the average number of trains unloaded per unit time. Define

$$\pi(i, j) = \lim_{t \to \infty} P\{(X_1(t), X_2(t)) = (i, j)\}.$$

Express the previous performance measures as a function of the known parameters and $\{\pi(i, j)\}$. **(c)** Actual data for the problem were collected for the first ten months of 1977. The time for a train to complete a trip from the unloader to the mines and back was 108 hours. There were fifteen unloader breakdowns during the period. Approximately four hundred trains were unloaded. The average unloading time was 9.5 hours per train. The data also showed that the unloader was out of service for 42.5 days. Find the parameters μ_1, μ_2, λ_1, and λ_2. **(d)** For $K = 5$ and the parameters given in Part c, compute the four performance measures defined in Part b. What is the relation between T and V?

13 **Modeling Two Coal Unloaders with Breakdowns**   Consider a variant of Problem 12 in which the plant operates two unloaders. When two unloaders are broken, two crews work independently on them. Assume that the two unloaders have the identical set of parameters: λ_2, μ_1, and μ_2. Then the process $\{(X_1(t), X_2(t)), t \geq 0\}$ can again be modeled as a continuous-time Markov chain but now with state space $\{(i, j), 0 \leq i \leq K$ and $j = 0, 1, 2\}$. Construct the transition diagram reflecting the addition of one extra unloader. Identify the transient state(s).

14 The sales department of a computer manufacturer has a fleet of thirty cars for use by its sales representatives for visiting clients. Times between breakdowns for each car follow an exponential distribution with mean one hundred days. There are three auto mechanics maintaining this fleet. Each can repair, on average, one car per day. Assume that repair times follow an exponential distribution. When the system is in steady state, what is the expected number of cars that are working on a given day?

15 **Deadlocks at an Automatic Manufacturing System**   Consider an automatic manufacturing system. A machine is producing parts at a rate of μ units per unit time. The processing time is assumed to follow an exponential distribution. Raw parts arrive from an input conveyor according to a Poisson process with rate λ. A robot picks up a raw part from the input conveyor as soon as it arrives from the input conveyor and loads it into the machine if the machine is free or the buffer if the machine is busy. At a production completion epoch, the robot picks up the finished part and moves it to the output conveyor. Robot transfer times are negligible. Assume that the buffer size is N. When the machine is busy, the buffer is full, and the robot is holding a

raw part, the system is said to have experienced a deadlock. At time 0, the buffer, the machine, and the robot are all empty. **(a)** Define an absorbing continuous-time Markov chain to model the time-to-deadlock random variable by stating the state space and constructing the corresponding transition diagram. **(b)** Assume that $\lambda = 4$ per minute and $\mu = 5$ per minute. For $N = 5{:}10$, compute the mean and variance of the time to deadlock.

16 Deadlocks at an Automatic Manufacturing System: Hyperexponential Processing Time   Consider a variant of Problem 15 in which the processing time distribution follows a hyperexponential distribution defined in Example 5.5.3 with $m = 3$. Thus the processing time is a probabilistic mixture of three exponential distributions with the weights $\{\alpha_1, \alpha_2, \alpha_3\}$ and the three exponential parameters $\{\mu_1, \mu_2, \mu_3\}$. Following Equation 5.5.9, the processing time distribution is given by

$$F(x) = \sum_{i=1}^{m} \alpha_i (1 - e^{-\mu_i x}) \qquad x \geq 0,$$

where $\alpha_1 + \alpha_2 + \alpha_3 = 1$ and $0 < \alpha_i < 1$ for all i. This processing time distribution is of practical interest for cases in which parts are to be manufactured in multiple configurations, each requiring a different processing time, and α_i is the fraction of parts to be made of configuration i. **(a)** Define an absorbing continuous-time Markov chain to model the time-to-deadlock random variable by stating the state space and constructing the corresponding transition diagram. **(b)** Assume again that $\lambda = 4$ per minute. The processing time parameters are $\{\alpha_1, \alpha_2, \alpha_3\} = \{1/2, 1/4, 1/4\}$ and $\{\mu_1, \mu_2, \mu_3\} = \{10, 5, 2.5\}$. It is easy to verify that the mean processing time is the same as in Problem 15, that is, $1/5$ minutes. Find the mean and variance of the time to deadlock for $N = 5{:}10$. **(c)** Problems 15 and 16 have the same mean processing times; however, the distributional forms are different. By comparing the results obtained from the numerical computation, what can you say about the effects of the processing time distribution on the time-to-deadlock statistics?

17 Consider a cafeteria with a seating capacity of C persons. Assume that there are two types of customers—those with a group size of g_1 and those with a group size g_2. For modeling purposes, we may assume that one type-i customer takes up g_i seats. When a type-i customer arrives and the cafeteria has enough seats to accommodate all members of the group, all members immediately receive service and depart together after an exponential sojourn in the cafeteria with parameter μ_i. On the other hand, if there are not enough seats to hold the entire group, the customers simply leave the cafeteria without receiving any service. Assume that the arrival process of type-i customers is Poisson with rate λ_i. Our objective is to calculate the fraction of customers lost for each type of customer. The model to be developed here can be used in communication systems in which a customer type represents a media type of a given size (voice, data, or video packets) and the system can process a maximum of C packets simultaneously. **(a)** Model the system as a continuous-time Markov chain so that we will be able to compute the loss probabilities. Define the state space and transition rates involved and the expression to compute the loss probabilities. **(b)** Solve the problem with $\lambda_1 = 10$, $\lambda_2 = 8$, $\mu_1 = 1$, $\mu_2 = 0.5$, $g_1 = 2$, $g_2 = 5$, and $C = 70$.

18 Consider a conveyor system for assembly CD players. Arrivals of chassis follow a Poisson process with parameter λ. Work stations 1 and 2 are sequential along the conveyor as shown in the following figure.

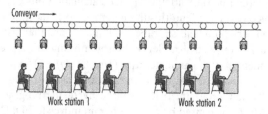

Work station 1   Work station 2

Work station i has n_i identical exponential servers, each with a service rate of μ_i and no storage spaces to hold waiting chassis. When an incoming chassis passes by work station 1, an available worker will pick it up and work on it. Once it is done, it leaves the system. If work station 1 is full, then a passing chassis will be worked on by any available idle worker at work station 2. If both stations are full, passing chassis will leave the system unassembled. We want to compute the fraction of chassis left unassembled. **(a)** Model the system as a continuous-time Markov chain so that we will be able to compute the loss probabilities. Define the state space and transition rates involved and the expression to compute the lost probabilities. **(b)** Solve the problem with $\lambda = 10$, $\mu_1 = 3$, $\mu_2 = 4$, $n_1 = 4$, and $n_2 = 3$.

19 An ISDN (Integrated Services Digital Networks) Protocol Consider a communication channel that can serve either voice or data transmissions. A transmitter contains a buffer of size C for data packets. Arrivals of voice call follow a Poisson process with rate λ_1. A voice call seizes the channel if no data packets are in the buffer. The service time for each voice call is exponential with rate μ_1. Data packets arrive in accordance with a Poisson process with rate λ_2 and are served one packet at a time at a rate of μ_2. If a voice call is being transmitted, arriving data packets can only be buffered. Use a continuous-time Markov chain to model the state of the system at any time by defining the state space and presenting a transition diagram.

20 A Computer with Virtual Memory We consider a partitioned computer system with virtual memory shown in the following figure.

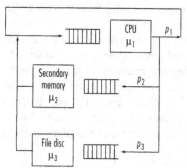

The system contains three components: the central processing unit (CPU), the secondary memory (SM), and the file disc (FD), which performs the input and output operations. The service times SM and FD follow independent exponential distributions with rates μ_2 and μ_3, respectively. Execution of a program on the CPU is terminated by the realization of the first of the three exponential random variables: t_1, the calculation time; t_2, the time for the occurrence of a page fault (insufficient memory); and t_3, the time for an I/O operation. The parameters for the last three exponential distributions are $1/c$, $1/q$, and $1/r$, respectively. Hence the sojourn time for a program in the CPU is exponential with parameter $\mu_1 = (1/c) + (1/q) + (1/r)$. If the termination of a program is caused by t_1, the program leaves the system and is replaced by a new program; if it is caused by t_2, the program is put in a queue for SM; if it is caused by t_3, the program moves to FD. One value for the parameter q suggested in computer engineering literature is

$$q = (0.01)\left(\frac{M}{K}\right)^{1.5},$$

where M is the total memory reserved for program execution and K is the degree of multiprogramming—namely, the number of programs in the system at any time. We see that the system can be modeled as a closed queueing network with state space $S = \{(n_1, n_2, n_3), \; n_1 \geq 0, \; n_2 \geq 0, \; n_3 \geq 0, \; n_1 + n_2 + n_3 = K\}$. **(a)** Find the limiting joint queue length distribution of the system in terms of the given parameters. In doing so, set the parameters involved in such a way that $\rho_1 = 1$. Also find the utilizations at each server. **(b)** Consider a numerical example with $1/\mu_2 = 5$ ms (milliseconds), $1/\mu_3 = 30$ ms, $c = 800$ ms, $r = 20$ ms, $M = 128$ pages, and $K = 6$. Compute the marginal queue length distributions at each server and their respective utilizations.

21 Consider a pure birth process $X = \{X(t), t \geq 0\}$ where $X(t)$ denotes the state of the system at time t. Assume that $X(0) = 0$ and

$$P\{\text{a birth occurs in } (t, \; t + h)|X(t) = \text{odd}\} = \lambda h + o(h)$$

$$P\{\text{a birth occurs in } (t, \; t + h)|X(t) = \text{even}\} = \mu h + o(h),$$

where $o(h)$ is the little-oh function. Find the following probabilities: $P\{X(t) = \text{odd}\}$ and $P\{X(t) = \text{even}\}$.

22 A small casino in Vegas has five slot machines. The time-to-failure rate of each machine follows an exponential distribution with a mean of one day. The casino has three repairers to maintain its slot machines. When a machine fails, one available repairer will work on the machine. The repair time follows an exponential distribution with a mean of one-half day. Assume that times to failure and repair times are all mutually independent random variables. The casino manager now has an option to hire a super-duper repairer who can work three times as fast but charges three times as much. Should the manager replace the three repairers by this super-duper person if he uses the mean number of failed machines at any time as the criterion for decision making?

*23 The H_2 Distribution with Balanced Means   A hyperexponential distribution of order 2, denoted by H_2, has the density

$$f(x) = p_1\mu_1 e^{-\mu_1 x} + p_2\mu_2 e^{-\mu_2 x} \qquad x \geq 0,$$

where $p_1 > 0$, $p_2 > 0$, and $p_1 + p_2 = 1$. Recall that the coefficient of variation c_X of a random variable X is defined as the standard deviation of X divided by the mean of X. If X follows H_2, then $c_X > 1$. **(a)** State the phase representation of the distribution. **(b)** Find $E[X]$, $Var[X]$, and c_X^2 using Equations 5.5.6 and 5.5.7. **(c)** The distribution H_2 has three unknown parameters: p_1, μ_1, and μ_2. If we impose the following condition: $p_1/\mu_1 = p_2/\mu_2 = r$, the distribution is said to have balanced means. In this case the number of unknown parameters reduces to two, for example, p_1 and μ_1. If we know the mean and coefficient of variation of random variable X, then the two parameters are uniquely determined. Show that

$$p_1 = \frac{1}{2}\left(1 + \sqrt{\frac{c_X^2 - 1}{c_X^2 + 1}}\right) \qquad \text{and} \qquad \mu_1 = \frac{2p_1}{E[X]}.$$

The H_2 distribution is frequently used to model the distribution of a random variable whose coefficient of variation is greater than 1.

*24 A Mixture of Two Erlang Distributions: The $E_{k-1,k}$ Distribution   Let X be a random variable whose density is a probabilistic mixture of two Erlang distributions, one with parameters $(k - 1, \mu)$ and the other with parameters (k, μ). This distribution is known as the $E_{k-1,k}$ distribution. It is useful to model densities whose coefficients of variation are in the interval $[0, 1]$. Let $f^e(s)$ denote the Laplace transform of the density $f(\cdot)$ of X. Then we have

$$f^{(e)}(s) = p\left(\frac{\mu}{s+\mu}\right)^{k-1} + q\left(\frac{\mu}{s+\mu}\right)^k,$$

where $q = 1 - p$ **(a)** Find the phase representation (α, T) of the density $f(\cdot)$. What is the order of this representation? **(b)** Find $E[X]$, $Var[X]$, and c_X^2. **(c)** Show that

$$\mu = \frac{k-p}{E[X]}, \qquad p = \frac{1}{1+c_X^2}\left[kc_X^2 - \{k(1+c_X^2) - k^2 c_X^2\}^{\frac{1}{2}}\right], \qquad \text{and} \quad \frac{1}{k} \leq c_X^2 \leq \frac{1}{k-1}.$$

We remark that, knowing $E[X]$, and c_X^2, we are able to use the previous relations to determine the parameters k, p, and μ and consequently the phase representation of $f(\cdot)$.

25 The service time statistics in minutes collected at an airline ticket counter are given as follows.

4.00, 1.44, 4.44, 1.74, 1.16, 4.20, 3.59, 2.14, 3.54, 2.56,
5.53, 2.02, 3.06, 1.66, 3.23, 4.84, 7.99, 3.07, 1.24, 3.40,
5.01, 2.78, 1.62, 5.19, 5.09, 3.78, 1.52, 3.94, 1.96, 6.20,
3.67, 3.37, 1.84, 1.60, 1.31, 5.64, 0.99, 3.06, 1.24, 3.11,
4.67, 0.90, 2.78, 1.64, 2.43, 5.26, 2.11, 4.27, 3.36, 4.76.

(a) Propose a phase representation of the service time distribution. (b) Assess the adequacy of using your phase representation to model the service time statistics.

26 Bayou City Imaging & Diagnostic Center has one mammogram machine and one X-ray machine to perform diagnostic checks for patients. The center operates on a walk-in basis. Patients arriving for mammograms follow a Poisson process with a mean of four per hour; patients arriving for X-rays follow another independent Poisson process with a mean of six per hour. The service times for the two procedures follow two independent exponential distributions with means twenty minutes and twelve minutes, respectively. Due to the health department's ordinance, the center can accommodate twelve patients at most. In other words, the total number of patients (of both types) in the center at any time cannot exceed twelve. When an arrival shows up and the system is full, the arrival is turned away and will have no effect on the future arrival streams. We are interested in finding the expected numbers of patients of either type who are turned away because of the capacity constraint and the marginal (limiting) queue length distributions of patients of either type. (a) Solve the problem by modeling it as a two-dimensional continuous-time Markov chain. (b) Solve the problem by using the results given in Example 5.8.5.

27 Consider an open central server model of the flexible manufacturing system shown in the following graph.

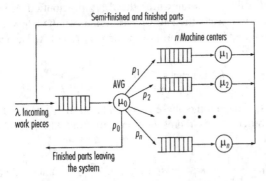

The system has n heterogeneous machine centers, each representing an exponential server with a buffer that can hold an unlimited number of incoming work pieces. The service rate at machine center i is μ_i, $i = 1, \ldots,$ n. The arrivals of workpiece from outside follow a Poisson process with rate λ. The transport mechanism is an automated guided vehicle (AGV). It is the central server of the system because every work piece must be handled by the AGV before and after each operation. This server, denoted as Server 0, is an exponential server with service rate μ_0. Each raw or semi-finished part will have probability p_i of requiring processing at machine center i, and with probability p_0 it will leave the system as a finished part. Let N_k denote the number of work pieces at Server k in steady state. Find the joint limiting probability distribution with $N_0 = i_0, N_1 = i_1, \ldots, N_m = i_m$.

28 Consider a production system consisting of four heterogeneous independent exponential servers forming an open Jackson network. The buffer sizes at each server are unlimited. The external arrival rates to the four servers are 5, 2, 3, and 0, respectively. The service rates are 8, 6, 4, and 6, respectively. The routing matrix is given as follows:

$$P = \begin{matrix} & \begin{matrix} 1 & 2 & 3 & 4 \end{matrix} \\ \begin{matrix} 1 \\ 2 \\ 3 \\ 4 \end{matrix} & \begin{bmatrix} & .3 & & .4 \\ .5 & & .1 & .2 \\ & .2 & & .5 \\ & & & \end{bmatrix} \end{matrix}.$$

(a) Find the mean queue lengths at each server. **(b)** Find the expected total sojourn time in the network of an arriving work piece. **(c)** Find the expected total sojourn time in the network if the arriving work piece starts from server i. **(d)** Conditioning on the server at the first entrance to the system, recompute the expected total sojourn time in the network W using the results obtained in Part c and hence verify the result found in Part b.

29 Consider a production system consisting of two work stations arranged in tandem. Raw parts arriving at Station 1 from outside of the system follow a Poisson process with a rate of ten per hour. Processing times at Station 1 follow an exponential distribution with a mean of four minutes. If a unit is found defective and reworkable, then the unit will be moved by an automated guided vehicle (AGV) to the incoming queue for Station 1 once more. Experience shows that 20 percent of the items inspected are reworkable and 10 percent are not reworkable. Nonreworkable work pieces are moved out of the system by the AGV and discarded. Nondefective items are moved to the waiting queue for Station 2 by the AGV. Processing times at Station 2 follow another independent exponential distribution with a mean of six minutes. All work pieces processed by Station 2 leave the system without the aid of the AGV. The AGV is an independent exponential queue with a mean of two minutes. Assume that all queues are unlimited in size. **(a)** Find the mean queue lengths at the two work stations and AGV. **(b)** Find the mean times for a nondefective item and a reject to go through the system.

30 Consider a flexible manufacturing system with two independent exponential servers in tandem as shown in following figure. Server i has a service rate of μ_i, $i = 1$ and 2. Each server performs one required operation. The system uses a pallet that holds N fixtures, each holding one work piece. A fixtured work piece on a pallet moves about the system. As soon as a work piece finishes the two operations, a new work piece takes its place in the fixture. Assume that the buffer in front of each server can hold all waiting work pieces; hence the system is a closed queueing network with N customers. Let $\pi(n, N - n)$ denote the stationary probability that there are n work pieces in Station 1 and $N - n$ work pieces in Station 2. Define $v = \mu_2/\mu_1$. **(a)** Express $\pi(n, N - n)$ in terms of v and n. **(b)** We call the average time spent by a work piece in the system the average cycle time. Let $E(C)$ denote the average cycle time. What is the rate of input to Station 1 if we know the value of $E(C)$? **(c)** Express $E(C)$ in terms of N, μ_1, μ_2, and $\pi(\cdot, \cdot)$. (Hint: The rate of input to Station 1 must be equal to the rate of output from Station 2.)

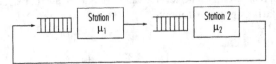

31 Consider a Gordon-Newell network. If the total number of customers in the network is N, for server i, we let $L_i(N)$ denote the mean queue length, $U_i(N)$ the utilization, and $W_i(N)$ the mean waiting time. The three performance measures are given, respectively, by Equations 5.8.20, 5.8.24, and 5.8.21. **(a)** Use Equations 5.8.20 and 5.8.21 to show algebraically that

$$L_i(N) = U_i(N)\mu_i\left[1 + L_i(N-1)\right]\frac{1}{\mu_i}.$$

(b) Apply Little's law to argue that the expression for $W_i(N)$ is given by Equation 5.8.24. This establishes Equation 5.8.24 without invoking the arrival theorem.

32 Consider a closed queueing network with s independent heterogeneous exponential servers and N customers. For server i, the service rate is μ_i. Assume that there are no limits on queue lengths in front of each server. A customer leaving a given server will move to any one of the other $s-1$ servers with equal probability. Let n_i denote the queue length at server i. Consider a continuous-time Markov chain with state definition $\boldsymbol{n} = (n_1, \ldots, n_s)$, where $n_i = 0, 1, \ldots, N$, for all i, and $n_1 + \cdots + n_s = N$. Speculate the form of the stationary joint queue length distribution, show that this chain is time reversible, and find the stationary distribution.

33 **A Fork Join Queue** Consider a parallel computer with two processors. Jobs arrive at the computer in accordance with a Poisson process with rate λ. Each arrival places exactly one task in front of each processor (the fork primitive). Assume that the computer has a buffer that can hold a total of N waiting tasks of either type. An arriving job, finding the buffer cannot accommodate the two additional tasks, will be lost. The processing of the tasks by each processor is done on a FIFO basis one at a time and is independent of the tasks in the other processor. For $i = 1, 2$, the processing time at processor i follows an exponential distribution with rate μ_i. A job leaves the system when its two tasks are complete (the join primitive). One performance measure of interest is the number of jobs in the system, namely, the queue length L. **(a)** Develop a continuous-time Markov chain for finding the expected queue length $E[L]$. Specifically, define the state space, give the transition probabilities, and present an expression for computing $E[L]$. **(b)** With $\lambda = 0.75$, $\mu_1 = 1$, $\mu_2 = 2$, and $N = 3$, compute $E[L]$.

34 Consider a variant of Problem 33. Assume that each processor has its own buffer that can hold M waiting tasks. When either of the two processors reaches its buffer capacity, any arriving job will be lost. For each processor, the system constitutes an $M/M/1/M + 1$ queue. Do the two parts in Problem 33 with $M = 4$.

*35 **A Continuous-Time Markov Chain of QBD Type** We consider an irreducible, positive-recurrent continuous-time Markov chain X whose infinitesimal generator is of block tridiagonal form:

$$Q = \begin{bmatrix} B_1 & A_0 & & & \\ A_2 & A_1 & A_0 & & \\ & A_2 & A_1 & A_0 & \\ & & \ddots & \ddots & \ddots \\ & & & \ddots & \ddots \end{bmatrix},$$

where all submatrices are of size $m \times m$. Let $S = \{(i, j), i \geq 0, i \leq j \leq m\}$ denote the state space of the chain X. We call the subset $\{(n, j), 1 \leq j \leq m\}$ of the state space *level* $l(n)$. Thus the state space is partitioned into levels $l(0)$, $l(1)$, ... This is a continuous-time *quasi birth and death* process. Its discrete-time counterpart was introduced as Problem 31 in Chapter 4. **(a)** Using the results obtained in Problem 31 of Chapter 4 and the idea of uniformization introduced in Section 5.6, show that the R and G matrices for the continuous-time QBD process satisfy the following conditions: $A_0 + RA_1 + R^2 A_2 = 0$ and $A_2 + A_1 G + A_2 G^2 = 0$. **(b)** Will R_{ij} and G_{ij}, as defined in Problem 31 of Chapter 4, retain the same interpretations in the context of a continuous-time QBD process? **(c)** Define U_{ij} as the transition rate associated with the first return to level $l(n + 1)$ via state $(n + 1, j)$ given that X starts from $(n + 1, j)$ and the taboo set is $l(n)$. Argue that $U = A_1 + A_0 G$. **(d)** Argue that we can construct a censored continuous-time Markov chain with state space $\{l(0), l(1), \ldots, l(n)\}$ and transition probability matrix

$$Q_1 = \begin{bmatrix} B_1 & A_0 & & & & \\ A_2 & A_1 & A_0 & & & \\ & A_2 & A_1 & A_0 & & \\ & & \ddots & \ddots & \ddots & \\ & & & A_2 & A_1 & A_0 \\ & & & & A_2 & U \end{bmatrix}.$$

(e) Show that $G = (-U)^{-1} A_2$ and $R = A_0 (-U)^{-1}$. **(f)** What interpretation can be given to the (i, j)th elements of $(-U)^{-1}$. **(g)** Let row vector $\pi_n = \{\pi(n, 1), \pi(n, 2), \ldots, \pi(n, m)\}$ denote the stationary probabilities associated with states in $l(n)$. Show that $\pi_n = \pi_0 R^n$, $n \geq 0$. Derive the equations for finding π_0. **(h)** Let $A = A_0 + A_1 + A_2$. We consider a continuous-time Markov chain with state space $(1, \ldots, m)$ and infinitesimal generator A. Assume that A is irreducible. We let x denote the stationary probability vector for this chain. We see that x is the marginal limiting distribution of the second state variable (index) of the original chain that contains two state indices i and j. In many applications, this second state variable denotes the "environmental" state of the system. Give interpretations for the terms $xA_2 e$ and $xA_0 e$. It can be shown that a sufficient condition for the positive recurrence of the QBD process is $xA_2e > xA_0e$.

*36 A Continuous-Time Markov Chain of QBD Type—A Variant   We consider an irreducible, positive-recurrent continuous-time Markov chain X whose infinitesimal generator is of block tridiagonal form:

$$
Q = \begin{bmatrix}
B_1 & B_0 & & & \\
B_2 & A_1 & A_0 & & \\
 & A_2 & A_1 & A_0 & \\
 & & \ddots & \ddots & \ddots
\end{bmatrix},
$$

where all submatrices are of size $m \times m$. Let $S = \{(i, j), i \geq 0, 1 \leq j \leq m\}$ denote the state space of the chain X. We call the subset $\{(n, j), 1 \leq j \leq m\}$ of the state space *level* $l(n)$. This is another continuous-time quasi birth and death process. Let row vector $\pi_n = \{\pi(n, 1), \pi(n, 2), \ldots, \pi(n, m)\}$ denote the stationary probabilities associated with states in $l(n)$. **(a)** For the process, show that $\pi_n = \pi_1 R^{n-1}$, $n \geq 1$. **(b)** Express π_1 in terms of π_0. **(c)** How do you find π_0? **(d)** When the dimensions of B_0, B_1, and B_2 are of sizes $m_1 \times m$, $m_1 \times m_1$, and $m \times m_1$, where $m_1 \neq m$, argue that the above results still hold.

*37 When the QBD Process Has a Structure   Consider Problem 36 again. Assume now that

$$
A_2 = v \cdot \alpha, \qquad \text{and} \qquad \alpha \cdot e = 1,
$$

where α is a row vector and v is a column vector. Show that the matrix R is given by

$$
R = -A_0 Z^{-1},
$$

where $Z = A_1 + A_0 e \cdot \alpha$. Thus the R matrix can be determined without using iterative methods.

*38 The M/PH/1 Queue   The M/PH/1 queue is a special case of an M/G/1 queue in which the service time distribution is $PH(\beta, S)$ of order m (see Section 5.5). Again, we let λ denote the arrival rate. Define a state by a doublet (i, j), where i denotes the number of the customer in the system at any time and j the state of the service phase. We see that the service phase can be considered an *environmental variable* for the system. The state space is then $\{0\} \cup \{(i, j), i \geq 1, 1 \leq j \leq m\}$. **(a)** Give a brief justification for the following infinitesimal generator of the underlying continuous-time Markov chain:

$$
Q = \begin{bmatrix}
-\lambda & \lambda\beta & & \\
S^0 & S - \lambda I & \lambda I & \\
 & S^0 \cdot \beta & S - \lambda I & \lambda I \\
 & & & \ddots & \ddots
\end{bmatrix}.
$$

(b) Derive a closed-form expression for the matrix R. **(c)** Show that $\pi_0 = 1 - \rho$, where $\rho = \lambda/\mu$. **(d)** Show that the mean queue length is given by $L = (1 - \rho)\beta R(I - R)^{-2}e$.

*39 Consider an *M/PH/*1 queue with an arrival rate of one customer per minute. The service time distribution follows the phase-type distribution described in Example 5.5.5. Note that the phase-type distribution given there is an approximation of a Weibull density with mean 0.8862. **(a)** Find the matrices *R* and *G* using successive approximation. **(b)** Use the closed-form expression derived in Part b of Problem 38 to find matrix *R*. **(c)** Find the matrix *U*. **(d)** Find π_0 and π_1 using the matrices *U* and *R*. **(e)** What is the mean queue length *L*?

*40 **The Limiting Phase-Type Excess-Life Distribution** If *F* is *PH*(α, *T*) with mean $\mu_1 = -\alpha T^{-1}e$ (see Equation 5.5.6), then show that the limiting excess-life distribution (see Equation 3.3.5)

$$F_E(x) = \frac{1}{\mu_1}\int_0^x [1 - F(u)]du$$

is *PH*(π, *T*) where π is the stationary probability vector of the irreducible generator $Q = T + T^0 \cdot \alpha$ (see Example 5.5.6). This problem presents another closure property of the phase-type distribution. (Hint: Argue that πT^0 represents the limiting aggregate rate of absorption of the phase-type distribution *F* and thus $\pi T^0 = 1/\mu_1$ (see Equation 5.5.3).)

*41 **A *PH/M/c* Queue** Consider a queue with *c* identical exponential servers working in parallel. The service rate of each server is μ. The arrival times are i.i.d. random variables, each following a phase-type distribution *PH*(α, *T*) of order *m*. The arrival rate is denoted by λ. Arrivals join a common waiting line and are served by the first idle server. This multiserver queue can be modeled by a continuous-time QBD process with state space $\{(i, j), i \geq 0, 1 \leq j \leq m\}$, where *i* denotes the number of customers in the system and *j* the phase of the arrival process. **(a)** Construct the infinitesimal generator of this system by defining the submatrices A_0, A_1, and A_2 and the submatrices for the boundary levels $l(0), \ldots, l(c)$ (the definition of level was given in Problem 35). **(b)** What is the stability condition for this queue? (Hint: See Part **h** of Problem 35.) **(c)** Let row vector $\pi_i = (\pi_{(i,1)}, \ldots, \pi_{(i,m)})$ denote the stationary probability vector for level $l(i)$. Show that $\pi_n = \pi_{c-1}R^{n-c+1}$ for $n \geq c - 1$. **(d)** How do you solve for the stationary probability vectors $\pi_0, \pi_1, \ldots, \pi_{c-1}$ associated with the boundary levels? **(e)** The time a fictitious customer would have to wait if he or she arrives at an arbitrary point is called the virtual waiting time. This is the work load at an arbitrary epoch. Show that the stationary distribution $\tilde{W}(\cdot)$ for the virtual waiting time is given by

$$\tilde{W}(x) = 1 - \pi_{c-1}R(I - R)^{-1}\exp[-c\mu x(I - R)]e \qquad x \geq 0.$$

(f) Let q_i denote the stationary probability that a customer will see that the queue length is *i* upon arrival. Use Equation 2.7.4 to express $\{q_i\}$ in terms of $\{\pi_i\}$. **(g)** Let *W*(*x*) denote the stationary distribution for the waiting time in queue of an arriving customer. We call this the actual waiting time distribution. Show that

$$W(x) = 1 - \rho^{-1}\pi_{c-1}R^2(I - R)^{-1}\exp[-c\mu x(I - R)]e \qquad x \geq 0.$$

*42 **A** *PH/M/c* **Queue—A Numerical Exercise**   Consider a *PH/M/c* queue with five servers. The service rate of each server is one. The interarrival time S follows a hyperexponential distribution with balanced means (see Example 1.3.5). Specifically, we assume that $E[S] = 0.25$ and $Var[S]/E^2[S] = 2$. (a) Find the R matrix. (b) Compute the virtual and actual waiting time distributions in the interval $[0, 10]$ and plot them. (c) Compute the queue length distribution and plot it.

*43 **A Two-Echelon Repairable Item Inventory System**   Consider a repair depot supporting two air force bases for the maintenance of aircraft engines. At Base i, we let BS_i denote the number of engines allocated to it and MS_i denote the desired number of operational engines at the base. Each engine which fails at a given Base i can be repaired at the base with probability α_i or must be repaired at the depot with probability $1 - \alpha_i$. We assume that base-repairable engines are always repaired at the bases. When a failed engine is sent to the depot for repair and the depot has a spare available, the spare is immediately sent to the base for replacement. For Base i at any point in time, we define $\#BU_i$ as the number of operational engines at the base, $\#BR_i$ as the number of engines in or awaiting repair at the base, and $\#DB_i$ as the number of engines from Base i backordered at the depot. At Base i, the inventory balance equation is then $BS_i = \#BU_i + \#BR_i + \#DB_i$. Since BS_i is a given parameter, one of the last three variables is redundant. Assume that we keep $(\#BR_1, \#DB_1, \#BR_2, \#DB_2)$ as a part of state definition. At Base i, the actual number of operational engines *in use* is then $\min(\#BU_i, MS_i)$ (when $\#BU_i > MS_i$, only MS_i operational engines will be in use and the balance are held as spares at the base). We assume that the time to failure distribution of each operational engine in use follows an exponential distribution with rate λ_i. At Base i, we let BC_i denote the number of repair channels at the base repair shop. The number of engines in repair in Base i's repair shop is $\min(BC_i, \#BR_i)$. Repair times are i.i.d. random variables and each follows an exponential distribution with rate μ_i. To summarize, the parameters for Base i are $BS_i, MS_i, MC_i, \alpha_i, \lambda_i,$ and μ_i.

At the depot, we let DS denote the number of depot spares and DC denote the number of depot repair channels. For the depot at any time, we define $\#DU$ as the number of operational engines at the depot and $\#DR$ as the number of engines in or awaiting repair at the depot. The inventory balance equation at the depot is then $\#DU + \#DR = DS + DB_1 + DB_2$. It is clear that when $\#DU > 0$, we have $DB_1 = 0$ and $DB_2 = 0$. In other words, when the depot has spares, backorders at bases cannot occur. This suggests that when $\#DU = 0$, the vector $(\#BR_1, \#DB_1, \#BR_2, \#DB_2)$ is sufficient to represent the state of the system. On the other hand, when $\#DU > 0$, the vector $(\#DU, \#BR_1, \#BR_2)$ describes the state of the system. The number of engines in repair at the depot at any time is $\min(DC, \#DR)$. We assume that repair times at the depot are i.i.d. random variables and each follows an exponential distribution with rate μ_D. At a depot repair completion epoch, the depot's strategy for filling backorders is a function of $\#DB_1$ and $\#DB_2$. Let BOS_i be the probability that a repaired item at the depot is sent to Base i given ($\#DB_1$, $\#DB_2$). For Base i, we define $v_i(DB_i) = w_i \cdot \#DB_i/BS_i$, where w_i is a given weighting parameter. Thus $v_i(\#DB_i)$ is the weight assigned to Base i in competing for the depot repaired item. The base with a higher weight receives the depot repaired item, for example, $BOS_i = 1$ if $v_i(\#DB_i) > v_j(\#DB_j)$. In the event of a tie, we let $BOS_i = \frac{1}{2}$ for each i. To summarize, the parameters for the depot are DS, DC, μ_D, w_1, and w_2. (a) Describe the state space for the underlying continuous-time Markov chain. (b) What is the size of the state space? (c) Assume that the system is in state ($\#BR_1$, $\#DB_1$, $\#BR_2$, $\#DB_2$). When a failure occurs at Base 1 and the failure is base repairable, what is the state to which a transition will occur? What is the corresponding transition rate? (d) Assume that the system is in state ($\#BR_1$, $\#DB_1$, $\#BR_2$, $\#DB_2$). When a failure occurs at Base 1 and the failure is depot repairable, what is the state to which a transition will occur? What is the corresponding transition rate? (e) Assume that the system is in state ($\#DU$, $\#BR_1$, $\#BR_2$). When a failure occurs at Base 1 and the failure is base repairable, what is the state to which a transition will occur? What is the corresponding transition rate? (f) Assume that the system is in state ($\#DU$, $\#BR_1$, $\#BR_2$). When a failure occurs at Base 1 and the failure is depot repairable, what is the state to which a transition will occur? What is the corresponding transition rate? (g) What are the possible transitions when a repair completion occurs at Base 1? What are the corresponding transition rates? (h) What are the possible transitions when a repair completion occurs at the depot? What are the corresponding transition rates? (i) Consider a numerical problem with (BS_1, MS_1, BC_1, α_1, $\lambda_1, \mu_1) = (1, 1, 1, 0.6, 0.6, 0.8)$, ($BS_2$, MS_2, BC_2, $\alpha_2, \lambda_2, \mu_2) = (2, 1, 1, 0.5, 0.6, 0.8)$, and ($DS$, DC, w_1, w_2, $\mu_D) = (1, 1, 0.5, 0.5, 0.8)$. For each base, compute the availability, the average number of operational engines, average number of engines in or awaiting base repair, and the average number of depot backorders. The availability at a base is defined as the probability that there is at least the desired number of operational engines at the base.

Bibliographic Notes

In Howard (1971), a Laplace transform approach to continuous-time Markov chains is given. A good introduction to continuous-time Markov chains can be found in Asmussen (1987), Çinlar (1975), Heyman and Sobel (1982), Karlin and Taylor (1975), Ross (1983), Taylor and Karlin (1994), and Wolff (1989). Applications of the continuous-time Markov chain to manufacturing systems are discussed in Viswanadham and Narahari (1992) and at a more advanced level in Buzacott and Shanthikumar (1993). For applications in telecommunications and computer networks, the books by Daigle (1992), Schwartz (1987), and Robertazzi (1994) provide a good introduction. Theory of queueing networks occupies a prominent position in the applications of the continuous-time Markov chain in many diverse fields; the books by Kelly (1979) and Walrand (1988) are excellent references. Matrix-geometric paradigms are useful modeling vehicles for algorithmic solutions of Markovian service systems. The two books by Neuts (1981, 1989) provide a good account of the theory and applications involved and contain extensive lists of references.

Example 5.1.5 is based on Tijms (1986, Example 2.6). Example 5.2.1 is typical of the examples shown in Howard (1971). The Yule process given as Example 5.2.4 is related to Ross (1983, Example 5.3 (a)), but in this text we use the transform approach instead. The formulas for finding variances of time to absorption in an absorbing Markov chain are based on Rust (1978). The discussions in Example 5.4.2 are related to those given in Taylor and Karlin (1994, 349–51). The materials on phase-type distribution in Section 5.5 are based on Neuts (1981). Example 5.5.5 is from Kao (1988). Additional references about fitting arbitrary distribution functions by phase-type distributions can be found in the survey paper by Neuts (1994). In Kao and Smith (1992), more closure properties were derived for phase-type renewal processes. The idea of uniformization for modeling and computation of continuous-time Markov chains has its origin in Jensen (1953). References on randomization can be found in a survey paper by Grassmann (1990). Materials concerning the Markov jump process with rewards are based on Serfozo (1979). Examples 5.7.2 and 5.7.3 are from Wood (1988) and Tijms (1986, Example 2.7), respectively. The main sources for Section 5.8 are Kelly (1979) and Ross (1983). Open Jackson networks and Gordon-Newell closed networks are also discussed in detail in Gross and Harris (1985) and Walrand (1988). Buzen's algorithm for finding the normalizing constant for the closed networks was presented in his 1973 paper. Problem 8 is based on Gershwin and Berman (1981). Problems 12 and 13 are based on the work by Chelst, Tilles, and Pipis (1981). Problems 15 and 27 are from Viswanadham and Narahari (1992). Problem 20 is related to Gelenbe and Pujolle (1987, Section 2.7). The H_2 and $E_{k-1,k}$ distributions used in Problems 23 and 24 were proposed in Tijms (1986). The data set given in Problem 25 was taken from Gross and Harris (1985). For further reading about materials relating to Problems 35, 36, 38, 40 and 41, a good source will be Neuts (1981). The explicit solution of the R matrix shown in Problem 37 was found by Gillent and Latouche (1983). Problem 43 about a two-echelon repairable inventory system is based on Gross, Kioussis, and Miller (1987). For references about closed queueing networks with blocking, readers may consult a survey paper by Onvural (1990) and a book by Perros (1994).

References

Asmussen, S. 1987. *Applied Probability and Queue.* New York: John Wiley & Sons.

Buzacott, J. A., and J. G. Shanthikumar. 1993. *Stochastic Models of Manufacturing Systems.* Englewood Cliffs, NJ: Prentice-Hall.

Buzen, J. P. 1973. Computational Algorithms for Closed Queueing Networks with Exponential Servers. *Communications of the ACM* 16(9):527–31.

Chelst, K., A. Z. Tilles, and J. S. Pipis. 1981. A Coal Unloader: A Finite Queueing System with Breakdowns. *Interfaces* 11(5):12–24.

Çinlar, E. 1975. *Introduction to Stochastic Processes.* Englewood Cliffs, NJ: Prentice-Hall.

Daigle, J. N. 1992. *Queueing Theory for Telecommunications.* Reading, MA: Addison-Wesley.

Gelenbe, E., and G. Pujolle. 1987. *Introduction to Queueing Networks.* New York: John Wiley & Sons.

Gershwin, S. B., and O. Berman. 1981. Analysis of Transfer Lines Consisting of Two Unreliable Machines with Random Processing Times and Finite Storage Buffers. *AIIE Transactions* 13(1):2–11.

Gillent, F., and G. Latouche. 1983. Semi-Explicit Solutions for *M/PH/1*—Like Queueing Systems. *European Journal of Operational Research* 18:151–60.

Grassmann, W. K. 1990. Computational Methods in Probability Theory. In *Handbook in Operations Research and Management Science, Volume 2; Stochastic Models.* Edited by D. P. Heyman and M. J. Sobel. Amsterdam: North Holland.

Gross, D., and C. M. Harris. 1985. *Fundamentals of Queueing Theory.* New York: John Wiley & Sons.

Gross, D., L. C. Kioussis, and D. R. Miller. 1987. A Network Decomposition Approach for Approximating the Steady-State Behavior of Markovian Multi-Echelon Repairable Item Inventory Systems. *Management Science* 33:1453–68.

Heyman, D. P., and M. J. Sobel. 1982. *Stochastic Models in Operations Research, Volume I.* New York: McGraw-Hill.

Howard, R. A. 1971. *Dynamic Probabilistic Systems, Volume II.* New York: John Wiley & Sons.

Jensen, A. 1953. Markoff Chains as an Aid in the Study of Markoff Processes. *Skandinavisk Aktuarietidskrift* 36:87–91.

Kao, E. P. C. 1988. Computing the Phase-Type Renewal and Related Functions. *Technometrics* 30(1):87–93.

Kao, E. P. C., and M. S. Smith. 1992. On Excess-, Current-, and Total-Life Distributions of Phase-Type Renewal Processes. *Naval Research Logistics* 32:789–99.

Karlin, S., and H. M. Taylor. 1975. *A First Course in Stochastic Processes.* New York: Academic Press.

Kelly, F. P. 1979. *Reversibility and Stochastic Networks.* New York: John Wiley & Sons.

Neuts, M. F. 1981. *Matrix-Geometric Solutions in Stochastic Models.* Baltimore: The Johns Hopkins University Press.

Neuts, M. F. 1989. *Structure Stochastic Matrices of M/G/1 Type and Their Applications.* New York: Marcel Dekker.

Neuts, M. F. 1995. Matrix-Analytic Methods in Queueing Theory. In *Advances in Queueing: Theory, Methods, and Open Problems.* Edited by Jewgeni Dshalaow. Boca Raton, FL: CRC Press.

Onvural, R. O. 1990. Survey of Closing Queueing Networks with Blocking. *ACM Computing Surveys* 22(2):83–121.

Perros, H. G. 1994. *Queueing Networks with Blocking.* New York: Oxford University Press.

Robertazzi, T. G. 1994. *Computer Networks and Systems: Queueing Theory and Performance Evaluation.* New York: Springer-Verlag.

Ross, S. M. 1983. *Stochastic Processes.* New York: John Wiley & Sons.

Rust, P. F. 1978. The Variance of Duration of Stay in an Absorbing Markov Chain. *Journal of Applied Probability* 15:420–25.

Schwartz, M. 1987. *Telecommunication Networks: Protocols, Modeling and Analysis.* Reading, MA: Addison-Wesley.

Serfozo, R. 1979. An Equivalence between Continuous and Discrete Time Markov Decision Processes. *Operations Research* 27:616–20.

Taylor, H. M., and S. Karlin. 1994. *An Introduction to Stochastic Modeling.* New York: Academic Press.

Tijms, H. C. 1986. *Stochastic Modelling and Analysis: A Computational Approach.* New York: John Wiley & Sons.

Viswanadham, N., and Y. Narahari. 1992. *Performance Modeling of Automated Manufacturing Systems.* Englewood Cliffs, NJ: Prentice-Hall.

Walrand, J. 1988. *An Introduction to Queueing Networks.* Englewood Cliffs, NJ: Prentice-Hall.

Wolff, R. W. 1989. *Stochastic Modeling and the Theory of Queues.* Englewood Cliffs, NJ: Prentice-Hall.

Wood, A. P. 1988. Optimal Maintenance Policies for Constantly Monitored Systems. *Naval Research Logistics* 35:461–71.

Appendix

Chapter 5: Section 2

Example 5.2.2 The use of MATLAB's matrix exponential routine **expm** makes the computing of time-dependent state probabilities rather simple as illustrated in this example. The output matrix contains the data for plotting Figure 5.4.

```
» [Z]=e522;
» function [Z]=e522
%
%    Part 4 of the barber shop story
%
Q=[-5 5 0 0 0; 2 -7 5 0 0; 0 4 -9 5 0;0 0 4 -9 5; 0 0 0 4 -4];
p0=[0 0 1 0 0];
x0=[]; x1=[]; x2=[]; x3=[]; x4=[]; t=[]; n=0;
while n <= 20
        x=n*0.1; t=[t x]; pt=p0*expm(Q*x);
        x0=[x0 pt(1,1)]; x1=[x1 pt(1,2)]; x2=[x2 pt(1,3)];
        x3=[x3 pt(1,4)]; x4=[x4 pt(1,5)]; n=n+1;
end
Z=zeros(21,6);
Z(:,2)=x0'; Z(:,3)=x1'; Z(:,4)=x2'; Z(:,5)=x3'; Z(:,6)=x4';
Z(:,1)=t';
```

Example 5.2.3 This example represents another application of MATLAB routine **expm**. The output matrix X contains the data for constructing Figure 5.5.

```
» [X]=e523;
» function [X]=e523
%
%   The Emergency Service Example
%
Q=zeros(9,9); IA=zeros(9,3); IA(1:3,1)=ones(3,1);
IA(4:6,2)=ones(3,1); IA(7:9,3)=ones(3,1);
lamda1=0.1; lamda2=0.125; bv=[];
mu11=0.5; mu12=0.25; mu21=0.2; mu22=0.4;
Q(1,3)=lamda2; Q(1,4)=lamda1;
Q(2,1)=mu21; Q(2,5)=lamda1; Q(2,8)=lamda2;
Q(3,1)=mu22; Q(3,6)=lamda1; Q(3,9)=lamda2;
Q(4,1)=mu11; Q(4,5)=lamda1; Q(4,6)=lamda2;
Q(5,2)=mu11; Q(5,4)=mu21;
Q(6,3)=mu11; Q(6,4)=mu22;
Q(7,1)=mu12; Q(7,8)=lamda1; Q(7,9)=lamda2;
Q(8,2)=mu12; Q(8,7)=mu21;
Q(9,3)=mu12; Q(9,7)=mu22;
h=-sum(Q'); QZ=diag(h); Q=Q+QZ;
p0=zeros(1,9); p0(1,1)=1;
z=zeros(9,1); z(5,1)=1; z(6,1)=1; z(8,1)=1; z(9,1)=1;
for t=0:1:24
    pt=p0*expm(Q*t); v=pt*IA; h=pt*z; v=[v h]; bv=[bv; v];
    end
t=0:1:24; X=[t' bv];
```

Chapter 5: Section 4

Example 5.4.1 The following program finds the results reported in the text.

```
» function  e541
%
%    Example 5.4.1
%    The trauma center example
%
p=[.7 .2 .1]; la=0.5;  lp=la*p; mu=.4;
R=zeros(5,3); V=zeros(5,5);
R(3,1)=lp(1,3); R(4,1:2)=lp(1,2:3); R(5,1:3)=lp(1,1:3);
R
V(1,1:4)=[-la lp(1,1:3)]; V(2,:)=[mu -(la+mu) lp(1,1:3)];
V(3,2:5)=[2*mu -(la+2*mu) lp(1,1:2)];
V(4,3:5)=[3*mu -(la+3*mu) lp(1,1)];
V(5,4:5)=[4*mu -(la+4*mu)];
N=inv(V); NI=-N; NI
sum(NI')
SI=NI*R
I=eye(5); V=(2*N*(N.*I))-(N.*N)
```

Chapter 5: Section 7

Example 5.7.1 This example illustrates accounting with the Markov jump process with rewards.

```
» e571
The equivalent CTMC

Q =

   -1.6000    1.4000    0.2000
    8.0000  -10.0000    2.0000
    1.2000    1.6000   -2.8000

  ri =     8.60      6.95     14.11

The Markov chain subordinated to a Poisson process

Q1 =

    0.8400    0.1400    0.0200
    0.8000         0    0.2000
    0.1200    0.1600    0.7200

r_hat =     1.79      6.95      5.73

 discrete discount factor    0.9901
  the discounted reward vector

g =

  299.6489
  306.2640
  312.9176
```

```
» function e571
%
%  Example 5.7.1
%
P1=[.2 .7 .1;.8 0 .2; .3 .4 .3];
R1=[5 2 1; 2 0 4; 7 3 11];
R2=[1 4 2];
R3=[12 4 7; 3 0 9; 5 3 15];
vi=[2 10 4]; beta=0.1; c=[1 1 1]./(beta+vi);
E=[0 1 1; 1 0 1; 1 1 0]; QE=P1.*E; qi=-sum(QE');
QE=QE+diag(qi); Q=[vi; vi; vi]'.*QE;
fprintf('The equivalent CTMC \n'); Q
ra=sum((P1.*R1)'); rb=R2.*c; c=vi.*c;
rc=sum((P1.*R3)'); rc=c.*rc; r=ra+rb+rc;
fprintf('  ri = %8.2f  %8.2f  %8.2f \n',r);
%
%  Convert the equivalent ctmc with generator of Equation 5.6.6 to
%  a Markov chain subordinated to a Poisson process
%
v=diag(Q); vmax=max(abs(v)); Q1=(1/vmax)*Q; v1=1+diag(Q1);
Q2=diag(v1); Q1=Q2+Q1.*E;
fprintf('The Markov chain subordinated to a Poisson process \n');
Q1
r2=(beta+vi)/(beta+vmax); r2=r2.*r;
fprintf('r_hat = %8.2f  %8.2f  %8.2f \n',r2);
alpha=vmax/(beta+vmax);
fprintf(' \n discrete discount factor %8.4f \n',alpha);
Q3=alpha*Q1; I=eye(3); g=(I-Q3)\r2';
fprintf('  the discounted reward vector \n'); g
```

Example 5.7.3 The following results and program are for the cash balance problem.

```
» e573
Q =
```

-0.4000	0.1600	0.1200	0.0800	0.0400	0	0
0.5000	-0.9000	0.1600	0.1200	0.0800	0.0400	0
0.4000	0.1000	-0.9000	0.1600	0.1200	0.0800	0.0400
0.2500	0.1500	0.1000	-0.9000	0.1600	0.1200	0.1200
0.1000	0.1500	0.1500	0.1000	-0.9000	0.1600	0.2400
0	0.1000	0.1500	0.1500	0.1000	-0.9000	0.4000
0	0	0.1000	0.1500	0.1500	0.1000	-0.5000

```
p =
```

0.3637	0.1190	0.1221	0.1120	0.0915	0.0618	0.1300

```
q =
```

1.4500	1.0000	0.6500	0.4700	0.4800	0.6500	0.9000

```
   long-run average cost under (4,10)   0.9794
» function e573
%
%  Example 5.7.3
```

```
%
d=[.4 .3 .2 .1]; w=[.2 .3 .3 .2];
lm=.5; mu=.4; alpha=4; beta=10; n=beta-alpha+1;
nd=length(d); nw=length(w);
dp=mu*d; wd=lm*w;
Q=zeros(n,n);
for i=1:n-1
    sc=i+1; sd=i+nd;
    if sd <= n
    Q(i,sc:sd)=dp;
    else
    m=n-sc+1; Q(i,sc:n)=dp(1:m);
    Q(i,n)=Q(i,n)+sum(dp(m+1:nd));
    end
end
wdf=fliplr(wd);
for i=n:-1:2
    sc=i-1; sd=i-nd;
    if sd >= 1
    Q(i,sd:sc)=wdf;
    else
    m=nw-sc+1; Q(i,1:sc)=wdf(m:nw);
    Q(i,1)=Q(i,1)+sum(wdf(1:m-1));
    end
end
s1=-sum(Q'); Q1=diag(s1); Q=Q+Q1;
Q
[p]=mc_limsr(Q); q=[]; c1=1; c2=.5; c3=.05;
p
for i=alpha:beta
    s1=0; s2=0;
    for j=0:alpha-1
      df=i-j;
      if df > 0
        if  df <= nw
        s1=s1+wd(df)*(alpha-j);
        end
      end
    end
    for j=beta+1:beta+nd
      df=j-i;
      if df >0
        if df <= nd
        s2=s2+dp(df)*(j-beta);
        end
      end
    end
    sm=s1*c1+s2*c2+c3*i; q=[q sm];
end
q
g=sum(p.*q);
fprintf(' long-run average cost under (4,10) %8.4f \n',g);
```

Chapter 5: Section 8

Example 5.8.9 For the closed queueing network example, the output matrix H contains the marginal queue length distributions. The results are plotted in Figure 5.17.

```
»  [H]=e589;
   no. of states  =            66
   G(N) =            7231882.9120

»  function [H]=e589
%
%   Example 5.8.9: Gordon-Newell Network
%
mu=[.25 .48 .08]; p=.75; q=1-p;
N=10; s=3;
%
%   Number of states in the state space
%
ns=binomial(N+s-1,N);
fprintf(' no. of states  = %10.0f \n',ns);
a=1; lm=[a a*p a*q]; r=lm./mu;
%
%   Buzen's Algorithm
%
N1=N+1; C=zeros(3,N1); C(2,1)=1; C(3,1)=1;
for i=1:N1
   C(1,i)=r(1)^(i-1);
end
for i=2:3
   for k=2:N1
   C(i,k)=C(i-1,k)+r(i)*C(i,k-1);
   end
end
GN=C(3,N1); Z=zeros(ns,4);
fprintf('   G(N) = %20.4f \n',GN);
%
%   End of Buzen's Algorithm
%
tm=0;
for i=0:N
  for j=0:N;
       k=N-(i+j);
       if k >= 0
       tm=tm+1; pb=(r(1)^i)*(r(2)^j)*(r(3)^k); pb=pb/GN;
       Z(tm,:)=[i j k pb];
       end
  end
end
%
%   Compute the marginal distribution
%
H=zeros(N1,4); v=ones(1,N1); v=cumsum(v)-1; H(:,1)=v';
for i=1:3
   v=[];
   for j=0:N
      sm=0;
      for k=1:ns
         if Z(k,i) == j
            sm=sm+Z(k,4);
         end
      end
      v=[v sm];
   end
   H(:,i+1)=v';
end
```

Markov Renewal and Semi-Regenerative Processes

Tips for Chapter 6

■ This is a relatively demanding chapter. The reward for plowing through the materials covered in this chapter is being able to see the unified framework Markov renewal processes supply for most stochastic models considered in the earlier chapters. Moreover, a Markov renewal paradigm enables us to handle many stochastic modeling problems beyond those of Markovian or exponential varieties.

■ Section 6.1 provides readers some basic ideas about Markov renewal processes and how they are related to the stochastic processes introduced earlier.

■ One of the most powerful ideas in stochastic processes is the notion of a semi-regenerative process. Our development culminates in the expression for the limiting state probability given in Equation 6.4.4. This expression can be pivotal in spawning other results of theoretical and applied interests.

■ Readers with an interest in queueing applications may use Examples 6.4.7 and 6.4.8 as supplementary reading.

■ We use Riemann-Stieltjes integrals and Riemann-Stieltjes transforms throughout this chapter. Hence the results obtained in this chapter are applicable to cases in which holding times are continuous, discrete, or continuous with discrete components.

6.0 Overview

In a Markov renewal process, the transitions between states of the system are governed by the transition probability matrix of the embedded Markov chain. Before a transition takes place, the sojourn time in a given state is a random variable following an arbitrary distribution, called the holding time distribution. This distribution is a function of the starting and destination states of the transition. When the state space of the process contains only one state, the resulting counting process is a renewal process and the holding time distribution corresponds to the interarrival time distribution. In addition, if the holding time follows an exponential distribution, then the resulting counting process is a Poisson process. When all holding times assume a given constant, the resulting process is a discrete-time Markov chain. When the holding time distributions are exponential and virtual transitions are not allowed, then the resulting process is a continuous-time Markov chain; if we allow virtual transitions, however, the resulting process is a Markov jump process. Therefore, all the stochastic processes considered earlier are special cases of a Markov renewal process. In a semi-regenerative process, the process *restarts* at the renewal times of an underlying renewal process. The notion of a regenerative process is generalized in this chapter to the semi-regenerative process. The idea of *process restart* depends not only on the renewal times but also on the observed states of the embedded Markov chain at the renewal times. This generalization makes the resulting stochastic process even more potent as a paradigm for problem solving.

In Section 6.1, a Markov renewal process is formally defined. The key development of Markov renewal theory given in Section 6.2 parallels that of renewal theory. First, we have a Markov renewal type of equation, Equation 6.2.14. This is obtained by conditioning on the transition time and destination state resulting from the first transition. We then have the time-dependent solution of Equation 6.2.15 and limiting solution of Equation 6.2.21 of the Markov renewal equation. An application of these results to the study of the various properties of a semi-Markov process is presented in Section 6.3. A semi-Markov process is one in which we track the state of the system at all times; we also introduce the related reward processes. The last section is about semi-regenerative process. The examples shown there should shed some light on applications of semi-regenerative process and its potency as a paradigm for stochastic modeling and analysis.

6.1 Introduction

Consider a stochastic process $(X, T) = \{(X_n, T_n), n = 0, 1, \ldots\}$, where random variables $\{X_n\}$ take on value in the state space $S = \{0, 1, \ldots\}$ and random variables $\{T_n\}$ take on values in $[0, \infty)$ (even though the process (X, T) is represented by a doublet, we refer to S as the state space of the process). Random variable X_n is the state of the system immediately after the nth transition and random variable T_n can be viewed as the epoch at which the nth transition occurs (other interpretations can be given to T_n, for example, see Example 6.1.2). Let $0 = T_0 < T_1 < \cdots$.

The stochastic process (X, T) is a *Markov renewal process* if

$$P\{X_{n+1} = j, T_{n+1} - T_n \le t | X_0, \ldots, X_n, T_0, \ldots, T_n\}$$
$$= P\{X_{n+1} = j, T_{n+1} - T_n \le t | X_n\}, \tag{6.1.1}$$

for all n, $t \ge 0$, and $j \in S$.

The process (X, T) is *time homogeneous* if for all i, j, and t, we have

$$P\{X_{n+1} = j, T_{n+1} - T_n \le t | X_n = i\} = Q_{ij}(t), \tag{6.1.2}$$

independent of n. In the sequel, we will focus on time-homogeneous Markov renewal processes.

For a fixed t, let square matrix $Q(t) = \{Q_{ij}(t)\}$, where the dimension of the matrix equals the size of the state space. The matrix function $\{Q(t), t \ge 0\}$ is called the *semi-Markov kernel* over the state space S. Define

$$p_{ij} = \lim_{t \to \infty} Q_{ij}(t). \tag{6.1.3}$$

We see that $\{p_{ij}\}$ are the one-step transition probabilities defined at the end of a transition epoch. The matrix P is the transition probability matrix of a Markov chain embedded at the transition epochs. Define

$$G_{ij}(t) = P\{T_{n+1} - T_n \le t | X_n = i, X_{n+1} = j\}. \tag{6.1.4}$$

The preceding gives the distribution function for the holding time in state i before a transition out of the state is made, given that we know that the next transition is to state j. Let τ_{ij} denote its mean. We call $G_{ij}(t)$ the *holding time* distribution.

From Equation 6.1.2, we find that the elements of the semi-Markov kernel are given by

$$Q_{ij}(t) = \frac{P\{X_{n+1} = j, T_{n+1} - T_n \le t, X_n = i\}}{P\{X_{n+1} = j, X_n = i\}} \frac{P\{X_{n+1} = j, X_n = i\}}{P\{X_n = i\}} = p_{ij} G_{ij}(t).$$

$$\tag{6.1.5}$$

EXAMPLE **6.1.1** **Movement of Coronary Patients in a Hospital** Based on a study of the patient records of 555 patients treated for heart attacks at the Hospital of St. Raphael, New Haven, Connecticut, a Markov renewal process has been constructed to model the movement of these patients over various patient care units. There, the state space contains the states: ECF, HOME, DIED, CCU, PCCU, ICU, MED, SURG, and AMB. We use integers 1, ..., 9 to denote them, respectively. The transition probability matrix of the embedded Markov chain is given by

$$P = \begin{bmatrix} 1 & & & & & & & & \\ & 1 & & & & & & & \\ & & 1 & & & & & & \\ & .0063 & .0962 & & .7447 & .0084 & .1339 & .0042 & .0063 \\ .0577 & .8298 & .0495 & .0192 & & .0137 & .0247 & .0027 & .0027 \\ & .1667 & & & .5833 & & .1667 & .0833 & \\ .0811 & .7028 & .1216 & & .0135 & .0405 & & .0135 & .0270 \\ & 1 & & & & & & & \\ & 1 & & & & & & & \end{bmatrix}.$$

We note that the matrix P is stated in the canonical form defined by Equation 4.3.1 and the embedded Markov chain is absorbing—eventually patients will be in one of the absorbing states 1, 2, or 3. The holding time distributions associated with those transitions for which $p_{ij} > 0$ are approximated by five Weibull distributions indexed by a, b, c, d, and f shown in Figure 6.1. Recalling Equation 4.5.4, the Weibull density is given by

$$f(t) = \left(\frac{\beta}{v}\right)\left(\frac{t}{v}\right)^{\beta-1} e^{-\left(\frac{t}{v}\right)^{\beta}} \qquad t > 0.$$

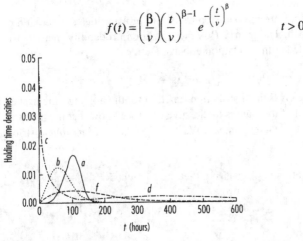

FIGURE
6.1   The five Weibull densities.

In Table 6.1, we tabulate the parameters, the means (in hours), and variances (in hours squared) of the five distributions for future reference. The assignments of the five distributions to model the holding time distributions $\{G_{ij}(t)\}$ for the process are as follows:

$$\{G_{ij}(t)\} = \begin{array}{c} \\ 4 \\ 5 \\ 6 \\ 7 \\ 8 \\ 9 \end{array} \begin{array}{cccccccccc} 1 & 2 & 3 & 4 & 5 & 6 & 7 & 8 & 9 \\ \left[\begin{array}{ccc|cccccc} & b & c & & a & a & a & a & b \\ d & & f & d & & a & d & a & a \\ & & c & & & d & & a & a \\ d & d & f & & d & d & & d & d \\ & d & & & & & & & \\ & d & & & & & & & \end{array}\right] \end{array}.$$

	Distribution	β	ν	Mean	Variance
TABLE **6.1** The parameters for the five Weibull distributions	a	4.74	109.33	100.07	577.92
	b	2.21	76.86	68.07	1060.20
	c	0.77	39.40	46.10	3708.80
	d	2.30	405.74	359.46	27,397.00
	f	1.62	182.92	163.79	10,695.70

In Figure 6.1, we plot the five Weibull densities. We observe that the five densities have distinguishing characteristics ranging from one that resembles an exponential density to one that is relatively flat over a long interval. Based on the structure of the embedded Markov chain, we see that the process eventually will move to one of the three absorbing states because patients eventually will leave the hospital. ∎

EXAMPLE **6.1.2** **Modeling Road Traffic** The highway department is studying the traffic pattern of a given thoroughfare (think of the times you drive through a street that has a rubber hose lying across it). They set up a sensor at a given spot along the road, monitoring the passing vehicles. There are three vehicle types of interest: cars, pickups, and trucks. For the nth passing vehicle, let X_n denote the type of vehicle, with 1, 2, and 3 representing a car, a pickup, and a truck, respectively. Let $T_{n+1} - T_n$ denote the headway between the $n + 1$ and the nth passing vehicle measured in feet as they pass by the sensor. The stochastic process $(X, T) = \{(X_n, T_n), n \geq 0\}$ is a Markov renewal process with state space $S = \{1, 2, 3\}$, transition probability matrix of the embedded Markov chain $P = \{p_{ij}\}$, and holding time distributions $\{G_{ij}(t)\}$, where $p_{ij} = P\{X_{n+1} = j | X_n = i\}$ and $G_{ij}(t) = P\{T_{n+1} - T_n \leq t | X_{n+1} = j, X_n = i\}$. We see that $G_{ij}(t)$ gives the probability distribution of the headway measured in feet when a vehicle of type j follows a vehicle of type i. Using Equation 6.1.5, we can construct the semi-Markov kernel $\{Q_{ij}(t)\}$ that characterizes the process. In contrast with the previous example, the embedded Markov chain here is assumed to be irreducible and recurrent; also, random variables $\{T_n\}$ are measured in feet (as opposed to time units). ∎

EXAMPLE **6.1.3** **An M/G/1/1 Queue** Consider a service system in which arrivals follow a Poisson process with parameter λ. The service times are i.i.d. random variables with a common distribution G and are independent of the arrival process. At an arrival epoch, if the customer sees that the server is busy, she departs without receiving any service and will have no influence on the future of the system. Let X_n denote the number of customers in the system at the nth epoch of state change. Since the system can only be idle or have one customer in it, the state space is given by $S = \{0, 1\}$. Let $\{T_n\}$ be the successive epochs of state change. Assume that $T_0 = 0$

and $X_0 = 0$. The stochastic process $(X, T) = \{(X_n, T_n), n \geq 0\}$ is a Markov renewal process with a semi-Markov kernel

$$Q(t) = \begin{bmatrix} 0 & 1-e^{-\lambda t} \\ 1 & G(t) & 0 \end{bmatrix} \qquad t \geq 0,$$

where $Q(t) = \{Q_{ij}(t)\}$. Even though the service time distribution is not memoryless, we look at the process only at the epoch of state change. Given that we know the state of the system at such an epoch, the future of the process becomes independent of the past. This enables us to assert that the process (X, T) is a Markov renewal process. ∎

Define the interarrival time between the nth state change and the $(n-1)$st state change as

$$D_n = T_n - T_{n-1} \qquad n = 1, 2, \ldots,$$

where $T_0 = 0$ (the term "state change" includes the possibility of virtual transitions). Then

$$T_n = \sum_{k=1}^{n} D_k$$

denotes the arrival time of the nth state change. For simplicity, we call $\{D_n\}$ the interrenewal times and $\{T_n\}$ the renewal times. One way to obtain the semi-Markov kernel is by Equation 6.1.5. Another way is by using the following:

$$Q_{ij}(t) = P\{X_{n+1} = j, \ T_{n+1} - T_n \leq t | X_n = i\} = P\{X_{n+1} = j, \ D_{n+1} \leq t | X_n = i\}$$

$$= \int_0^t dP\{X_{n+1} = j, \ D_{n+1} \leq u | X_n = i\}$$

$$= \int_0^t P\{X_{n+1} = j | D_{n+1} = u, \ X_n = i\} dP\{D_{n+1} \leq u | X_n = i\}. \qquad (6.1.6)$$

EXAMPLE
6.1.4

An M/G/1 Queue The assumptions underlying an $M/G/1$ queue are identical to those of an $M/G/1/1$ queue studied in Example 6.1.3; however, we now have an unlimited number of waiting spaces. In Example 4.4.2, we have studied the limiting behaviors of an $M/G/1$ queue. Let X_n denote the number of customers left behind by the nth departure and T_n the time of the nth departure. We use Equation 6.1.6 to find the semi-Markov kernel and this in turn shows that the stochastic process $(X, T) = \{(X_n, T_n), n \geq 0\}$ is a Markov renewal process with state space $S = \{0, 1, \ldots\}$.

For notational convenience, we use $pos(n; \lambda, t)$ to denote the probability that there are n Poisson arrivals in an interval of length t. If $X_n = i$ and $i \geq 1$, then at a departure epoch there is at least one customer still in the system and so another service of length S can start immediately, where S follows distribution function

G. This implies that $D_{n+1} = S$. For $i \geq 1$ and $j - i + 1 \geq 0$, we use Equation 6.1.6 to obtain

$$Q_{ij}(t) = P\{X_{n+1} = j, \ S \leq t | X_n = i\}$$

$$= \int_0^t P\{X_{n+1} = j | S = u, \ X_n = i\} dP\{S \leq u | X_n = i\}$$

$$= \int_0^t pos(j - i + 1; \ \lambda, \ u) dG(u). \tag{6.1.7}$$

When $X_n = 0$, it means that an idle period I will elapse before the next arrival occurs, where $I \sim exp(\lambda)$. With $X_n = 0$, we use $X_n' = 1$ to indicate that the system size becomes one at the end of the idle period. Hence

$$Q_{0j}(t) = P\{X_{n+1} = j, \ D_{n+1} \leq t | X_n = 0\}$$

$$= \int_0^t dP\{X_{n+1} = j, \ I \leq s, \ D_{n+1} \leq t | X_n = 0\}$$

$$= \int_0^t P\{X_{n+1} = j, \ S \leq t - s | X_n = 0, \ I = s\} dP\{I \leq s | X_n = 0\}$$

$$= \int_0^t P\{X_{n+1} = j, \ S \leq t - s | X_n' = 1\} \lambda e^{-\lambda s} ds = \int_0^t Q_{1j}(t - s) \lambda e^{-\lambda s} ds. \tag{6.1.8}$$

In establishing the preceding, we note that at the end of an idle period the system behaves similarly to $X_n = 1$. This yields the last equality. Finally, in an interrenewal interval, exactly one service completion occurs. It implies that for $i \geq 2$, we have $p_{ij} = 0$ and $Q_{ij}(t) = 0$ for $j < i - 1$ and all $t \geq 0$. For $n \geq 0$, we let

$$q_n(t) = \int_0^t pos(n; \ \lambda, \ u) dG(u) \quad \text{and} \quad p_n(t) = \int_0^t \lambda e^{-\lambda u} q_n(t - u) du.$$

Then the semi-Markov kernel $Q(t)$ has the form

$$
Q(t) =
\begin{array}{c}
 \\
0 \\
1 \\
2 \\
\vdots
\end{array}
\begin{array}{cccc}
0 & 1 & 2 & \\
\left[\begin{array}{cccc}
p_0(t) & p_1(t) & p_2(t) & \cdots \\
q_0(t) & q_1(t) & q_2(t) & \cdots \\
& q_0(t) & q_1(t) & \cdots \\
& & & \ddots
\end{array}\right].
\end{array}
$$

If we set t equal to ∞ in the previous matrix, we obtain the transition probability matrix of the embedded Markov chain described in Example 4.3.2. ∎

EXAMPLE
6.1.5 **Departures from an M/M/1/3 Queue** Consider a single-server service system with Poisson arrivals, exponential service times, and two waiting spaces (so the capacity of the system is 3). Let λ be the arrival rate and μ the service rate. As usual, we assume that service times are independent of the arrival process. Let X_n denote the number of customers left behind by the nth departure and T_n the time of the nth departure. Again, we use Equation 6.1.6 to find the semi-Markov kernel and hence establish that the resulting stochastic process $(X, T) = \{(X_n, T_n), n \geq 0\}$ is

a Markov renewal process. Even though the system capacity is 3, its maximal possible occupancy is 2 at a departure epoch. The state space is thus given by $S = \{0, 1, 2\}$.

As in Example 6.1.4, we use $pos(n; \lambda, t)$ to denote the probability that there are n Poisson arrivals in an interval of length t. Since each interrenewal time yields exactly one service completion, when $X_n = 2$ we have $p_{20} = 0$ and, consequently, $Q_{20}(t) = 0$ for all $t \geq 0$.

Consider the cases with $(X_n, X_{n+1}) = (i, j)$, where $i = 1, 2, j = 0, 1$, and $j - i + 1 \geq 0$. In these cases a service of length S can start immediately after a departure and the arrivals during S will not cause X_{n+1} to reach its maximal possible value of 2. For these cases, Equation 6.1.7 is readily applicable with $S - exp(\mu)$ and

$$Q_{ij}(t) = \int_0^t pos(j-i+1; \lambda, z)\mu e^{-\mu z}dz \qquad (i,j) \in \{(1,\ 0),(1,\ 1),(2,\ 1)\}.$$

For the cases in which $i = 1$ and 2 and $j = 2$, we need to take account of arrival overflows during S. This implies

$$Q_{ij}(t) = \sum_{k=2}^{\infty} \int_0^t pos(k-i+1; \lambda, z)\mu e^{-\mu z}dz \qquad (i,j) \in \{(1,\ 2),(2,\ 2)\}.$$

When $X_n = 0$, the system will experience an idle period I followed by a service period S, where $I \sim exp(\lambda)$ and $S \sim exp(\mu)$. Thus $D_{n+1} = I + S$. Consider first the case in which $X_{n+1} = 0$ or 1. With $S \sim exp(\mu)$, for $j = 0$ and 1, we use Equation 6.1.8 to write

$$Q_{0j}(t) = \int_0^t Q_{1j}(t-s)\lambda e^{-\lambda s}ds$$

$$= \int_0^t \int_0^{t-s} pos(j; \lambda, z)\mu e^{-\mu z}dz\lambda e^{-\lambda s}ds$$

$$= \int_0^t pos(j; \lambda, z)\mu e^{-\mu z}\int_0^{t-z} \lambda e^{-\lambda s}dsdz \qquad \text{by interchanging the order of integration}$$

$$= \int_0^t pos(j; \lambda, z)[1 - e^{-\lambda(t-z)}]\mu e^{-\mu z}dz.$$

When $X_{n+1} = 2$, we again take account of the arrival overflows during S and find

$$Q_{02}(t) = \sum_{k=2}^{\infty} \int_0^t pos(k; \lambda, z)[1 - e^{-\lambda(t-z)}]\mu e^{-\mu z}dz.$$

This concludes the construction of the semi-Markov kernel. By letting $t \to \infty$ in $\{Q_{ij}(t)\}$, we obtain the transition probability matrix P of the embedded Markov chain defined at each departure epoch, where

$$P = \begin{bmatrix} \dfrac{1}{1+\rho} & \dfrac{\rho}{(1+\rho)^2} & \dfrac{\rho^2}{(1+\rho)^2} \\[2mm] \dfrac{1}{1+\rho} & \dfrac{\rho}{(1+\rho)^2} & \dfrac{\rho^2}{(1+\rho)^2} \\[2mm] & \dfrac{1}{1+\rho} & \dfrac{\rho}{1+\rho} \end{bmatrix}$$

and $\rho = \lambda/\mu$. ∎

EXAMPLE
6.1.6

A GI/M/1 Queue The *GI/M*/1 queue was first studied in Example 4.3.3. Consider a single-server service system with an unlimited number of waiting rooms. The inter-arrival times of successive customers are i.i.d. random variables following a common distribution G. Service times are i.i.d. random variables, each following an exponential distribution with parameter μ. The service times are independent of the arrival process. Let $\{T_n\}$ be the epochs of successive arrivals and X_n denote the number of customers in the system just before the nth arrival. Then stochastic process $(X, T) = \{(X_n, T_n), n \geq 0\}$ is a Markov renewal process with state space $S = \{0, 1, \ldots\}$.

When $X_{n+1} = 0$, we know that the interval $D_{n+1} = T_{n+1} - T_n = I$ contains two segments: a period in which the server completes the service of all $X_n + 1$ customers and a subsequent period in which the server is idle, where I is an interarrival time interval. When $X_{n+1} \geq 1$, then we know that the server is busy throughout the interarrival time interval I. For all $i \geq 0$ and all $j \geq 1$, we use Equation 6.1.6 to write

$$Q_{ij}(t) = \int_0^t P\{X_{n+1} = j | I = s, X_n = i\} dP\{I \leq s | X_n = i\}$$

$$= \int_0^t pos(i+1-j; \mu, s) dG(s),$$

where we use $pos(k; \mu, s)$ to denote the probability that, for the Poisson process with parameter μ, k events have occurred in an interval of length s. To find $Q_{i0}(t)$ for $i \geq 0$, we see that

$$Q_{i0}(t) + \sum_{j=1}^{\infty} Q_{ij}(t) = G(t).$$

This gives

$$Q_{i0}(t) = G(t) - \sum_{j=1}^{i+1} \int_0^t pos(i+1-j; \mu, s) dG(s) = G(t) - \sum_{k=0}^{i} \int_0^t pos(k; \mu, s) dG(s).$$

Let

$$q_n(t) = \int_0^t pos(n; \mu, s) dG(s) \qquad \text{and} \qquad r_n(t) = G(t) - \sum_{k=0}^{n} q_k(t);$$

the semi-Markov kernel then reads

$$
Q(t) = \begin{array}{c} 0 \\ 1 \\ 2 \\ \vdots \end{array}
\begin{array}{ccc} 0 & 1 & 2 \end{array}
\left[\begin{matrix} r_0(t) & q_0(t) & & & \\ r_1(t) & q_1(t) & q_0(t) & & \\ r_2(t) & q_{2(t)} & q_1(t) & q_0(t) & \\ \vdots & \vdots & \vdots & \vdots & \ddots \\ \vdots & & & & \end{matrix} \right] .
$$

■

EXAMPLE
6.1.7

The Machine Reliability Problem Consider a machine containing M parts. If one part fails, so does the machine. Let $\{T_n\}$ denote the times of successive machine failures and X_n the part that causes the nth machine failure. The interrenewal time (to be specific, the interfailure time) $D_{n+1} = T_{n+1} - T_n$ is the sum of the repair time of the part that fails at T_n and a machine functioning interval, L, following the repair. For $i = 1, ..., M$, we assume that the lifetime of part i follows an exponential distribution with parameter μ_i and the repair time of part i, V_i, follows distribution G_i. Moreover, we assume that lifetimes are mutually independent and independent of repair times. The stochastic process $(X, T) = \{(X_n, T_n), n \geq 0\}$ is a Markov renewal process with state space $S = \{1, 2, ..., M\}$. We observe that the probability density that a failure of part j occurs at time s while other parts have not failed by time s is given by $\mu_j e^{-\mu_j s} \prod_{i \neq j} e^{-\mu_i s} = \mu_j e^{-\mu s}$, where $\mu = \mu_1 + \cdots + \mu_M$. The semi-Markov kernel is

$$
\begin{aligned}
Q_{ij}(t) &= P\{X_{n+1} = j, D_{n+1} \leq t | X_n = i\} = \int_0^t dP\{X_{n+1} = j, V_i \leq s, D_{n+1} \leq t | X_n = i\} \\
&= \int_0^t P\{X_{n+1} = j, D_{n+1} \leq t | V_i = s, X_n = i\} dP\{V_i \leq s | X_n = i\} \\
&= \int_0^t P\{X_{n+1} = j, L \leq t - s\} dG_i(s) \\
&= \int_0^t \int_0^{t-s} dP\{X_{n+1} = j, L = \tau\} d\tau dG_i(s) = \int_0^t \int_0^{t-s} \mu_j e^{-\mu\tau} d\tau dG_i(s) \\
&= \int_0^t \mu_j e^{-\mu\tau} \int_0^{t-\tau} dG_i(s) d\tau = \int_0^t \mu_j e^{-\mu\tau} G_i(t-\tau) d\tau = \frac{\mu_j}{\mu} \int_0^t G_i(t-\tau) \mu e^{-\mu\tau} d\tau.
\end{aligned}
$$

When $t \to \infty$, the last integral is one since it is the convolution of G_i and an exponential distribution with parameter μ. Hence $p_{ij} = \mu_j/\mu$ for all i. ▪

To conclude this section, in the next three examples we show that renewal processes, discrete-time Markov chains, and continuous-time Markov chains are all special cases of Markov renewal processes.

EXAMPLE 6.1.8 **Renewal Processes** Consider a Markov renewal process whose state space contains only a single state, say state 1. In this case, $X_n = 1$ for all n and $\{T_n\}$ constitute a renewal process with interrenewal times $\{D_n\}$. ▪

EXAMPLE 6.1.9 **Discrete-Time Markov Chains** Consider a Markov renewal process whose holding time distributions have the following form:

$$
G_{ij}(t) = \begin{cases} 1 & \text{if } t \geq 1 \\ 0 & \text{if } t < 1 \end{cases}
$$

for all i and j. This means each transition takes exactly one time unit. The resulting process is a discrete Markov chain whose transition probability matrix is given by $\{p_{ij}\}$. ▪

EXAMPLE   **Continuous-Time Markov Chains**   Consider a Markov renewal process whose semi-
6.1.10   Markov kernel is given by

$$Q_{ij}(t) = p_{ij}(1 - e^{-\mu_i t}) \qquad t > 0$$

for all i and j. We see that the holding time in state i follows an exponential distribution with parameter μ_i. If $p_{ii} = 0$ for each i, then the resulting process is a continuous-time Markov chain; otherwise, it is the continuous-time Markov process with a discrete state space presented in Example 5.6.3. ∎

6.2   Markov Renewal Functions and Equations

In a renewal process, we have the renewal function $M(t) = E[N(t)]$ denoting the expected number of renewals by time t. Generalizing the idea to Markov renewal processes, we now have the Markov renewal functions $\{R_{ij}(t)\}$, where $R_{ij}(t)$ represents the expected number of visits (renewals) to state j in the interval $[0, t]$ given the process starts from state i. Moreover, we generalize the concept of renewal-type equations to Markov renewal equations.

Define $Q_{ij}^{(n)}(t) = P\{X_n = j, T_n \le t | X_0 = i\}$. Using the law of total probability, we have

$$
\begin{aligned}
Q_{ij}^{(n)}(t) &= \sum_{k=0}^{\infty} \int_0^t dP\{X_n = j, T_n \le t, X_1 = k, T_1 \le s | X_0 = i\} \\
&= \sum_{k=0}^{\infty} \int_0^t P\{X_n = j, T_n \le t | X_1 = k, T_1 = s, X_0 = i\} dP\{X_1 = k, T_1 \le s | X_0 = i\} \\
&= \sum_{k=0}^{\infty} \int_0^t P\{X_{n-1} = j, T_{n-1} \le t - s | X_0 = k\} dQ_{ik}(s) \\
&= \sum_{k=0}^{\infty} \int_0^t dQ_{ik}(s) Q_{kj}^{(n-1)}(t - s).
\end{aligned}
\tag{6.2.1}
$$

In matrix notation, the preceding reads

$$Q^{(n)}(t) = \int_0^t dQ(s) Q^{(n-1)}(t - s) = (Q * Q^{(n-1)})(t), \tag{6.2.2}$$

where we use an asterisk (*) to denote matrix convolution and define $Q^{(0)}(t) = I$ for all $t \ge 0$. Define the Laplace-Stieltjes transform of $Q(t)$ (in matrix notation)

$$Q^e(s) = \int_0^{\infty} e^{-st} dQ(t).$$

From Equation 6.2.2, we obtain

$$Q^{e(n)}(s) = Q^e(s) Q^{e(n-1)}(s) = \cdots = \left(Q^e(s)\right)^n \qquad n \ge 0. \tag{6.2.3}$$

Define the following two indicator random variables

$$1_j(X_n) = \begin{cases} 1 & \text{if } X_n = j \\ 0 & \text{otherwise} \end{cases}$$

and

$$I_{[0,t]}(T_n) = \begin{cases} 1 & \text{if } T_n \in [0,\ t] \\ 0 & \text{otherwise.} \end{cases}$$

Let $R_{ij}(t)$ denote the expected number of times $X_n = j$ during the interval $[0,\ t]$ given that $X_0 = i$. We call $R_{ij}(t)$ the Markov renewal function. For the sample path shown in Figure 6.2, we see that $R_{12}(t) = 3$. The Markov renewal function can be found from $\{Q_{ij}^{(n)}(t)\}$ as follows:

$$R_{ij}(t) = E\left[\sum_{n=0}^{\infty} 1_j(X_n)I_{[0,t]}(T_n)\Big|X_0 = i\right] = \sum_{n=0}^{\infty} E\left[1_j(X_n)I_{[0,t]}(T_n)\Big|X_0 = i\right]$$

$$= \sum_{n=0}^{\infty} P\{X_n = j,\ T_n \le t | X_0 = i\} = \sum_{n=0}^{\infty} Q_{ij}^{(n)}(t).$$

In matrix notation, the Markov renewal function can be stated as

$$R(t) = \sum_{n=0}^{\infty} Q^{(n)}(t). \tag{6.2.4}$$

Since $Q_{ii}^{(0)}(t) = 1$ for all i and $t \ge 0$, in Equation 6.2.4 when $i = j$ the transition $X_0 = i$ at time T_0 is counted as a part of the Markov renewal function. Conversely, in renewal theory the initial renewal at the time origin is generally not included in the renewal function (and hence $M(t) = \sum_{n=1}^{\infty} F_n(t)$). We write

$$R(t) = I + Q(t) + Q^{(2)}(t) + Q^{(3)}(t) + \cdots$$

$$= I + Q(t) + Q * \left[Q(t) + Q^{(2)}(t) + Q^{(3)}(t) + \cdots\right] = I + (Q * R)(t). \tag{6.2.5}$$

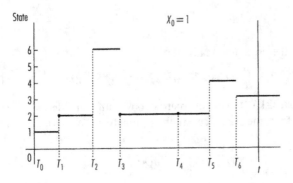

FIGURE

6.2   A sample path of a Markov renewal process.

The preceding shows that the Markov renewal function $R(t)$ is a solution to the system of integral equations (Equation 6.2.5). In component form, Equation 6.2.5 reads

$$R_{ij}(t) = \delta_{ij} + \sum_{k=0}^{\infty} \int_0^t dQ_{ik}(s)R_{kj}(t-s), \qquad (6.2.6)$$

where $\delta_{ij} = 1$ if $i = j$ and 0 otherwise. In renewal theory, $m(t) = dM(t)/dt$ is the renewal density and $dM(t) = m(t)dt$ represents the probability that a renewal occurs in $(t, t+dt)$. A similar interpretation can be given to $dR_{ij}(t)$. We can think of $dR_{ij}(t)$ as the probability that a transition into state j occurs in $(t, t+dt)$ for a process that starts in state i at time $T_0 = 0$. An equivalent way to write Equation 6.2.6 is

$$R_{ij}(t) = \delta_{ij} + \sum_{k=0}^{\infty} \int_0^t Q_{ik}(t-s)dR_{kj}(s). \qquad (6.2.7)$$

It is instructive to consider the case in which the process has only one state and hence becomes a renewal process. Dropping all subscripts, Equation 6.2.7 is reduced to

$$R(t) - 1 = \int_0^t Q(t-s)dR(s) = Q(t) + \int_{0^+}^t Q(t-s)dR(s).$$

The term $Q(t)$ surfaces in the last equality because of the jump of size 1 of the renewal function occurring at $s = 0$ (since $R(0) = 1$). The last equality is precisely the renewal function derived in Example 3.2.1.

Define the Laplace-Stieltjes transform of $R(t)$ (in matrix notation)

$$R^e(s) = \int_0^{\infty} e^{-st} dR(t).$$

Then Equation 6.2.6 can be written in the transform domain as

$$R^e(s) = I + Q^e(s)R^e(s). \qquad (6.2.8)$$

When the state space S is finite, the Laplace-Stieltjes transform of $R(t)$ is given by

$$R^e(s) = \left[I - Q^e(s)\right]^{-1}. \qquad (6.2.9)$$

If S is not finite, then it can be shown that $R(s)$ is the minimal nonnegative solution M of the system of equations $[I - Q^e(s)]M = I$.

EXAMPLE
6.2.1
Consider a Markov renewal process with state space $S = \{1, 2\}$ whose semi-Markov kernel has the following components:

$$P = \begin{bmatrix} .6 & .4 \\ .2 & .8 \end{bmatrix} \quad \text{and} \quad \left\{\frac{d}{dt} G(t)\right\} = \begin{bmatrix} 2e^{-2t} & e^{-t} \\ 2(e^{-t} - e^{-2t}) & te^{-t} \end{bmatrix}.$$

From the holding time densities $\{dG_{ij}(t)/dt\}$, we find the Laplace transform of $G(t)$

$$G^e(s) = \begin{bmatrix} \dfrac{2}{(s+2)} & \dfrac{1}{(s+1)} \\ \dfrac{2}{(s+1)(s+2)} & \dfrac{1}{(s+2)^2} \end{bmatrix}.$$

Since $Q^e(s) = P \square G^e(s)$, the Laplace transform of the semi-Markov kernel is given by

$$Q^e(s) = \begin{bmatrix} \dfrac{1.2}{(s+2)} & \dfrac{0.4}{(s+1)} \\ \dfrac{0.4}{(s+1)(s+2)} & \dfrac{0.8}{(s+1)^2} \end{bmatrix}.$$

Using Equation 6.2.9, we find

$$R^e(s) = [I - Q^e(s)]^{-1} = \begin{bmatrix} \dfrac{s+0.8}{(s+2)} & \dfrac{-0.4}{(s+1)} \\ \dfrac{-0.4}{(s+1)(s+2)} & \dfrac{s^2+2s+0.2}{(s+1)^2} \end{bmatrix}^{-1}$$

$$= \begin{bmatrix} \dfrac{(s^2+2s+0.2)(s+2)}{s(s+1)(s+1.8)} & \dfrac{0.4(s+2)}{s(s+1.8)} \\ \dfrac{0.4}{s(s+1.8)} & \dfrac{(s+0.8)(s+1)}{s(s+1.8)} \end{bmatrix}.$$

To find the Markov renewal function $R(t)$, we perform partial fraction expansions on $(1/s)R^e(s)$ and obtain

$$\frac{1}{s}R^e(s) = \begin{bmatrix} \dfrac{2/9}{s^2} + \dfrac{161/81}{s} - \dfrac{1}{s+1} + \dfrac{1/81}{s+1.8} & \dfrac{4/9}{s^2} - \dfrac{2/81}{s} + \dfrac{2/81}{s+1.8} \\ \dfrac{2/9}{s^2} - \dfrac{10/81}{s} + \dfrac{10/81}{s+1.8} & \dfrac{4/9}{s^2} + \dfrac{61/81}{s} + \dfrac{20/81}{s+1.8} \end{bmatrix}.$$

We invert the previous transform and find the Markov renewal function

$$R(t) = \begin{bmatrix} \dfrac{2}{9}t + \dfrac{161}{81} - e^{-t} + \dfrac{1}{81}e^{-1.8t} & \dfrac{4}{9}t - \dfrac{2}{81} + \dfrac{2}{81}e^{-1.8t} \\ \dfrac{2}{9}t - \dfrac{10}{81} + \dfrac{10}{81}e^{-1.8t} & \dfrac{4}{9}t + \dfrac{61}{81} + \dfrac{20}{81}e^{-1.8t} \end{bmatrix}.$$

In Figure 6.3, we plot the four Markov renewal functions found in the Appendix for comparison. We see that the renewal functions approach their respective linear asymptotes rapidly as t increases. Moreover, the slopes of the linear asymptotes depend on the destination states and are independent of the starting states at time 0. ∎

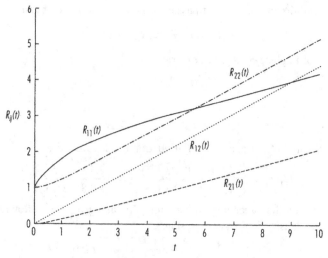

FIGURE

6.3   The Markov renewal functions $\{R_{ij}(t)\}$.

For a given state j, we let $\{S_n^j\}$ denote the successive times at which state j is revisited. For the sample path shown in Figure 6.2, with $j = 2$ we have $S_0^2 = T_1$, $S_1^2 = T_3$, $S_2^3 = T_4$, and so on. If we consider each return to state j a renewal, we have an ordinary renewal process with interarrival times $\{S_{n+1}^j - S_n^j\}$, $n = 0$, $1, 2, \ldots$ provided the Markov renewal process starts from state j at time 0 (in this case, $P\{S_0^j = 0\} = 1$). Let $F_{jj}(t)$ be the distribution function of these interarrival times and μ_{jj} denote its mean. On the other hand, if the Markov renewal process starts from state $i \neq j$, then we have a delayed renewal process with an initial inter-arrival time distribution $F_{ij}(t)$ and $S_0^j \sim F_{ij}(t)$. We note that $F_{ij}(t)$ is the distribution of the first passage time from state i to state j. Let μ_{ij} denote its mean.

From renewal theory, we know that the renewal function is given by $\sum_{n=1}^{\infty} F_n(t)$, where $F_n(t)$ is the n-fold convolution of F with itself (as indicated by subscript n). Since the Markov renewal function $R_{jj}(t)$ includes the initial renewal at time 0, we have

$$R_{jj}(t) = \sum_{n=0}^{\infty} F_{jj}^{(n)}(t) \tag{6.2.10}$$

(here, superscript n is used to indicate the n-fold convolution of F_{jj} with itself). For any i and j, we define the Laplace-Stieltjes transform as

$$F_{ij}^e(s) = \int_0^\infty e^{-st} dF_{ij}(t).$$

Then Equation 6.2.10 can be written in terms of Laplace-Stieltjes transforms as

$$R_{jj}^e(s) = \frac{1}{1 - F_{jj}^e(s)}. \tag{6.2.11}$$

For $i \neq j$, conditioning on the first passage time from i to j, we obtain

$$R_{ij}(t) = \int_0^t dF_{ij}(u)R_{jj}(t-u),$$

or in terms of Laplace-Stieltjes transforms,

$$R_{ij}^e(s) = F_{ij}^e(s)R_{jj}^e(s). \qquad (6.2.12)$$

EXAMPLE 6.2.2   Consider Example 6.2.1 again. Using Equation 6.2.12, we write

$$F_{12}^e(s) = \frac{R_{12}^e(s)}{R_{22}^e(s)} = \frac{0.4(s+2)}{(s+0.8)(s+1)} = \frac{2.4}{s+0.8} - \frac{2}{s+1}.$$

An inversion of the preceding to the time domain gives the following first-passage-time density from state 1 to state 2

$$f_{12}(t) = 2.4e^{-0.8t} - 2e^{-t} \qquad t \geq 0.$$

The mean of the first passage time, μ_{12}, is 1.75. ∎

In Section 3.5, we have defined a transient renewal process as one with a defective interarrival time distribution F, that is, $F(\infty) < 1$. In a Markov renewal process, the same definition applies to state j where $F_{jj}(t)$ corresponds to the interarrival time distribution of the renewal process with interarrival times $\{S_{n+1}^j - S_n^j\}$. In other words, state j is recurrent if $F_{jj}(\infty) = 1$ and transient if $F_{jj}(\infty) < 1$. Moreover, the periodicity of state j is determined by the periodicity of the distribution $F_{jj}(t)$.

In renewal theory, the renewal-type equation, Equation 3.2.1, plays a prominent role. The equation, written in the notation of a Laplace-Stieltjes integral, reads

$$g(t) = h(t) + \int_0^t dF(x)g(t-x),$$

where h is given, F is the interarrival time distribution, and g is unknown. We recall that the solution of the preceding is given by

$$g(t) = h(t) + \int_0^t dM(x)h(t-x), \qquad (6.2.13)$$

where $M(x)$ is the renewal function of the underlying renewal process with interarrival time distribution F.

Generalizing the idea to a Markov renewal process, we have the *Markov renewal equation*

$$g_i(t) = h_i(t) + \sum_{j \in S} \int_0^t dQ_{ij}(x)g_j(t-x), \qquad (6.2.14)$$

where $\{h_i(t)\}$ are given, $\{Q_{ij}(s)\}$ constitute the semi-Markov kernel, and $\{g_j(t)\}$ are unknown.

In matrix and convolution notations, Equation 6.2.14 can be written as $g(t) = h(t) + (Q*g)(t)$, where $g(t)$ and $h(t)$ are two column vectors. Let $g^e(s)$ and $h^e(s)$ be the Laplace transform of the respective functions. Then $g^e(s) = h^e(s) + Q^e(s)g^e(s)$ or $[I - Q^e(s)]g^e(s) = h^e(s)$. When the state space is finite, the last expression is given by

$$g^e(s) = \left[I - Q^e(s)\right]^{-1} h^e(s) = R^e(s)h^e(s),$$

where the final equality is due to Equation 6.2.9. Inverting the previous equation to the time domain, we obtain the *time-dependent* solution to the Markov renewal equation $g(t) = (R*h)(t)$ or, in component form,

$$g_i(t) = \sum_{j \in S} \int_0^t dR_{ij}(x)h_j(t - x). \tag{6.2.15}$$

When the Markov renewal process has only one state, after dropping all subscripts, Equation 6.2.15 reduces to

$$g(t) = h(t) + \int_{0^+}^t dR(x)h(t - x).$$

The term $h(t)$ appears because of the jump of size 1 of the renewal function at $x = 0$ (since $R(0) = 1$). The previous expression is actually Equation 6.2.13 (even though we have used the Laplace-Stieltjes integral in Equation 6.2.13, the integration is still from 0 [vis à vis 0^+] onward because the renewal function $M(t)$ defined in renewal theory does not include the renewal at time 0). We mention that under some regularity conditions, Equation 6.2.15 still holds even when the state space is not finite.

Before looking into the *limiting* solution of the Markov renewal equation, we examine the renewal functions $\{R_{ij}(t)\}$ for large t. If state j is transient and $i \neq j$, then

$$\lim_{t \to \infty} R_{ij}(t) = F_{ij}(\infty)R_{jj}(\infty). \tag{6.2.16}$$

If state j is recurrent and $F_{ij}(\infty) = 1$, then from Equation 3.5.4 we conclude that

$$\lim_{t \to \infty} R_{ij}(t) - R_{ij}(t - a) = \frac{a}{\mu_{jj}}, \tag{6.2.17}$$

where μ_{jj} is the mean recurrence time of state j and

$$\mu_{jj} = \int_0^\infty t \, dF_{jj}(t).$$

We observe that $1/\mu_{jj}$ represents the rate of entrance into state j.

For a Markov renewal process, we now consider the case in which the embedded Markov chain $X = \{X_n\}$ is irreducible and recurrent. Let $\{\pi_i\}$ denote the limiting probabilities of the embedded chain as defined by Equation 4.3.6 and τ_i the mean unconditional holding time in state i (unconditional on the destination states), where

$$\tau_i = \int_0^\infty \left[1 - \sum_k Q_{ik}(t)\right] dt = \sum_j p_{ij}\tau_{ij}.$$

Define a cycle as the time between successive entrances into state j. In Equation 4.3.7, we have established that π_i/π_j represents the expected number of visits to state i in a cycle. Then the mean cycle length μ_{jj} is given by

$$\mu_{jj} = \tau_j + \sum_{i \neq j} \frac{\pi_i}{\pi_j}\tau_i = \tau_j \frac{\pi_j}{\pi_j} + \frac{1}{\pi_j}\sum_{i \neq j} \pi_i\tau_i = \frac{1}{\pi_j}\sum_i \pi_i\tau_i. \tag{6.2.18}$$

In other words, the entrance rate to state j is given by

$$\frac{1}{\mu_{jj}} = \frac{\pi_j}{\sum_i \pi_i \tau_i} \tag{6.2.19}$$

Another way to find $\{\mu_{ij}\}$ is by solving the system of linear equations $\mu_{ij} = \tau_i + p_{ij}\mu_{jj}$ for all i and j.

Let P_j denote the long-run fraction of time the Markov renewal process is in state j; then a regenerative type of argument implies that

$$P_j = \frac{\tau_j}{\mu_{jj}} = \frac{\pi_j \tau_j}{\sum_i \pi_i \tau_i}. \tag{6.2.20}$$

EXAMPLE 6.2.3　Consider Example 6.2.1 once more. The mean holding times are

$$\{\tau_{ij}\} = \left\{-\frac{d}{ds} G_{ij}^e(s)\Big|_{s=0}\right\} = \begin{bmatrix} 0.5 & 1 \\ 1.5 & 2 \end{bmatrix}.$$

Using $\tau_i = \sum_j p_{ij}\tau_{ij}$, we find that $\tau_1 = 0.7$ and $\tau_2 = 1.9$. The limiting probability vector of the embedded chain X is $(1/3, 2/3)$. This gives $\sum_i \pi_i \tau_i = (1/3)(7/10) + (2/3)(19/10) = 3/2$. From Equation 6.2.19, we obtain the entrance rates $1/\mu_{11} = (1/3)/(3/2) = 2/9$ and $1/\mu_{22} = (2/3)/(3/2) = 4/9$. These entrance rates are the asymptotic rates of renewal shown in Equation 6.2.17, they are the slopes of the renewal functions found in Example 6.2.1. Using Equation 6.2.20, we find the long-run fractions of time the process spends in states 1 and 2. These fractions are $P_1 = 7/45$ and $P_2 = 38/45$. ∎

We now return to an important result in Markov renewal theory—namely, the limiting solution of the Markov renewal equation. Again, we assume that the embedded Markov chain X is irreducible and recurrent, the state space is finite, the states are aperiodic, and $\{h_i(t)\}$ are directly Riemann integrable functions (see Section 3.2). Taking the limit of Equation 6.2.15, we have

$$\lim_{t\to\infty} g_i(t) = \lim_{t\to\infty} \sum_{j\in S} \int_0^t dR_{ij}(x) h_j(t-x) = \lim_{t\to\infty} \sum_{j\in S} (R_{ij} * h_j)(t)$$

$$= \lim_{t\to\infty} \sum_{j\in S} (F_{ij} * R_{jj} * h_j)(t) \qquad \text{by Equation 6.2.12}$$

$$= \lim_{t\to\infty} \sum_{j\in S} (R_{jj} * F_{ij} * h_j)(t).$$

Define $H_j(t) = (F_{ij} * h_j)(t)$ (for the moment, we suppress the argument "i" in $H_j(t)$ for notational simplicity). Then the preceding can be written as

$$\lim_{t\to\infty} g_i(t) = \lim_{t\to\infty} \sum_{j\in S}(R_{ij} * H_j)(t) = \lim_{t\to\infty}\sum_{j\in S}\int_0^t dR_{ij}(x)H_j(t-x)$$

$$= \sum_{j\in S}\lim_{t\to\infty}\int_0^t dR_{ij}(x)H_j(t-x), \qquad \text{since } S \text{ is finite}$$

$$= \sum_{j\in S}\frac{1}{\mu_{jj}}\int_0^\infty H_j(t)dt \qquad \text{by Equation 3.2.3.}$$

Define $\eta_j = \int_0^\infty H_j(t)dt$. We see that

$$\eta_j = \int_0^\infty\int_0^t dF_{ij}(s)h_j(t-s)dt = \int_0^\infty dF_{ij}(s)\int_s^\infty h_j(t-s)dt$$

$$= F_{ij}(\infty)\int_0^\infty h_j(t)dt = \int_0^\infty h_j(t)dt.$$

Using Equation 6.2.19, we conclude that the limiting solution of the Markov renewal equation is given by

$$\lim_{t\to\infty} g_i(t) = \frac{\displaystyle\sum_{j\in S}\pi_j\eta_j}{\displaystyle\sum_{j\in S}\pi_j\tau_j}. \qquad (6.2.21)$$

We observe that the limiting solution of the Markov renewal equation is independent of the starting state i. Under some regularity conditions, we mention that the previous result is applicable to cases in which the state space is not finite. The uses of Equation 6.2.21 will be demonstrated repeatedly in the remaining part of this chapter.

6.3   Semi-Markov Processes and Related Reward Processes

Consider a Markov renewal process (X, T) with state space S and semi-Markov kernel $\{Q(t), t\ge 0\}$. We are now interested in the stochastic process $V = \{V(t), t\ge 0\}$, where

$$V(t) = X_n \qquad \text{if } T_n \le t < T_{n+1}.$$

The stochastic process V is called a *semi-Markov process*. The process retains the Markov property at renewal times $\{T_n\}$ and is thus called semi-Markov.

Define transition function $P_{ij}(t) = P\{V(t) = j | X_0 = i\}$. Let S_j denote the unconditional holding time in state j. Then we recall $E[S_j] = \tau_j$ and $P\{S_j > t\}$ $= 1 - \sum_{k\in S}Q_{jk}(t)$. Define $h_j(t) = P\{S_j > t\}$. Conditioning on the time T_1 and the destination X_1 of the first transition out of $X_0 = i$, we write

$$P_{ij}(t) = P\left\{V(t) = j, \ T_1 > t \middle| X_0 = i\right\} + P\left\{V(t) = j, \ T_1 \le t \middle| X_0 = i\right\}$$

$$= \delta_{ij} P\{S_i > t\} + \underset{X_1, T_1}{E} \left\{[T_1 < t]P\{V(t) = j | X_1, \ T_1\}\right\}$$

$$= \delta_{ij} h_i(t) + \sum_{k \in S} \int_0^t dQ_{ik}(s) P_{kj}(t - s),$$

where $[A] = 1$ if A is true and 0 otherwise. The preceding is a Markov renewal type of equation of the of form of Equation 6.2.14. It can be written in matrix form as

$$P(t) = H(t) + \int_0^t dQ(s)P(t - s), \tag{6.3.1}$$

where $P(t) = \{P_{ij}(t)\}$ and $H(t)$ is a diagonal matrix whose jth diagonal element is $h_j(t)$. Define Laplace transforms

$$P_{ij}^e(s) = \int_0^\infty e^{-st} dP_{ij}(t) \qquad \text{and}$$

$$h_j^e(s) = \int_0^\infty e^{-st} h_j(t) dt = \int_0^\infty e^{-st} \left[1 - \sum_{k \in S} Q_{jk}(t)\right] dt = \frac{1}{s}\left[1 - \sum_{k \in S} Q_{jk}^e(s)\right]$$

and let $P^e(s) = \{P_{ij}^e(s)\}$ and $H^e(s)$ be a diagonal matrix whose jth diagonal element is $h_j^e(s)$. We state Equation 6.3.1 in terms of Laplace transforms

$$\frac{1}{s}P^e(s) = H^e(s) + \frac{1}{s}Q^e(s)P^e(s) \qquad \text{or} \qquad \frac{1}{s}\left[I - Q^e(s)\right]P^e(s) = H^e(s).$$

Using Equation 6.2.9, the last expression can be stated as

$$\frac{1}{s}P^e(s) = [I - Q^e(s)]^{-1}H^e(s) = R^e(s)H^e(s) \qquad \text{or} \qquad \frac{1}{s}P_{ij}^e(s) = R_{ij}^e(s)h_j^e(s). \tag{6.3.2}$$

Inverting the last Laplace transform to the time domain, we obtain

$$P_{ij}(t) = \int_0^t dR_{ij}(u)h_j(t - u). \tag{6.3.3}$$

We now give an intuitive interpretation of Equation 6.3.3. As mentioned earlier, $dR_{ij}(u)$ can be viewed as the probability that a transition into state j occurs in $(u, u + du)$ given that the process starts from state i at time 0. Knowing that the process enters state j in $(u, u + du)$, it will still be in the state at time t if the holding time in the state is longer than $t - u$. Combining the two observations and considering all possible values of u enable us to write the right side of Equation 6.3.3.

Assume now that the embedded Markov chain X is irreducible and recurrent. Having defined $h_j(t) = P\{S_j > t\}$, we note that the right side of Equation 6.3.3 is actually the right side of Equation 6.2.15 without the summation sign.

Following steps similar to those leading to Equation 6.2.21, we find the limiting state probabilities

$$\lim_{t \to \infty} P_{ij}(t) = \frac{\pi_j \eta_j}{\sum_{k \in S} \pi_k \tau_k} = \frac{\pi_j \tau_j}{\sum_{k \in S} \pi_k \tau_k}, \tag{6.3.4}$$

where we use $\eta_k = \int_0^\infty h_k(t)dt = \int_0^\infty P\{S_k > t\}dt = E[S_k] = \tau_k$

The previous result is identical to Equation 6.2.20 obtained by a different approach.

EXAMPLE
6.3.1

We return to Example 6.2.1. First, we find that

$$h_1^e(s) = \frac{1}{s}\left[1 - \left(\frac{1.2}{s+2} + \frac{0.4}{s+1}\right)\right] = \frac{s+1.4}{(s+1)(s+2)}$$

$$h_2^e(s) = \frac{1}{s}\left[1 - \left(\frac{0.4}{(s+1)(s+2)} + \frac{0.8}{(s+1)^2}\right)\right] = \frac{s^2+4s+3.8}{(s+1)^2(s+2)}.$$

In Example 6.2.1, we have already obtained $R^e(s)$. Applying Equation 6.3.2 yields

$$\frac{1}{s}P^e(s) = \begin{bmatrix} \dfrac{(s+1.4)(s^2+2s+0.2)}{s(s+1.8)(s+1)^2} & \dfrac{0.4(s^2+4s+3.8)}{s(s+1.8)(s+1)^2} \\[4mm] \dfrac{0.4(s+1.4)}{s(s+1)(s+1.8)(s+2)} & \dfrac{(s+0.8)(s^2+4s+3.8)}{s(s+1)(s+1.8)(s+2)} \end{bmatrix}$$

$$= \begin{bmatrix} \dfrac{7/45}{s} - \dfrac{1/18}{s+1.8} + \dfrac{9/10}{s+1} + \dfrac{2/5}{(s+1)^2} & \dfrac{38/45}{s} + \dfrac{1/18}{s+1.8} - \dfrac{9/10}{s+1} - \dfrac{2/5}{(s+1)^2} \\[4mm] \dfrac{7/45}{s} - \dfrac{1/5}{s+1} - \dfrac{5/9}{s+1.8} + \dfrac{3/5}{s+2} & \dfrac{38/45}{s} + \dfrac{1/5}{s+1} + \dfrac{5/9}{s+1.8} - \dfrac{3/5}{s+2} \end{bmatrix}.$$

Inverting the preceding transform, we obtain the transition function

$$P(t) = \begin{bmatrix} \dfrac{7}{45} - \dfrac{1}{18}e^{-1.8t} + \dfrac{9}{10}e^{-t} + \dfrac{2}{5}te^{-t} & \dfrac{38}{45} + \dfrac{1}{18}e^{-1.8t} - \dfrac{9}{10}e^{-t} - \dfrac{2}{5}te^{-t} \\[4mm] \dfrac{7}{45} - \dfrac{1}{5}e^{-t} - \dfrac{5}{9}e^{-1.8t} + \dfrac{3}{5}e^{-2t} & \dfrac{38}{45} + \dfrac{1}{5}e^{-t} + \dfrac{5}{9}e^{-1.8t} - \dfrac{3}{5}e^{-2t} \end{bmatrix}.$$

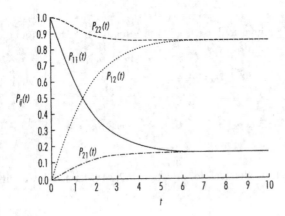

FIGURE
6.4   The transition function.

In Figure 6.4, we plot $\{P_{ij}(t)\}$, obtained in the Appendix, over the interval $(0, 10)$. We observe that the limiting probabilities in states 1 and 2 are independent of the starting states and converge to their respective values found in Example 6.2.3. ▪

Let $U_{ij}(t)$ denote the expected sojourn time in state j during the interval $[0, t]$ given that $X_0 = i$. Define indicator random variable $1_j(V(t)) = 1$ if $V(t) = j$ and 0 otherwise. From Equation 6.3.1, we find that

$$U_{ij}(t) = E\left[\int_0^t 1_j(V(s))ds \Big| X_0 = i\right] = \int_0^t E\left[1_j(V(s))ds \Big| X_0 = i\right]$$

$$= \int_0^t P_{ij}(s)ds = \int_0^t \int_0^s dR_{ij}(u)h_j(s-u)ds$$

$$= \int_0^t dR_{ij}(u)\int_u^s h_j(s-u)ds = \int_0^t dR_{ij}(u)\int_0^{s-u} h_j(z)dz. \qquad (6.3.5)$$

The Laplace-Stieltjes transform of $U_{ij}(t)$ can then be stated in terms of $h_j^e(s)$ as

$$U_{ij}^e(s) = \int_0^\infty e^{-st}dU_{ij}(t) = \int_0^\infty e^{-st}P_{ij}(t)dt = \int_0^\infty e^{-st}\int_0^t dR_{ij}(u)h_j(t-u)dt$$

$$= \int_0^\infty e^{-su}dR_{ij}(u)\int_u^\infty e^{-s(t-u)}h_j(t-u)dt = \int_0^\infty e^{-su}dR_{ij}(u)\int_0^\infty e^{-sz}h_j(z)dz$$

$$= R_{ij}^e(s)h_j^e(s). \qquad (6.3.6)$$

With $(1/s)U_{ij}^e(s) = \int_0^\infty e^{-st}U_{ij}(t)dt$, we use the final value property of the Laplace transform to study the limiting behavior of $U_{ij}(t)$. We observe that

$$\lim_{t\to\infty} U_{ij}(t) = \lim_{s\to 0} s\left(\frac{1}{s}U_{ij}^e(s)\right) = \lim_{s\to 0} U_{ij}^e(s) = \lim_{s\to 0} R_{ij}^e(s)h_j^e(s)$$

$$= \lim_{s\to 0} s\left(\frac{1}{s}R_{ij}^e(s)\right)h_j^e(0) = R_{ij}(\infty)\tau_j, \qquad (6.3.7)$$

where we use $h_j^e(0) = \int_0^\infty h_j(t)dt = \tau_j$. If state j is transient, then $R_{ij}(\infty)$ is finite and Equation 6.3.7 implies that the expected time in j in an infinite horizon is equal to the expected number of transitions into j in an infinite horizon times the mean holding time in j per transition into j—a result intuitively expected. When states i and j are in the class of recurrent states, that is, $F_{ij}(\infty) = 1$, then $R_{ij}(\infty)$ and $U_{ij}(\infty)$ will both be infinite.

Consider again the Markov renewal process whose embedded Markov chain X is irreducible and recurrent. Recall that Equation 6.3.4 is a point-wise limit. Since the point-wise limit gives the Cesàro limit (or the time average, see Section 1.4), we have

$$\lim_{t\to\infty} \frac{U_{ij}(t)}{t} = \lim_{t\to\infty} \frac{\int_0^t P_{ij}(u)du}{t} = \frac{\pi_j \tau_j}{\sum_{k\in S} \pi_k \tau_k}. \tag{6.3.8}$$

In the remainder of this section, we will consider the reward process generated by a semi-Markov process. Suppose that $f_j(t)$ is the reward rate at time t when the state of the semi-Markov process $V(t)$ is equal to j. Let $M_i(t)$ denote the expected reward received in the interval $[0, t]$ given that $X_0 = i$. Then

$$M_i(t) = E\left[\int_0^t f_{V(s)}(s)ds \Big| X_0 = i\right] = \int_0^t \sum_{j\in S} P_{ij}(s)f_j(s)ds = \sum_{j\in S} \int_0^t P_{ij}(s)f_j(s)ds$$

$$= \sum_{j\in S} \int_0^t \int_0^s dR_{ij}(u)h_j(s-u)f_j(s)ds = \sum_{j\in S} \int_0^t dR_{ij}(u)\int_u^t h_j(s-u)f_j(s)ds$$

$$= \sum_{j\in S} \int_0^t dR_{ij}(u)\int_0^{t-u} h_j(z)f_j(z+u)dz. \tag{6.3.9}$$

When $f_j(t) = g_j$ for all t, the discounted reward over an infinite horizon has a simple form. We let D_i denote the expected discounted reward received in the interval $[0, \infty)$, given that the process starts from state i, and β the factor for continuous discounting. Then we have

$$D_i = E\left[\int_0^\infty e^{-\beta t} f_{V(t)}(t)dt \Big| X_0 = i\right] = \int_0^\infty e^{-\beta t} E\left[f_{V(t)}(t)\Big| X_0 = i\right]dt$$

$$= \int_0^\infty e^{-\beta t} \sum_{j\in S} g_j \int_0^\infty P_{ij}(t)dt = \sum_{j\in S} g_j \int_0^\infty e^{-\beta t} P_{ij}(t)dt = \sum_{j\in S} g_j U_{ij}^e(\beta), \tag{6.3.10}$$

where we use Equation 6.3.6 to establish the last equality.

Consider again the case in which the embedded Markov chain is irreducible and recurrent.

When $f_j(t) = g_j$ for all t, we use Equation 6.3.8 to compute the (undiscounted) expected reward rate g as follows

$$g = \lim_{t\to\infty} \frac{\sum_{j\in S} g_j \int_0^t P_{ij}(u)du}{t} = \frac{\sum_{j\in S} g_j \pi_j \tau_j}{\sum_{k\in S} \pi_k \tau_k} = \frac{\sum_{j\in S} \pi_j r_j}{\sum_{k\in S} \pi_k \tau_k}, \tag{6.3.11}$$

where we define $r_j = \tau_j g_j$.

The quantity r_j represents the expected reward per each sojourn in state j. We see that the numerator of Equation 6.3.11 gives the expected reward per renewal and the denominator the expected time per renewal. The asymptotic reward rate g is simply the ratio of the two. While Equation 6.3.11 has been established for the case in which rewards are generated at a constant rate of g_j, when the process is in state j, the result actually applies to any reward generation pattern as long as r_j represents the expected reward per each occupancy of state j. Moreover, under some regularity conditions, Equation 6.3.11 also works when the state space is not finite.

EXAMPLE 6.3.2

A Repair Limit Replacement Method In a repair limit replacement model, when a piece of equipment requires repair it is first inspected and the repair cost is estimated. If the estimated cost is less than or equal to a "repair limit," then it is repaired; otherwise it is replaced. In this example, we consider a specific case in which equipment will be kept for at most two years and the repair limits are age specific. Let R_i be the repair limit for i-year-old equipment.

For i-year-old equipment, we assume that the time between successive failures follows an exponential density f_i whose parameter μ_i depends on i; the repair cost density h_i is also exponential with parameter λ_i. Let A denote the cost of a replacement. For a specific pair of repair limits (R_0, R_1), we want to compute the long-run expected average cost per year g. The underlying Markov renewal process has a state space $S = \{0, 1, 2\}$ whose elements denote the ages of the equipment. To find g, we need to obtain the transition probabilities of the embedded chain $\{p_{ij}\}$, the mean holding times $\{\tau_i\}$, and the expected single transition costs $\{r_i\}$.

For i-year-old equipment, we let P_i denote the probability that a repair cost X_i exceeds the repair limit R_i. Then

$$P_i = P\{X_i > R_i\} = 1 - H_i(R_i),$$

where H_i is the repair cost distribution. Also, we let U_i denote the mean repair cost given that $X_i \le R_i$. Define the partial expectation $M_i = \int_0^{R_i} uh_i(u)du$. Then

$$U_i = \int_0^{R_i} u\frac{h_i(u)}{H_i(R_i)}du = \frac{\int_0^{R_i} uh_i(u)du}{H_i(R_i)} = \frac{M_i}{H_i(R_i)}.$$

For i-year-old equipment, let $v_i(t)$ denote the failure rate at time t. Conditioning on the first failure time during the year, we write the following renewal-type equation

$$v_i(t) = f_i(t) + \int_0^t v_i(t-u)(1-P_i)f_i(u)du.$$

Define the Laplace transform

$$v_i^e(s) = \int_0^\infty e^{-st}v_i(t)dt.$$

The renewal-type equation can now be stated in the transform domain as

$$v_i^e(s) = \frac{\mu_i}{s+\mu_i} + (1-P_i)\left(\frac{\mu_i}{s+\mu_i}\right)v_i^e(s).$$

Upon simplification, we obtain $v_i^e(s) = \mu_i/(s + P_i\mu_i)$. This gives the failure rate at time t

$$v_i(t) = \mu_i e^{-P_i\mu_i t} \qquad t \geq 0.$$

For i-year-old equipment, let random variable Z_i denote the equipment replacement time. We see that the density function for Z_i is given by $P_i v_i(t)$. In other words, we have

$$f_{Z_i}(t) = P_i\mu_i e^{-P_i\mu_i t} \qquad t \geq 0.$$

For i-year-old equipment, the holding time in state i is $\min\{Z_i, 1\}$ and its mean is given by

$$\tau_i = \int_0^1 P\{Z_i > t\}dt = \int_0^1 e^{-P_i\mu_i t}dt = \frac{1}{P_i\mu_i}\left[1 - e^{-P_i\mu_i}\right], \qquad i = 0 \text{ and } 1.$$

In addition, we have $\tau_2 = 0$. The probability that the equipment will last to the next year is given by

$$P_{i,i+1} = \int_1^\infty f_{Z_i}(t)dt = e^{-P_i\mu_i} \qquad i = 0 \text{ and } 1.$$

The transition probability matrix of the embedded Markov chain has the form

$$
\begin{array}{c c c c}
 & 0 & 1 & 2 \\
P = \begin{array}{c} 0 \\ 1 \\ 2 \end{array} &
\left[\begin{array}{ccc}
1 - P_{01} & P_{01} & \\
1 - P_{12} & & P_{12} \\
1 & &
\end{array}\right].
\end{array}
$$

For i-year-old equipment, we let $c_i(t)$ denote the expected cost rate at time t. Conditioning on the type of action taken at the failure, we have

$$c_i(t) = \left[U_i(1-P_i) + AP_i\right]\mu_i e^{-P_i\mu_i t} = (M_i + AP_i)\mu_i e^{-P_i\mu_i t}.$$

Thus the expected single transition cost for state i is given by

$$r_i = \int_0^1 c_i(t)dt = \left[M_i + AP_i\right]\left(\frac{1}{P_i}\right)\left[1 - e^{-P_i\mu_i}\right]$$

$$= \left[\frac{M_i}{P_i} + A\right]\left[1 - e^{-P_i\mu_i}\right] \qquad i = 0 \text{ and } 1$$

and $r_2 = A$.

We now look at a numerical example with replacement cost $A = \$400$, mean repair costs $1/\lambda_0 = \$100$ and $1/\lambda_1 = \$150$, and failure rates $\mu_0 = 2$ failures per year and $\mu_1 = 3$ failures per year. Consider the case in which we set the repair limits $(R_0, R_1) = (293, 125)$. Then we find that $p_{01} = 0.8987$ and $p_{12} = 0.2715$, the limiting probabilities of the embedded chain $(\pi_0, \pi_1, \pi_2) = (0.4667, 0.4194, 0.1139)$, the mean holding times $(\tau_0, \tau_1, \tau_2) = (0.9485, 0.5588, 0)$, and the expected single transition costs $(r_0, r_1, r_2) = (190.40, 342.50, 400)$. Using Equation 6.3.11, we obtain the expected average cost per year $g = \$410.73$. If we use a policy of replacing the equipment every two years, then the expected average cost per year will be

$$\frac{(\$100 \times 2 + \$150 \times 3) + \$400}{2} = \$525.$$

If we replace the equipment every year, then the annual cost will be $(\$100 \times 2) + \$400 = \$600$. In Figure 6.5, we plot the expected average costs per year, found in the Appendix, for $R_0 = 100:50:400$ and $R_1 = 100:50:400$. The figure indicates that the repair limit replacement policy $R_0, R_1) = (293, 125)$ is in the region in which the minimal cost solution lies. ∎

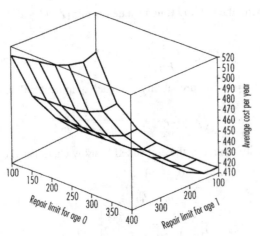

6.5   The expected average costs per year under various repair limit policies.

EXAMPLE   **A Discrete-Time Inventory System with Batch Demands**   Consider an inventory system in
6.3.3   which the times between successive demand occurrences are i.i.d. random variables following a common probability mass function $P_T(\cdot)$. At each demand occurrence, the number of units demanded follows another independent probability mass function $p_D(\cdot)$. Assume that when demand exceeds the stock on hand the unmet demands are lost. The inventory system is under continuous review and delivery lead times are negligible.

Let h denote the holding cost per unit of inventory in stock per unit time and π the penalty cost per unit of lost sales. The ordering cost consists of two components: a fixed cost of K per each order placed and a variable cost c per unit purchased.

The inventory replenishment policy under consideration is as follows: at the end of each demand occurrence, if the remaining inventory is below s, an order is placed so that there will be N units on hand after ordering. Consider a Markov renewal process in which the embedded Markov chain is defined at the epochs immediately after each demand occurrence or the receipt of items ordered. We call these epochs the transaction epochs. Then the state of the system represents the inventory level at these transaction epochs and $S = \{0, 1, ..., N\}$. Let X_n denote the inventory level at transaction epoch n. We now set up this problem as a semi-Markov reward process to find the expected average cost per unit time g.

When $X_n = i \geq s$, the transition probabilities of the embedded Markov chain are given by

$$p_{ij} = \begin{cases} p_D(i-j) & \text{if } 1 \leq j < i \\ \sum_{k \geq i} p_D(k) & \text{if } j = 0 \\ 0 & \text{otherwise,} \end{cases}$$

the mean holding times are

$$\tau_i = \sum_k k p_T(k) = E[T],$$

and the expected single transition costs are

$$r_i = ihE[T] + \sum_{k>i} \pi(k-i)p_D(k) = ihE[T] + L_i,$$

where L_i represents the expected cost of lost sales given that the inventory on hand is i. When $X_n = i < s$, an inventory replenishment is ordered and thus we have

$$p_{ij} = \begin{cases} 1 & \text{if } j = N \\ 0 & \text{otherwise,} \end{cases}$$

$\tau_i = 0$, and $r_i = K + c(N - i)$. Using these data and Equation 6.3.11, we can compute the expected average cost per unit time.

Consider a numerical example with demand distribution $\{p_D(1), p_D(2), p_D(3), p_D(4)\} = \{0.2, 0.5, 0.2, 0.1\}$ and interdemand occurrence time distribution $\{p_T(1), p_T(2), ..., p_T(7)\} = \{0.05, 0.05, 0.10, 0.15, 0.50, 0.10, 0.05\}$, cost parameters $h = 1$, $\pi = 20$, $K = 25$, and $c = 4$, and order-up-to-quantity $N = 6$. To determine the optimal s to minimize the expected average cost per unit time, we compute the g values for $s = 0{:}1{:}6$. The results found in the Appendix are plotted in Figure 6.6. We see that the value of s that minimizes the expected cost rate is 1. This implies that we should order only when the inventory level reaches 0. For this problem, we remark that under some s values the corresponding embedded Markov chains contain transient states. As an example, when $s = 6$, we place an order at each demand occurrence. Since the maximal number of units demanded at each demand occurrence is 4, states 0, 1, 2 will not be reached in the long run and are transient states. However, the presence of transient states does not affect our solution because the respective limiting probabilities are simply zero. ∎

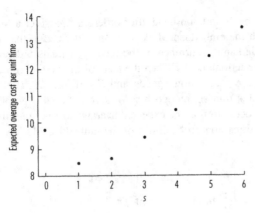

FIGURE
6.6   The expected average cost rates for different reorder points s.

6.4 Semi-Regenerative Processes

In Section 3.6, we have introduced the idea of a regenerative process. In a regenerative process, the process "restarts" at the renewal times of an *underlying* renewal process in the sense that the future of the process becomes a probabilistic replica of the process itself. In a semi-regenerative process, the idea of *process restart* depends not only on the renewal times $\{T_n\}$ but also on the states $\{X_n\}$ of the embedded Markov chain at the renewal times, where the *underlying* process is now a Markov renewal process $(X, T) = \{(X_n, T_n)\}$. If the Markov renewal process has only one state, the semi-regenerative process reduces to a regenerative process.

Before we define a semi-regenerative process formally, we first review the concept of a stopping time (Section 3.2). Consider a stochastic process $Z = \{Z(t), t \geq 0\}$. A random variable T is called a stopping time for Z if the occurrence or nonoccurrence of the event $\{T < t\}$ depends only on the history $\{Z(u), u \leq t\}$ of the process up to t.

The stochastic process $Z = \{Z(t), t \geq 0\}$ with state space E is called a *semi-regenerative process* if there exists a Markov renewal process (X, T) such that (i) $\{T_n\}$ are the stopping times for Z, (ii) for each n, X_n is determined by the history of Z up to T_n, and (iii) for each n, $m \geq 1$, $0 \leq t_1 < t_2 < \cdots < t_m$ and any nonnegative function f defined on E^m, we have

$$E\left[f\big(Z(T_n + t_1), \ldots, Z(T_n + t_m)\big) \mid Z(u),\ u \leq T_n \right] = E\left[f\big(Z(t_1), \ldots, Z(t_m)\big) \mid \hat{X}_0 \right].$$

(6.4.1)

where we use $\hat{X}_0 = X_n$ to denote the initial state of the embedded Markov chain of the Markov renewal process $\{(\hat{X}, \hat{T})\}$ with the same semi-Markov kernel and whose time origin is at T_n.

Upon observing X_n at T_n, Equation 6.4.1 says that the future of the process Z is (conditionally) independent of the history of Z before T_n.

EXAMPLE
6.4.1 **Excess Lives**   We return to the excess-life random variable of a renewal process with interarrival times $\{X_n\}$ considered in Section 3.3. Recall that the arrival times $\{S_n\}$ are defined by $S_n = X_1 + \cdots + X_n$, $n = 1, 2, \ldots$, and the excess life at t is defined by $Y(t) = S_{N(t)+1} - t$, where $N(t)$ is the number of renewals by time t. A sample path of the stochastic process

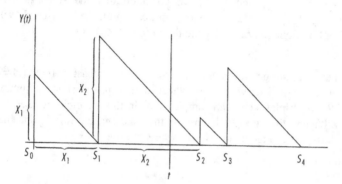

FIGURE
6.7   A sample path of the excess-life process $\{Y(t),\ t \geq 0\}$.

$Y = \{Y(t),\ t \geq 0\}$ is depicted in Figure 6.7. In the figure, we see that $\{S_n\}$ is a sequence of stopping times with respect to Y because the occurrence or nonoccurrence of $\{S_n < t\}$ depends only on the history of Y up to t (in other words, it does not depend on the process Y after t). This implies that the first condition for Y being semi-regenerative is met. Following the semi-regenerative paradigm, we construct a Markov renewal process $(W, S) = \{(W_n, S_n),\ n \geq 0\}$, where we have $W_n = 1$ for all n reflecting the fact that in a renewal process there is only a single state to which transitions can be made. Thus the second condition for Y being semi-regenerative is trivially met. The future of the process Y starts afresh at each renewal time S_n. Hence Equation 6.4.1 is satisfied and the third defining condition for being a semi-regenerative process is met. We conclude that stochastic process Y is semi-regenerative; of course, we could have done the same by simply noting Y is a regenerative process. This exercise is a prelude to subsequent examples. ▪

EXAMPLE
6.4.2 **The Yule Process Revisited**   We return to the Yule process studied in Example 5.2.4. Recall that the Yule process is a version of a pure birth process in which the birth rate at any time t is given by $j\lambda$, where j is the population size $Z(t)$ at time t and λ is the birth rate of each member of the population. We consider the case in which $Z(0) = 1$. Let T_n denote the time of the nth change of population size and X_n the

population at the completion of the nth population size change. We see that $X_{n+1} = X_n + 1$, $n = 0, 1, \ldots$, and $X_0 = 1$. Let $p_{ij} = P\{X_{n+1} = j | X_n = i\}$. Then

$$p_{ij} = \begin{cases} 1 & \text{if } i \geq 1 \text{ and } j = i + 1, \\ 0 & \text{otherwise.} \end{cases}$$

Using a result related to competing exponential processes (see Example 1.3.2), we find that $P\{T_{n+1} - T_n \leq t | X_n = n + 1\} = 1 - e^{-\lambda(n+1)t}$. This in turn gives

$$G_{ij} = \begin{cases} 1 - e^{-i\lambda} & \text{if } i \geq 1 \text{ and } j = i + 1 \\ 0 & \text{otherwise.} \end{cases}$$

Having specified $\{p_{ij}\}$ and $\{G_{ij}\}$, we conclude that the stochastic process $(X, T) = \{(X_n, T_n)\}$ is a Markov renewal process with the semi-Markov kernel $Q(t)$ defined by Equation 6.1.5. By the definition of $Z(t)$, we have

$$Z(t) = X_n \qquad T_n \leq t < T_{n+1}.$$

Based on the preceding relation, it is easy to see that stochastic process $\mathbf{Z} = \{Z(t), t \geq 0\}$ is a semi-regenerative process whose sample paths are step functions. In Figure 6.8, we display one such sample path. In this example, we see that the process $\mathbf{Z}$ beyond time t depends only on the state change time T_n and the state X_n if the last state change before time t occurs at time T_n. ∎

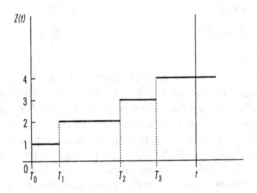

FIGURE 6.8   A sample path of $\{Z(t), t \geq 0\}$.

<table>
<tr><td>EXAMPLE
6.4.3</td><td>**Queue Lengths of an M/G/1 Queue**   In Example 4.3.2, we considered an $M/G/1$ queue and obtained the limiting probabilities of queue length at departure epochs. In Example 3.6.3, we used the regenerative approach to compute the limiting queue length distribution at any time. Lastly, in Example 6.1.4, we constructed a Markov renewal process to model the $M/G/1$ queue. Continuing with these examples, we define $Z(t)$ as the queue length at time t (the number of customers in the system at time t) and show that stochastic process $\{Z(t), t \geq 0\}$ is a semi-regenerative process.</td></tr>
</table>

Let T_n denote the time of the nth departure from the queue, X_n the number of customers left behind by the nth departure, S_n the service time of the nth customer, and I the length of an idle period. We see that

$$T_{n+1} - T_n = I \times [X_n = 0] + S_{n+1},$$

where $[A] = 1$ if A is true and 0 otherwise. Hence $P\{T_{n+1} - T_n \leq t | X_n\}$ is the service time distribution G if $X_n > 0$ and is the convolution of an exponential distribution with parameter λ and distribution G if $X_n = 0$. The one-step transition probabilities $\{p_{ij}\}$, where $p_{ij} = P\{X_{n+1} = j | X_n = i\}$, have been specified in Example 4.3.2. Combining these results in Equation 6.1.5, we can construct the semi-Markov kernel $Q(t)$ associated with the Markov renewal process $(X, T) = \{(X_n, T_n)\}$. Under the current approach we build the kernel from its components $\{p_{ij}\}$ and $\{G_{ij}\}$, while in Example 6.1.4 we constructed $Q(t)$ directly.

The queue length at time t is given by

$$Z(t) = X_n \qquad T_n \leq t < T_{n+1}.$$

The stochastic process $\mathbf{Z} = \{Z(t), t \geq 0\}$ represents the time-dependent queue length process with a state space $S = \{0, 1, 2, \ldots\}$. A sample path of $\mathbf{Z}$ was shown in Figure 3.23. We see that $\{T_n\}$ is a sequence of stopping time with respect to $\mathbf{Z}$ in the sense that the occurrence or nonoccurrence of $\{T_n < t\}$ depends only on the history of $\mathbf{Z}$ up to time t. Also, the probabilistic behaviors of $\mathbf{Z}$ beyond t depend only on T_n and the state of the system X_n if the last departure before t occurs at T_n. This implies that stochastic process $\mathbf{Z}$ is a semi-regenerative process. ∎

A regenerative process is a powerful tool for stochastic modeling because (i) by conditioning on the time of occurrence of the first event of interest, we can derive a renewal-type equation, (ii) Equation 3.6.3 provides its time-dependent solution, and (iii) Equation 3.6.4 provides its limiting solution. For semi-regenerative processes, we now present results parallel to those of Equations 3.6.3 and 3.6.4. These generalizations make the semi-regenerative process even more useful for solving problems in stochastic processes.

Let $\mathbf{Z}$ be a semi-regenerative process with state space F. Let (X, T) be a Markov renewal process embedded in $\mathbf{Z}$. The state space, the semi-Markov kernel, and the Markov renewal kernel for (X, T) are S, Q, and R, respectively. For any subset A of F, we define

$$u_i(A, t) = P\left\{Z(t) \in A, T_1 > t | X_0 = i\right\}$$

and

$$f_i(A, t) = P\left\{Z(t) \in A | X_0 = i\right\}.$$

Using the law of total probability, we write

$$f_i(A, t) = P\left\{Z(t) \in A, T_1 > t | X_0 = i\right\} + P\left\{Z(t) \in A, T_1 \leq t | X_0 = i\right\}$$

$$= u_i(A, t) + \sum_{k \in S} \int_0^t dQ_{ik}(s) f_k(A, t - s), \qquad \text{(6.4.2)}$$

where the last equality is obtained by conditioning on the time T_1 and state X_1 of the first transition out of state $X_0 = i$. Equation 6.4.2 is a Markov renewal type of equation defined by Equation 6.2.14. Following 6.2.15, its (time-dependent) solution is given by

$$f_i(A, t) = \sum_{j \in S} \int_0^t dR_{ij}(s) u_j(A, t - s). \tag{6.4.3}$$

We find the limiting solution of Equation 6.4.2 by invoking Equation 6.2.21

$$\lim_{t \to \infty} f_i(A, t) = \frac{1}{\sum_{i \in S} \pi_i \tau_i} \left\{ \sum_{j \in S} \pi_j \int_0^\infty u_j(A, t) dt \right\} = \sum_{j \in S} \frac{1}{\mu_{jj}} \int_0^\infty u_j(A, t) dt,$$
$$\tag{6.4.4}$$

where, as in Equation 6.2.21, $\{\pi_i\}$ are the limiting probabilities of the Markov chain embedded in (X, T), $\{\tau_i\}$ are the mean unconditional holding times in each state of S, and μ_{jj} is the mean recurrence time of state j defined by Equation 6.2.19.

EXAMPLE
6.4.4   **The Machine Reliability Problem Revisited**   We return to Example 6.1.7 to study the following two random variables:

$$Y_t = \text{the part that causes the last failure before time } t$$
$$W_t = \text{the status of the machine at time } t,$$

where $W_t = 1$ denotes that the machine is working at time t and 0 otherwise. Using an argument similar to those in the last three examples, we conclude that $(Y, W) = \{(Y_t, W_t)\}$ is a semi-regenerative process with state space $F = \{(Y_t, W_t) | Y_t \in \{1, \dots, M\}, W_t \in \{0, 1\}\}$, where the embedded Markov renewal process (X, T) has been defined in Example 6.1.7.

Let $A = \{Y_t = j, W_t = 0\}$. Then A represents the event that at time t the machine is not working and the part causing the last machine failure before t is j. Following the semi-regenerative paradigm, we define $f_i(A, t) = P\{Y_t = j, W_t = 0 | X_0 = i\}$ and $u_i(A, t) = P\{Y_t = j, W_t = 0, T_1 > t | X_0 = i\}$, where T_1 denotes the time of the first transition out of state $X_0 = i$. Since only when $X_0 = j$ it is possible that $Y_t = j$, $W_t = 0$, and $T_1 > t$, we have $u_i(A, t) = \delta_{ij}[1 - G_j(t)]$.

Applying Equation 6.4.3, we find the time-dependent solution

$$P\{Y_t = j, W_t = 0 | X_0 = i\} = \sum_{k=1}^{M} \int_0^t dR_{ik}(s) u_k(A, t - s)$$

$$= \sum_{k=1}^{M} \int_0^t dR_i(s) \delta_{kj} [1 - G_j(t - s)]$$

$$= \int_0^t dR_{ij}(s) [1 - G_j(t - s)]. \tag{6.4.5}$$

In Example 6.1.7, we found that $p_{ij} = \mu_j/\mu$ for all i. This implies that the transition probability of the embedded Markov chain has identical rows and $\pi_j = \mu_j/\mu$ for all j, where $\{\pi_j\}$ are the limiting probabilities of the embedded chain. The mean unconditional holding time in state j, τ_j, is $E[V_j] + (1/\mu)$— namely, the sum of the mean repair time of part j plus the mean time for a failure to occur after the repair of part j. Also, we see that

$$\int_0^\infty u_k(A, t)dt = \delta_{kj}\int_0^\infty [1 - G_j(t)]dt = \delta_{kj}E[V_j].$$

Using these results in Equation 6.4.4, we conclude that

$$\lim_{t\to\infty} P\{Y_t = j, W_t = 0\} = \frac{\pi_j E[V_j]}{\sum_{i=1}^M \pi_i \tau_i} = \frac{\pi_j E[V_j]}{\sum_{i=1}^M \frac{\mu_i}{\mu}\left[\frac{1}{\mu} + E[V_i]\right]} = \frac{\mu_j E[V_j]}{1 + \sum_{i=1}^M \mu_i E[V_i]}. \tag{6.4.6}$$

It is instructive to look at the problem of finding the limiting probability from the vantage point of regenerative theory. Let a regeneration cycle be the time between successive failures. Conditioning on the part causing the failure, the mean cycle length is $(1/\mu) + \sum(\mu_i/\mu)E[V_i]$, which is just the denominator of the third term of Equation 6.4.6. The expected time that event A will occur in a regeneration cycle is $(\mu_j/\mu)E[V_j]$. This is the numerator of the third term. Thus the regenerative approach produces the identical result—as expected. ∎

EXAMPLE 6.4.5 **The Machine Reliability Problem: A Numerical Example** Consider a piece of equipment containing two parts. As in Example 6.4.4, for Part i, $i = 1, 2$, the time-to-failure distribution is exponential with parameter μ_i and the repair time distribution is G_i. We assume that $G_1 \sim PH(\mathbf{a}, T)$ and $G_2 \sim PH(\mathbf{b}, S)$, where $\mathbf{a} = (1, 0)$, $\mathbf{b} = (b_1, b_2)$, $b_1 > 0$, $b_2 > 0$, $b_1 + b_2 = 1$, and

$$T = \begin{bmatrix} -\lambda_1 & \lambda_1 \\ & -\lambda_2 \end{bmatrix} \quad \text{and} \quad S = \begin{bmatrix} -\beta_1 & \\ & -\beta_2 \end{bmatrix}.$$

We see that G_1 is a generalized Erlang distribution, G_2 is a hyperexponential distribution, and the Laplace transforms of the two densities are

$$g_1^e(s) = \prod_{i=1}^2 \frac{\lambda_i}{s + \lambda_i} \quad \text{and} \quad g_2^e(s) = \sum_{i=1}^2 b_i \frac{\beta_i}{s + \beta_i}.$$

Define $\pi_i = \mu_i/\mu$, $i = 1, 2$. Example 6.1.7 suggests that $Q_{ij}(t) = K_i(t)\pi_j$ for i, $j = 1, 2$, where $K_i = [G_i^* E](t)$, the asterisk (*) is the convolution operator, and $E \sim exp(\mu)$. Since E is $PH(\theta, V)$, where $\theta = [1]$ and $V = [-\mu]$, using the formulas given in Example 5.5.1, we find that $K_i \sim PH(\gamma_i, L_i)$, $i = 1, 2$, where $\gamma_1 = (1, 0, 0)$, $\gamma_2 = (b_1, b_2, 0)$,

$$L_1 = \begin{bmatrix} -\lambda_1 & \lambda_1 & \\ & -\lambda_2 & \lambda_2 \\ & & -\mu \end{bmatrix} \quad \text{and} \quad L_2 = \begin{bmatrix} -\beta_1 & & \beta_1 \\ & -\beta_2 & \beta_2 \\ & & -\mu \end{bmatrix}.$$

Without resorting to the formulas, the preceding phase-type representations could also have been obtained by looking at the structures of the transition diagrams involved. Using Equation 5.5.4, or by inspection, we find the Laplace transform of the densities h_1 and h_2,

$$g_1^e(s) = \frac{\mu}{s+\mu} \prod_{i=1}^{2} \frac{\lambda_i}{s+\lambda_i} \equiv \psi\Pi \quad \text{and} \quad g_2^s(s) = \frac{\mu}{s+\mu} \sum_{i=1}^{2} b_i \frac{\beta_i}{s+\beta_i} \equiv \psi\Sigma,$$

where we define $\psi = \mu/(s+\mu)$, and Π and Σ are defined accordingly. From Example 6.1.7, the Laplace-Stieltjes transform of $Q(t)$ is

$$Q^e(s) = \begin{bmatrix} \pi_1 g_1^e(s) & \pi_2 g_1^e(s) \\ \pi_1 g_2^e(s) & \pi_2 g_2^e(s) \end{bmatrix}.$$

As before, we define the matrix of Laplace-Stieltjes transforms of Markov renewal function $R(t)$ as $R^e(s) = \int_0^\infty e^{-st} dR(t)$. Following Equation 6.2.9, we obtain

$$R^e(s) = [I - Q^e(s)]^{-1} = [1 - (\pi_1\psi\Pi + \pi_2\psi\Sigma)]^{-1} \begin{bmatrix} 1 - \pi_2\psi\Sigma & \pi_2\psi\Pi \\ \pi_1\psi\Sigma & 1 - \pi_1\psi\Pi \end{bmatrix}.$$

To find the time-dependent probability $P\{Y_t = j, \ W_t = 0 | X_0 = i\}$, from Equation 6.4.5 we see that all we have to do is invert the Laplace transform $R_{ij}^e(s)[(1/s)(1 - g_j^e(s))]$. We illustrate the solution procedure in the following numerical example.

Consider a case in which the mean times to failure for the two parts are 1500 hours and 500 hours. The repair-time distributions are of phase type with the parameters $a = (1, 0)$, $b = (0.05, 0.95)$,

$$T = \begin{bmatrix} -.1 & .1 \\ & -.08 \end{bmatrix} \quad \text{and} \quad S = \begin{bmatrix} -.008 & \\ & -.2 \end{bmatrix}.$$

This set of parameters yields $E[V_1] = 22.50$, $Sdv[V_1] = 16.01$, $E[V_2] = 11$, and $Sdv[V_2] = 38.59$. As expected, we note that the repair time for Part 2 has a considerably larger variability.

The limiting probability $P\{Y_t = j, \ W_t = 0\}$ can be found from Equation 6.4.6. For this set of data, we obtain

$$\lim_{t\to\infty} P\{Y_t = 1, \ W_t = 0\} = 0.0145 \quad \text{and} \quad \lim_{t\to\infty} P\{Y_t = 2, \ W_t = 0\} = 0.0212.$$

To compute the time-dependent probability, we use numerical inversion of a Laplace transform (see Section 1.3). For $P\{Y_t = 1, \ W_t = 0 | X_0 = 1\}$ and $0 \le t \le 150$, the time-dependent probabilities, found in the Appendix, are plotted in Figure 6.9. The figure shows that after one hundred hours of operation the probabilities become quite close to the limiting value of 0.0145. ∎

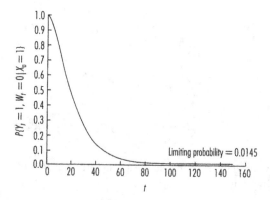

FIGURE
6.9 The time-dependent probability $P\{Y_t = 1, W_t = 0 | X_0 = 1\}$.

EXAMPLE
6.4.6*   **Excess Lives and Ages of a Semi-Markov Process**   Let (X, T) be a Markov renewal process and V the associated semi-Markov process as defined in Section 6.3. Let T_n be the nth transition time of (X, T). We define excess life and age of the Markov renewal process at time t as we did in a renewal process, namely,

$$Y(t) = T_{n+1} - t \qquad \text{and} \qquad A(t) = t - T_n,$$

where $T_n \le t < T_{n+1}$. Using an argument analogous to that given in Example 6.4.1, we can establish that $(V, Y) = (V(t), Y(t))$, $(V, A) = (V(t), A(t))$ and (V, Y, A) $= (V(t), Y(t), A(t))$ are all semi-regenerative processes.

Looking first at semi-regenerative process (V, Y) with state space $F = \{\{0, 1, ...\} \cup (0, \infty)\}$, we define $A = \{V(t) = j, Y(t) > y\}$, $u_i(A, t) = P\{V(t) = j, Y(t) > y, T_1 > t | X_0 = i\}$, and $f_i(A, t) = P\{V(t) = j, Y(t) > y | X_0 = i\}$. If the first transition out of state i occurs after time t, then it is possible that $V(t) = j$ only if $i = j$. Moreover, $Y(t) > y$ and $T_1 > t$ jointly imply $T_1 > t + y$. These observations lead to

$$u_i(A, t) = \delta_{ij} P\left\{T_1 > t + y | X_0 = i\right\}$$

$$= \delta_{ij} P\{S_j > t + y\} = \delta_{ij} h_j(t + y) = \delta_{ij} P\left\{1 - \sum_{k \in S} Q_{jk}(t + y)\right\},$$

where S_j denotes the unconditional holding time in state j and $h_j(y) = P\{S_j > y\}$. Applying Equation 6.4.3, we find

$$P\left\{V(t) = j, Y(t) > y | X_0 = i\right\} = \sum_{k \in S} \int_0^t dR_{ik}(s) u_k(A, t - s)$$

$$= \int_0^t dR_{ij}(ds) h_j(t + y - s). \qquad (6.4.7)$$

Assume that the Markov chain embedded under (X, T) is ergodic and hence the limiting probabilities $\{\pi_i\}$ exist. Using Equation 6.4.4, we obtain the limiting probability

$$\lim_{t \to \infty} P\{V(t) = j, \ Y(t) > y | X_0 = i\} = \frac{1}{\sum_{i \in S} \pi_i \tau_i} \sum_{k \in S} \pi_k \int_0^\infty \delta_{kj} h_j(t+y) dt$$

$$= \frac{1}{\sum_{i \in S} \pi_i \tau_i} \pi_j \int_0^\infty h_j(t+y) dt$$

$$= \frac{1}{\sum_{i \in S} \pi_i \tau_i} \pi_j \int_y^\infty h_j(\xi) d\xi = \frac{1}{\mu_{jj}} \int_y^\infty h_j(\xi) d\xi, \quad (6.4.8)$$

where μ_{jj} is the mean recurrence time of state j. An insightful way to look at Equation 6.4.8 is by rewriting it as

$$\lim_{t \to \infty} P\{V(t) = j, \ Y(t) > y\} = \frac{\tau_j}{\mu_{jj}} \frac{1}{\tau_j} \int_y^\infty h_j(\xi) d\xi = P_j \frac{\int_y^\infty P\{S_j > \xi\} d\xi}{\tau_j} = P_j h_{j,e}(y),$$

$$(6.4.9)$$

where P_j denotes the long-run fraction of time the process (X, T) is in state j (see Equation 6.2.20) and $h_{j,e}(y)$ is the *equilibrium* complementary cumulative distribution of h_j (see Equation 3.3.4). In steady state, we see that $V(t)$ and $Y(t)$ are independent.

We now turn our attention to the semi-regenerative process (V, Y, A) with state space $F = \{\{0, 1, \ldots\} \cup (0, \infty) \cup (0, t)\}$. We define $A = \{V(t) = j, \ Y(t) > y, \ A(t) > x\}$, $u_i(A, \ t) = P\{V(t) = j, \ Y(t) > y, \ A(t) > x, \ T_1 > t | X_0 = i\}$, and $f_i(A, \ t) = P\{V(t) = j, \ Y(t) > y, \ A(t) > x | X_0 = i\}$. By taking into consideration the added event $A(t) > x$, we modify our earlier result and obtain

$$u_i(A, \ t) = \delta_{ij} P\{S_j > t + y\}[t > x],$$

where $[B] = 1$ if B is true and 0 otherwise. The previous equality implies that t must be greater than x so that $A(t) > x$ and $T_1 > t$. Using Equation 6.4.3, we obtain the time-dependent probability

$$P\{V(t) = j, \ Y(t) > y, \ A(t) > x | X_0 = i\} = \sum_{k \in S} \int_0^t dR_{ik}(s) u_k(A, \ t-s)$$

$$= \int_0^t dR_{ij}(s) P\{S_j > t + y - s\}[t - s > x]$$

$$= \int_0^t dR_{ij}(s) h_j(t + y - s)[s < t - x]$$

$$= \int_0^{t-x} dR_{ij}(s) h_j(t + y - s). \quad (6.4.10)$$

For the limiting probability, we first compute

$$\int_0^\infty u_i(A, \ t) dt = \delta_{ij} \int_0^\infty h_j(t+y)[t > x] dt = \delta_{ij} \int_x^\infty h_j(t+y) dt = \delta_{ij} \int_{x+y}^\infty h_j(\xi) d\xi.$$

The rest is entirely identical to the derivation of Equation 6.4.8. Thus we find the joint limiting distribution

$$\lim_{t \to \infty} P\{V(t) = j, \ Y(t) > y, \ A(t) > x | X_0 = i\} = \frac{1}{\mu_{jj}} \int_{x+y}^\infty h_j(\xi) d\xi. \quad \blacksquare \quad (6.4.11)$$

EXAMPLE
6.4.7*

Queue Lengths of an M/G/1 Queue Revisited In Example 6.4.3, $Z(t)$ denotes the queue length at time t. There we showed that $Z = \{Z(t), t \geq 0\}$ is a semi-regenerative process with state space $S = \{0, 1, 2, ...\}$ and an embedded Markov renewal process (X, T).

In this example, we are interested in the time-dependent and limiting probabilities of the event $Z(t) = j$. Define $A = \{Z(t) = j\}$, $u_i(A, t) = P\{Z(t) = j, T_1 > t | X_0 = i\}$, and $f_i(A, t) = P\{Z(t) = j | X_0 = i\}$. Define $K_{ij}(t) = u_i(A, t)$ so that j is shown explicitly. For $i = 0$ and $j = 0$, we see that $K_{ij}(t) = P\{Z(t) = 0, T_1 > t | X_0 = 0\}$ $= P\{$the first arrival occurs after time $t\} = e^{-\lambda t}$, where λ is the arrival rate. For $i = 0$ and $j > 0$, we condition on the length of the first arrival and the length of the first service time. The scenario is shown in Figure 6.10. Given that

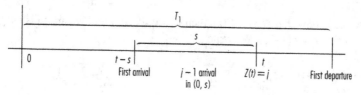

FIGURE
6.10 The scenario with $X_0 = 0$ and $Z(t) = j > 0$.

the first departure occurs after time t and $Z(t) = j > 0$, the first arrival must occur before time t, say at time $t - s$, where $s > 0$. The service time of the first customer must be at least s and during $(t - s, t)$ there must be exactly $j - 1$ arrivals. Combining these observations, we obtain

$$K_{ij}(t) = \int_0^t \lambda e^{-\lambda(t-s)}[1 - G(s)]pos(j-1; \lambda, s)ds \qquad i = 0, j > 0. \quad \text{(6.4.12)}$$

When $X_0 = i > 0$, T_1 is actually the length of the service time of the first customer who is present at time 0. We have $Z(t) = j \geq i$ if there are $j - i$ arrivals in $(0, t)$. This gives

$$K_{ij}(t) = [1 - G(t)]pos(j - i; \lambda, t) \qquad i > 0, j \geq i. \quad \text{(6.4.13)}$$

With $u_i(A, t) = K_{ij}(t)$ so defined, the time-dependent queue length distributions can be found by Equation 6.4.3.

For the limiting queue length distribution, we consider the case in which the traffic intensity $\rho = \lambda/\mu < 1$, where $1/\mu$ is the mean service time. In this case the queue is stable and the limiting queue length probabilities $\{\pi_j\}$ defined at departure epochs are strictly positive. From Equation 6.4.4, we find

$$\lim_{t \to \infty} P\{Z(t) = j | X_0 = i\} = \sum_{k \in S} \frac{1}{\mu_{kk}} \int_0^\infty u_k(A, t)dt = \sum_{k=0}^j \frac{1}{\mu_{kk}} \int_0^\infty u_k(A, t)dt.$$

The last equality is due to $u_k(A, t) = 0$ when $k > j$. We will now show that the previous queue length probability at any time in steady state is π_j. To begin, we note that

$$\sum_{i=0}^{\infty} \pi_i \tau_i = \pi_0 \left[\frac{1}{\lambda} + \frac{1}{\mu} \right] + (1 - \pi_0) \frac{1}{\mu} = \pi_0 \frac{1}{\lambda} + \frac{1}{\mu} = \left(1 - \frac{\lambda}{\mu} \right) \left(\frac{1}{\lambda} \right) + \frac{1}{\mu} = \frac{1}{\lambda},$$

where we use the relation $\pi_0 = 1 - \rho$. With Equation 6.2.19, we see that $1/\mu_{kk} = \lambda \pi_k$ and

$$v_j \equiv \lim_{t \to \infty} P\{Z(t) = j | X_0 = i\} = \sum_{k=0}^{j} \lambda \pi_k \int_0^{\infty} u_k(A, t) dt.$$

Define probability generating functions

$$H(z) = \sum_{j=0}^{\infty} v_j z^j \quad \text{and} \quad G(z) = \sum_{j=0}^{\infty} \pi_j z^j.$$

If we can show that $H(z) = G(z)$, then $\pi_j = v_j$ for all j and we are finished.

Define $a = \lambda(1 - z)$. We write

$$(1 - z)H(z) = (1 - z) \sum_{j=0}^{\infty} \left[\sum_{k=0}^{j} \lambda \pi_k \int_0^{\infty} u_k(A, t) dt \right] z^j = a \sum_{k=0}^{\infty} \pi_k \sum_{j=k}^{\infty} z^j \int_0^{\infty} K_{kj}(t) dt. \quad (6.4.14)$$

Let $g^e(s)$ denote the Laplace transform of the service time density. We now consider the right side of Equation 6.4.14 in three separate parts:

(i) For $k = 0$ and $j = 0$, we have

$$a\pi_0 \int_0^{\infty} K_{00}(t) dt = a\pi_0 \int_0^{\infty} e^{-\lambda t} dt = (1 - z)\pi_0 \int_0^{\infty} \lambda e^{-\lambda t} dt = (1 - z)\pi_0.$$

(ii) For $k = 0$ and $j \geq 1$, using Equation 6.4.12, we have

$$a\pi_0 \sum_{j=1}^{\infty} z^j \int_0^{\infty} \left[\int_0^t \lambda e^{-\lambda(t-s)} [1 - G(s)] e^{-\lambda s} \frac{(\lambda s)^{j-1}}{(j-1)!} ds \right] dt$$

$$= a\pi_0 z \int_0^{\infty} \left[\int_0^t \lambda e^{-\lambda(t-s)} [1 - G(s)] e^{-\lambda s} e^{\lambda sz} ds \right] dt$$

$$= a\pi_0 z \int_0^{\infty} [1 - G(s)] e^{-as} \left[\int_s^{\infty} \lambda e^{-\lambda(t-s)} dt \right] ds$$

$$= a\pi_0 z \int_0^{\infty} [1 - G(s)] e^{-as} ds = \pi_0 z \int_0^{\infty} [1 - G(s)] a e^{-as} ds = \pi_0 z [1 - g^e(a)].$$

(iii) For $k \geq 1$ and $j \geq k$, using Equation 6.4.13, we have

$$a \sum_{k=1}^{\infty} \pi_k \sum_{j=k}^{\infty} z^j \int_0^{\infty} [1 - G(t)] e^{-\lambda t} \frac{(\lambda t)^{j-k}}{(j-k)!} dt = a \sum_{k=1}^{\infty} \pi_k z^k \int_0^{\infty} [1 - G(t)] e^{-\lambda t} e^{\lambda tz} dt$$

$$= \sum_{k=1}^{\infty} \pi_k z^k \int_0^{\infty} [1 - G(t)] a e^{-at} dt$$

$$= \sum_{k=1}^{\infty} \pi_k z^k [1 - g^e(a)].$$

Combining these three parts into the right side of Equation 6.4.14, we find

$$(1-z)H(z) = \pi_0\left[1-z+z\left(1-g^e(a)\right)\right] + \sum_{k=1}^{\infty}\pi_k z^k\left(1-g^e(a)\right)$$

$$= \pi_0\left[1-z+z\left(1-g^e(a)\right)\right] + \left[G(z)-\pi_0\right]\left(1-g^e(a)\right)$$

$$= \pi_0(1-z)g^e(a) + G(z)\left(1-g^e(a)\right). \tag{6.4.15}$$

To show that $(1-z)G(z)$ is equal to Equation 6.4.15, we first note that, from Example 4.3.2,

$$\sum_{j=0}^{\infty}a_j z^j = \sum_{j=0}^{\infty}z^j\int_0^{\infty}e^{-\lambda x}\frac{(\lambda x)^j}{j!}g(x)dx$$

$$= \int_0^{\infty}e^{-\lambda x}e^{\lambda xz}g(x)dx = \int_0^{\infty}e^{-ax}g(x)dx = g^e(a).$$

Using the results of Example 4.3.2, we write

$$(1-z)G(z) = G(z) - zG(z) = G(z) - \sum_{j=0}^{\infty}\pi_j z^{j+1}$$

$$= G(z) - \sum_{j=0}^{\infty}\left[\pi_0 a_j + \sum_{i=1}^{j+1}\pi_i a_{j-i+1}\right]z^{j+1}$$

$$= G(z) - z\pi_0 g^e(a) - \sum_{j=0}^{\infty}z^{j+1}\sum_{i=1}^{j+1}\pi_i a_{j-i+1}$$

$$= G(z) - z\pi_0 g^e(a) - \sum_{i=1}^{\infty}\pi_i\sum_{j=i-1}^{\infty}z^{j+1}a_{j-i+1}$$

$$= G(z) - z\pi_0 g^e(a) - \sum_{i=1}^{\infty}\pi_i\sum_{k=0}^{\infty}z^{k+i}a_k$$

$$= G(z) - z\pi_0 g^e(a) - [G(z)-\pi_0]g^e(a)$$

$$= \pi_0(1-z)g^e(a) + G(z)\left(1-g^e(a)\right).$$

The last expression is simply Equation 6.4.15. Hence we show that, for an *M/G/*1 queue, the limiting queue length distribution at any time, including that of an arrival epoch, is equal to the limiting queue length distribution at departure epochs (see PASTA discussed in Section 2.7). ∎

EXAMPLE **6.4.8***
Queue Lengths of a GI/M/1 Queue   In Example 4.3.3, we defined the *GI/M/*1 queue and obtained the limiting queue length distribution *at arrival epochs.* Continuing with the same paradigm, we now study the limiting queue length distribution *at any time.* As opposed to the *M/G/*1 queue, we will see that for a *GI/M/*1 queue the two distributions are different.

Let T_n denote the arrival epoch of the *n*th customer and X_n the number of customers in the system just before the *n*th arrival. In Example 6.1.6, we showed that

$(X, T) = \{(X_n, T_n)\}$ is a Markov renewal process with state space $S = \{0, 1, \dots\}$ and semi-Markov kernel $Q(t)$ described there. Let $Z(t)$ denote the queue length at time t. An argument similar to that given in Example 6.4.3 will show that $Z = \{Z(t), t \geq 0\}$ is a semi-regenerative process with the embedded Markov renewal process (X, T).

As in Example 6.4.7, we define $A = \{Z(t) = j\}$, $u_i(A, t) = P\{Z(t) = j$, $T_1 > t | X_0 = i\}$, $f_i(A, t) = P\{Z(t) = j | X_0 = i\}$, and $K_{ij}(t) = u_i(A, t)$ so that j is shown explicitly. For $j > 0$ and $i \geq j - 1$, there must be exactly $i + 1 - j$ service completions in the initial interval $(0, t)$ and so

$$K_{ij}(t) = P\{Z(t) = j, T_1 > t | X_0 = i\} = P\{Z(t) = j | T_1 > t, X_0 = i\}P\{T_1 > t | X_0 = i\}$$

$$= pos(i + 1 - j; \mu, t)[1 - G(t)].$$

For $j = 0$ and $i \geq 0$, the system has $i + 1$ service completions in the initial interval $(0, t)$ and hence

$$K_{ij}(t) = \sum_{k=i+1}^{\infty} pos(k; \mu, t)[1 - G(t)].$$

With $K_{ij}(t)$ so defined, we will be able to compute the time-dependent queue length probability $P\{Z(t) = j | X_0 = i\}$ using Equation 6.4.3.

Let $\rho = 1/r = \lambda/\mu$ denote the traffic intensity of the queue. In Example 4.3.3, we established that when $\rho < 1$, the limiting queue length distribution at arrival epochs exists and is given by $\pi_j = (1 - a)a_j$, $j = 0, 1, \dots$, where $0 < a < 1$ is the unique solution of $a = G^e(\mu(1 - a))$. To find the limiting queue length distribution at any time, we consider the case in which the interarrival time has a density g (when interarrival times are periodic, the limiting distribution will be periodic and the treatment will be somewhat involved). From Equation 6.4.4, we see that

$$\lim_{t \to \infty} P\{Z(t) = j | X_0 = i\} = \lambda \sum_{k=0}^{\infty} (1 - a)a^k \int_0^{\infty} K_{kj}(t)dt,$$

where we use $\sum \pi_i \tau_i = (1/\lambda)\sum \pi_i = 1/\lambda$. Before we proceed, we first establish the following identity:

$$\int_0^{\infty} [1 - G(t)]e^{-\mu(1-a)t} dt = \frac{1}{\mu(1 - a)} - \frac{1}{\mu(1 - a)} G^e(\mu(1 - a))$$

$$= \frac{1}{\mu(1 - a)}(1 - a) = \frac{1}{\mu}, \tag{6.4.16}$$

where we use $a = G^e(\mu(1 - a))$ in establishing the second equality. For $j = 0$, we see that

$$\lim_{t \to \infty} P\{Z(t) = j | X_0 = i\} = \lambda \sum_{k=0}^{\infty} (1 - a)a^k \int_0^{\infty} [1 - G(t)] \sum_{n=k+1}^{\infty} pos(n; \mu, t)dt$$

$$= \lambda \int_0^{\infty} dt [1 - G(t)]e^{-\mu t} \sum_{n=1}^{\infty} (1 - a)\frac{(\mu t)^n}{n!} \sum_{k=0}^{n-1} a^k$$

$$= \lambda \int_0^{\infty} dt [1 - G(t)]e^{-\mu t} \sum_{n=1}^{\infty} \frac{(\mu t)^n}{n!}(1 - a^n)$$

$$= \lambda \int_0^\infty [1 - G(t)]e^{-\mu t} \Big[(e^{\mu t} - 1) - (e^{\mu t a} - 1) \Big] dt$$

$$= \lambda \int_0^\infty [1 - G(t)][1 - e^{-\mu(1-a)t}] dt$$

$$= \lambda \left[\int_0^\infty [1 - G(t)] dt - \int_0^\infty [1 - G(t)][e^{-\mu(1-a)t}] dt \right] = \lambda \left(\frac{1}{\lambda} - \frac{1}{\mu} \right) = 1 - \rho,$$

where the second to the last equality is due to Equation 6.4.16. Finally, for $j > 0$, we have

$$\lim_{t \to \infty} P\{Z(t) = j | X_0 = i\} = \lambda \sum_{k=i-1}^\infty (1 - a)a^k \int_0^\infty [1 - G(t)] pos(k + 1 - j; \mu, t)$$

$$= \lambda(1 - a) \int_0^\infty dt [1 - G(t)] e^{-\mu t} a^{j-1} \sum_{k=j-1}^\infty a^{k+1-j} \frac{(\mu t)^{k+1-j}}{(k+1-j)!} dt$$

$$= \lambda(1 - a)a^{j-1} \int_0^\infty [1 - G(t)] e^{-\mu(1-a)t} dt$$

$$= \lambda(1 - a)a^{j-1} \left(\frac{1}{\mu} \right) = \rho(1 - a)a^{j-1}.$$

Again, we have used Equation 6.4.16 in the previous derivation. It should be clear that the limiting queue length distribution at any time of a *GI/M/*1 queue is different from that defined at arrival epochs. We summarize the given useful result as follows:

$$\lim_{t \to \infty} P\{Z(t) = j | X_0 = i\} = \begin{cases} 1 - \rho & j = 0 \\ \rho(1 - a)a^{j-1} & j > 0. \end{cases} \quad \blacksquare \qquad (6.4.17)$$

EXAMPLE **6.4.9**   **The GI/M/1 Queue: Numerical Illustration**   We give two numerical examples to illustrate the computation of queue length distribution for a *GI/M/*1 queue. In both cases, the service rates μ are set at 1 and the arrival rates λ are set at 0.95 (hence the mean interarrival times are both 1.0526) so that the traffic intensity is $\rho = 0.95$. The two interarrival time distributions are the two phase-type distributions defined in Example 6.4.5 with parameters $a = (1, 0)$, $(\lambda_1, \lambda_2) = (95/81, 1/5)$, $b = (0.05, 0.95)$ and $(\beta_1, \beta_2) = (1/10, 361/210)$. Using Equation 5.5.7, we find that the two variances of the interarrival time are 0.7670 and 9.5349. The hyperexponential interarrival time distribution exhibits a considerably larger variability.

For reference, we call the queue whose interarrival time distribution is generalized Erlang Queue 1 and the one whose interarrival time distribution is hyperexponential Queue 2. For Queue 1, the equation $a = G^e[\mu(1 - a)]$ reduces to

$$a = \left[\frac{\lambda_1}{(1 - a) + \lambda_1} \right] \left[\frac{\lambda_2}{(1 - a) + \lambda_2} \right],$$

and for Queue 2 it becomes

$$a = \left[\frac{b_1 \beta_1}{(1-a) + \beta_1} \right] + \left[\frac{b_2 \beta_2}{(1-a) + \beta_2} \right].$$

Solving each one of the two equations for a, we find that the two values are 0.9410 and 0.9891. In Figure 6.11, for the two queues, we plot the queue length distributions defined at arrival epochs and any time over the range of queue length from 0 to 10 to highlight the regions where the most noticeable discrepancies between the two occur. The details of computation are given in the Appendix. It is obvious that the two queue length distributions for Queue 2 behave differently from their counterparts for Queue 1. ∎

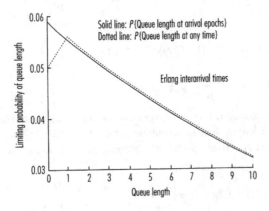

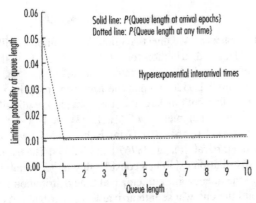

FIGURE 6.11 The queue length distributions for the two queues.

Problems

1 Consider a Markov renewal process with state space $S = \{1, 2\}$ and semi-Markov kernel

$$Q(t) = \begin{bmatrix} 0.7 - 0.7e^{-2t} & 0.3 - 0.3e^{-t} \\ 0.2 - 0.2e^{-2t} - 0.4te^{-2t} & 0.8 - 0.8e^{-t} - 0.8te^{-t} \end{bmatrix}.$$

(a) What is the transition matrix of the embedded Markov chain? **(b)** What are holding time densities $\{g_{ij}(t)\}$, where $g_{ij}(t) = dG_{ij}(t)/dt$?

2 Consider a Markov renewal process with state space $S = \{1, ..., n\}$ and semi-Markov kernel $Q(t)$. Let $T_0 = 0$ and define $D_n = T_n - T_{n-1}, n = 1, 2, ...,$ and $v_i = P\{X_0 = i$ for all i. **(a)** Derive an expression for computing the joint probability $P\{X_1 = i_1, D_1 \le t_1, X_2 = i_2, D_2 \le t_2\}$ as a function of $\{v_i\}$ and $\{Q_{ij}(t)\}$. **(b)** Derive an expression for computing the conditional probability $P\{D_1 \le t_1, D_2 \le t_2 | X_0 = i_0, X_1 = i_1, X_2 = i_2\}$ as a function of the holding time distributions $\{G_{ij}(t)\}$. **(c)** Use the result obtained in Part b to find $P\{D_1 \le x, D_2 \le y | X_0 = 1, X_1 = 2, X_2 = 1\}$ for the Markov renewal process described in Problem 1.

3 Consider a Markov renewal process with state space $S = \{1, ..., n\}$ and semi-Markov kernel $Q(t)$. Let $T_0 = 0$ and define $D_n = T_n - T_{n-1}, n = 1, 2, ...,$ and $v_i = P\{X_0 = i\}$ for all i. For the embedded Markov chain, as usual, we let $p_{ij}^{(n)} = P\{X_n = j | X_0 = i\}$, row vector $s = \{s_j(n)\}$, where $s_j(n) = P\{X_n = j | v\}$, where $v = \{v_i\}$. Use the law of total probability to establish $P\{D_{m+1} \le t\} = s(n)Q(t)e$, where e is a column vector of ones.

4 Consider a stationary Markov renewal process with state space $S = \{1, ..., n\}$ and semi-Markov kernel $Q(t)$. Let $\pi = \{\pi_i\}$ be the stationary probability vector of the embedded Markov chain (that is, it satisfies $\pi = \pi P$ and $\pi e = 1$, where e is a column vector of ones). Show that **(a)** $P\{D_{m+1} \le t\} = \pi Q(t)e$ and **(b)** $P\{D_{m+1} \le t_1, D_{m+2} \le t_2\} = \pi Q(t_1)Q(t_2)e$.

5 Consider a Markov renewal process with state space $S = \{1, 2\}$ and semi-Markov kernel

$$Q(t) = \begin{bmatrix} 0 & 1 - e^{-\lambda t} \\ 1 - e^{-\mu t} & 0 \end{bmatrix}.$$

Another way to look at this stochastic process is that we have an alternating renewal process with two alternating exponential interarrival time distributions with parameters λ and μ, respectively. **(a)** Find $P\{D_{m+1} \le t\}$ in steady state. **(b)** Find $P\{D_{m+1} \le t_1, D_{m+2} \le t_2\}$ in steady state. **(c)** Find $\{dR_{ij}(t)/dt\}$. **(d)** Find $\{F_{ij}(t)\}$. **(e)** Let $V = \{V(t), t \ge 0\}$ be the semi-Markov process associated with the Markov renewal process and $Y(t)$ and $A(t)$ the excess life and age of the Markov renewal process at time t. Find $P\{V(t) = j, Y(t) > y, A(t) > x | X_0 = i\}$ for all i and j. **(f)** Use Equation 6.4.11 to find the limiting probabilities of $P\{V(t) = j, Y(t) > y, A(t) > x | X_0 = i\}$ for all i and j.

6 Consider a Markov renewal process with state space $S = \{1, 2\}$ and semi-Markov kernel

$$Q(t) = \begin{bmatrix} 1 - e^{-\lambda t} & 0 \\ a(1 - e^{-\mu t}) & (1 - a)(1 - e^{-\mu t}) \end{bmatrix},$$

where $a > 0$. Assume that $P\{X_0 = 2\} = 1$. **(a)** What is the transition probability matrix of the embedded chain? **(b)** What is the n-step transition probability matrix of the embedded chain? **(c)** What kind of embedded Markov chain is this? **(d)** What are the holding time distributions in states 1 and 2? **(e)** Derive $P\{D_{m+1} \leq t\}$ for any finite m.

7 Use Equation 6.2.5 to establish Equation 6.2.8: $R^e(s) = I + Q^e(s)R^e(s)$.

8 Consider a piece of equipment whose performance at any time can be characterized by one of the four states *initial*, *good*, *marginal*, and *failed*, denoted by 0, 1, 2, and 3, respectively. The state of the system is always known with certainty. The changes of state can be described by the following transition probability matrix

$$P = \begin{bmatrix} 0 & 0.80 & 0.15 & 0.05 \\ & & 0.97 & 0.03 \\ & & & 1 \\ & & & 1 \end{bmatrix}.$$

When the equipment reaches state 3, it must be replaced. A replacement will return the equipment to state 0 again. When the equipment is in state i, the mean sojourn time in the state is τ_i. Here, we assume that $\tau_0 = 6.017$, $\tau_1 = 4.6188$, and $\tau_2 = 5.72$. The mean replacement time is 3. The cost parameters include A_i, the cost per unit time in state i, the fixed cost of replacement R, and the equipment down time of D per unit time. We are given $A_0 = 1$, $A_1 = 2$, $A_2 = 3$, $R = 5$, and $D = 3$. **(a)** Find the long-run average cost per unit time for having the equipment. **(b)** Consider another replacement policy. The policy calls for replacing the equipment when it reaches either state 2 or 3 for the first time. The fixed cost of replacement in state 2 is 4, the variable cost of down time in state 2 is 2 per unit time, and mean time to replace the equipment when it is done in state 2 is 2. Find the long-run average cost per unit time for following this replacement policy.

9 A single product is subject to random demand. The times between successive demands are i.i.d. random variables with a common mean $1/\lambda$. The delivery lead time is zero. No stockouts are permitted. There is a fixed cost of ordering of A and a carrying cost of h per unit of inventory per unit time. Since the delivery lead time is zero, ordering can take place when a demand occurs and there is nothing on hand. Assume that whenever we order, the order size is $Q + 1$ so that the inventory on hand will move from Q, $Q - 1, \ldots, 2, 1$, and back to Q again. **(a)** Model the inventory system as a semi-Markov process with rewards by defining the state space, stating the transition probability matrix of the embedded Markov chain, expected cost per sojourn in each state, and expected length of sojourn time in each state.

(b) Find the long-run average cost per unit time under the replenishment policy. **(c)** Derive an optimal Q that minimizes the long-run average cost.

10 A personal computer has two key components: a microprocessor and a memory board populated with memory chips. We call them parts 1 and 2, respectively. The time of failure distribution for part i is exponential with parameter λ_i, $i = 1$ and 2. The two lifetimes are assumed to be independent. The computer fails when either part fails. When part 1 fails, it is replaced with a new part. The time to replace part 1 follows an Erlang distribution with parameters $n = 2$ and μ so that its means is $2/\mu$. When part 2 fails, with probability a it requires a repair and with probability $1 - a$ it requires a replacement. The time to repair part 2 follows an exponential distribution with parameter v and the time to replace part 2 follows an Erlang distribution with parameters $n = 3$ and γ so that its mean is $3/\gamma$. Let $V(t)$ denote the state of the system at time t and the state space $\{0, 1, 2\}$, where 0 denotes the computer is working, 1 denotes part 1 is under replacement, and 2 denotes part 2 is under either replacement or repair. **(a)** Formulate $V = \{V(t), t \geq 0\}$ as a semi-Markov process by stating the Laplace transform of the semi-Markov kernel, $Q^e(s)$. **(b)** Describe how you will compute the time-dependent probabilities that $P\{V(t) = i | X_0 = 0\}$, $i = 1, 2, 3$.

11 Consider a numerical example based on Problem 10. The parameters for the problem are $\lambda_1 = 0.004$, $\lambda_2 = 0.0012$, $\mu = 2$, $v = 1$, $\gamma = 8$, and $a = 0.2$. When the computer is working, it generates a return of \$18 per unit time; when it fails, it costs \$700 to replace part 1 and \$475 to replace part 2. If a repair of part 2 is needed, the cost is \$50 per unit time. What is the long-run average return per unit time generated by the computer?

*12 Consider an automatic manufacturing system with a flexible (programmable) manufacturing cell producing two different parts for two separate assembly lines producing two types of final product (for $i = 1, 2$, part i is for product i). The cell is only capable of fabricating one part at a time. When the cell completes a part of a given type, it places the completed part in the buffer of the assembly line for the type, and its controller (for example, a microprocessor) determines the type of part to produce next. Assume that the two assembly lines each have a buffer to hold three waiting parts. When both buffers are full, the cell stops its manufacturing operation and will restart as soon as there is at least one space available at either buffer. Each assembly line will take one part from its buffer at a time as long as there are parts in the buffer. Assume that product assembly time at line i follows an exponential distribution with parameter μ_i and the part production time follows an exponential distribution with parameter λ_i. Assume also that all processing times are mutually independent. Denote the state of the system at any time by a doublet (i, j), where i represents the number of parts at line 1 and j represents the number of parts at line 2 (if $i > 1$ ($j > 1$), then there are $i - 1$ ($j - 1$) parts in the buffer for part 1 (2)). When assembly line i is idle, the system incurs a lost opportunity cost of C_i. The part selection protocol residing at the controller is shown in the following table. For example, if the system is in state (3,4), then the cell will produce part 1 next.

Part 2

		0	1	2	3	4
	0	1	1	1	1	1
	1	2	1	1	1	1
Part 1	2	2	2	1	1	1
	3	2	2	2	2	1
	4	2	2	2	2	

To model the problem by a semi-Markov reward process, we see that a Markov renewal process can be constructed where the transition epoch are the production completion epochs of a part of each type, and the epochs at which the manufacturing cell becomes active again (that is, the system leaves the state (4,4) due to one assembly completion). **(a)** Find the transition probability matrix of the embedded Markov chain. **(b)** Find the expected holding times $\{\tau_{(i,j)}\}$ in each state. **(c)** Find the expected single transition costs $\{r_{(i,j)}\}$ in each state.

13 **An *M/G*/1 Queue with Group Arrivals**   Consider a variant of Example 6.1.4. In this variant, at each arrival epoch the number of arriving customers is not one but follows a probability distribution $\{p_k\}$, where $p_k = P$ {there are k arriving customers at an arrival epoch} and $k \geq 1$. Customers are served one at a time. As in Example 6.1.4, we let X_n denote the number of customers left behind by the nth departure and T_n the time of the nth departure. We see that the stochastic process $\{(X_n, T_n), n \geq 0\}$ is a Markov renewal process. **(a)** Let $\phi_v(t)$ denote the probability that there are a total of v arriving customers in $(0, t]$. Express $\phi_v(t)$ in terms of λ and $\{p_k\}$. **(b)** Specify the semi-Markov kernel $\{Q(t), t \geq 0\}$ for this process.

14 **An *M/G*/1 Queue with Special Service after an Idle Period**   Consider another variant of Example 6.1.4. In this variant, the service time for the first customer who enters an empty system (in other words, the customer who initiates a busy period) follows a different distribution than $G(\cdot)$. We call this special service time distribution $G_1(\cdot)$. In some applications, the special service time may include a set-up time for the server. How should we modify the semi-Markov kernel described in Example 6.1.4 to account for the special service time distribution?

15 **An *M/G*/1 Queue under an *N* Policy**   In Example 3.6.4, we described the use of an N policy to control an *M/G*/1 queue. Under the policy, as soon as the system becomes idle it is closed down. It will be reopened as soon as there are N customers present. How should we modify the semi-Markov kernel described in Example 6.1.4 to account for this change of system start-up policy?

Bibliographic Notes

The three primary sources of references used in this chapter are Çinlar (1975), Disney (1982), and Howard (1971). Çinlar (1975) provides a formal account of Markov renewal theory; even though it was published more than twenty years ago, it remains an invaluable source of reference. Disney's (1982) tutorial provides a nice introduction to the subject and includes a host of references dealing with real-world applications of Markov renewal theory. Howard (1971) presents the subject from the vantage point of transform analysis. The Markov renewal theory is a key prerequisite for studying stochastic systems beyond those of Markovian models. In Disney and Kiessler (1987), queueing networks are analyzed through the use of the Markov renewal approach. The large number of references cited in Neuts (1981, 1989, 1994) provide a huge and rich reservoir for readers who are interested in carrying out an in-depth study of the subject.

Examples 6.1.1 and 6.1.2 are from Kao (1974a), and Disney (1982), respectively. Examples 6.1.4 and 6.1.6 are based on Çinlar (1975). Example 6.1.5 is found in Disney (1982). Example 6.1.7 follows Çinlar (1975, Example 6.23). Example 6.2.1 is based on Howard (1971, 188, Problem 11.3). Examples 6.3.2 and 6.3.3 are based on Hastings (1969) and Kao (1975), respectively. Examples 6.4.3, 6.4.4, 6.4.7, and 6.4.8 all have their origins in Çinlar (1975). The excess lives and ages of a semi-Markov process studied in Example 6.4.6 are discussed in Çinlar (1975, Example 6.18), Disney (1982), and Ross (1983). In this chapter, we skip a deliberation of a *transient* semi-Markov process similar to that given in Chapters 4 and 5. For the first two moments of time to absorption in such a process, see Kao (1974b). Problems 2, 3, 4, 5, and 6 are related to materials discussed in Disney and Kiessler (1987). Problem 8 is based on Kao (1973). Problem 10 was developed in the spirit of Çinlar (1975, Problem 8.4). Problem 12 is from Seidman and Schweitzer (1984) and Tijms (1986). Problems 13–15 are based on Neuts (1989).

References

Çinlar, E. 1975. *Introduction to Stochastic Processes*. Englewood Cliffs, NJ: Prentice-Hall.

Disney, R. L. 1982. A Tutorial on Markov Renewal Theory, Semi-Regenerative Processes, and Their Applications. In *Advanced Techniques in the Practice of Operations Research*. Edited by H. J. Greenberg, F. H. Murphy, and S. H. Shaw. Amsterdam: North Holland.

Disney, R. L., and P. C. Kiessler. 1987. *Traffic Processes in Queueing Network: A Markov Renewal Approach*. Baltimore: The Johns Hopkins University Press.

Hastings, N. A. J. 1969. The Repair Limit Replacement Method. *Operational Research Quarterly* 20:337–49.

Howard, R. A. 1971. *Dynamic Probabilistic Systems, Volume II*. New York: John Wiley & Sons.

Kao, E. P. C. 1973. Optimal Replacement Rules When Changes of State Are Semi-Markovian. *Operations Research* 21(6):1231–49.

Kao, E. P. C. 1974a. Modeling the Movement of Coronary Patients within a Hospital by Semi-Markov Processes. *Operations Research* 22(4):683–99.

Kao, E. P. C. 1974b. A Note on the First Two Moments of Times in Transient States in a Semi-Markov Process. *Journal of Applied Probability* 11(1):193–98.

Kao, E. P. C. 1975. A Discrete-Time Inventory Model with Arbitrary Interval and Quantity Distributions of Demand. *Operations Research* 23(6):1132–42.

Neuts, M. F. 1981. *Matrix-Geometric Solutions in Stochastic Models*. Baltimore: The Johns Hopkins University Press.

Neuts, M. F. 1989. *Structured Stochastic Matrices of M/G/1 Type and Their Applications*. New York: Marcel Dekker.

Neuts, M. F. 1995. Matrix-Analytic Methods in Queueing Theory. In *Advances in Queueing: Theory, Methods, and Open Problems*. Edited by Jewgeni Dshalaow. Boca Raton, FL: CRC Press.

Ross, S. M. 1983. *Stochastic Processes*. New York: John Wiley & Sons.

Seidman, A., and P. J. Schweitzer. 1984. Part Selection Policy for a Flexible Manufacturing Cell Feeding Several Production Lines. *IEE Transactions* 16:355–62.

Tijms, H. C. 1986. *Stochastic Modelling and Analysis: A Computational Approach*. New York: John Wiley & Sons.

Appendix

Chapter 6: Section 2

Example 6.2.1 The following program computes the four renewal functions for the Markov renewal process specified in the example.

```
» [R]=e621;
» function [R]=e621
%
%    Example 6.2.1
%
a=[];
b=[2/9 161/81 -1 1/81]; a=[a; b];
b=[4/9 -2/81 0 2/81]; a=[a; b];
b=[2/9 -10/81' 0 10/81]; a=[a; b];
b=[4/9 61/81 0 20/81]; a=[a; b];
t=0:0.1:10; R=t';
for i=1:4
    c1=a(i,1); c2=a(i,2); c3=a(i,3); c4=a(i,4); x=[];
    for t=0:0.1:10
        r=(c1*t)+c2+(c3*exp(-t))+(c4*exp(-1.8*t)); x=[x r];
    end
    R=[R x'];
end
```

Chapter 6: Section 3

Example 6.3.1 The following program computes the four transition functions for the Markov renewal process specified in the example

```
» [X]=e631;
» function [X]=e631
%
%    Example 6.3.1
%
a=[7/45 -1/18 9/10 2/5 0];
b=[38/45 1/18 -9/10 -2/5 0];
c=[7/45 -5/9 -1/5 0 3/5];
```

```
d=[38/45 5/9 1/5 0 -3/5];
coe=[a; b; c; d]; t=0:0.2:10; X=[t'];
for t=0:0.2:10
    x=zeros(1,4);
    for i=1:4
    x(1,i)=coe(i,1)+coe(i,2)*exp(-1.8*t)+coe(i,3)*exp(-t);
    x(1,i)=x(1,i)+coe(i,4)*(t*exp(-t))+coe(i,5)*exp(-2*t);
    end
    Y=[Y; x];
end
X=[X Y];
```

Example 6.3.2   The following program computes the long-run average costs for the set $\{(R_0, R_1), R_0 = 100:50:400, R_1 = 100:50:400\}$. The results are plotted in Figure 6.5.

```
>> [X]=632;
>> function [X]=e632
%
%   Example 6.3.2
%
mu0=2; mu1=3; A=400; lm0=1/100; lm1=1/150; R0=293; R1=125;
X=[];
for R0=100:50:400
    for R1=100:50:400
P0=exp(-R0*lm0); P1=exp(-R1*lm1);
t0=lm0*R0; M0=(1/lm0)*gammainc(t0,2);
t1=lm1*R1; M1=(1/lm1)*gammainc(t1,2);
P=zeros(3,3); P01=exp(-P0*mu0); P12=exp(-P1*mu1);
P(1,2)=P01; P(1,1)=1-P01; P(2,3)=P12; P(2,1)=1-P12; P(3,1)=1;
p=mc_limsr(P);
p0=P0*mu0; tau0=(1/p0)*(1-exp(-p0));
p1=P1*mu1; tau1=(1/p1)*(1-exp(-p1));
tau=[tau0 tau1 0];
r0=((M0/P0)+A)*(1-exp(-p0));
r1=((M1/P1)+A)*(1-exp(-p1));
r=[r0 r1 A];
g=(p*r')/(p*tau');
x=[R0 R1 g]; X=[X; x];
    end
end
```

Example 6.3.3   Program **e633** computes the long-run average costs for the discrete-time inventory problem with batch demands. The output matrix X contains the data for constructing Figure 6.6.

```
>> [X]=e633;
>> function [X]=e633
%
%   Example 6.3.3
%
pt=[.05 .05 .10 .15 .50 .10 .05];
pd=[.2 .5 .2 .1]; h=1; K=25; c=4; pen=20; N=6; NS=N+1;
n=length(pt); t=cumsum(ones(1,n)); ET=t*pt';
m=length(pd); Li=[];
for i=0:m-1
    suma=0;
```

```
    i1=max(i,1);
    for k=m:-1:i1
        suma=suma+(k-i)*pd(k);
    end
    suma=suma*pen; Li=[Li suma];
end
cost=[]; sx=0:N; X=zeros(NS,2); X(:,1)=sx';
for s=0:N
P=zeros(NS,NS); tau=[]; r=[];
for i=s:N
    for j=1:i
        ij=i-j;
        if ij > 0
        if ij < 5
        P(i+1,j+1)=pd(ij);
        end
        end
    end
tau=[tau ET]; r1=i*h*ET;
if i < m
r1=r1+Li(i+1);
end
P(i+1,1)=1-sum(P(i+1,:)); r=[r r1];
end
for i=s-1:-1:0
    P(i+1,NS)=1; tau=[0 tau]; r1=K+c*(N-i); r=[r1 r];
end
IP=eye(NS)-P; IP(:,1)=ones(NS,1); IP=inv(IP); pi=IP(1,:);
up=pi*r'; down=pi*tau';
g=up/down;
cost=[cost g];
end
X(:,2)=cost';
```

Chapter 6: Section 4

Example 6.4.5 Program **e645_a** is for computing $P\{Y_t = j, W_t = 0\}$, $j = 1, 2$, using Equation 6.4.6. Program **phase_mv** is for computing the mean and variance of a phase-type distribution. Its listing is given under Example 7.2.1 in the Appendix of Chapter 7. Program **e645_b** gives the Laplace transform of $P\{Y_t = 1, W_t = 0 | X_0 = 1\}$. We numerically invert the Laplace transform using **e645c** to obtain the time-dependent probabilities shown in Figure 6.9.

```
>> e645_a;
 mean =        22.50 sdv =        16.01
 mean =        11.00 sdv =        38.59
P(Yt=1, Wt=0)=    0.0145  P(Yt=2, Wt=0)=    0.0212

>> [X]=e645_c;

>> function E645_a
%
%  Example 6.4.5
%
lm1=0.1; lm2=0.08; bt1=0.008; bt2=.2;
a=[1 0]; T=[-lm1 lm1; 0 -lm2];
b=[.05 .95]; S=[-bt1 0; 0 -bt2];
```

```
[m1,v1]=phasemv(a,T);
fprintf(' mean = %10.2f sdv = %10.2f \n',m1,sqrt(v1));
[m2,v2]=phasemv(b,S);
fprintf(' mean = %10.2f sdv = %10.2f \n',m2,sqrt(v2));
%
%   Limiting behavior
%
mu1=1/1500; mu2=1/500; mu=mu1+mu2; pi1=mu1/mu; pi2=mu2/mu;
EV1=m1; EV2=m2; mv=[mu1 mu2]; EV=[EV1 EV2];
deno=1+sum(mv.*EV); p=(mv.*EV)/deno;
fprintf('P(Yt=1, Wt=0)= %8.4f  P(Yt=2, Wt=0)= %8.4f \n',p);

» function [z]=e645_b(x,y)
%
s=x+y*i; lm1=0.1; lm2=0.08; bt1=0.008; bt2=0.2;
mu1=1/1500; mu2=1/500; mu=mu1+mu2;
pi1=mu1/mu; pi2=mu2/mu; b1=.05; b2=0.95;
summ=(b1*(bt1/(s+bt1)))+(b2*(bt2/(s+bt2)));
prod=(lm1/(s+lm1))*(lm2/(s+lm2));
pp1=pi1*mu/(s+mu); pp2=pi2*mu/(s+mu);
deno=1-pp2*summ-pp1*prod;
t1=(1-pp2*summ)*(1-prod)/(deno*s); z=real(t1);

» function [X]=e645_c
%
%   The main calling program for Example 6.4.5
%
tx=[0]; qx=[1];
for t=1:1:150
tx=[tx t];
q=invt_laq('e645_b',t); qx=[qx q];
end
ns=length(tx); X=zeros(ns,2); X(:,1)=tx'; X(:,2)=qx';
```

Example 6.4.9 Program **e649** uses the "root-finding" approach for computing the limiting queue length distributions for the two cases—one with generalized Erlang interarrival time distribution and the other with hyperexponential interarrival time distribution. Programs **e649_a** and **e649_b** supply the two equations, each of the form $G^2[\mu(1-a)] - a = 0$ for root finding for the respective cases.

```
» [X]=e649;
  lamdas:    1.17283951   5.00000000
   betas:    0.10000000   1.71904762
  m1 =    1.0526  v1 =     0.7670
  m2 =    1.0526  v2 =     9.5349
root for generalized Erlang = 0.94100691
root for hyperexponential   = 0.98909858

aa =

    0.9410    0.9891

» function [X]=e649
%
%   Example 6.4.9
%
```

```
global lm1 lm2 bt1 bt2 b
r=1/.95; rho=0.95;
lm1=1/(r-.20); lm2=1/0.20;
a=[1 0]; T=[-lm1 lm1; 0 -lm2];
[m1,v1]=phasemv(a,T);
fprintf('  lamdas: %12.8f  %12.8f \n',lm1,lm2);
b=[.05 .95]; bt1=1/10; bt2=.95/(r-0.5);
fprintf('   betas: %12.8f  %12.8f \n',bt1,bt2);
S=[-bt1 0; 0 -bt2];
[m2,v2]=phasemv(b,S);
fprintf('  m1 = %10.4f  v1 = %10.4f \n',m1,v1);
fprintf('  m2 = %10.4f  v2 = %10.4f \n',m2,v2);
%
%   find the root of generalized Erlang
%
a1=fzero('e649_a',0.5);
fprintf(' root for generalized Erlang = %10.8f \n',a1);
%
%   find the root of hyperexponential
%
a2=fzero('e649_b',0.1);
fprintf(' root for hyperexponential  = %10.8f \n',a2);
aa=[a1 a2]
%
%   Queue length distribution at arrival epochs
%
X=zeros(101,5); z=ones(1,101); z=cumsum(z)-1; X(:,1)=z';
for i=1:2
    a=aa(i); x=1-a; v=[x];
    for j=1:100
    x=x*a; v=[v x];
    end
    X(:,1+i)=v';
end
%
%   Queue length distribution at any time
%
for i=1:2
    a=aa(i); x=1-rho; v=[x]; x=rho*(1-a); v=[v x];
    for j=1:99
    x=x*a; v=[v x];
    end
    X(:,3+i)=v';
end

» function [y]=e649_a(x)
%
global lm1 lm2 bt1 bt2 b
y=(lm1/((1-x)+lm1))*(lm2/((1-x)+lm2))-x;

» function [y]=e649_b(x)
%
global lm1 lm2 bt1 bt2 b
y=((b(1,1)*bt1/((1-x)+bt1))+(b(1,2)*bt2/((1-x)+bt2)))-x;
```

7

Brownian Motion and Other Diffusion Processes

Tips for Chapter 7

■ This chapter provides an introductory exposition of a rather difficult but very important subject, Brownian motion and related processes. To cover the various topics well, we need mathematical prerequisites beyond those stated in the preface. Thus our development will be necessarily nonrigorous. For serious inquiries about the subject, readers will find many helpful references at the end of the chapter.

■ Continuous-time finance is a fertile ground for applications of Brownian motion and related processes. Its principle tool is Ito's formula. In Sections 7.3 and 7.4, we give a self-contained exposition of the formula and its variants and review a few classical examples in continuous-time finance.

■ The last section (Section 7.5) is a brief introduction to stochastic optimal control. The section provides readers who are curious about how optimization is done over a system of stochastic differential equations a glimpse of this intrinsically challenging subject.

7.0 Overview

The Brownian motion process was studied originally by English botanist Robert Brown for modeling the motion of a small particle immersed in a liquid subject to molecular collisions. A mathematical model of the process was first derived by Einstein and the underlying theory was subsequently perfected by Fokker, Planck, and Wiener, among others.

Brownian motion is a Markov process with a continuous state space and a continuous index set. The process and its many generalizations occupy a central role in applied probability. A prominent generalization is the class of processes known as Ito processes, which are functions of the Brownian motion process. Since publication of the seminal paper of Black and Scholes in 1973, Ito processes have remained in the center stage of continuous-time finance. In operations research, Brownian motion is useful in approximations of stochastic service systems. Coupled with optimal control, the process is pivotal in modeling and analysis of many problems in engineering, economics, and management science. The mathematics involved in studying Brownian motion is relatively intricate. Our presentation in this chapter is necessarily rudimentary. Readers who are intrigued by the subject are referred to the References given at the end of the chapter for further studies.

In Section 7.1, we look at Brownian motion from the perspective of a limiting process of a random walk. We also derive the diffusion equation underlying the Brownian motion process. The derivation is based on applying a Chapman-Kolmogorov type of argument to a Markov process. We show that the time-dependent solution of the probability density that the process is in a given state assumes a normal distribution. The next section complements the first one. It also presents a direct generalization of the diffusion process in which the resulting normal density has a time-dependent mean and variance. In Section 7.3, we introduce Ito's calculus and construct many Ito processes of interest. The following section extends Ito's calculus to multiple dimensions. The optimal control of Ito processes is given in Section 7.5. In our exposition, we present many examples in queueing approximations and continuous-time finance to illustrate the use of various ideas in applications.

7.1 Introduction

Before introducing Brownian motion processes, we first look at a random walk. Let $\{X_i\}$ be i.i.d. random variables with

$$P\{X_i = k\} = p \quad \text{and} \quad P\{X_i = -k\} = 1 - p = q,$$

where X_i denotes the size of the ith step, with probability p that the walk is toward the positive direction and with probability q toward the negative direction. It is easy to verify that $E[X_i] = (p - q)k$ and $Var[X_i] = 4pqk^2$. Let $Z_0 = 0$, and for $n = 1, 2, \ldots$, let $Z_n = X_1 + \cdots + X_n$. Then Z_n denotes the position of the random walk after n steps on a Z-n plane. The stochastic process $\{Z_n, n \geq 0\}$ is called a *random walk* process. We now extend this idea to continuous time. An interval of length t is divided into subintervals of length Δt. Let

$$Z(t) = X_1 + \cdots + X_{\left\lceil \frac{t}{\Delta t} \right\rceil}.$$

Assume t is large so that the term in the brackets can be approximated by an integer. Then we have

$$E[Z(t)] = \left(\frac{t}{\Delta t}\right)(p - q)k \quad \text{and} \quad Var[Z(t)] = \left(\frac{t}{\Delta t}\right)4pqk^2.$$

We now ask the question: How do we choose the values of p and k so that as $\Delta t \to 0$,

$$E[Z(t)] \to \mu t \quad \text{and} \quad Var[Z(t)] \to \sigma^2 t$$

or, equivalently,

$$\frac{(p-q)k}{\Delta t} \to \mu \quad \text{and} \quad \frac{4pqk^2}{\Delta t} \to \sigma^2.$$

The following choices provide the answer to our question:

$$p = \frac{1}{2}\left(1 + \frac{\mu\sqrt{\Delta t}}{\sigma}\right), \quad q = \frac{1}{2}\left(1 - \frac{\mu\sqrt{\Delta t}}{\sigma}\right), \quad \text{and} \quad k = \sigma\sqrt{\Delta t}.$$

Since $Z(t)$ is the sum of a large number of i.i.d. random variables, by the central limit theorem we know that $Z(t)$ follows a normal distribution for large t. By the nature of the random walk, we infer that for $u > t > 0$, the increments $Z(t) - Z(0)$ and $Z(u) - Z(t)$ are independent and the increment $Z(u) - Z(t)$ is asymptotically normal with mean $\mu(u - t)$ and variance $\sigma^2(u - t)$.

EXAMPLE   To visualize the limiting process described in the preceding exposition, we con-
7.1.1   sider the case with $\mu = 0$ and $\sigma^2 = 1$. This corresponds to a symmetric random walk with $p = 0.5$. The step size is governed by $k = \sigma\sqrt{\Delta t}$. In Figure 7.1, we plot the two simulated sample paths generated from the program shown in the Appendix, one with $\Delta t = 0.01$ and the other with $\Delta t = 0.001$. We see that the two sample paths are close to being continuous. As Δt decreases, the sample path becomes more "rugged." ∎

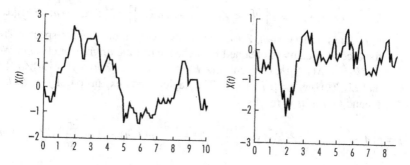

FIGURE
7.1   Random walks as approximations to standard Brownian motion processes (the right side of the figure has $\Delta t = 0.01$ and the left side $\Delta t = 0.001$).

The limiting process $\mathbf{Z} = \{Z(t),\ t \geq 0\}$ constructed in the previous manner is called a *Brownian motion process* or *Wiener process*. Brownian processes inherit the properties stated earlier in the context of being the limiting process of a random walk, namely (i) $Z(0) = 0$, (ii) $\mathbf{Z}$ has stationary and independent increments, and (iii) for every $t \geq 0$, $Z(t) \sim N(\mu t, \sigma^2 t)$. The parameters μ and σ^2 are called the *drift* and *variance parameters* (or *diffusion coefficient*) of the process, respectively. The sample path of Brownian motion $Z(t)$ is a continuous function of t, but $Z(t)$ is nowhere differentiable. To demonstrate the latter, we consider a Brownian motion process with $\mu = 0$ and $\sigma^2 = 1$. Then the random variable $Z(t + \Delta t) - Z(t)$ is normal with mean 0 and variance Δt. The random variable $[Z(t + \Delta t) - Z(t)]/\Delta t$ represents the rate of change of $Z(t)$ at t. We now see that

$$E\left[\frac{Z(t+\Delta) - Z(t)}{\Delta t}\right] = 0 \quad \text{and} \quad Var\left[\frac{Z(t+\Delta) - Z(t)}{\Delta t}\right] = \frac{1}{(\Delta t)^2}\Delta(t) = \frac{1}{\Delta t}$$

As $\Delta t \to 0$, the rate of change will have an infinite variance. In other words, the sample paths of $\mathbf{Z}$ are never smooth.

The Diffusion Equations for Brownian Motion

For discrete-time Markov chains, the multistep transition probabilities $\{p_{ij}^{(n)}\}$ govern the law of motion; for continuous-time Markov chains, the transition function $\{P_{ij}(t)\}$ does the job. The former is characterized by the Chapman-Kolmogorov equations and the latter by the Kolmogorov differential equations. We will now develop a similar system of equations, called the diffusion equations, for defining the law of motion of Brownian motion.

Let $f(x, t; y)$ denote the probability density of $Z(t)$ given $Z(0) = y$. In other words, we have $f(x, t; y)dx = P\{x < Z(t) < x + dx | Z(0) = y\}$. Using the law of total probability, we write $f(x, t; y)dx = E[P\{x < Z(t) < x + dx | Z(0) = y, Z(h)\}]$. By the independent and stationary increment properties, the last expression can be reduced to

$$f(x, t; y)dx = E\left[P\{x < Z(t) < x + dx | Z(h)\}\right] = E\left[f(x, t - h; Z(h))\right]dx$$

or $f(x, t; y) = E[f(x, t - h; Z(h))]$, where the expectation is with respect to $Z(h)$ and the second equality is obtained by a change of time origin from 0 to h. We note that $Z(h) \sim N(y + \mu h, \sigma^2 h)$ and $E[Z(h) - y] = \mu h$, $Var[Z(h) - y] = Var[Z(h)] = \sigma^2 h$, and $E[(Z(h) - y)^2] = \mu^2 h^2 + \sigma^2 h$. The Taylor series expansion of $f(x, t - h; Z(h))$ around $f \equiv f(x, t; y)$ results in

$$f = E\left[f - \frac{\partial f}{\partial t}h + \frac{\partial f}{\partial y}[Z(h) - y] + \frac{1}{2}\frac{\partial^2 f}{\partial t^2}h^2 + \frac{1}{2}\frac{\partial^2 f}{\partial y^2}[Z(h) - y]^2 - \frac{\partial^2 f}{\partial t\partial y}[Z(h) - y]h + \cdots\right]$$

$$= f - \frac{\partial f}{\partial t}h + \frac{\partial f}{\partial y}\mu h + \frac{1}{2}\frac{\partial^2 f}{\partial t^2}h^2 + \frac{1}{2}\frac{\partial^2 f}{\partial y^2}\left(\mu^2 h^2 + \sigma^2 h\right) - \frac{\partial^2 f}{\partial t\partial y}\mu h^2$$

$$= f - \frac{\partial f}{\partial t}h + \frac{\partial f}{\partial y}\mu h + \frac{1}{2}\frac{\partial^2 f}{\partial y^2}\sigma^2 h + o(h).$$

Dividing both sides of the last equality by h and taking the limit as $h \to 0$, we find the following backward diffusion equation

$$\frac{\partial}{\partial t} f(x, t; y) = \mu \frac{\partial}{\partial y} f(x, t; y) + \frac{1}{2}\sigma^2 \frac{\partial^2}{\partial y^2} f(x, t; y),$$

or in abbreviated form,

$$f_t = \mu f_y + \frac{1}{2}\sigma^2 f_{yy}, \tag{7.1.1}$$

where we use subscripts to indicate the respective partial derivatives.

As in the case of Markov chains, we call Equation 7.1.1 the backward diffusion equation since in its derivation the infinitesimal time segment h appears in the backward part of the overall time segment t. The forward diffusion equation can be derived similarly (see Problem 10). It is given by

$$\frac{\partial}{\partial t} f(x, t; y) = -\mu \frac{\partial}{\partial x} f(x, t; y) + \frac{1}{2}\sigma^2 \frac{\partial^2}{\partial x^2} f(x, t; y), \tag{7.1.2}$$

or in abbreviated form,

$$f_t = -\mu f_x + \frac{1}{2}\sigma^2 f_{xx}.$$

For the forward diffusion equation (Equation 7.1.2), we now perform a change of variable with $z = (x - y - \mu t)/\sigma$. Define q such that $f(x, t; y) = f(\sigma z + y + \mu t, t; y) = q(z, t)$. Using the chain rule, we write

$$f_x = q_z \frac{\partial z}{\partial x} = \frac{1}{\sigma} q_z$$

$$f_{xx} = \frac{1}{\sigma}\frac{\partial}{\partial x} q_z = \frac{1}{\sigma}\left[\frac{\partial^2 q}{\partial z^2}\frac{\partial z}{\partial x}\right] = \left(\frac{1}{\sigma^2}\right)q_{zz}$$

$$f_t = \frac{\partial}{\partial t} q(z, t) = \frac{\partial q}{\partial z}\frac{\partial z}{\partial t} + \frac{\partial q}{\partial t}\frac{\partial t}{\partial t} = -\frac{\mu}{\sigma} q_z + q_t.$$

Upon substituting the previous results into Equation 7.1.2, we find

$$-\frac{\mu}{\sigma} q_z + q_t = -\mu\left(\frac{1}{\sigma}\right)q_z + \frac{1}{2}\sigma^2\left(\frac{1}{\sigma}\right)^2 q_{zz} \qquad \text{or}$$

$$\frac{1}{2}\frac{\partial^2 q(z, t)}{\partial z^2} = \frac{\partial q(z, t)}{\partial t}.$$

The preceding partial differential equation is called the Fokker-Planck equation or the heat equation in physics; it occupies a prominent position in partial differential equations. Our next step is to solve the given equation for $q(z, t)$. For a fixed t, we define the moment generating function

$$M(s; t) = E[e^{sZ(t)}] = \int_{-\infty}^{\infty} q(z, t)e^{sz}dz,$$

where $Z(t) \sim q(z, t)$. We see that

$$\frac{\partial M(s; t)}{\partial t} = \frac{\partial}{\partial t}\int_{-\infty}^{\infty} q(z, t)e^{sz}dz = \int_{-\infty}^{\infty}\frac{\partial}{\partial t}q(z, t)e^{sz}dz = \int_{-\infty}^{\infty}\frac{1}{2}\frac{\partial^2}{\partial z^2}q(z, t)e^{sz}dz$$

$$= \frac{1}{2}\int_{-\infty}^{\infty}e^{sz}d\left(\frac{\partial}{\partial z}q(z, t)\right) = \frac{1}{2}e^{sz}\frac{\partial}{\partial z}q(z, t)\Big|_{-\infty}^{\infty} - \frac{1}{2}\int_{-\infty}^{\infty}\frac{\partial}{\partial z}q(z, t)se^{sz}dz$$

$$= 0 + \left\{-\frac{s}{2}\int_{-\infty}^{\infty}e^{sz}d(q(z, t))\right\}$$

$$= -\frac{s}{2}e^{sz}q(z, t)\Big|_{-\infty}^{\infty} + \frac{s}{2}\int_{-\infty}^{\infty}q(z, t)se^{sz}dz = \frac{1}{2}s^2M(s; t),$$

where the third equality is due to $q_t = (1/2)q_{zz}$ and the fifth and seventh equalities are resulting from integration by parts. In the preceding derivation, the first terms on the right side of the fifth and seventh equalities vanish because $q(z, t)$ is actually the standard normal density. The equation

$$\frac{\partial M(s; t)}{\partial t} = \frac{1}{2}s^2M(s; t)$$

has the initial condition $M(s; 0) = 1$. For a fixed s, we define the Laplace transform

$$M^e(s; v) = \int_0^{\infty} M(s; t)e^{-vt}dt.$$

Taking the previous Laplace transform, we obtain $vM^e(s; v) - M(s; 0) = (1/2)s^2M^e(s, v)$, $(v - (1/2)s^2)M^e(s; v) = 1$, or $M^e(s; v) = 1/[(v - (1/2)s^2]$. Upon inverting the last expression to the time domain, we find that the moment generating function for $q(z, t)$ is

$$M(s; t) = e^{(1/2)s^2t}$$

It is the moment generating function of the normal density with mean 0 and variance t (see Equation 1.3.10). Hence, we conclude

$$q(z, t) = \frac{1}{\sqrt{2\pi t}}e^{-\frac{z^2}{2t}}.$$

Performing a transformation of variable back to argument x using $z = (x - y - \mu t)/\sigma$, we find the solution to the forward diffusion equation (Equation 7.1.2)

$$f(x, t; y) = \frac{1}{\sqrt{2\pi\sigma^2 t}}e^{-\left\{\frac{(x-y-\mu t)^2}{2\sigma^2 t}\right\}}. \tag{7.1.3}$$

The preceding result implies that $Z(t) - Z(0) \sim N(\mu t, \sigma^2 t)$. We see that normality is a consequence of the independent and stationary increment properties of the process **Z**.

First Passage Time

Given that $Z(0) = 0$, we now study the first passage time T_x that Brownian motion **Z** reaches absorbing barrier x (with $x > 0$). Let $g(x) \equiv g^e(s) = E[e^{-sT_x}]$ denote the Laplace transform of T_x. By the independent and stationary properties of **Z**, we observe

$$E\left[e^{-sT_{x+y}}\right] = E\left[e^{-s[(T_{x+y}-T_y)+T_y]}\right] = E\left[e^{-s(T_{x+y}-T_y)}\right]E\left[e^{-T_y}\right] = E\left[e^{-sT_x}\right]E\left[e^{-sT_y}\right].$$

Thus, we have a functional equation $g(x + y) = g(x)g(y)$ with its solution given by $g(x) = e^{-cx}$ for some $c > 0$. To find c, we condition on $Z(h)$, where h is an initial time segment. Assuming that the probability that the first passage to x in the initial time segment h is negligible, we obtain

$$g(x) = E_{Z(h)}\left[E\left[e^{-sT_x}|Z(h)\right]\right] = E_{Z(h)}\left[E\left[e^{-s(h+T_{x-Z(h)})}\right]\right]$$

$$= e^{-sh}E_{Z(h)}\left[E\left[e^{-s(T_{x-Z(h)})}\right]\right] = e^{-sh}E_{Z(h)}\left[g(x - Z(h))\right]. \qquad (7.1.4)$$

Using a Taylor series, we expand $g(x - Z(h))$ around $g(x)$ and obtain

$$g(x - Z(h)) = g(x) - Z(h)g'(x) + \frac{1}{2}Z(h)^2 g''(x) + \cdots.$$

With $e^{-sh} = 1 - sh + o(h)$, $E[Z(h)] = \mu h$, and $E[Z(h)^2] = \sigma^2 h + \mu^2 h^2$, we rewrite Equation 7.1.4 as

$$g(x) = (1 - sh)\left[g(x) - \mu h g'(x) + \frac{1}{2}\sigma^2 h g''(x)\right] + o(h)$$

$$= g(x) - shg(x) - \mu h g'(x) + \frac{1}{2}\sigma^2 h g''(x) + o(h).$$

Dividing the last equality by h and taking the limit as $h \to 0$, we obtain the following second-order linear differential equation:

$$sg(x) = -\mu g'(x) + \frac{1}{2}\sigma^2 g''(x).$$

Since $g(x) = e^{-cx}$, we have $g'(x) = -ce^{-cx}$ and $g''(x) = c^2 e^{-cx}$. Substituting these expressions in the previous differential equation and dividing the result so obtained by e^{-cx}, we find the following quadratic equation defining c:

$$\sigma^2 c^2 + 2\mu c - 2s = 0.$$

The root of the quadratic equation is given by

$$c = \frac{-\mu \pm \sqrt{\mu^2 + 2s\sigma^2}}{\sigma^2}.$$

Consider the case in which the drift μ is strictly positive. Since $s > 0$, we have

$$c = \frac{-\mu + \sqrt{\mu^2 + 2s\sigma^2}}{\sigma^2} > 0.$$

Summarizing the preceding derivation, for a Brownian motion process with drift $\mu > 0$ and variance parameter σ^2 and $Z(0) = 0$, the Laplace transform of T_x is given by

$$E[e^{-sT_x}] = e^{-\frac{x}{\sigma^2}\left[\sqrt{\mu^2 + 2s\sigma^2} - \mu\right]},$$

(7.1.5)

where T_x is the first passage time to barrier $x > 0$.

An inversion of Equation 7.1.5 will yield the density $g(t|0, x)$ for the first passage time T_x with

$$g(t|0, x) = \frac{x}{\sigma\sqrt{2\pi t^3}} e^{-\frac{(x-\mu t)^2}{2\sigma^2 t}} \qquad t > 0.$$

(7.1.6)

Using the Laplace transform (Equation 7.1.5), we find

$$E[T_x] = \frac{x}{\mu} \qquad \text{and} \qquad Var[T_x] = \frac{x\sigma^2}{\mu^3}.$$

(7.1.7)

Given $Z(0) = 0$ and the barrier x, we observe that the mean and variance of the first passage time depend only on the mean and variance of the Brownian motion process. From the vantage point of random walk, an intuitive interpretation of Equation 7.1.7 could have been constructed using the asymptotic mean (Equation 3.5.1) and variance (Equation 3.5.2) of $N(t)$ associated with a renewal process. If the Brownian motion process starts with $Z(0) = y$ ($y < x$ with x as the barrier), then we construct a related Brownian motion $Z_1(t) = Z(t) - y$. The two processes $Z_1 = \{Z_1(t), t \geq 0, Z_1(0) = 0\}$ and $Z = \{Z(t), t \geq 0, Z(0) = y\}$ are probabilistically equivalent and depicted in Figure 7.2.

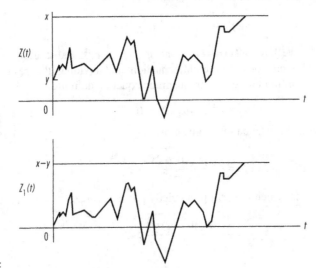

FIGURE

7.2   Two probabilistic equivalent Brownian motion processes.

They differ only by a translation. Z_1 reaching x is equivalent to Z reaching $x - y$. Let T_x^1 denote the first passage time to x for the process Z_1. Then Equations 7.1.5, 7.1.6, and 7.1.7 hold with $y - x$ replacing x on the respective right sides.

EXAMPLE
7.1.2

A Database Degradation Model Consider a database that can hold a maximum of M records. There are three types of requests made to the database: add, delete, or query a record. In our model, we assume that when a record is deleted its space becomes available for subsequent use (this is known as on-the-fly maintenance— as opposed to simply tagging the record as being deleted). For a request, we let p denote the probability that it is for addition, q for deletion, and r for query. Let Z_n denote the number of records in the database after the nth request and X_n the change in the number of records contained in the database resulting from the nth request. Then we see that $Z_n = Z_0 + X_1 + \cdots + X_n$, where Z_0 is the initial size of the database and

$$P\{X_n = +1\} = p, \qquad P\{X_n = 0\} = r, \qquad \text{and} \qquad P\{X_n = -1\} = q.$$

The stochastic process $\{Z_n, n \geq 0\}$ is a random walk on the set of integers $\{0, 1, ..., M\}$. Let $M + 1$ denote the state of database overflow. The time to database overflow is the first passage time from state Z_0 to state $M + 1$. The problem can be solved by modeling the process as an absorbing Markov chain with absorbing state $M + 1$. When the value of M is large, we now use Brownian motion to obtain approximate solutions.

With $E[X_n] = (+1)p + (-1)q = p - q$, $E[X_n^2] = (+1)^2 p + (-1)^2 q = p + q$, and $Var[X_n] = (p + q) - (p - q)^2$, we set $\mu = p - q$ and $\sigma^2 = p + q - (p - q)^2$. We then approximate the random walk $\{Z_n, n \geq 0\}$ by the Brownian motion process $\{Z(t), t \geq 0\}$ with drift μ and variance parameter σ^2, where $Z(t)$ denotes the number of records in the database at time t. With $Z(0) = Z_0$ and an absorbing barrier $x = M + 1$, the results obtained earlier enable us to compute the time to database overflow statistics.

We now consider a numerical example. Suppose that 90 percent of all requests do not add or delete records and 90 percent of the remaining requests add records. This gives $p = 0.09$, $q = 0.01$, and $r = 0.9$. Consider a database that can hold a maximum of six hundred records and starts with 50 percent of the space occupied. In this case, we have $Z(0) = y = 300$ and an absorbing barrier $x = 601$ or, equivalently, with $x = 601 - 300 = 301$ in Equations 7.1.5 through 7.1.7. Let T denote the database overflow time (in units of number of requests). Then we find

$$E[T] = \frac{301}{0.08} = 3762.5 \qquad \text{and} \qquad Var[T] = \frac{(301)(0.0936)}{(0.08)^3} = 55026.56.$$

This shows that the variance of T is substantially larger than its mean. Using Equation 7.1.6, we compute the probability density of $g_T(t)$ and plot the results in Figure 7.3. Suppose that the database processes, on average, five hundred requests per day (the figure suggests that it is likely the database overflow will occur in about nine days). ∎

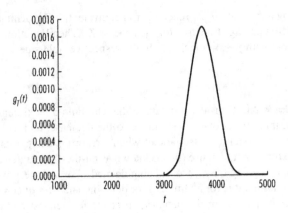

FIGURE
7.3   Time to database overflow density.

EXAMPLE
7.1.3

The Worth of Perpetual Warrants A warrant is an option to buy a fixed number of shares of a given stock at a stated price during a specified interval. In this example, we consider the case in which warrants do not have expiration dates. These warrants are called perpetual warrants. For simplicity, we assume that the stated price in the warrant is one. Let $Y(t)$ denote the market price of the stock at time t. If the holder of such an option decides to exercise the option at time t, the resulting profit would be $Y(t) - 1$. One strategy for the holder of the warrant is to exercise the option the first time $Y(t)$ reaches a prespecified level x. This produces a profit of $x - 1$ for the holder. It is obvious that we need only to consider the case for which $x > 1$. Let T_x denote the first time the market price of the stock reaches x and β the discount factor to be used by the holder of the option. Then discounted profit to the holder of the option is given by $e^{-\beta T_x}[Y(T_x) - 1] = e^{-\beta T_x}[x - 1]$.

We now describe the laws of motion for stochastic process $Y = \{Y(t), t \geq 0\}$. We assume that $Y(t) = e^{Z(t)}$, where $\{Z(t), t \geq 0\}$ is a Brownian motion process with drift μ, variance parameter σ^2, and $Z(0) = \ln y$. Process Y is a geometric Brownian motion process. More will be said about this process in the sequel. Since $Z(t) - Z(0) \sim N(\mu t, \sigma^2 t)$, by setting the argument of the moment generating function (Equation 1.3.10) of the corresponding normal density to one, we find

$$E[e^{Z(t)-Z(0)}] = e^{\left(\mu + \frac{1}{2}\sigma^2\right)t} \quad \text{and} \quad E[e^{Z(t)}] = E[e^{Z(t)-Z(0)+Z(0)}] = y e^{\left(\mu + \frac{1}{2}\sigma^2\right)t}.$$

This implies

$$E[Y(t)|Y(0) = y] = E[e^{Z(t)}|Z(0) = \ln y] = y e^{\left(\mu + \frac{1}{2}\sigma^2\right)t}.$$

In the last expression, we see that the stock is increasing in value at a rate of $\mu + (1/2)\sigma^2$ per unit time. For owning the option to be sensible, we would require that the holder's discount rate β be greater than $\mu + (1/2)\sigma^2$.

By definition, T_x is the first time $Y(t)$ reaches x. It is also the first time $\ln Y(t)$ reaches $\ln x$. Since $Z(t) = \ln Y(t)$, T_x is also the first time $Z(t)$ reaches $\ln x$ given that $Z(0) = \ln y$. Moreover, T_x is probabilistically equivalent to the first passage time to $\ln x - \ln y = \ln(x/y)$ for Brownian motion $\{Z'(t), t \geq 0\}$ with drift μ, variance parameter σ^2, and $Z'(0) = 0$. Using Equation 7.1.5, we find that

$$E[e^{-\beta T_x}] = e^{-\ln\left(\frac{x}{y}\right)z} = \left(\frac{y}{x}\right)^z,$$

where

$$z = \frac{1}{\sigma^2}\left[\sqrt{\mu^2 + 2\beta\sigma^2} - \mu\right].$$

The expected discounted profit is given by

$$g(x, y) = (x-1)E\left[e^{-\beta T_x} \,\middle|\, Y(0) = y\right] = (x-1)\left[\frac{y}{x}\right]^z.$$

The value of x that maximizes the expected discounted profit can be found from the solution of

$$\frac{\partial}{\partial x}g(x, y) = \left(\frac{y}{x}\right)^z - (x-1)\left(\frac{z}{x}\right)\left(\frac{y}{x}\right)^z = 0.$$

This gives the optimizing value $x^* = z/(z-1)$. Simple algebra will show that $z > 1$ or, equivalently, $(1/\sigma^2)[\sqrt{\mu^2 + 2\beta\sigma^2} - \mu] > 1$, is guaranteed by the condition $\beta > \mu + (1/2)\sigma^2$ described earlier. If the current stock price is y, then the worth of the warrant is given by

$$g(x^*, y) = \left(\frac{1}{z-1}\right)\left(\frac{y(z-1)}{z}\right)^z.$$

We now consider a numerical example. Assume that the price $Y(t)$ of a perpetual warrant at time t follows a geometric Brownian motion process with $Y(t) = \exp(Z(t))$, where $Z(t) - Z(0) \sim N(\mu t, \sigma^2 t)$, $\mu = 0.07$, and $\sigma^2 = 0.02$, and $Y(0) = 0.7$. In the Appendix, we simulate the price movement of this warrant over the interval $[0, 10]$ and plot the results in Figure 7.4. We address the numerical valuation of this particular warrant in Problem 13. ■

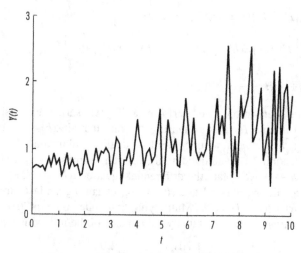

FIGURE
7.4 The price of a perpetual warrant (a simulated geometric Brownian motion process).

Brownian Motion with Two Barriers

Consider Brownian motion $Z = \{Z(t), t \geq 0\}$ with drift $\mu \neq 0$ and variance parameter σ^2. Assume that $Z(0) = y$ and there are two barriers $x > 0$ and $z > 0$. We are interested in finding the probability $H(y)$ that $Z(t)$ reaches x before reaching $-z$ given that $Z(0) = y$. Such a scenario is depicted in Figure 7.5. Let T_x denote the first passage time to x and T_z the first passage time to $-z$ and $T_{xz} = \min\{T_x, T_z\}$. Then we have $H(y) = P\{Z(T_{xz}) = x | Z(0) = y\}$. Conditioning on the value $Z(h)$ occurring in the initial infinitesimal time segment h, we write

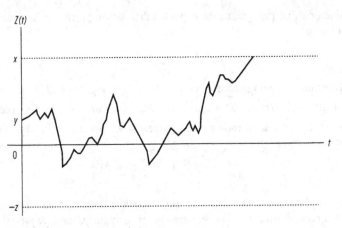

FIGURE
7.5 A Brownian motion process with two barriers.

$$H(y) = E_{Z(h)}\left[P\{Z(T_{xz}) = x | Z(0) = y, Z(h)\}\right] = E_{Z(h)}\left[P\{Z(T_{xz}) = x | Z(h)\}\right]$$

$$= E_{Z(h)}\left[H(Z(h))\right] = E_{Z(h)}\left[H\left(y + (Z(h) - y)\right)\right]$$

$$= E_{Z(h)}\left[H(y) + [Z(h) - y]H_y + \frac{1}{2}[Z(h) - y]^2 H_{yy} + \cdots\right]$$

$$= H(y) + \mu h H_y + \frac{1}{2}\sigma^2 h H_{yy} + o(h). \tag{7.1.8}$$

In the previous derivation, we perform a Taylor series expansion of $H(y + (Z(h) - y))$ around $H(y)$. We then use the fact that $E[Z(h) - y] = \mu h$ and $E[(Z(h) - y)^2] = \mu^2 h^2 + \sigma^2 h$ in writing the last equality. Subtracting $H(y)$ from both sides of Equation 7.1.8 and dividing the resulting expression by h and taking the limit as $h \to 0$, we obtain the differential equation $\mu H_y + (1/2)\sigma^2 H_{yy} = 0$ with boundary conditions $H(x) = 1$ and $H(-z) = 0$. Integrating the last expression once yields $(2\mu/\sigma^2)H(y) + H_y = C_1$. Multiplying both sides by $\exp(2\mu y/\sigma^2)$, we note that the resulting left side is actually the left side of the following equality

$$\frac{d}{dy} H(y) \exp\left(\frac{2\mu y}{\sigma^2}\right) = C_1 \exp\left(\frac{2\mu y}{\sigma^2}\right).$$

Upon integrating the preceding, we find

$$H(y)\exp\left(\frac{2\mu y}{\sigma^2}\right) = C_1 \exp\left(\frac{2\mu y}{\sigma^2}\right) + C_2 \quad \text{or} \quad H(y) = C_1 + C_2 \exp\left(-\frac{2\mu y}{\sigma^2}\right).$$

Using the two boundary conditions, finding constants C_1 and C_2 is straightforward. This gives the desired result

$$H(y) = \frac{\exp\left(-\dfrac{2\mu y}{\sigma^2}\right) - \exp\left(\dfrac{2\mu z}{\sigma^2}\right)}{\exp\left(-\dfrac{2\mu x}{\sigma^2}\right) - \exp\left(\dfrac{2\mu z}{\sigma^2}\right)}. \tag{7.1.9}$$

For the special case in which $Z(0) = 0$ and $\sigma^2 = 1$, Equation 7.1.9 reduces to

$$H(0) = \text{Prob}\{\text{the process hits } x \text{ before going down to } -z\} = \frac{1 - e^{2\mu z}}{e^{-2\mu x} - e^{2\mu z}}. \tag{7.1.10}$$

If $\mu < 0$, then $e^{2\mu z} \to 0$ as $z \to \infty$. Thus Equation 7.1.10 implies

$$\text{Prob}\{\text{the process reaches } x\} = e^{2\mu x}. \tag{7.1.11}$$

Therefore, if the process has a negative drift, the probability that it will never reach x is given by $1 - e^{2\mu x}$.

7.2 Diffusion Processes

Consider a continuous-time stochastic process $X = \{X(t), t \geq 0\}$ with a state space S. A *stopping time* T of X satisfies the property that for each t, the occurrence or nonoccurrence of the event $\{T \leq t\}$ is determined by the random variables $\{X(u); u \leq t\}$. If the stochastic process X possesses the Markov property at stopping times, then the process is said to possess a *strong Markov property*. In such cases, we have

$$P\{X(T+s) \in A \mid X(u); \ u \leq T\} = P\{X(s) \in A \mid X(0) = X_T\},$$

where X_T denotes the state of the system at stopping time T and A is any subset of S.

> A stochastic process $X = \{X(t), t \geq 0\}$ is a *diffusion process* if (i) the sample paths of X are continuous and (ii) X possesses the strong Markov property.

The most fundamental diffusion process is Brownian motion (the Wiener process). A continuous-time continuous-state Markov process is also a diffusion process. For such a Markov process $X = \{X(t), t \geq 0\}$, we define the transition probability $F(x, t; y) = P\{X(t) \leq x \mid X(0) = y\}$. Starting with the Chapman-Kolmogorov equation, the following forward Kolmogorov equation can be derived for the Markov process X:

$$\frac{\partial}{\partial t} F(x, t; y) = -\mu \frac{\partial}{\partial x} F(x, t; y) + \frac{1}{2}\sigma^2 \frac{\partial^2}{\partial x^2} F(x, t; y) \tag{7.2.1}$$

with the boundary condition $F(x, 0; y)$ for $x < y$ and 1 if $x \geq y$ (since $X(0) = y$). We see that $F(x, t; y)$ satisfies the forward diffusion equation (Equation 7.1.3) with the stated boundary condition. In the sequel, Brownian motion and Markov processes will occasionally be referred to as diffusion processes.

The diffusion process $\{|X(t)|, t \geq 0\}$ is said to have a reflecting barrier. Its law of motion again follows Equation 7.2.1 but the boundary condition is given by $F(0+, t; y) = 0$ for all $y > 0$ and $t > 0$. For $\mu \neq 0$, the time-dependent solution of Equation 7.2.1 is

$$F(x, t; y) = \Phi\left(\frac{x - y - \mu t}{\sqrt{\sigma^2 t}}\right) - e^{\frac{2x\mu}{\sigma^2}} \Phi\left(\frac{-x - y - \mu t}{\sqrt{\sigma^2 t}}\right), \qquad (7.2.2)$$

where $\Phi(x)$ denotes the standard normal distribution evaluated at x. For the limiting probability $F(x) = \lim\limits_{t \to \infty} F(x, t; y)$, the left side of Equation 7.2.1 vanishes. Thus, for $x > 0$, the differential equation characterizing the limiting distribution is given by

$$0 = -\mu \frac{\partial}{\partial x} F(x) + \frac{1}{2}\sigma^2 \frac{\partial^2}{\partial x^2} F(x). \qquad (7.2.3)$$

When $\mu < 0$, the limiting distribution F can be found by noting $\Phi(\infty) = 1$ in Equation 7.2.2 and hence

$$F(x) = 1 - e^{\frac{2x\mu}{\sigma^2}} = 1 - e^{-\left(\frac{2|\mu|}{\sigma^2}\right)x}. \qquad (7.2.4)$$

The preceding shows that when the diffusion process is moved about positive values and with drift $\mu < 0$, the asymptotic state density is exponential with mean

$$\lim_{t \to \infty} E[X(t)|X(0) = y] = \frac{\sigma^2}{2|\mu|}. \qquad (7.2.5)$$

EXAMPLE 7.2.1

Diffusion Approximation of Queue Length of a GI/G/1 Queue under Heavy Traffic Consider a single-server system in which the interarrival times $\{X_i\}$ of customers follow a renewal process with $E[X] = 1/\lambda$ and $Var[X] = \sigma_a^2$, and service times $\{S_i\}$ are i.i.d. random variables and follow a common distribution G with $E[S] = 1/\mu_s$ and $Var[S] = \sigma_s^2$. Assume that there is an unlimited number of waiting spaces, and $\{X_i\}$ and $\{S_i\}$ are mutually independent.

Let $X(t)$ denote the number of customers in the system at time t, $A(t)$ the number of arrivals by time t, and $D(t)$ the number of departures by time t. We assume that $X(0) = 0$. Then we have $X(t) = A(t) - D(t)$. Under the heavy traffic assumption that the traffic intensity ρ is close to one, $X(t)$ does not reach 0, the interdeparture time approximately follows the service time distribution G, and $\{A(t), t \geq 0\}$ and $\{D(t), t \geq 0\}$ are mutually independent. Following Equations 3.5.1 and 3.5.2, we see that for large t, $E[A(t)] = \lambda t$, $Var[A(t)] = \sigma_a^2\lambda^3 t$, $E[D(t)] = \mu_s t$, $Var[D(t)] = \sigma_s^2\mu_s^3 t$, and $A(t)$ and $D(t)$ are asymptotically normal. This implies that for a fixed but large t, $X(t)$ is normal with $E[X(t)] = (\lambda - \mu_s)t$ and $Var[X(t)] = (\sigma_a^2\lambda^3 + \sigma_s^2\mu_s^3)t$. We now conclude that $\{X(t), t \geq 0\}$ can be approximated by a diffusion process with drift

$\mu = \lambda - \mu_s$ and variance parameter $\sigma^2 = \sigma_a^2 \lambda^3 + \sigma_s^2 \mu_s^3$. Assume that $\mu_s > \lambda$ so that the queue is stable. This means that the drift $\mu < 0$. Of course, under the heavy traffic assumption, $|\mu|$ will be very small. The model so constructed satisfies the conditions leading to Equation 7.2.4. Thus we conclude that the limiting distribution for $X(t)$ is exponential with mean given by Equation 7.2.5, where

$$\lim_{t \to \infty} E[X(t)] \equiv \frac{1}{\tau} = \frac{\sigma_a^2 \lambda^3 + \sigma_s^2 \mu_s^3}{2(\mu_s - \lambda)}.$$

The previous expression is the mean queue length at any time. Let $\hat{p}_n$ denote the approximated value of $\lim_{t \to \infty} P\{X(t) = n\}$. For $n = 0, 1, \ldots$, we discretize the limiting distribution F and find

$$\hat{p}_n = F(n+1) - F(n) = (1 - e^{-(n+1)\tau}) - (1 - e^{-n\tau}) = e^{-n\tau}(1 - e^{-\tau}) = \hat{\rho}^n(1 - \hat{\rho}), \quad (7.2.6)$$

where $\hat{\rho} = e^{-\tau}$.

Let $\rho = \lambda/\mu_s$ denote the traffic intensity of the queue. From queueing theory, it is known that $p_0 = 1 - \rho$. An improvement over Equation 7.2.6 is obtained by letting $\hat{p}_n = c\hat{\rho}^n(1 - \hat{\rho})$, $n = 1, 2, \ldots$ and summing $\hat{p}_1 + \hat{p}_2 + \cdots$ to obtain the normalizing constant $c = \rho/\hat{\rho}$. This finetunes the approximation and yields

$$\hat{p}_0 = 1 - \rho \quad \text{and} \quad \hat{p}_n = \rho\hat{\rho}^{n-1}(1 - \hat{\rho}) \qquad n = 1, 2, \ldots. \quad (7.2.7)$$

We now look at a numerical example. Assume that the interarrival time follows a phase-type distribution with representation (α, T) and the service time follows a phase-type distribution with representation (β, S), where $\alpha = (1\ 0\ 0\ 0\ 0)$, $\beta = (0.7\ 0\ 0\ 0\ 0.3\ 0)$,

$$T = \begin{bmatrix} -4.75 & 4.75 & & & \\ & -4.75 & 4.75 & & \\ & & -4.75 & 4.75 & \\ & & & -4.75 & 4.75 \\ & & & & -4.75 \end{bmatrix}$$

$$S = \begin{bmatrix} -6.6778 & 6.6778 & & & & \\ & -6.6778 & 6.6778 & & & \\ & & -6.6778 & 6.6778 & & \\ & & & -6.6778 & & \\ & & & & -1.0332 & 1.0332 \\ & & & & & -1.0332 \end{bmatrix}.$$

We see that $PH(\alpha, T)$ is Erlang $(5, 4.75)$ and $PH(\beta, S)$ is a probabilistic mixture of Erlang $(4, 6.6778)$ and Erlang $(2, 1.0332)$. Using Equations 5.5.6 and 5.5.7, we find $E[X] = 1.0526$, $Var[X] = 0.2216$, $E[S] = 1$, $Var[S] = 1$. This gives traffic intensity $\rho = 0.95$. By rescaling the parameter λ of Erlang $(5, \lambda)$ of the arrival process, we construct another set of data with $\rho = 0.98$. For these two $PH/PH/1$ queues, the limiting queue length distributions at any time t can be computed exactly as shown in the Appendix. In Figure 7.6 we plot the actual distributions against the diffusion approximations. The figures show that the approximations are adequate. Moreover, as the traffic intensity increases, the approximation improves. For example, the maximal difference $p_n - \hat{p}_n$ in each case occurs at $n = 1$. When $\rho = 0.95$, the difference is 0.0144; when ρ moves to 0.98, the difference reduces to 0.0061. ∎

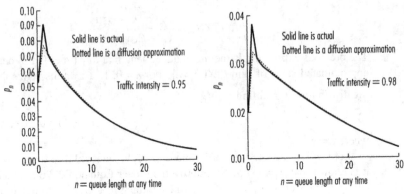

FIGURE
7.6   Limiting queue length distributions: actual versus diffusion approximations.

EXAMPLE
7.2.2   **Diffusion Approximation of Virtual Delay of a GI/G/1 Queue under Heavy Traffic**   We consider the $GI/G/1$ queue once more. The virtual delay at any time t is the delay in queue a customer must experience if the customer arrives at time t. Let $L(t)$ denote the amount of work brought to the server in $(0, t]$. The virtual delay at time t is the server's remaining work at time t if the service discipline is FIFO. In Figure 7.7, we display a typical sample path of the stochastic process $\{V(t), t \geq 0\}$. The figure shows that at the absence of an arrival the work is depleting at a rate of 1 (the slope of the sample path when $V(t) > 0$ and excluding the arrival epochs is -1).

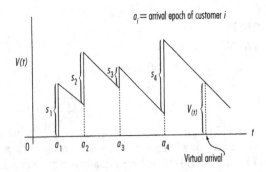

FIGURE
7.7   A sample path of $\{V(t), t \geq 0\}$.

To derive a diffusion model to approximate the virtual delay distribution, we note that $L(t) = S_1 + \cdots + S_{A(t)}$. Since $\{S_i\}$ are i.i.d. random variables, for large t, we have

$$E[L(t)] = E[S]E[A(t)] = \frac{1}{\mu_s}\lambda t = \rho t.$$

The formula $Var[L(t)] = E[Var[L(t)]|A(t)] + Var[E[L(t)]|A(t)]$ enables us to find the variance of $L(t)$. Since $Var[L(t)|A(t)] = A(t)\sigma_s^2$, we have $E[Var[L(t)|A(t)]] = E[A(t)]\sigma_s^2 = \lambda t\sigma_s^2$ for large t. Also, since $E[L(t)|A(t)] = A(t)E[S] = A(t)(1/\mu_s)$, we have $Var[E[L(t)|A(t)]] = (1/\mu_s)^2$ and $Var[A(t)] = (1/\mu_s)^2\sigma_a^2\lambda^3 t = \lambda\rho^2\sigma_a^2 t$. Combining the two parts, we find that for large t,

$$Var[L(t)] = \lambda\left(\sigma_s^2 + \rho^2\sigma_a^2\right)t \equiv \sigma^2 t.$$

The preceding results suggest

$$\lim_{\Delta t \to 0}\left\{\frac{E[L(t+\Delta t)] - E[L(t)]}{\Delta t}\right\} = \rho \quad \text{and}$$

$$\lim_{\Delta t \to 0}\left\{\frac{Var[L(t+\Delta t)] - Var[L(t)]}{\Delta t}\right\} = \sigma^2$$

We observe that work is accumulating at a rate of ρ and depleting at a rate of 1. Therefore the remaining work $V(t)$ is being generated at a rate of $\rho - 1$. Moreover, the variance associated with $V(t)$ is induced only by $L(t)$. Hence we conclude that the process $\{V(t), t \geq 0\}$ can be modeled as a diffusion process with drift $\mu = \rho - 1$ and variance parameter σ^2. The stability condition $\rho < 1$ ensures that $\mu < 0$. Consequently, Equation 7.2.4 applies and we conclude that the virtual delay follows an exponential distribution with its mean given by Equation 7.2.5.

Let $F(x)$ denote the diffusion approximation of virtual delay distribution with its mean $1/\gamma$ given by Equation 7.2.5. We can finetune the approximation by noting that the actual distribution has a probability mass of $1 - \rho$ at 0. Therefore a modified approximation has

$$\hat{F}(x) = \begin{cases} 1 - \rho & x = 0 \\ 1 - \rho e^{-\gamma x} & x > 0. \end{cases}$$

Using the data set described in Example 7.2.1, we compute the actual virtual waiting time distributions and their respective diffusion approximations in the Appendix. The results are displayed in Figure 7.8. ∎

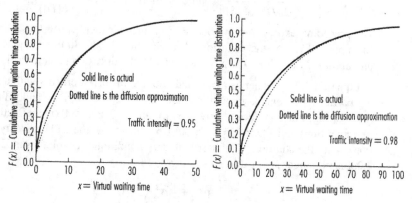

FIGURE

7.8 Limiting virtual waiting time distribution: actual versus diffusion approximations.

EXAMPLE
7.2.3 **Diffusion Approximation of a Multiprogramming System**   Consider a cyclic queueing system consisting of two queues in tandem (see Figure 7.9). The first queue is a CPU (central processing unit) and the second a DTU (data transfer unit). The system is under heavy traffic and can accommodate J programs. Each program goes through both queues in sequence and then returns to the first queue. When a program completes the service at the CPU, it moves to the tail of the DTU queue. After the DTU service, it moves back to the tail of the CPU queue. Both queues use a FIFO discipline. The service times at the CPU are i.i.d. random variables with mean $E[S]$ and variance $Var[S]$ and the service times at the DTU are i.i.d. random variables with mean $E[D]$ and variance $Var[D]$. Moreover, the two sets of service times are mutually independent.

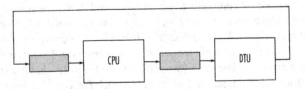

FIGURE
7.9   A multiprogramming system.

Let $A(t)$ denote the number of arrivals at the CPU in $(0, t]$ and $D(t)$ the number of CPU service completions in $(0, t]$. Under heavy traffic, the interarrival times at the CPU are the service times at the DTU. Following the argument given in Example 7.2.1, we conclude that $A(t) \sim N(t/E[D], tVar[D]/(E[D])^3)$ and $D(t) \sim N(t/E[S], tVar[S]/(E[S])^3)$ for large t. Let $X(t)$ denote the queue length at the CPU at time t. With $X(0) = 0$, we have $X(t) = A(t) - D(t)$. Then X is approximately normally distributed with

$$E[X(t)] = \mu t = \left[\frac{1}{E[D]} - \frac{1}{E[S]} \right] t \quad \text{and} \quad Var[X(t)] = \sigma^2 t = \left[\frac{Var[D]}{(E[D])^3} + \frac{Var[S]}{(E[S])^3} \right] t.$$

The preceding expressions are derived in a manner identical to those shown in Example 7.2.1. For the diffusion approximation of the limiting queue length probability $F(x) = \lim_{t \to \infty} F(x, t; y)$ at the CPU, the diffusion equation is identical to Equation 7.2.3—which is associated with the diffusion process with a reflecting barrier. The boundary conditions are $F(x, 0; y) = 0$ for $x < y$ and 1 for $x \geq 1$ (the starting state condition), $F(0+, t; y)$ for all $y > 0$ and $t > 0$ (the reflecting barrier condition) and $F(J, t; y) = 1$ for all $y > 0$ and $t > 0$ (the CPU capacity condition). With the addition of the last boundary condition, the solution of Equation 7.2.4 is no longer applicable. In solving

$$-\mu F_x + \frac{1}{2}\sigma^2 F_{xx} = 0, \tag{7.2.8}$$

we mimic the approach leading to Equation 7.1.8 except that after integrating Equation 7.2.8 once, we multiply both sides of the resulting equation by

$\exp(-2\mu x/\sigma^2)$ (as opposed to by $\exp(2\mu x/\sigma^2)$ because of the negative sign in Equation 7.2.8). This yields the solution $F(x) = A[1 - B \exp(2\mu x/\sigma^2)]$. The upper boundary condition requires $1 = A[1 - B \exp(2\mu J/\sigma^2)]$ or $A = [1 - B \exp(2\mu J/\sigma^2)]^{-1}$. Thus we obtain

$$F(x) = \frac{1 - B\exp(2\mu x/\sigma^2)}{1 - B\exp(2\mu J/\sigma^2)}, \tag{7.2.9}$$

where the last unknown constant B is to be determined by the lower boundary condition $F(0+)$. We consider three cases.

When $J = 1$, we consider a regeneration cycle whose length is the time between two successive entrances to CPU. Applying the limit theorem for regenerative processes, we have

$$F(0+) = P\{CPU \text{ is idle}\} = \frac{E[D]}{E[D] + E[S]}.$$

With $x = 0$ in the right side of Equation 7.2.9 and the right side of the last expression replacing the left side of Equation 7.2.9, we can easily find the constant B.

When $J = 2$, we let a regeneration point be the epoch at which one program starts from the CPU and the other program starts from the DTU simultaneously. The length of a regeneration cycle is then given by $\max\{S, D\}$. A glance at Figure 7.10 will show why the last assertion is true.

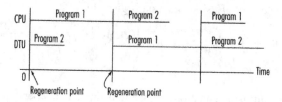

FIGURE
7.10 Regeneration cycles for $J = 2$.

The limit theorem of regenerative processes gives

$$1 - F(0+) = \frac{E[S]}{E[\max(S, D)]}.$$

The constant B can now be found. For a large J, the denominator of Equation 7.2.9 is approximately 1 (recall $\mu < 0$) and $F(x) = 1 - B \exp(2\mu x/\sigma^2)$. We use the argument given in Example 7.2.1 and propose $F(0+) = 1 - B = 1 - \rho$, where $\rho = E[S]/E[D] < 1$. Hence $B = \rho$. ∎

Generalizations of Diffusion Processes

The diffusion equations (Equations 7.1.1 and 7.1.2) can be traced back to the random walk in which the law of motion is governed by transition probabilities that are independent of state k and transition time i, namely, we have $P\{X_i = k\} = p$ and $P\{X_i = -k\} = q$. Starting with a random walk with state dependent transition probabilities, diffusion processes whose law of motion are state dependent can be constructed similarly. Define a diffusion process $X = \{X(t), t \geq 0\}$ with

$$\mu(x) = \lim_{\Delta t \to 0} \frac{E\left[\{X(t+\Delta t) - X(t)\}|X(t) = x\right]}{\Delta t} \tag{7.2.10}$$

and

$$\sigma^2(x) = \lim_{\Delta t \to 0} \frac{Var\left[\{X(t+\Delta t) - X(t)\}|X(t) = x\right]}{\Delta t}. \tag{7.2.11}$$

We see in the preceding generalization that instantaneous mean and variance depend on the state of the process $X(t) = x$, whereas in the diffusion model defined in Section 7.1 we have $\mu(x) = \mu$ and $\sigma^2(x) = \sigma^2$. Again, we let $f(x, t; y)$ denote the probability density of $X(t)$ given $X(0) = y$.

The forward diffusion equation associated with this diffusion process is given by

$$\frac{\partial}{\partial t} f(x, t; y) = -\frac{\partial}{\partial x}\{\mu(x)f(x, t; y)\} + \frac{1}{2}\frac{\partial^2}{\partial x^2}\{\sigma^2(x)f(x, t; y)\}. \tag{7.2.12}$$

Similarly, the backward diffusion equation is given by

$$\frac{\partial}{\partial t} f(x, t; y) = \mu(x)\frac{\partial}{\partial y} f(x, t; y) + \frac{1}{2}\sigma^2(x)\frac{\partial^2}{\partial y^2} f(x, t; y). \tag{7.2.13}$$

As expected, the two diffusion equations reduce to their state independent counterparts when the instantaneous mean and variance are state independent. The solutions of these diffusion equations depend on initial and boundary conditions. A further generalization of diffusion process is to allow the instantaneous mean and variance to be a function of state and time. In this case, the two parameters are represented by $\mu(x, t)$ and $\sigma^2(x, t)$ and the diffusion equations remain the same except $\mu(x, t)$ and $\sigma^2(x, t)$ now replace the respective roles of $\mu(x)$ and $\sigma^2(x)$.

Define the limiting density $f(x) = \lim_{t \to \infty} f(x, t; y)$. Setting the left side of the forward diffusion equation (Equation 7.2.12) equal to zero and replacing $f(x, t; y)$ by $f(x)$ on the right side of Equation 7.2.12, we obtain the differential equation defining the limiting density $f(x)$

$$0 = -\frac{\partial}{\partial x}\{\mu(x)f(x)\} + \frac{1}{2}\frac{\partial^2}{\partial x^2}\{\sigma^2(x)f(x)\}. \tag{7.2.14}$$

When there is a reflecting barrier at $x = 0$, the boundary condition is given by

$$\frac{1}{2}\frac{\partial^2}{\partial x^2}\{\sigma^2(x)f(x)\}\Big|_{x=0} = \mu(0)f(0). \tag{7.2.15}$$

The solution of Equation 7.2.14 with the boundary condition of Equation 7.2.15 is given by

$$f(x) = \frac{H}{\sigma^2(x)} \exp\left\{ 2\int_0^x \frac{\mu(y)}{\sigma^2(y)} dy \right\},$$
(7.2.16)

where H is a constant of integration. When $\mu(y) = \mu$ and $\sigma^2(y) = \sigma^2$ for all y, Equation 7.2.16 reduces to $f(x) = (H/\sigma^2)\exp\{2x\mu/\sigma^2\}$. Setting $H = 2\mu$ and recalling $\mu < 0$, we reclaim Equation 7.2.4 as expected upon integration.

EXAMPLE 7.2.4 **Diffusion Approximation of a GI/G/c Queue** Consider an extension of the $GI/G/1$ queue described in Example 7.2.1 to the case there are c identical servers, where $c \geq 1$. Under this system, the number of customers in the system at time t is equal to x, the service rate at time t is $\mu_s \min(x, c) \equiv (x \wedge c)\mu_s$. Using an argument identical to the one in Example 7.2.1, we obtain, for $x \geq 0$,

$$\mu(x) = \lambda - (x \wedge c)\mu_s \quad \text{and} \quad \sigma^2(x) = \sigma_a^2\lambda^3 + (x \wedge c)\sigma_s^2\mu_s^3,$$

or

$$\mu(x) = \begin{cases} \lambda - x\mu_s & \text{if } 0 \leq x \leq c \\ \lambda - c\mu_s & \text{if } x > c \end{cases} \quad \text{and} \quad \sigma^2(x) = \begin{cases} \sigma_a^2\lambda^3 + x\sigma_s^2\mu_s^3 & \text{if } 0 \leq x \leq c \\ \sigma_a^2\lambda^3 + c\sigma_s^2\mu_s^3 & \text{if } x > c. \end{cases}$$
(7.2.17)

For the $GI/G/c$ queue, the traffic intensity is given by $\rho = \lambda/(c\mu_s)$. Under the heavy traffic assumption, we expect that ρ is close to but less than one. This implies that $\mu(x) < 0$ for all $x \geq 0$. The difficulty of this problem lies with the discontinuities in the derivatives of $\mu(x)$ and $\sigma^2(x)$ at the point where $x = c$. To circumvent the difficulty, we define

$$f(x) = \begin{cases} f_1(x) = H_1 g_1(x) & 0 \leq x \leq c \\ f_2(x) = H_2 g_2(x) & x \geq c, \end{cases}$$

where we use the $\mu(x)$ and $\sigma^2(x)$ defined in Equation 7.2.17 for the respective regions of x to solve Equation 7.2.14. For each region, the solution again has the form of Equation 7.2.16. For $x \geq c$, it is straightforward to find that

$$g_2(x) = \exp\left\{ -2\left(\frac{(c\mu_s - \lambda)x}{\sigma_a^2\lambda^3 + c\sigma_s^2\mu_s^3} \right) \right\}.$$

For the region $0 \leq x \leq c$, deriving an explicit result from Equation 7.2.16 is somewhat cumbersome. After some tedious algebra, we obtain

$$g_1(x) = \left(\sigma_a^2\lambda^3 + x\sigma_s^2\mu_s^3 \right)^{u-1} \exp\left\{ -\frac{2x}{\sigma_s^2\mu_s^2} \right\},$$

where

$$u = \frac{2\lambda}{\sigma_s^2\mu_s^3} \left(\left(\frac{\sigma_a\lambda}{\sigma_s\mu_s} \right)^2 + 1 \right).$$

The two unknown constants H_1 and H_2 are determined by the normalization condition that

$$H_1 \int_0^c g_1(x)dx + H_2 \int_c^\infty g_2(x)dx = 1 \qquad (7.2.18)$$

and the continuity condition of $f(x)$ at $x = c$

$$H_1 g_1(c) = H_2 g_2(c). \qquad (7.2.19)$$

We now consider a numerical example. For simplicity, we use an *M/M/5* with traffic intensity 0.95. Setting $\mu_s = 1$, all other parameters are defined accordingly. Once the queue length density $f(x)$ is found, we discretize the queue length mass function using

$$P_n = \int_{n-0.5}^{n+0.5} f(x)dx \qquad n = 1, 2, \dots.$$

We truncate the computation at $n = 179$. In Figure 7.11, we plot the actual distribution and the diffusion approximation. We see that the two queue length distributions are indistinguishable in this numerical example. The mean and standard deviation for the queue length are 21.46 and 19.44, respectively, for the diffusion approximation. They compare favorably with 21.41 and 19.43 obtained from using the actual distributions. ∎

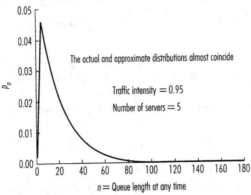

FIGURE
7.11 Limiting queue length distributions: actual versus diffusion approximations.

To prepare for the development of stochastic calculus, we now return to Brownian motion with drift $\mu = 0$ and variance parameter $\sigma^2 = 1$, called standard Brownian motion. Using standard Brownian motion processes as building blocks, we will construct more complex stochastic processes in the next section.

Standard Brownian Motion

A Brownian motion process $\mathbf{Z} = \{Z(t), t \geq 0\}$ with drift $\mu = 0$ and variance parameter $\sigma^2 = 1$ is called standard Brownian motion. Unless stated otherwise, we assume that $Z(0) = 0$. From Equation 7.1.3, we see that $Z(t) \sim N(0, t)$ and $E[Z(t)^2] = t$. Assuming that $t > s \geq 0$, we find the covariance function of **Z** as follows:

$$Cov[Z(t), Z(s)] = E[Z(t)Z(s)] - E[Z(t)]E[Z(s)] = E[Z(t)Z(s)]$$
$$= E\big[Z(s)[[Z(t) - Z(s)] + Z(s)]\big] = 0 + E[Z(s)^2] = s,$$

where we use the independent increment property of Brownian motion to establish the fourth equality. A stochastic process $\{X(t), t \geq 0\}$ is called a *Gaussian* process if $X(t_1), \ldots, X(t_n)$ follows a multivariate normal distribution for all $t_1, \ldots, t_n$. Since a multivariate normal distribution is completely characterized by the marginal means and the matrix of covariances, standard Brownian motion is also a Gaussian process with $E[Z(t)] = 0$ for all $t \geq 0$ and $Cov[Z(t), Z(s)] = t \wedge s$, where $a \wedge b = \min\{a, b\}$.

Let T_x denote the first passage time to x. Using the law of total probability, we write

$$P\{Z(t) \geq x\} = P\big\{Z(t) \geq x | T_x \leq t\big\}P\{T_x \leq t\} + P\big\{Z(t) \geq x | T_x > t\big\}P\{T_x > t\}.$$

If $T_x \leq t$, then by symmetry we have $P\{Z(t) \geq x | T_x \leq t\} = 1/2$. To confirm this, we observe that in Figure 7.12, from time T_x onward, there is equal probability that $Z(t)$ is above or below the darkened line (by noting that standard Brownian motion is a symmetric random walk in the limit). On the other hand, if $T_x > t$, then $P\{Z(t) \geq x | T_x > t\} = 0$. Hence we find

$$P\{T_x \leq t\} = 2P\{Z(t) \geq x\}. \tag{7.2.20}$$

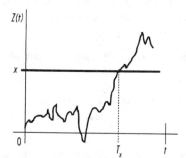

FIGURE
7.12   A sample path of standard Brownian motion.

Since $Z(t) \sim N(0, t)$, we obtain the following first-passage-time distribution

$$P\{T_x \leq t\} = 2P\left\{\frac{Z(t)}{\sqrt{t}} \geq \frac{x}{\sqrt{t}}\right\} = 2\frac{1}{\sqrt{2\pi}}\int_{\frac{x}{\sqrt{t}}}^{\infty} e^{-\frac{y^2}{2}} dy = \sqrt{\frac{2}{\pi}}\int_{\frac{x}{\sqrt{t}}}^{\infty} e^{-\frac{y^2}{2}} dy. \tag{7.2.21}$$

Differentiating the preceding with respect to t and applying Leibnitz's rule, we find the first-passage-time density

$$g_{T_x}(t) = \frac{x}{\sqrt{2\pi t^3}} e^{-\frac{x^2}{2t}} \qquad t > 0.$$

The preceding density is precisely Equation 7.1.6 with $\mu = 0$ and $\sigma^2 = 1$ (while Equation 7.1.6 was derived in the last section for the case with $\mu > 0$, the formula works also for $\mu = 0$).

7.3 Ito's Calculus and Stochastic Differential Equations

Let $Z = \{Z(t), t \geq 0\}$ be a standard Brownian motion process with drift μ, variance parameter σ^2, and $Z(0) = 0$. We know that $Z(t) \sim N(0, t)$. Let dt be an infinitesimal time interval. We define

$$dZ(t) = Z(t + dt) - Z(t).$$

Then we see that random variable $dZ(t) \sim N(0, dt)$ and $\{dZ(t), t \geq 0\}$ is a stochastic process. This stochastic process is sometimes referred to as *white noise*.

We first look at the central moments of $dZ(t)$. We note that $E[dZ(t)^k] = 0$ for $k = 1, 3, 5, \ldots$. The second moment $E[dZ(t)^2] = Var[dZ(t)] + E^2[dZ(t)] = dt$. For other even-order moments of $dZ(t)$, we use the identity

$$E[dZ(t_1) \cdots dZ(t_n)] = \sum E\left[dZ(t_{i_1})dZ(t_{i_2})\right] \cdots E\left[dZ(t_{i_{n-1}})dZ(t_{i_n})\right],$$

where the summation is over all possible ways of dividing the n points into $n/2$ combinations of pairs. Using the given identity, we find

$$E[dZ(t)^4] = \binom{4}{2}\left(E[dZ(t)^2]\right)^2 = 3dt^2 \qquad E[dZ(t)^6] = \binom{6}{2}\left(E[dZ(t)^2]\right)^3 = 15dt^3,$$

and so on. We define $dt^n = 0$ for $n > 1$. Then $Var[dZ(t)^2] = E[dZ(t)^4] - E^2[dZ(t)^2]$ $= 3dt^2 - dt^2 = 0$. For any random variable X, we have the relation

$$Var[X] = 0 \quad \Rightarrow \quad E[X] = X. \tag{7.3.1}$$

Applying Equation 7.3.1 to random variable $dZ(t)^2$, we obtain a useful identity

$$dZ(t)^2 = dt. \tag{7.3.2}$$

We see that $E[dZ(t)dt] = 0$ and $E[(dZ(t)dt)^2] = E[dZ(t)^2]dt^2 = 0$. This implies that $Var[dZ(t)dt] = 0$. Applying Equation 7.3.1 to random variable $dZ(t)dt$, we find

$$dZ(t)dt = 0. \tag{7.3.3}$$

Let Z be a standard Brownian motion process. Let $X = \{X(t), t \geq 0\}$ be a stochastic process defined by

$$dX(t) = \mu(X(t), t)dt + \sigma(X(t), t)dZ(t).$$

Suppressing the indexing variable t, the preceding can be stated as

$$dX = \mu(X, t)dt + \sigma(X, t)dZ. \tag{7.3.4}$$

This equation is called a *stochastic differential equation*.

Stochastic processes defined by Equation 7.3.4 are sometimes called *Ito processes*. The roles played by $\mu(X, t)$ and $\sigma(X, t)$ in Equation 7.3.4 are the generalizations of Equations 7.2.10 and 7.2.11 to their time-dependent counterparts, namely,

$$\mu(X(t), \ t) = \lim_{\Delta t \to 0} \frac{E\big[\{X(t+\Delta t) - X(t)\}|X(t), \ t\big]}{\Delta t}$$

and

$$\sigma^2(X(t), \ t) = \lim_{\Delta t \to 0} \frac{Var\big[\{X(t+\Delta t) - X(t)\}|X(t), \ t\big]}{\Delta t}.$$

The previous expressions show that $\mu(X(t), t) \Delta t$ is the mean rate of change in $X(t)$ in the interval $(t, \ t + \Delta t)$ and $\sigma^2(X(t), t)\Delta t$ is the variance of change in $X(t)$ in the same interval. We see that Equation 7.3.4 is a way to specify a diffusion process with state- and time-dependent drift and variance parameter using the notation of the differential.

In the sequel, we need the following two properties of dX to establish an important result in stochastic calculus:

$$dX dt = 0 \tag{7.3.5}$$

and

$$dX^2 = \sigma^2(X, t)dt, \tag{7.3.6}$$

where dX is defined by Equation 7.3.4. To establish the preceding identities, we observe

$$dX dt = [\mu(X, \ t)dt + \sigma(X, \ t)dZ]dt = \mu(X, \ t)dt^2 + \sigma(X, \ t)dZ dt = 0$$

and

$$\begin{aligned}
dX^2 &= [\mu(X, \ t)dt + \sigma(X, \ t)dZ][\mu(X, \ t)dt + \sigma(X, \ t)dZ] \\
&= \mu(X, \ t)^2 dt^2 + 2\sigma(X, \ t)\mu(X, \ t)dZ dt + \sigma^2(X, \ t)dZ^2 \\
&= \sigma^2(X, \ t)dZ^2 = \sigma^2(X, \ t)dt,
\end{aligned}$$

where we use Equations 7.3.2 and 7.3.3 in deriving the previous results.

EXAMPLE
7.3.1

Arithmetic Brownian Motion   Consider the process $dY = \mu dt + \sigma dZ$, where μ and σ are constants. This process is said to follow *arithmetic Brownian motion* with drift μ and volatility σ. This process is actually the diffusion process considered in Section 7.1 except that we now express the process in differential notations. The random variable dY is a linear transformation of a normal variate. Hence dY is normal with mean μdt and $Var[dY(t)] = \sigma^2 Var[dZ(t)] = \sigma^2 E[dZ(t)^2] = \sigma^2 dt$. The process can be used to model economic variables that grow at a linear rate and show increasing uncertainty over time. ∎

EXAMPLE
7.3.2

Geometric Brownian Motion Consider the process $dY = \mu Y dt + \sigma Y dZ$ or

$$\frac{dY}{Y} = \mu dt + \sigma dZ. \tag{7.3.7}$$

The previous expression suggests that the rate of change in $X(t)$ follows an arithmetic Brownian motion. For this process, we see that $\mu(Y, t) = \mu Y$ and $\sigma(Y, t) = \sigma Y$. It is called geometric Brownian motion with drift μ and volatility σ. The process is used frequently in modeling stock prices. This process has appeared in Example 7.1.3 and will reappear in the sequel. ∎

EXAMPLE
7.3.3

Merton's Portfolio Selection Problem In this example, we show the construction of $\mu(X, t)$ and $\sigma(X, t)$ for an Ito process arising in a portfolio selection problem. Let $X(t)$ denote the wealth of an investor at time t. The investor allocates a fraction w of the wealth in a risky asset and the remainder in a sure asset. The sure asset produces a rate of return s. The risky asset yields a rate of return μ ($\mu > s$) along with a variance of σ^2 per unit time. In other words, the risky asset earns a return $dr(t)$ in $(t, t + dt)$, where $dr = \mu dt + \sigma dZ$, and Z is a standard Brownian motion process. We see that

$$X(t + dt) = X(t) + s(1 - w)X(t)dt + dr(t)\big(wX(t)\big)$$
$$= X(t) + s(1 - w)X(t)dt + \big(\mu dt + \sigma dZ(t)\big)\big(wX(t)\big)$$
$$= X(t) + [\mu w + s(1 - w)]X(t)dt + \sigma wX(t)dZ(t).$$

Rearranging the terms, we obtain the Ito process

$$dX = \mu(X, t)dt + \sigma(X, t)dZ,$$

where $\mu(X, t) = [\mu w + s(1 - w)]X(t)$ and $\sigma(X, t) = \sigma wX(t)$. ∎

In many applications of stochastic differential equations, an important question is the following: If X is an Ito process defined by Equation 7.3.4 and $Y(t)$ is a function of $X(t)$ and t, then what is the stochastic process $Y = \{Y(t), t \geq 0\}$? The following lemma provides the means for finding the answer, which is the fundamental result in stochastic calculus.

Ito's Lemma

Using the Taylor series expansion of $Y(X, t) = f(X(t), t)$, we write

$$dY = Y_t dt + Y_X dX + \frac{1}{2}Y_{tt}dt^2 + \frac{1}{2}Y_{XX}dX^2 + 2Y_{Xt}dXdt + \cdots,$$

where, again, we use subscripts to denote the indicated partial differentiations.

Invoking Equations 7.3.5 and 7.3.6 and ignoring the terms of order $o(dt)$, we obtain *Ito's lemma*

$$dY = Y_t dt + Y_X dX + \frac{1}{2} Y_{XX} \sigma^2(X, t) dt. \qquad (7.3.8)$$

Using Equation 7.3.4, we can express the preceding formula in terms of **Z** as follows:

$$dY = \left(Y_t + Y_X \mu(X, t) + \frac{1}{2} Y_{XX} \sigma^2(X, t) \right) dt + Y_X \sigma(X, t) dZ. \qquad (7.3.9)$$

A special case of Equation 7.3.9 is when $\mu(X, t) = 0$ and $\sigma(X, t) = 1$ and consequently $dX = dZ$. In this case, $Y(t)$ is simply a function of $Z(t)$ and Equation 7.3.9 reduces to

$$dY = \left(Y_t + \frac{1}{2} Y_{ZZ} \right) dt + Y_Z dZ. \qquad (7.3.10)$$

EXAMPLE **7.3.4**   **Arithmetic Brownian Motion Revisited**   Consider the stochastic differential equation $dY = \mu dt + \sigma dZ$ once more. Comparing it against Equation 7.3.10 term by term yields two conditions for $Y(t)$ to satisfy

$$Y_t + \frac{1}{2} Y_{ZZ} = \mu \qquad \text{and} \qquad Y_Z = \sigma.$$

Since $Y_Z = \sigma$, we have $Y_{zz} = 0$, $Y_t = \mu$, and $dY = \sigma dZ$. Integrating the last expression produces $Y(t) = \sigma Z(t) + f(t)$. Differentiating the latter, we obtain $Y_t = f'(t) = \mu$. This implies that $f(t) = \mu t + c$ and $Y(t) = \sigma Z(t) + \mu t + c$. The initial condition $Y(0) = 0$ results in $c = 0$. Since

$$Y(t) = \mu t + \sigma Z(t),$$

for a fixed t, $Y(t)$ is indeed a linear transformation of the normal deviate $Z(t)$.

If $t > s$, $Y(t)$ is a future value of the process relative to time s. The distribution of Y_t, given X_s, is normal with mean $Y_s + \mu(t - s)$ and variance $\sigma^2(t - s)$. ∎

EXAMPLE **7.3.5**   **Geometric Brownian Motion Revisited**   We now consider the stochastic differential equation $dY = \mu Y dt + \sigma Y dZ$ again. Comparing this equation with Equation 7.3.10, we find the following two conditions:

$$Y_t + \frac{1}{2} Y_{ZZ} = \mu Y \qquad (7.3.11)$$

and

$$Y_Z = \sigma Y. \qquad (7.3.12)$$

We write Equation 7.3.12 as $\partial Y / \partial Z = \sigma Y$ or $\partial Y / Y = \sigma \partial Z$. Upon integration, we find $\ln Y = \sigma Z + f(t)$ or $Y = \exp(\sigma Z + f(t))$. Differentiating the last expression, we obtain $Y_t = Y f'(t)$, $Y_Z = Y \sigma$, and $Y_{ZZ} = Y \sigma^2$. Using these identities in Equation 7.3.11, we obtain $Y f'(t) + (1/2) Y \sigma^2 = \mu Y$. Since Y can never be zero (if it is, it will stay there forever), we divide the last expression by Y and obtain $f'(t) + (1/2) \sigma^2 = \mu$ or $f'(t) = \mu - (1/2) \sigma^2$. Upon integration, we have

$$f(t) = \left(\mu - \frac{1}{2} \sigma^2 \right) t + c,$$

where c is a constant of integration. This implies that

$$Y = e^{\sigma Z + \left(\mu - \frac{1}{2} \sigma^2 \right) t + c} \tag{7.3.13}$$

When $t = 0$, $\sigma Z(0) = 0$, and $[\mu - (1/2) \sigma^2] t = 0$, we have $Y(0) = e^c$ and

$$Y(t) = Y(0) e^{\left(\mu - \frac{1}{2} \sigma^2 \right) t + \sigma Z}$$

or

$$\ln Y(t) = \ln Y(0) + \sigma Z(t) + \left(\mu - \frac{1}{2} \sigma^2 \right) t.$$

The right-side terms of the preceding expression are all constants except $Z(t)$. We conclude that $\ln Y(t)$ is normal with

$$E[\ln Y(t)] = \ln Y(0) + \mu t - \frac{1}{2} \sigma^2 t \qquad \text{and} \qquad Var[\ln Y(t)] = \sigma^2 t. \tag{7.3.14}$$

Setting $a = [\mu - (1/2) \sigma^2]$ and $Y(0) = 1$, Equation 7.3.13 can be stated as

$$Y(t) = e^{X(t)}, \tag{7.3.15}$$

where $X(t) = at + \sigma Z$. Since $X(t)$ is Brownian motion with drift μ and variance parameter σ^2, the process $\{Y(t), t \geq 0\}$ defined by Equation 7.1.15 is called geometric Brownian motion. The process was used in Example 7.1.3 for modeling stock prices.

We now consider the computation of the conditional expectation $E[Y(t)|Y(u), 0 \leq u \leq s]$ for $t > s$. Before doing so, we recall that the moment generating function of a normal variate W with mean $E[W]$ and $Var[W]$ is given by $E[e^{Wr}] = \exp(\tau E[W] + (\tau^2 Var[W]/2))$. Letting $\tau = 1$ in the last expression, we find $E[Y(t)] = E\{\exp(X(t))\} = \exp\{E[X(t)] + (Var[X(t)]/2)\} = \exp\{[a + (\sigma^2)/2]t\} = \exp(\mu t)$. The conditional expectation is then given by

$$E[Y(t)|Y(u),\ 0 \leq u \leq s] = E\left[e^{X(t)} | X(u),\ 0 \leq u \leq s \right]$$

$$= E\left[e^{X(s) + X(t) - X(s)} | X(u),\ 0 \leq u \leq s \right]$$

$$= e^{X(s)} E\left[e^{X(t) - X(s)} | X(u),\ 0 \leq u \leq s \right]$$

(by independent increments of X)

$$= Y(s) E\left[e^{X(t-s)} \right]$$

(by independent and stationary increments of X)

$$= Y(s) e^{\mu(t-s)}.$$

The previous derivation shows that the conditional mean of $Y(t)$ given $Y(s)$ is $Y(s)e^{\mu(t-s)}$.

Geometric Brownian motion is frequently used to model stock prices. Stock prices are nonnegative and typically exhibit long-term exponential growth. Another insightful interpretation of such a usage is through the argument of percentage changes. To be specific, we let X_n denote the price of a stock at time n. Now we consider the ratio $Y_n = X_n/X_{n-1}$. In a perfect market, this ratio tends to stay in equilibrium and thus it is reasonable to assume $\{Y_n\}$ are i.i.d. random variables. We write

$$X_n = \frac{X_n}{X_{n-1}} \frac{X_{n-1}}{X_{n-2}} \cdots \frac{X_1}{X_0} X_0 = Y_n Y_{n-1} \cdots Y_1 X_0.$$

This implies that $\ln X_n = \sum_{i=1}^{n} Y_i + \ln X_0$. Since $\left\{\sum_{i=1}^{n} Y_i\right\}$ are approximately normal by the central limit theorem, so are $\{\ln X_n\}$. Consequently, $\{X_n\}$ can be approximated by geometric Brownian motion. ∎

EXAMPLE
7.3.6

Forward Contracts on a Non-Dividend-Paying Security   A forward contract is an agreement between two parties to buy or sell a security at a certain future time for a certain price, called the delivery price. The party to buy the security is said to assume a *long* position and the party to sell the security is said to assume a *short* position (the terms *long* and *short* are also used as verbs and adjectives with their respective meanings in security trading). Let t denote the current time and T the maturity date of the contract. Let $S(t)$ denote the price of the security at time t. Assume that $S(t)$ follows geometric Brownian motion with expected return μ and volatility σ. In other words, $S(t)$ is characterized by the stochastic differential equation $dS = \mu S dt + \sigma S dZ$. Let $Y(t)$ denote the forward price at time t. The forward price at any time is the delivery price that would make the contract have a zero value. Let r be the risk-free interest rate. We now consider the case in which the security does not yield any dividends.

For arbitrage opportunities to be absent, the forward and security prices must be related as follows: $Y(t) = S(t)e^{r(T-t)}$. Otherwise, if $Y(t) > S(t)e^{r(T-t)}$, an arbitrageur can borrow S dollars for a period of $T - t$ at the risk-free interest rate, buy the security, and take a short position in the forward contract. At time T, the security is sold for $Y(t)$. After paying the loan of $S(t)e^{r(T-t)}$, the arbitrageur nets a profit of $Y(t) - S(t)e^{r(T-t)}$. If the inequality goes the other direction, a similar scenario can be constructed.

Since $Y(t)$ is a function of Brownian motion $S(t)$, we use Ito's lemma to find the stochastic process Y. First, we see that

$$\frac{\partial Y}{\partial S} = e^{r(T-t)}, \qquad \frac{\partial^2 Y}{\partial S^2} = 0, \qquad \text{and} \qquad \frac{\partial Y}{\partial t} = -rSe^{r(T-t)}.$$

We then use Equation 7.3.9 to write

$$dY = \left(-rSe^{r(T-t)} + e^{r(T-t)}\mu S\right)dt + e^{r(T-t)}\sigma S dZ.$$

Using $Y = Se^{r(T-t)}$ in the previous expression, we obtain $dY = (\mu - r)Ydt + \sigma YdZ$. Therefore we conclude that $Y(t)$ is geometric Brownian motion with an expected growth rate of $\mu - r$ and volatility σ. ∎

EXAMPLE
7.3.7

The Black-Scholes Differential Equation A *derivative security* is a security whose value depends on the values of other variables. Derivative securities are also called *contingent claims*. Frequently, the variables underlying derivative securities are the prices of traded security. The forward contract considered in Example 7.3.6 provides one such example. A stock option represents another derivative security. We now derive the Black-Scholes differential equation for a non–dividend-paying derivative security.

Assume that the price of a stock $S(t)$ follows geometric Brownian motion

$$dS = \mu Sdt + \sigma SdZ. \tag{7.3.16}$$

Let $Y(t)$ be the price of a derivative security contingent on $S(t)$. Black and Scholes (1973) observed that a portfolio consisting of one derivative security short and $\partial Y/\partial S$ shares of stock long would be riskless in the sense that the gain from one instrument would always offset the loss from the other instrument so that the overall value of the portfolio at the end of a *short* period would be known with certainty.

Let f be the value of the portfolio. Then we see that

$$f = -Y + \frac{\partial Y}{\partial S} S = -Y + Y_S S \tag{7.3.17}$$

and

$$df = -dY + Y_S dS. \tag{7.3.18}$$

From Ito's lemma in Equation 7.3.9, we obtain

$$dY = \left(Y_t + Y_S \mu S + \frac{1}{2} Y_{SS} \sigma^2 S^2 \right) dt + Y_S \sigma SdZ. \tag{7.3.19}$$

Using Equations 7.3.16 and 7.3.19 in Equation 7.3.18, after some cancellations we find that

$$df = -\left(Y_t + \frac{1}{2} Y_{SS} \sigma^2 S^2 \right) dt. \tag{7.3.20}$$

In Equation 7.3.20, we observe that df does not involve dZ (the Wiener process has been eliminated) and the portfolio f must be riskless in the infinitesimal time interval.

Let r denote the risk-free interest rate. To exclude arbitrage opportunities, we must have

$$\frac{\partial f}{f} = rdt \quad \text{or} \quad \partial f = rfdt. \tag{7.3.21}$$

The argument leading to Equation 7.3.21 is similar to that presented in Example 7.3.6. Using Equations 7.3.17 and 7.3.20 in Equation 7.3.21, we obtain the Black-Scholes differential equation for option pricing,

$$Y_t + rSY_S + \frac{1}{2} Y_{SS} \sigma^2 S^2 = rY. \tag{7.3.22}$$

The solution of differential equation (Equation 7.3.22) depends on the boundary conditions imposed by the specific derivative security. As an example, we consider a European *call option* in which the owner of the option is entitled to purchase a share of a given stock at an exercise price c at a future time T. In this case the boundary condition is given by

$$Y(S(T),\ T) = \max\{S(T) - c,\ 0\}. \tag{7.3.23}$$

The price of the option at time $t < T$ will be its expected price at time T discounted back to t. Hence

$$Y(S(t),\ t) = e^{-r(T-t)} E\big[Y(S(T),\ T)\big]. \quad \blacksquare \tag{7.3.24}$$

EXAMPLE 7.3.8

Forward Contracts on a Non-Dividend-Paying Security Revisited   Let K be the delivery price of the security at maturity date T. We now consider two equally attractive portfolios. The first portfolio is a long forward contract on the security plus an amount of cash $Ke^{-r(T-t)}$ at time t and the other is one unit of the security at time t. Both portfolios ensure that the investor will possess one unit of the security at the maturity date T. Thus the two portfolios will be equally acceptable to the investor at time t and we have

$$Y(t) + Ke^{-r(T-t)} = S(t) \quad \text{or} \quad Y = S - Ke^{-r(T-t)}, \tag{7.3.25}$$

where $Y(t)$ is the price of the forward contract at time t. To show that $Y(t)$ defined by Equation 7.3.25 indeed satisfies the Black-Scholes differential equation (Equation 7.3.22), we find $Y_t = -rKe^{-r(T-t)}$, $Y_S = 1$, $Y_{SS} = 0$ and make the respective substitutions in Equation 7.3.22. This gives $-rKe^{-r(T-t)} + rS = rY$. Dividing the last expression by r establishes Equation 7.3.25. $\blacksquare$

EXAMPLE 7.3.9

European Call Options   In the last part of Example 7.3.7, we have shown that the option price at time t is given by Equation 7.3.24. A direct way to find $E[Y(S(T), T)]$ is by solving the partial differential equation (Equation 7.3.22) using the boundary condition of Equation 7.3.23. Another way is by assuming that there exist risk-neutral investors such that the price of the stock follows geometric Brownian motion

$$dS = rSdt + \sigma SdZ,$$

namely, with drift r replacing μ. For notational convenience, we define $X(T) = \ln S(T)$. From Equation 7.3.14, we conclude that $X(T)$ is normal with

$$E[X(T)] = \ln S(t) + \left(r - \frac{1}{2}\sigma^2\right)(T - t) \quad \text{and} \quad Var[X(T)] = \sigma^2(T - t).$$

We let $n_{X(T)}$ denote the normal density with the given mean and variance. Then $S(T) = e^{X(T)}$ and $E[Y(S(T),\ T)] = E[Y(e^{X(T)},\ T)] = E[\max\{e^{S(T)} - c,\ 0\}]$. This implies that

$$E\big[Y(S(T),\ T)\big] = \int_{\ln c}^{\infty} [e^u - c] n_{X(T)}(u)\,du. \tag{7.3.26}$$

Substituting Equation 7.3.26 into Equation 7.3.24 and simplifying, we obtain the solution

$$Y(S(t),\ t) = S(t)\Phi(z_1) - ce^{-r(T-t)}\Phi(z_2), \tag{7.3.27}$$

where

$$z_1 = \frac{\ln\left(\dfrac{S(t)}{c}\right) + \left(r + \dfrac{\sigma^2}{2}\right)(T-t)}{\sigma\sqrt{T-t}} \quad \text{and} \quad z_2 = \frac{\ln\left(\dfrac{S(t)}{c}\right) + \left(r - \dfrac{\sigma^2}{2}\right)(T-t)}{\sigma\sqrt{T-t}}.$$

We now illustrate the use of the option pricing equation (Equation 7.3.27) in a numerical example. Consider a non–dividend-paying stock whose current price is $80. The exercise price of the option is $75. The risk-free interest rate is 10 percent per annum and the volatility is 20 percent per annum. The option will expire in six months. Using Equation 7.3.27 with $r = 0.1$, $\sigma = 0.2$, $S(t) = 80$, $c = 75$, and $T - t = 0.5$, we find in the Appendix that the value of the call option is $9.92. If we purchase the stock now, we pay $80; if we purchase the call option and acquire the stock on the option's expiration date, we pay $75 then. Consequently, the stock price has to rise by ($75 + $9.92) − $80 = $ 4.92 for the purchaser of the call to break even. ∎

7.4 Multidimensional Ito's Lemma

Consider n Ito's processes $dX_1, \ldots, dX_n$ with

$$dX_i = \mu_i(\vec{X},\ t)dt + \sigma_i(\vec{X},\ t)dZ_i, \tag{7.4.1}$$

where $\vec{X} = \{X_1, \ldots, X_n\}$, μ_i and σ_i may be any function of $\{X_i(t)\}$ and t, and $\{Z_i(t)\}$ are Wiener processes. We assume that

$$dZ_i dZ_j = \rho_{ij}dt, \tag{7.4.2}$$

where ρ_{ij} is the instantaneous correlation coefficient between the Wiener processes dZ_i and dZ_j. When $i = j$, we have $\rho_{ij} = 1$ and Equation 7.4.2 reduces to Equation 7.3.2. Using Equations 7.3.3 and 7.4.2 and $dt^2 = 0$, we obtain the following useful result:

$$dX_i dX_j = \left[\mu_i(\vec{X},\ t)dt + \sigma_i(\vec{X},\ t)dZ_i\right]\left[\mu_j(\vec{X},\ t)dt + \sigma_j(\vec{X},\ t)dZ_j\right]$$

$$= \sigma_i(\vec{X},\ t)\sigma_j(\vec{X},\ t)\rho_{ij}dt. \tag{7.4.3}$$

As in the one-dimensional case, we now study the function of Ito processes. Let f be an arbitrary function and $Y(\vec{X},\ t) = f(\vec{X},\ t)$. We use the Taylor series expansion to write

$$dY = Y_t dt + \sum_{i=1}^{n} Y_{X_i} dX_i + \frac{1}{2}\sum_{i=1}^{n}\sum_{j=1}^{n} Y_{X_i X_j} dX_i dX_j + \cdots. \tag{7.4.4}$$

Ignoring the terms of order $o(dt)$ and using Equation 7.4.3, we obtain the *multidimensional Ito's lemma*

$$dY = Y_t dt + \sum_{i=1}^{n} Y_{X_i} dX_i + \frac{1}{2} \sum_{i=1}^{n} \sum_{j=1}^{n} Y_{X_i X_j} \sigma_i \sigma_j \rho_{ij} dt. \qquad (7.4.5)$$

When $n = 1$, Equation 7.4.5 becomes Equation 7.3.8—as expected. Substituting Equation 7.4.1 into Equation 7.4.5, we find another version of the lemma

$$dY = \left(Y_t + \sum_{i=1}^{n} Y_{X_i} \mu_i + \frac{1}{2} \sum_{i=1}^{n} \sum_{j=1}^{n} Y_{X_i X_j} \sigma_i \sigma_j \rho_{ij} \right) dt + \sum_{i=1}^{n} Y_{X_i} \sigma_i dZ_i. \qquad (7.4.6)$$

The preceding version is the multidimensional extension of Equation 7.3.6. In Equations 7.4.4 through 7.4.6, we again use subscripts X_i X_j, and t of Y to denote the indicated partial derivatives; μ_i and σ_i are shorted-handed notations for $\mu_i(\vec{X}, t)$ and $\sigma_i(\vec{X}, t)$, respectively.

EXAMPLE
7.4.1

Fischer's Model of Index Bonds Index bonds link investment contracts to the price level to lessen the impact of inflation. Consider the case in which the price level is modeled by the geometric Brownian motion process

$$\frac{dP}{P} = \pi dt + s dZ_1. \qquad (7.4.7)$$

Hence the (instantaneous) proportional change in the price level is normal with mean πdt and variance $s^2 dt$.

We consider an investor holding three assets in a portfolio—a real bond, a risky asset (equity), and a nominal bond. The values of these three assets at time t are denoted by $Q_1(t)$, $Q_2(t)$ and $Q_3(t)$, respectively. The real bond pays a nominal return r_1 plus the realized rate of inflation. Thus the Ito process governing the nominal return on the index bond is given by

$$\frac{dQ_1}{Q_1} = r_1 dt + \frac{dP}{P} = (r_1 + \pi)dt + s dZ_1 = R_1 dt + s_1 dZ_1, \qquad (7.4.8)$$

where $R_1 \equiv r_1 + \pi$ and $s_1 \equiv s$. The nominal return on equity is

$$\frac{dQ_2}{Q_2} = R_2 dt + s_2 dZ_2, \qquad (7.4.9)$$

where R_2 is the expected nominal return on equity per unit time and s_2^2 is the variance of the nominal return per unit time. The deterministic nominal return on nominal bonds is

$$\frac{dQ_3}{Q_3} = R_3 dt, \qquad (7.4.10)$$

where R_3 is the expected nominal return on the nominal bond per unit time. For $i = 1, 2, 3$, we see that $\mu_i = R_i Q_i$ and $\sigma_i = s_i Q_i$ (where we define $s_3 = 0$).

The *real* return for asset i can be measured by $d(Q_i/P)/(Q_i/P)$. Hence for each i, we define $Y^i(Q_i, P, t) = Q_i/P$. Applying the multidimensional Ito's lemma (Equation 7.4.6) with $n = 2$, for $i = 1, 2, 3$, we obtain

$$dY^i = \left[Y_t^i + Y_{Q_i}^i \mu_i + Y_P^i \mu_P + \frac{1}{2} Y_{Q_i Q_i}^i \sigma_i^2 + \frac{1}{2} Y_{PP}^i \sigma_P^2 + Y_{Q_i P}^i \sigma_i \sigma_P \rho_{iP} \right] dt$$
$$+ Y_{Q_i}^i \sigma_i dZ_i + Y_P^i \sigma_P dZ_1, \qquad (7.4.11)$$

where $\mu_P = \pi P$ and $\sigma_P = s_1 P$ (we emphasize that subscripts of Y^i denote their indicated partial derivatives and subscripts of other terms serve only as indices). The partial derivatives for use in Equation 7.4.11 are

$$Y_{Q_i}^i = \frac{1}{P} \qquad\qquad Y_{Q_i Q_i}^i = 0 \qquad\qquad Y_{Q_i P}^i = -\frac{1}{P^2}$$

$$Y_P^i = -\frac{Q_i}{P^2} \qquad\qquad Y_{PP}^i = \frac{2Q_i}{P^3} \qquad\qquad Y_t = 0.$$

Using the preceding in Equation 7.4.11 for $i = 1$, we obtain

$$dY^1 = \left[\left(\frac{1}{P}\right) R_1 Q_1 - \frac{Q_1}{P^2} \pi P + \frac{1}{2}\left(\frac{2Q_1}{P^3}\right)(s_1 P)^2 - \frac{1}{P^2}(s_1 Q_1)(s_1 P)\rho_{Q_1 P} \right] dt$$
$$+ \frac{1}{P}(s_1 Q_1) dZ_1 - \frac{Q_1}{P^2}(s_1 P) dZ_1$$
$$= \frac{Q_1}{P}\left[R_1 - \pi + s_1^2 - s_1^2 \right] dt = \frac{Q_1}{P} r_1 dt$$

or

$$\frac{d(Q_1/P)}{Q_1/P} = r_1 dt,$$

where we use $\rho_{Q_1 P} = 1$ (since the uncertainty for both processes is dZ_1). The previous result suggests that for the index bond, the real return is equal to the nominal return r_1 as we expect. Similarly, we find the real return on equity

$$\frac{d(Q_2/P)}{Q_2/P} = \left[R_2 - \pi + s_1^2 - \rho_{Q_2 P} s_1 s_2 \right] dt + s_2 dZ_2 - s_1 dZ_1 = r_2 dt + s_2 dZ_2 - s_1 dZ_1,$$

where $r_2 \equiv R_2 - \pi + s_1^2 - \rho_{Q_2 P} s_1 s_2$ is the expected real return on equity. We observe that the expected real return on equity is not the expected nominal return minus the expected rate of inflation. Finally, the real return on the nominal bond is given by

$$\frac{d(Q_3/P)}{(Q_3/P)} = (R_3 - \pi + s_1^2) dt - s_1 dZ_1 = r_3 dt - s_1 dZ_1,$$

where $r_3 \equiv R_3 - \pi + s_1^2$ is the expected real return on the nominal bond. In the presence of price uncertainty, the preceding suggests that for the nominal bond the expected real return r_3 is greater than the expected nominal return adjusted by the inflation. To shed some light on this anomaly, we consider the case of simply holding \$1 worth of cash (we may think of having a nominal bond with $R_3 = 0$). Suppose that the price, one period hence, is P. Then the real return is $1/P$. If P has uncertainty, then the real return is $E[1/P]$. On the other hand, if the increase in price is deterministic and of magnitude $E[P]$, then the real return is $1/E[P]$. Define $g(P) = 1/P$. For $P > 0$, the function g is convex. Invoking Jensen's inequality, we have $E[g(P)] \geq g(E[P])$ or $E[1/P] \geq 1/E[P]$. This implies that under price uncertainty the real return is in general larger than its deterministic counterpart. ∎

EXAMPLE
7.4.2

Capital Asset Pricing Let X be a freely traded security with no cash payouts. Risk uncorrelated with changes in X is not priced. We assume that X follows geometric Brownian motion with drift μ_1 and volatility σ_1. Hence

$$\frac{dX}{X} = \mu_1 dt + \sigma_1 dZ_1,$$

where dZ_1 is a standard Wiener process. Let Y be a security following the Ito process

$$dY = \mu_2 dt + \sigma_2 dZ_2,$$

where dZ_2 is another standard Wiener process, $\mu_2 = \mu_2(Y, t)$, $\sigma_2 = \sigma_2(Y, t)$, and $dZ_1 dZ_2 = \rho dt$. We define the *beta* of security Y through its correlation with X as

$$\beta = \frac{Cov\left(\dfrac{dX}{X}, \dfrac{dY}{Y}\right)}{Var\left(\dfrac{dX}{X}\right)} = \frac{\rho\left(\dfrac{1}{Y}\right)\sigma_1\sigma_2}{\sigma_1^2} = \frac{\rho\sigma_2}{Y\sigma_1}.$$

To elaborate a little more about the role played by β, we make a brief digression here. For a given period, we let Δ_1 denote the change in the value of \$1 invested in the market index and Δ_2 the change in the value of \$1 invested in a given security. We use a straight line to fit the relation between Δ_1 and Δ_2, specifically, $\Delta_2 = \alpha + \beta\Delta_1$, where α and β are two constants. The best linear predictor of the fit is given by $\beta = (\sigma_2/\sigma_1)\rho_{12}$. So we have

$$\beta = \frac{\sigma_2}{\sigma_1}\left(\frac{\sigma_{12}}{\sigma_1\sigma_2}\right) = \frac{\sigma_{12}}{\sigma_1^2} = \frac{Cov(\Delta_1, \Delta_2)}{Var(\Delta_1)}.$$

When $\beta = 1$, the return on the security tends to follow the return on the market; when $\beta = 2$, the return on the security moves twice as fast as the return on the market.

Returning to the asset pricing problem, we assume that the value of an asset V depends on the Ito process dY, the asset has a claim to continuous cash payment of $g(X, t)$, and the asset has a remaining life τ (for example, the asset can be an annuity that expires worthless after an interval of length τ). If we are currently at time t and T is the expiration date, then $\tau = T - t$, $t = T - \tau$, and $dt = -d\tau$. Thus we have the partial derivative $V_t = -V_\tau$. Since $V = V(Y, \tau)$, we use the one-dimensional Ito's lemma (Equation 7.3.9) to write

$$dV = \left(V_t + V_Y \mu_2 + \frac{1}{2} V_{YY} \sigma_2^2 \right) dt + V_Y \sigma_2 dZ_2$$

$$= \left(-V_\tau + V_Y \mu_2 + \frac{1}{2} V_{YY} \sigma_2^2 \right) dt + V_Y \sigma_2 dZ_2.$$

Another way to state the previous Ito process is

$$dV = \mu_V dt + \sigma_V dZ_V,$$

where $\mu_V = -V_\tau + V_Y \mu_2 + (1/2) V_{YY} \sigma_2^2$, $\sigma_V = V_Y \sigma_2$, and $dZ_V = dZ_2$.

We now form a portfolio P that is long one unit of V and h units of the index portfolio X. So we see that $P = P(V, X) = V + hX$ and the multidimensional Ito's lemma (Equation 7.4.5) is applicable. With $P_V = 1$, $P_{VV} = 0$, $P_X = h$, $P_{XX} = 0$, $P_{VX} = 0$, and $P_t = 0$, we find

$$dP = \left[\mu_V + h\mu_1 X \right] dt + \sigma_V dZ_V + h\sigma_1 X dZ_1$$

$$= \left[-V_\tau + V_Y \mu_2 + \frac{1}{2} V_{YY} \sigma_2^2 + h\mu_1 X \right] dt + V_Y \sigma_2 dZ_2 + h\sigma_1 X dZ_1.$$

Before we proceed, we cite a useful relation about a standard bivariate normal distribution. The relation says that dZ_1 and dZ_2 follow standard bivariate normal distribution with correlation ρ if and only if $dZ_2 = \rho dZ_1 + \sqrt{1-\rho^2} dZ_\epsilon$, where dZ_1 and dZ_ϵ are standard normal variables. We now use this relation to eliminate the term involving dZ_2 in the last equation. This gives

$$dP = \left[-V_\tau + V_Y \mu_2 + \frac{1}{2} V_{YY} \sigma_2^2 + h\mu_1 X \right] dt + \left[V_Y \sigma_2 \rho + h\sigma_1 X \right] dZ_1$$

$$+ V_Y \sigma_2 \sqrt{1-\rho^2} dZ_\epsilon.$$

We choose h to remove the term involving dZ_1 by setting $V_Y \sigma_2 \rho + h\sigma_1 X = 0$ and obtain

$$h = -\frac{V_Y \sigma_2 \rho}{\sigma_1 X} = -\frac{V_Y \beta Y}{X} \quad \text{or} \quad hX = -V_Y \beta Y.$$

The expected total return on the hedged portfolio has two parts: the expected capital gain on P, $E[dP]$, and the expected cash flows to P, $g(X, t)dt$. Let r be the risk-free interest rate. By noting that the residual risk dZ_ϵ is not priced, we set the expected return on the hedged portfolio equal to the risk-free interest rate and obtain

$$rPdt = r(V + hX) = r(V - V_Y \beta Y)dt$$

$$= \left[-V_\tau + V_Y \mu_2 + \frac{1}{2} V_{YY} \sigma_2^2 - V_y \beta Y \mu_1 \right] dt + g(X, t)dt.$$

Rearranging the last equality, we find the desired result

$$g(X,\ t) = rV + V_\tau - \left[\frac{1}{2}V_{YY}\sigma_2^2 + V_Y Y\left(\frac{\mu_2}{Y} - \beta(\mu_1 - r)\right)\right]. \quad \blacksquare$$

7.5 Control of Systems of Stochastic Differential Equations

We consider the following optimization problem

$$J(0,\ x_0) \equiv \max_D\left\{\int_0^T f(X,\ D,\ s)ds + B(X(T),\ T)\right\} \qquad (7.5.1)$$

subject to

$$dX = \mu(X,\ D,\ s)ds + \sigma(X,\ D,\ s)dZ \qquad \text{and} \qquad X(0) = x_0. \qquad (7.5.2)$$

In Equation 7.5.1, X is the state variable, D the decision variable, f the objective function, and B the function of terminal reward. For a given D, dX is defined by the stochastic differential equation (Equation 7.5.2). We see that $J(0,\ x_0)$ represents the optimal reward obtainable in $[0,\ T]$ given that at time 0 we are in state x_0. For any $0 < t < T$, we can define $J(t,\ x_t)$ similarly, namely, $J(t,\ x)$ represents the optimal reward obtainable in $[t,\ T]$ given that at time t we are in state $X(t) = x$. In other words, we have

$$J(t,\ x) \equiv \max_D\left\{\int_t^T f(X,\ D,\ s)ds + B(X(T),\ T)\right\} \qquad (7.5.3)$$

subject to

$$dX = \mu(X,\ D,\ s)ds + \sigma(X,\ D,\ s)dZ \qquad \text{and} \qquad X(t) = x. \qquad (7.5.4)$$

Applying the principle of optimality in dynamic programming, we can rewrite Equation 7.5.3 as

$$J(t,\ x) = \max_D E\left\{\int_t^{t+\Delta t} f(X,\ D,\ s)ds + J(t + \Delta t,\ x + \Delta X)\right\}. \qquad (7.5.5)$$

Equation 7.5.5 is Bellman's functional equation. The integral on the right side of Equation 7.5.5 is the reward received in $(t,\ t + \Delta t)$ and the last term of Equation 7.5.5 is the optimal reward from time $t + \Delta t$ onward, given that $X(t + \Delta t) = x + \Delta X$. We observe that

$$\int_t^{t+\Delta t} f(X,\ D,\ s)ds = f(x,\ D,\ t)\Delta t + o(\Delta t). \qquad (7.5.6)$$

A Taylor series expansion yields

$$J(t + \Delta t,\ x + \Delta X) = J(t,\ x) + J_t(t,\ x)\Delta t + J_x(t,\ x)\Delta X$$

$$+ \frac{1}{2}J_{tt}(t,\ x)\Delta t^2 + \frac{1}{2}J_{xx}(t,\ x)\Delta X^2$$

$$+ J_{xt}(t,\ x)\Delta t\Delta X + o(\Delta t). \qquad (7.5.7)$$

Substituting Equations 7.5.6 and 7.5.7 into Equation 7.5.5 and ignoring terms involving $o(\Delta t)$, we obtain

$$J(t, x) = \max_{D} E\{f(x, D, t)\Delta t + J(t, x) + J_t(t, x)\Delta t + J_x(t, x)\Delta X$$

$$+ \frac{1}{2}J_{xx}(t, x)\Delta X^2 + J_{xt}(t, x)\Delta t \Delta X\}$$

or

$$0 = \max_{D} E\left\{\begin{matrix} f(x, D, t)\Delta t + J_t(t, x)\Delta t + J_x(t, x)\Delta X \\ + \frac{1}{2}J_{xx}(t, x)\Delta X^2 + J_{xt}(t, x)\Delta t \Delta X\} \end{matrix}\right\}$$

$$= \max_{D}\left\{\begin{matrix} f(x, D, t)\Delta t + J_t(t, x)\Delta t + J_x(t, x)E[\Delta X] \\ + \frac{1}{2}J_{xx}(t, x)E[\Delta X^2] + J_{xt}(t, x)E[\Delta t \Delta X] \end{matrix}\right\}.$$

From Equations 7.3.4 and 7.3.6, we infer $E[\Delta X] = \mu(x, D, t)\Delta t$ and $E[\Delta X^2] = \sigma^2(x, D, t)\Delta t$, respectively. The last equality also implies that $E[\Delta X \Delta t] = E[\Delta X]\Delta t = \mu(x, D, t)\Delta t^2 = o(\Delta t)$. Using these observations, we obtain

$$0 = \max_{D}\left\{\begin{matrix} f(x, D, t)\Delta t + J_t(t, x)\Delta t + J_x(t, x)\mu(x, D, t)\Delta t \\ + \frac{1}{2}J_{xx}(t, x)\sigma^2(x, D, t)\Delta t \end{matrix}\right\}.$$

Dividing the previous equation by Δt, we find the following fundamental result in stochastic control theory:

$$-J_t(t, x) = \max_{D}\left\{\begin{matrix} f(x, D, t) + J_x(t, x)\mu(x, D, t) \\ + \frac{1}{2}J_{xx}(t, x)\sigma^2(x, D, t) \end{matrix}\right\}. \qquad (7.5.8)$$

Equation 7.5.8 is known as the *Hamilton-Jacobi-Bellman equation.*

The boundary condition for the preceding problem is $J(T, X(T)) = B(X(T), T)$.

The Hamilton-Jacobi-Bellman equation can be generalized to several state variables. To illustrate, we look at the case of two state variables X_1 and X_2 with

$$dX_i = \mu_i(\vec{X}, D, t)dt + \sigma_i(\vec{X}, D, t)dZ_i \qquad i = 1, 2,$$

where $\vec{X} = \{X_1, X_2\}$, D is the decision variable, and other components are defined as in Equation 7.4.1—specifically, $dZ_1 dZ_2 = \rho dt$. In the two-dimensional generalization, Equations 7.5.1 through 7.5.6 remain unchanged except that we now use $\vec{x}$, $\vec{\Delta X}$, and $\vec{X}$ to denote the two-dimensional analogs of x, ΔX, X. The only nonnotational modification occurs at Equation 7.5.7. From the multidimensional Ito's lemma Equation 7.4.7, we see that

$$J(t+\Delta t, \ \vec{x}+\vec{\Delta} X) = J(t+\vec{x}) + J_t(t, \ \vec{x})\Delta t + J_{x_1}(t, \ \vec{x})\Delta X_1 + J_{x_2}(t, \ \vec{x})\Delta X_2$$

$$+\frac{1}{2}J_{x_1 x_1}(t, \ \vec{x})\Delta X_1^2 + \frac{1}{2}J_{x_2 x_2}(t, \ \vec{x})\Delta X_2^2 + J_{x_1 x_2}(t, \ \vec{x})\Delta X_1 \Delta X_2$$

$$+ J_{x_1 t}(t, \ \vec{x})\Delta X_1 \Delta t + J_{x_1 t}(t, \ \vec{x})\Delta X_2 \Delta t + o(\Delta t).$$

Taking the expectation of the previous expression, using $E[\Delta X_i] = \mu_i(\vec{x}, \ D, \ t)\Delta t$, $E[\Delta X_i^2] = \sigma_i^2(\vec{x}, \ D, \ t)\Delta t$, $E[\Delta X_1 \Delta X_2] = \sigma_1(\vec{x}, \ D, \ t)\sigma_2(\vec{x}, \ D, \ t)\rho\Delta t$, and $E[\Delta X_i \Delta t]$ $= o(\Delta t)$ and ignoring terms involving $o(\Delta t)$, we obtain

$$E[J(t+\Delta t, \ \vec{x}+\vec{\Delta} X)] = J(t, \ \vec{x}) + J_t(t, \ \vec{x})\Delta t + J_{x_1}(t, \ \vec{x})\mu_1(\vec{x}, \ D, \ t)\Delta t$$

$$+ J_{x_2}(t, \ \vec{x})\mu_2(\vec{x}, \ D, \ t)\Delta t + \frac{1}{2}J_{x_1 x_1}(t, \ \vec{x})\sigma_1^2(\vec{x}, \ D, \ t)\Delta t$$

$$+\frac{1}{2}J_{x_2 x_2}(t, \ \vec{x})\sigma_2^2(\vec{x}, \ D, \ t)\Delta t$$

$$+ J_{x_1 x_2}(t, \ \vec{x})\sigma_1(\vec{x}, \ D, \ t)\sigma_2(\vec{x}, \ D, \ t)\rho\Delta t.$$

Using the given result, we mimic the steps leading to Equation 7.5.8 to find the two-dimensional Hamilton-Jacobi-Bellman equation. It is given by

$$-J_t(t, \ \vec{x}) = \max_D \{ f(\vec{x}, \ D, \ t) + J_{x_1}(t, \ \vec{x})\mu_1(\vec{x}, \ D, \ t) + J_{x_2}(t, \ \vec{x})\mu_2(\vec{x}, \ D, \ t)$$

$$+\frac{1}{2}J_{x_1 x_1}(t, \ \vec{x})\sigma_1^2(\vec{x}, \ D, \ t) + \frac{1}{2}J_{x_2 x_2}(t, \ \vec{x})\sigma_2^2(\vec{x}, \ D, \ t)$$

$$+ J_{x_1 x_2}(t, \ \vec{x})\sigma_1(\vec{x}, \ D, \ t)\sigma_2(\vec{x}, \ D, \ t)\rho. \tag{7.5.9}$$

We rewrite Equation 7.5.9 in a matrix-vector notation to make it suggestive for easy generalization to higher dimensions. First, we define

$$\vec{J}_1 = \begin{bmatrix} J_{x_1}(t, \ \vec{x}), & J_{x_2}(t, \ \vec{x}) \end{bmatrix}$$

$$\vec{\mu} = \begin{bmatrix} \mu_1(\vec{x}, \ D, \ t) \\ \mu_2(\vec{x}, \ D, \ t) \end{bmatrix}$$

$$J_2 = \begin{bmatrix} J_{x_1 x_1}(t, \ \vec{x}) & J_{x_1 x_2}(t, \ \vec{x}) \\ J_{x_2 x_1}(t, \ \vec{x}) & J_{x_2 x_2}(t, \ \vec{x}) \end{bmatrix}$$

$$\Sigma = \begin{bmatrix} \sigma_1^2(\vec{x}, \ D, \ t) & \rho\sigma_2(\vec{x}, \ D, \ t)\sigma_1(\vec{x}, \ D, \ t) \\ \rho\sigma_1(\vec{x}, \ D, \ t)\sigma_2(\vec{x}, \ D, \ t) & \sigma_2^2(\vec{x}, \ D, \ t) \end{bmatrix},$$

where Σ is the variance-covariance matrix. Recalling that the trace of a square matrix is the sum of its diagonal elements, we state Equation 7.5.9 in matrix-vector notation as follows:

$$-J_t(t, \ \vec{x}) = \max_D \left\{ f(\vec{x}, \ D, \ t) + \vec{J}_1 \vec{\mu} + \frac{1}{2} trace(J_2 \Sigma) \right\}. \tag{7.5.10}$$

Generalization of Equation 7.5.10 to cases in which the number of state variables exceeding two is now obvious.

We note that the various versions of the Hamilton-Jacobi-Bellman equation (Equations 7.5.8, 7.5.9, and 7.5.10) give necessary conditions for an extremum. If f is a concave function of its arguments, then they will produce a maximum and if f is a convex function of its arguments, then they will produce a minimum.

EXAMPLE 7.5.1

Consider the following optimization problem:

$$\min_{D} E\left[\int_0^T e^{-\beta t}(aX^2 + bD^2)dt \right]$$

subject to

$$dX = Ddt + \sigma X dZ,$$

where $a > 0$, $b > 0$, $\sigma > 0$, β is a discount factor, T is large, and dX is an Ito process. In this case, we have $f(x, D, t) = e^{-\beta t}(aX^2 + bD^2)$. Applying Equation 7.5.8, we obtain

$$-J_t = \min_{D}\left\{ e^{-\beta t}(ax^2 + bD^2) + J_x D + \frac{1}{2}J_{xx}\sigma^2 x^2 \right\} = \min_{D} C(D), \qquad (7.5.11)$$

where $C(D)$ denotes the term in the braces. Setting $dC(D)/dD = 0$ yields

$$2bDe^{-\beta t} + J_x = 0 \qquad \text{or} \qquad D(x) = \frac{-J_x e^{\beta t}}{2b}.$$

Substituting the last equality into Equation 7.5.11, we find

$$-J_t = e^{-\beta t}\left(ax^2 + b\frac{J_x^2 e^{2\beta t}}{4b^2} \right) - J_x\left(\frac{J_x e^{\beta t}}{2b} \right) + \frac{1}{2}J_{xx}\sigma^2 x^2$$

or

$$-e^{\beta t}J_t = ax^2 - \frac{J_x^2 e^{2\beta t}}{4b} + \frac{1}{2}e^{\beta t}J_{xx}\sigma^2 x^2. \qquad (7.5.12)$$

The preceding is a partial differential equation. We now conjecture that $J(t, x) = g(x)f(t)$ and, moreover, $g(x) = kx^2$ and $f(t) = e^{-\beta t}$. This gives

$$J_t = -\beta e^{-\beta t}kx^2 \qquad \text{or} \qquad -J_t e^{\beta t} = \beta kx^2$$

$$J_x = 2kxe^{-\beta t} \qquad \text{or} \qquad J_x e^{\beta t} = 2kx$$

$$J_{xx} = 2ke^{-\beta t} \qquad \text{or} \qquad J_{xx} e^{\beta t} = 2k.$$

Using the given expressions in Equation 7.5.12, we obtain the condition for optimality

$$\beta kx^2 = ax^2 - \frac{k^2 x^2}{b} + k\sigma^2 x^2 \qquad \text{or} \qquad \frac{k^2}{b} + (\beta - \sigma^2)k - a = 0.$$

Since f is a convex function of its arguments, we conclude that the k obtained from solving the previous quadratic equation yields the minimal expected discounted cost. The solution is given by

$$k = \frac{b\left[(\sigma^2 - \beta) + \sqrt{(\beta - \sigma^2)^2 - 4a/b}\right]}{2},$$

where we pick the value of k such that $J(t, x) \geq 0$. The optimal decision is linear as $D(x) = -kx/b$ and the optimal value of the objective function is $J(t, x) = e^{-\beta t}kx^2$. ∎

An Infinite Horizon, Discounted, Time-Homogeneous Problem

To consider a discounted infinite horizon problem, we assume that the objective function f, drift μ, and volatility σ are independent of time. Hence we have $f(X, D)$, $\mu(X, D)$, and $\sigma(X, D)$ replacing $f(X, D, t)$, $\mu(X, D, t)$, and $\sigma(X, D, t)$. This version of the problem is called the time-homogeneous problem. The time-homogeneity assumption eliminates the time variable from the resulting partial differential equation and simplifies its solution. We state the problem of interest to us as follows:

$$V(x) = \max_D E\left\{\int_t^\infty e^{-\beta s} f(X, D)ds\right\} \tag{7.5.13}$$

subject to

$$dX = \mu(X, D)dt + \sigma(X, D)dZ \quad \text{and} \quad X(t) = x. \tag{7.5.14}$$

The roles played by X, D, and f in Equation 7.5.13 are identical to those defined in Equation 7.5.1. Similarly, Equation 7.5.14 is defined as in Equation 7.5.2 except that now we assume time homogeneity in μ and σ.

At time 0, the present value of the objective function is given by $J = e^{-\beta t}V(x)$. We see that $J_t = -\beta e^{-\beta t}V$, $J_x = e^{-\beta t}V_x$, and $J_{xx} = e^{-\beta t}V_{xx}$. Using these relations in Equation 7.5.8, we obtain

$$\beta e^{-\beta t}V(x) = \max_D\left\{e^{-\beta t}f(x, D) + e^{-\beta t}V_x(x)\mu(x, D) + \frac{1}{2}e^{-\beta t}V_{xx}(x)\sigma^2(x, D)\right\}$$

$$\tag{7.5.15}$$

or

$$\beta V(x) = \max_D\left\{f(x, D) + V_x(x)\mu(x, D) + \frac{1}{2}V_{xx}(x)\sigma^2(x, D)\right\}, \tag{7.5.16}$$

where we observe that the immediate return at time t, $f(x, D)$, is also being discounted back to its present value. Equation 7.5.16 is called the present-value version of the Hamilton-Jacobi-Bellman equation. We can attach economical interpretation to Equation 7.5.16. We can view $f(x, D)dt$ as the expected cash flow and, from Ito's lemma (Equation 7.3.8), $[V_x(x)\mu(x, D) + (1/2)V_{xx}(x)\sigma^2(x, D)]dt$ as the expected capital gain. The term on the right side of Equation 7.5.16 is then the total return per unit time of the optimally managed asset J. Such a return must be equivalent to the risk-free return βV in a risk-neutral economy and hence the equality of the two sides of Equation 7.5.16.

EXAMPLE 7.5.2

We return to Example 7.5.1 with an explicit assumption of an infinite horizon. Following Equation 7.5.17, the Hamilton-Jacobi-Bellman equation reads

$$\beta V(x) = \max_D \left\{ ax^2 + bD^2 + V_x(x)D + \frac{1}{2}V_{xx}(x)\sigma^2 X^2 \right\} = \min_D C(D), \qquad (7.5.17)$$

where $C(D)$ denotes the term in the braces. Setting $dC(D)/dD = 0$, we obtain the necessary condition for optimality $D(x) = -V_x(x)/2b$. Using the last expression in Equation 7.5.17, we obtain the following differential equation:

$$\beta V(x) = ax^2 - \frac{[V_x(x)]^2}{4b} + \frac{1}{2}V_{xx}(x)\sigma^2 x^2.$$

We conjecture that the solution is of the form $V(x) = kx^2$. Upon substitution and dividing both sides by x^2, we obtain

$$\beta k = a - \frac{k^2}{b} + k\sigma^2 \qquad \text{or} \qquad \frac{k}{b} + (\beta - \sigma^2)k - a = 0.$$

The quadratic equation characterizing k is identical to that obtained in Example 7.5.1. We see that the present approach makes it easier to find an optimizing D. ∎

EXAMPLE 7.5.3

Consider an asset that pays a perpetual cash flow at a rate of $X(t)dt$, where X follows arithmetic Brownian motion process with drift D and volatility σ. The cash flow may become negative. When this occurs, the firm must raise equity to meet the cash requirements. A manager may influence the growth rate D of X at a cost of D^2dt to the shareholders. The problem is to find the maximal present value of the cash flows net of management expenses obtainable. To state the problem in the form of Equation 7.5.13, we have

$$V = \max_D E\left\{ \int_0^\infty e^{-\beta t}(X - D^2)dt \right\}$$

subject to

$$dX = Ddt + \sigma dZ.$$

Using Equation 7.5.16, we write the Hamilton-Jacobi-Bellman equation as follows:

$$\beta V = \max_D \left\{ x - D^2 + DV_x + \frac{1}{2}\sigma^2 V_{xx} \right\} = \max_D C(D), \qquad (7.5.18)$$

where $C(D)$ is the term enclosed by the braces. Setting $dC(D)/dD = 0$, we find $V_x = 2D$. Since $C(D)$ is concave in D, the solution gives the maximum. The solution of $V_x = 2D$ is $V = 2Dx + E$, where E is a constant. So we have $V_{xx} = 0$ (here, we assume that $D[x]$ is a constant for all x). Using these results in Equation 7.5.18, we obtain

$$\beta(2Dx + E) = x - D^2 + 2D^2 \quad \text{or} \quad 2\beta Dx + \beta E = x + D^2.$$

Matching the coefficients of the last polynomial in x, we require $2\beta D = 1$ and $\beta E = D^2$. This gives $D = 1/(2\beta)$ and $E = 1/(4\beta^3)$. The present value of the cash flows net management expenses is

$$V = \frac{x}{\beta} + \frac{1}{4\beta^3},$$

where $X(t) = x$, and the optimal decision is to set $D = 1/(2\beta)$. ∎

EXAMPLE
7.5.4*

Merton's Portfolio Selection Problem Revisited   We return to one of the classical problems in financial economics of allocating personal wealth among consumption, investment in a single risk asset, and investment in a risk-free security. In addition to the notations introduced in Example 7.3.3, we let c denote the consumption rate. Following the development of Example 7.3.3, the wealth $X(t)$ of the investor is described by the Ito process

$$dX = [\mu w X + s(1 - w)X - c] + \sigma w X dZ. \tag{7.5.19}$$

In Equation 7.5.19, we see that the growth rate is reduced by consumption by an amount c. Assume that the investor has a utility function of the form $f(c) = c^\gamma / \gamma$, where $\gamma < 1$, and the investor is interested in maximizing the expected utility over an infinite time horizon. The objective function of the problem is

$$\max_{c,w} E\left\{ \int_0^\infty e^{-bt} f(c)dt \right\} = \max_{c,w} E\left\{ \int_0^\infty e^{-bt} \frac{c(t)^\gamma}{\gamma} dt \right\}, \tag{7.5.20}$$

where we recall that decision variable w represents the fraction of wealth invested in risky asset. In Equation 7.5.20, we have two decision variables, c and w. The constraint is given by Equation 7.5.19. While Equation 7.5.16 was developed for a single decision variable, it is equally applicable to the case of two decision variables. Hence the Hamilton-Jacobi-Bellman equation reads

$$\beta V(x) = \max_{c,w}\left\{ \frac{c^\gamma}{\gamma} + V_x(x)[\mu w x + s(1-w)x - c] + \frac{1}{2}V_{xx}(x)\sigma^2 x^2 w^2 \right\}$$

$$= \max_{c,w} C(c, w), \tag{7.5.21}$$

where $C(c, w)$ represents the term enclosed by the braces. We now find the first-order conditions

$$\frac{\partial}{\partial c} C(c,w) = c^{\gamma-1} - V_x(x) = 0 \qquad \Rightarrow \qquad c = \left[V_x(x)\right]^{\frac{1}{\gamma-1}}$$

$$\frac{\partial}{\partial w} C(c,w) = V_x(x)[\mu - s]x + \sigma^2 x^2 V_{xx} w = 0 \qquad \Rightarrow \qquad w = \frac{V_x(x)[s - \mu]}{\sigma^2 x V_{xx}(x)}.$$

Substituting the preceding results into Equation 7.5.21, we obtain the necessary condition for optimality

$$\beta V(x) = \left[V_x(x)\right]^{\frac{\gamma}{\gamma-1}} + sxV_x(x) - \frac{(s-\mu)^2[V_x(x)]^2}{\sigma^2 V_{xx}(x)} + \frac{(s-\mu)^2[V_x(x)]^2}{2\sigma^2 V_{xx}(x)}$$

$$= \left[V_x(x)\right]^{\frac{\gamma}{\gamma-1}} + sxV_x(x) - \frac{(s-\mu)^2[V_x(x)]^2}{2\sigma^2 V_{xx}(x)}. \tag{7.5.22}$$

The objective function is concave with respect to c. It can be shown that the given condition is also sufficient for optimality. We conjecture that the solution to the preceding nonlinear second-order differential equation is of the form $V(x) = AX^{\gamma}$. Using the last equality, we obtain $V_x(x) = \gamma A x^{\gamma-1}$ and $V_{xx}(x) = \gamma(\gamma-1)Ax^{\gamma-2}$. Upon substituting these identities into Equation 7.5.22 and dividing both sides by x^{γ}, we find

$$\beta A = (\gamma A)^{\frac{\gamma}{\gamma-1}} + s\gamma A + \frac{(s-\mu)^2 \gamma A}{2\sigma^2(1-\gamma)} \quad \text{or} \quad (\gamma A)^{\frac{\gamma}{\gamma-1}} = \left[\beta - s\gamma - \frac{(s-\mu)^2 \gamma}{2\sigma^2(1-\gamma)}\right]A.$$

We rewrite the last equation as

$$(\gamma A)^{\frac{\gamma}{\gamma-1}} = \left[\frac{\beta - s\gamma - \dfrac{(s-\mu)^2 \gamma}{2\sigma^2(1-\gamma)}}{\gamma}\right]\gamma A \quad \text{or} \quad (\gamma A)^{\frac{1}{\gamma-1}} = \left[\frac{\beta - s\gamma - \dfrac{(s-\mu)^2 \gamma}{2\sigma^2(1-\gamma)}}{\gamma}\right].$$

We conclude that

$$\gamma A = \left[\frac{\beta - s\gamma - \dfrac{(s-\mu)^2 \gamma}{2\sigma^2(1-\gamma)}}{\gamma}\right]^{\gamma-1}. \tag{7.5.23}$$

Using Equation 7.5.23, the optimal consumption rate can be found from

$$c(x) = (\gamma A x^{\gamma-1})^{\frac{1}{\gamma-1}} = x(\gamma A)^{\frac{1}{1-\gamma}}.$$

We see that the optimal solution calls for the investor to consume a constant fraction of wealth at each moment. The optimal fraction depends on all the parameters—β, μ, γ, s, and σ^2. The optimal division of wealth between the two kinds of investment is determined by

$$w(x) = \frac{(\mu - s)}{(1-\gamma)\sigma^2}.$$

We note that the optimal fraction of investment on risky asset is independent of the total wealth. It varies directly with the expected return of the risky assets and inversely with its variance. ∎

Problems

1 Consider the following stochastic differential equation

$$dY = 3Y^{1/3}dt + 3Y^{2/3}dZ,$$

where $Y(0) = 8$. Solve for $Y(t)$.

2 Let $Z = \{Z(t), t \geq 0\}$ be a standard Brownian motion process. For $0 \leq s < t$, find $Cov[Z(s), Z(t)]$.

3 Let $Z = \{Z(t), t \geq 0\}$ be a standard Brownian motion process. For all $t \geq 0$ and some fixed $s \geq 0$, we define $Y(t) = Z(t + s) - Z(s)$. Show that $Y = \{Y(t), t \geq 0\}$ is also a standard Brownian motion process.

4 Let $Z = \{Z(t), t \geq 0\}$ be a standard Brownian motion process. Define $Y(0) = 0$ and let $Y(t) = tZ(1/t)$. (a) Show that $Y = \{Y(t), t \geq 0\}$ is also a standard Brownian motion process. (b) For $0 \leq s < t$, find $Cov[Y(s), Y(t)]$.

5 Let $Z = \{Z(t), t \geq 0\}$ be a standard Brownian motion process. Define

$$Y(t) = \int_0^t Z(u)du.$$

The stochastic process $Y = \{Y(t), t \geq 0\}$ is called *integrated Brownian motion*. The definition implies the rate of change, $dY(t)/dt$, follows Brownian motion. (a) Find $E[Y(t)]$. (b) Find $E[(Y(t))^2]$. (c) For $0 \leq s < t$, find $Cov[Y(s), Y(t)]$. (d) Is Y a Gaussian process? (e) Is Y a Markov Process? Elaborate.

6 Suppose we want to model the cash balance of a bank by a stochastic differential equation. Assume that the average rate at which cash is demanded at time t is $(1.02)^t$, the variance in the demand for cash is proportional to the average rate at which cash is demanded (that is, $Var[D] = \sigma^2(1.02)^t$), and the bank receives a constant inflow of cash at a rate of C per unit time. Write a stochastic differential equation that would provide an appropriate description of the bank's cash balance.

7 **Ornstein-Uhlenbeck Processes**   Let $Z = \{Z(t), t \geq 0\}$ be a standard Brownian motion process. We have nonnegative constants μ, σ, and η. Define the following stochastic differential equation

$$X = \eta(\mu - X)dt + \sigma dZ$$

so that $\mu(X, t) = \eta(\mu - X)$ and $\sigma(X, t) = \sigma$. The previous stochastic differential equation is a special case of mean-reverting processes for modeling the prices of raw materials such as aluminum or oil. In the short run the price of a raw material might fluctuate in response to ongoing events. In the long run the price ought to revert back to the marginal cost of production μ. The expected change in X depends on the difference between X and μ. Define

$$Y = (X - \mu)e^{\eta(t - t_0)}.$$

(a) Use Ito's lemma to establish

$$dY = e^{\eta(t - t_0)}dZ.$$

(b) From the last expression, what can you say about the distributional form of X?

8 We write Equation 7.1.3 as

$$f(x, t; y) = \varphi(x, \sigma^2 t, y + \mu t),$$

where

$$\varphi(x, t, y) = \frac{1}{\sqrt{2\pi t}} e^{-\frac{(x-y)^2}{2t}} \qquad t > 0, \quad -\infty < x < \infty, \quad -\infty < y < \infty$$

and φ is called the *Gauss kernel*. **(a)** Show that Equation 7.1.3 satisfies the backward diffusion equation (Equation 7.1.1)

$$f_t = \mu f_y + \frac{1}{2} \sigma^2 f_{yy}$$

by differentiating $f(x, t; y)$ with respect to y and t using the chain rule (note that φ satisfies the heat equation $\varphi_t = (1/2)\varphi_{yy}$). **(b)** We now rewrite Equation 7.1.3 as

$$f(x, t; y) = \varphi(x, \sigma^2 t, y + \mu t).$$

Show that Equation 7.1.3 satisfies the forward equation (Equation 7.1.2)

$$f_t = -\mu f_x + \frac{1}{2} \sigma^2 f_{xx}.$$

9 Let $Z = \{Z(t), t \geq 0\}$ be a standard Brownian motion process. Define the following stochastic differential equation

$$dX = \left(\frac{1}{4}\right) dt + \sqrt{X} dZ,$$

with $X(0) = 1$. Show that this stochastic differential equation has the solution

$$X(t) = \left(1 + \frac{1}{2} Z(t)\right)^2.$$

*10 The Forward Diffusion Equation   Consider a Brownian motion process $Z = \{Z(t), t \geq 0\}$ with drift parameter μ and variance parameter σ^2. Define

$$f(x, t; y) dx = P\{x < Z(t) < x + dx | Z(0) = y\}.$$

Derive the forward diffusion equation (Equation 7.1.2)

$$\frac{\partial}{\partial t} f(x, t; y) + \mu \frac{\partial}{\partial x} f(x, t; y) = \frac{1}{2} \sigma^2 \frac{\partial^2}{\partial x^2} f(x, t; y).$$

(Hint: Write $Z(t) = [Z(t) - Z(t-h)] + Z(t-h)$ and note that $Z(t) - Z(t-h)$ and $Z(t-h)$ are independent. Moreover, $Z(t) - Z(t-h)$ follows $N(\mu h, \sigma^2 h)$. By conditioning on $Z(t) - Z(t-h)$, use the law of total probability to express $f(x, t; y)$ in terms of $f(x-w, t-h; y)$ and the normal density $n(w; \mu h, \sigma^2 h)$, where $w = Z(t) - Z(t-h)$.)

*11 Consider a Brownian motion process with drift μ and variance parameter σ^2. Use the Laplace transform of the first-passage-time density of Equation 7.1.5 to verify that the mean and variance of the first passage time are given by Equation 7.1.7.

12 Consider a Brownian motion process $Z = \{Z(t), t \geq 0\}$ with drift $\mu = 5$, variance parameter $\sigma^2 = 100$, and $Z(0) = 0$. Let T_x denote the first passage time to reach x. Consider the case in which $x = 10$. **(a)** Compute $P\{T_x > 10\}$ by numerically inverting the Laplace transform. **(b)** Compute $P\{T_x > 10\}$ by numerically integrating the density of T_x.

13 We return to the numerical problem considered in Example 7.1.3. Assume that the discount factor $\beta = 0.1$ **(a)** Does it make sense to have the warrant under the circumstances? **(b)** If the answer to Part a is affirmative, what is the market price of the stock x^* at which it is optimal to exercise the option to buy? **(c)** What is the expected profit under Part b? **(d)** Find the expected time to reach x^*, $E[T_{x^*}]$, and the corresponding variance $Var[T_{x^*}]$. **(e)** Plot the expected profits $g(x, y)$ under market prices $x = 0.1{:}0.1{:}10$.

14 Let $Z = \{Z(t), t \geq 0\}$ be a standard Brownian motion process with $Z(0) = 0$. Define $X(t) = |Z(t)|$. Thus $X = \{X(t), t \geq 0\}$ is a standard Brownian motion process reflected at the origin. **(a)** Show that

$$P\{X(t) \leq x\} = \frac{2}{\sqrt{2\pi t}} \int_{-\infty}^{x} e^{-\frac{\mu^2}{2t}} du.$$

(b) Find the density of $X(t)$, $f_{X(t)}(x)$. **(c)** Show that

$$E[X(t)] = \sqrt{\frac{2t}{\pi}} \quad \text{and} \quad Var[X(t)] = \left(1 - \frac{2}{\pi}\right)t.$$

Bibliographic Notes

Brownian motion is covered in many texts on stochastic processes. The list includes Cox and Miller (1967), Heyman and Sobel (1982), Karlin and Taylor (1975), Parzen (1964), Resnick (1992), and Ross (1983). For an exposition at an intermediate level, Karlin and Taylor (1981) and Harrison (1985) are good sources. The first section follows the development of Cox and Miller (1967). Examples 7.1.2 and 7.1.3 are based on Heyman (1982) and Karlin and Taylor (1975, 363–65), respectively. The generalizations of diffusion process described in Section 7.2 are related to materials shown in Kleinrock (1976, Chapter 2). Examples 7.2.1, 7.2.2, 7.2.3, and 7.2.4 are from Kobayashi (1974), Heyman (1975), Gaver and Shedler (1973), and Halachmi and Franta (1978), respectively. The exposition on Ito's calculus is influenced by Hull (1993), Ingersoll (1987, Chapter 16), Malliaris and Brock (1982), Shimko (1992), and Winston (1981). The expression for finding even-numbered higher moments of a normal process is given in Parzen (1964, 95, Example 4E). Example 7.3.3 is based on Merton (1971). Examples 7.3.6, 7.3.7, and 7.3.8 are discussed at length in Hull (1993). The derivation of Equation 7.3.27 in Example 7.3.9 can be found in Chow (1979). Example 7.4.1 is from Fischer (1972). Additional materials on the capital asset pricing model of Example 7.4.2 can be found in Merton (1994) and the numerous references given there. The last section is based on Winston (1981). The survey paper by Karatzas

(1989) provides additional references on the applications of optimal control of stochastic differential equations in finance. The book by Harrison (1985) also addresses the optimal control of Brownian motion process. It contains several interesting applications in the last two chapters. The use of dynamic programming in optimization over Ito's processes is discussed in Dixit and Pindyck (1994). Problems 1 and 6 are from Winston (1981). Additional reading about Problem 7 can be found in Ingersoll (1987, 351). Part a of Problem 8 is based on Karlin and Taylor (1981, 217, Example 2). Problem 9 is from Åström (1970, 76–77).

References

Åström, K. J. 1970. *Introduction to Stochastic Control Theory*. New York: Academic Press.

Black, F., and M. Scholes. 1973. The Pricing of Options and Corporate Liabilities. *Journal of Political Economy* 81:637–54.

Chow, G. C. 1979. Optimum Control of Stochastic Differential Equation Systems. *Journal of Economic Dynamics and Control* 1:143–75.

Cox, D. R., and H. D. Miller. 1967. *The Theory of Stochastic Processes*. New York: John Wiley & Sons.

Dixit, A. K., and R. S. Pindyck. 1994. *Investment under Uncertainty*. Princeton, NJ: Princeton University Press.

Fischer, S. 1972. The Demand for Index Bonds. *Journal of Political Economy* 83:509–34.

Gaver, D. P., and G. S. Shedler. 1973. Processor Utilization in Multiprogramming Systems via Diffusion Approximations. *Operations Research* 21(2):569–76.

Halachmi, B., and W. R. Franta. 1978. A Diffusion Approximation to the Multi-Server Queue. *Management Science* 24(3):522–29.

Harrison, J. M. 1985. *Brownian Motion and Stochastic Flow Systems*. New York: John Wiley & Sons.

Heyman, D. P. 1975. A Diffusion Approximation for $GI/G/1$ Queue in Heavy Traffic. *The Bell System Technical Journal* 54:1637–46.

Heyman, D. P. 1982. Mathematical Models of Database Degradation. *ACM Transactions on Database Systems* 7(4): 615–31.

Heyman, D. P., and M. J. Sobel. 1982. *Stochastic Models in Operations Research, Volume 1: Stochastic Processes and Operating Characteristics*. New York: McGraw-Hill.

Hull, J. 1993. *Options, Futures, and Other Derivative Securities*. Englewood Cliffs, NJ: Prentice-Hall.

Ingersoll, J. E. 1987. *Theory of Financial Decision Making*. Totowa, NJ: Rowman & Littlefield.

Karatzas, I. 1989. Optimization Problems in the Theory of Continuous Trading. *SIAM Journal on Control and Optimization* 27(6):1221–59.

Karlin, S., and H. M. Taylor. 1975. *A First Course in Stochastic Processes*. 2nd ed. New York: Academic Press.

Karlin, S., and H. M. Taylor. 1981. *A Second Course in Stochastic Processes*. New York: Academic Press.

Kleinrock, L. 1976. *Queueing Systems, Volume II: Computer Applications*. New York: John Wiley & Sons.

Kobayashi, H. 1974. Application of the Diffusion Approximation to Queueing Networks I: Equilibrium Queue Distributions. *Journal of the Associate for Computing Machinery* 21(2):316–28.

Malliaris, A. G., and W. A. Brock. 1982. *Stochastic Methods in Economics and Finance*. Amsterdam: North Holland.

Merton, R. C. 1971. Optimal Consumption and Portfolio Rules in a Continuous Time Model. *Journal of Economic Theory* 3:373–413.

Merton, R. C. 1994. *Continuous-Time Finance*. Cambridge, MA: Blackwell.

Parzen, E. 1964. *Stochastic Processes*. San Francisco: Holden-Day.

Resnick, S. I. 1992. *Adventures in Stochastic Processes*. Boston: Birkhauser.

Ross, S. M. 1983. *Stochastic Processes*. New York: John Wiley & Sons.

Shimko, D. C. 1992. *Finance in Continuous Time: A Primer*. Miami, FL: Kolb Publishing Company.

Winston, W. L. 1981. *Notes on Brownian Motion, Stochastic Calculus, and Stochastic Control*. Bloomington: Graduate School of Business, Indiana University.

Appendix

Chapter 7: Section 1

Example 7.1.1   The program **e711** uses random walks to simulate a Brownian motion. The variable dt corresponds to the value of Δt. The output matrix contains the data used in constructing Figure 7.1.

```
»  [X]=e711;
»  function [X]=e711
%
%   Simulate a standard Wiener process
%
sigma=1; mu=0; dt=0.001; inc=0.1; Lt=10;
k=sigma*sqrt(dt);
p=0.5*(1+mu*sqrt(dt)/sigma); q=1-p; sumx=0; Xt=[sumx];
for bt=0.1:inc:Lt
    nx=inc/dt;
    ns=binornd(nx,p); nf=nx-ns; net=ns-nf;
    sumx=sumx+(net*k); Xt=[Xt sumx];
end
t=0.1:inc:Lt; t=[0 t]; n=length(t);
[X]=zeros(n,2); X(:,1)=t'; X(:,2)=Xt';
```

Example 7.1.2   Equation 7.1.6 is used to compute the first passage density for this example.

```
»  [X]=e712;
»  function [X]=e712_c
%
%   Pdf of the first passage time of Example 7.1.2
%
p=0.09; q=0.01; mu=p-q; var=p+q-mu^2; bx=301;
tx=1000:50:5000;
up=(bx-mu*tx).^2; dm=2*var*tx; qo=-up./dm;
ep=exp(qo);
dm=2*pi*(tx.^3); dm=dm.^0.5; dm=sqrt(var)*dm;
s=bx./dm; qx=s.*ep;
[m,n]=size(tx); X=zeros(n,2);
X(:,1)=tx'; X(:,2)=qx';
```

Example 7.1.3   This example illustrates the use of MATLAB supplied subroutine **normrnd** to simulate a geometric Brownian motion process. The output matrix D contains the data for constructing Figure 7.3.

```
» [D]=e713;
» function [D]=e713
%
%   Example 7.1.3: Simulate the geometric BM
%
mu=0.07; sigma2=0.02; y=0.7; z0=log(y); a=[0]; b=[y];
for t=0.1:0.1:10
  mut=mu*t; sigmat=sqrt(sigma2*t); z=normrnd(mut,sigmat);
  z=exp(z0+z); a=[a t]; b=[b z];
end
D=[a; b]';
```

Chapter 7: Section 2

Example 7.2.1   In the following, **e721** is the main calling program. Programs **e721a** and **e721b** contain the parameters for the interarrival time and service time distributions, respectively. Program **phaseak** is for generating the submatrices of the infinitesimal generator of a continuous-time Markov chain for the underlying queue. The exact solution for the queue length distribution is obtained using **gph1_sra.** Program **phasemv** computes the mean and variance of a phase-type distribution.

```
» [X]=e721;
  elapsed time =        0  N =      11
  elapsed time          0 seconds
  no. of iters          9  no. of flops        92793
»
function [a,T]=e721a
%
%   The interarrival time distribution:   Ph(a,T)
%   traffic intensity is   rho
%
rho=0.95; k=5*rho;
a=[1 0 0 0 0]; v=k*ones(1,5); v1=v(1,1:4);
T=diag(-v)+diag(v1,1);

function [b,S]=e721b
%
%   The service time distribution:   phase(b,S)
%
b=zeros(1,6); b(1,1)=.7; b(1,5)=.3;
v=6.67778512*ones(1,4); v1=v(1,1:3);
P=diag(-v)+diag(v1,1); P1=zeros(4,2); P=[P P1];
Q=zeros(2,4); v=1.033237*ones(1,2); v1=v(1,1);
Q1=diag(-v)+diag(v1,1); Q=[Q Q1]; S=[P;Q];

function [A]=phaseak(a,T,b,S)
%
%   Find matrices   Ak   for   k = 0, 1, ...
%
```

```
%   Input:   the interrenewal distribution  PH(a,T)
%            the distribution H(u) with phase representation  PH(b,S)
%            tolerance for termination  epsi
%
%   For PH/PH/1 queue, service time is PH(a,T), arrival time is
%   PH(b,S)
%
%   Reference:  P. 244, Theorem 5.1.5, of Neuts' 1989 M/G/1 Book
%
epsi=1.0e-8; time=clock;
[m,k]=size(T); [n,k]=size(S);
In=eye(n); Im=eye(m); em=ones(m,1); en=ones(n,1);
T0=-T*em; S0=-S*en;
Ib=dirpro(Im,b); Is=dirpro(Im,S0); Ti=dirpro(T0,In);
Ai=dirpro(a,In);
TI=dirpro(T,In); IS=dirpro(Im,S); TS=inv(TI+IS);
A=-Ib*TS*Is; C=-Ai*TS*Ti; D=-Ib*TS*Ti; E=-Ai*TS*Is; Ak=D*E;
er=max(max(Ak)); Ck=C; k=1;
%
while er >= epsi
      A=[A Ak]; k=k+1;
      Ak=D*Ck*E;
      Ck=Ck*C;
      er=max(max(Ak));
end
time=etime(clock,time); k = k-1;
fprintf(' elapsed time = %10.0f  N = %6.0f \n',time,k);

function [u1,v]=phasemv(a,T)
%
%   Find mean (u1) and variance (v) of phase-type distribution
%
[m,n]=size(T);
e=ones(n,1);
u1=-a*(inv(T))*e;
u2=2*a*(inv(T)*inv(T))*e;
v=u2-(u1*u1);

function [R]=gph1_sra(A)
%
%   Find  R  matrix  by the state reduction method
%   Ref:  Kao, ORSA Journal on Computing, Vol 3, No. 3, 1991
%
time=clock; flops(0); [m,n]=size(A); nb=n/m;
%
%   Construct the initial state reduction box
%
iz=1+(nb-1)*m; AN=A(:,iz:n); GA=AN; i=nb-1;
while i >= 1
    is=1+(i-1)*m; ie=i*m; GA=[GA A(:,is:ie)]; i=i-1;
end
GU=GA; GU(:,[1:m])=[]; GD=GA; GD(:,[iz:n])=[];
GD(:,iz-m:n-m)=GD(:,iz-m:n-m)+A(:,1:m); Z=[GU; GD];
it=0; epsi=1.0e-8; er=1.0; m1=m+1; En=zeros(2*m,m); R=zeros(m,m);
%
%   Do the state reduction
%
```

```
while  er > epsi
   it=it+1; Eo=En; j=1; Ro=R;
   while j <= m
       [nr,nc]=size(Z); nc1=1:nc-1; nr1=1:nr-1;
       c=Z(nr,nc1); cs=sum(c); d=Z(nr1,nc)./cs;
       En(nr1,m1-j)=d; Z=Z(nr1,nc1)+d*c; j=j+1;
   end
%  Find  R
   E0=En(1:m,:); E1=En(m1:2*m,:); R=zeros(m,m); R(:,1)=E0(:,1);
j=2;
   while j <= m
   i=1; R(:,j)=E0(:,j);
       while i < j
       R(:,j)=R(:,j)+E1(i,j)*R(:,i); i=i+1;
       end
   j=j+1;
   end
   er=max(max(abs(R-Ro))); Z=[AN Z]; Z=[GU; Z];
end
%
time=etime(clock,time); flop=flops;
fprintf(' elapsed time %10.0f seconds  \n',time);
fprintf(' no. of iters %10.0f  no. of flops %12.0f  \n',it,flop);
```

Example 7.2.2 This example uses programs **e721a** and **e721b** to obtain the input parameters. The output matrix X contains the data for constructing Figure 7.8.

```
» [X]=e722;
   elapsed time =       0  N =    11
   elapsed time        0 seconds
   no. of iters    .   9  no. of flops        92793

» function [X]=e722
%
%    Find virtual delay distribution of PH/PH/1 queue
%    Compare the results with a diffusion approximation
%
[a,T]=e721a; [b,S]=e721b;
[A]=phaseak(b,S,a,T);
[R]=gph1_sra(A);
[ma,va]=phasemv(a,T); [ms,vs]=phasemv(b,S);
rho=ms/ma; mu=abs(rho-1);
s2=(1/ma)*(vs+rho*rho*va); md=s2/(2*mu); mdi=1/md; y=[1-rho];
m=50;
for z=1:1:m
   x=1-rho*exp(-mdi*z); y=[y x];
end
%
% Exact computation
%
[n1,n2]=size(R); e=ones(n1,1); St=-S*e; SB=St*b;
TT=S+R*SB; b1=rho*b; y1=[];
for z=0:1:m
   x=1-sum(b1*expm(z*TT)); y1=[y1 x];
end
X=zeros(m+1,3); X(:,1)=(0:1:m)'; X(:,2)=y1'; X(:,3)=y';
```

Example 7.2.4 The main calling program **e724** does the diffusion approximation of a GI/G/c queue. The program calls **e724b** and **e724c** for the functions $g_2(x)$ and $g_1(x)$, respectively. The output matrix X contains the data for constructing Figure 7.11.

```
» [X]=e724a;
   mean queue length      =     21.46
  var of queue length     =     19.44
   mean queue length (actual)    =    21.41
  var of queue length (actual)   =    19.43

» function [X]=e724a
%
%    For Example 7.2.4
%
%    Find queue length distribution of M/M/c queue at
%    arbitrary time t - A diffusion approximation
%
mq=180;
global T1 T2 T3 T4 T5 T6
c=5; rho=0.95; mu=1;
lm=rho*c*mu; ma=1/lm; va=1/lm^2; ms=1/mu; vs=1/mu^2;
T1=va*(lm^3); T2=vs*(mu^3); T3=mu*sqrt(vs);
T4=(lm*sqrt(va)/T3)^2; T4=(2*lm/T2)*(T4+1);
T5=c*mu-lm; T6=T1+c*T2;
g2=quad('e724b',c,mq);
g1=quad('e724c',0,c);
H=[g1 e724c(c); g2 -e724b(c)]; HI=inv(H); h=HI(1,:); px=[];
cum_c=h(1,1)*g1; cum_o=0;
for k=1:mq
   k=k-0.5;
   if k < c
   y=h(1,1)*quad('e724c',0,k); z=y-cum_o; cum_o=y;
   px=[px z];
   else
   y=h(1,2)*quad('e724b',c,k); y=cum_c+y; z=y-cum_o;
   cum_o=y; px=[px z];
   end
end
%
x=ones(1,mq); x=cumsum(x)-1; EQ=sum(x.*px);
EX2=sum(px.*(x.^2)); VQ=sqrt(EX2-EQ^2);
fprintf('   mean queue length  = %8.2f \n',EQ);
fprintf('  var of queue length = %8.2f \n',VQ);
%
% Exact computation for M/M/c  queue
% Reference Gross and Harris, p. 87 (cf. Reference, Chapter 5)
%
r=lm/mu; sm=0;
for n=0:c-1
   if n==0
   nf=1;
   else
   ns=1:n; nf=cumprod(ns); nf=nf(1,n);
   end
   sm=sm+(r^n)/nf;
end
ns=1:c; nf=cumprod(ns); nf=nf(1,c); st=(c*r^c)/(nf*(c-r));
p0=1/(sm+st); pa=[p0];
for n=1:mq-1
```

```
    if n <= c
        ns=1:n; ns=cumprod(ns); ns=ns(1,n);
        x=p0*(lm^n)/(ns*mu^n); pa=[pa x];
    else
        x=p0*(lm^n)/((c^(n-c))*nf*(mu^n)); pa=[pa x];
    end
end
X=zeros(mq,3); X(:,1)=(1:mq)'; X(:,2)=px'; X(:,3)=pa';
x=ones(1,mq); x=cumsum(x)-1; EQ=sum(x.*pa);
EX2=sum(pa.*(x.^2)); VQ=sqrt(EX2-EQ^2);
fprintf('    mean queue length (actual)   = %8.2f \n',EQ);
fprintf('    var of queue length (actual)  = %8.2f \n',VQ);

» function [y]=e724b(x)
%
%    g2(x)
%
global T1 T2 T3 T4 T5 T6
[y]=exp((-2*T5*x)/T6);

» function [y]=e724c(x)
%
%    g1(x)
%
global T1 T2 T3 T4 T5 T6
y=(T1+x*T2).^(T4-1); y=y.*exp(-(2*x/(T3^2)));
```

Chapter 7: Section 9

Example 7.3.9   The program **e739** computes the value of a European call option.

```
» e739
 The value of the call option =     9.92
» function e739
%
%    Example 7.3.9 - European Option
%
St=80; c=75; Tt=0.5; sigma=0.2; r=0.1;
t1=log(St/c); t2=sigma*sqrt(Tt); t3=(r+sigma^2/2)*Tt;
t4=(r-sigma^2/2)*Tt;
z1=(t1+t3)/t2; z2=(t1+t4)/t2;
d1=normcdf(z1); d2=normcdf(z2);
x=St*d1-c*exp(-r*Tt)*d2;
fprintf(' The value of the call option = %8.2f \n',x);
```

Getting Started with MATLAB

This is a brief introduction to MATLAB with special emphasis on its use in the context of this text. The objectives are to give readers with no prior exposure to MATLAB a quick start and highlight some issues that may be confusing for novices. If you want a more detailed exposition on MATLAB, read Chapter 2, entitled "Tutorial," of *MATLAB User's Guide* (The Math-Works, Inc., Natick, MA, 1993). Another option is Chapter 5, entitled "MATLAB Tutorial," of *The Student Edition of MATLAB, Version 4, User's Guide* (Prentice-Hall, Englewood Cliffs, NJ, 1995). Of course, a more comprehensive source is *MATLAB Reference Guide* (The MathWorks, Inc., Natick, MA, 1992). To know more about MATLAB, you may want to visit the home page of MathWorks on the World Wide Web (http://www.mathworks.com).

MATLAB is available for various computing platforms ranging from work stations to personal computers. It works under operating systems such as UNIX, Microsoft Windows, Macintosh, and Open VMS. If you access MATLAB in a network environment, it is up to your system administrator to set up the software correctly; otherwise, you need to make the proper installation yourself. MATLAB is an interpretative language. Thus, you do not need to do any program compilation. From now on, we assume that MATLAB is in place and ready to work.

To launch MATLAB, you invoke the command **matlab** under a UNIX system or click the MATLAB icon in Microsoft Windows or the Macintosh desktop to get into the MATLAB environment. To exit from MATLAB, you enter the command **quit**. To interrupt an execution of a MATLAB run (for example, if you run into a nonterminating loop), use CTRL-C in a UNIX or Macintosh system or CTRL-BREAK in Microsoft Windows.

Preliminaries

MATLAB is case sensitive. Thus **A** and **a** refer to two different objects. For example, if you enter the following three lines, you get the answer that follows.

```
A =[1  2  3  4];
a =[4  5  6  7];
c =A+a

c =
      5     7     9    11
```

Note the roles played by the semicolons. If you use a semicolon at the end of a statement, the result of the computation is not displayed. This is illustrated as follows:

```
E=[1  2  3  4]; F=[4  5  6  7]; E=E+F;
```

If you need to know the answer, then enter

```
E

E =
    5    7    9    11
```

MATLAB is built on the idea of two-dimensional arrays (known as matrices). For example, if you enter

---

```
size(E)

ans =
    1    4
```

---

you see that the row vector E is a matrix of dimension 1×4. If you enter

---

```
length(E)

ans =
    4
```

---

then MATLAB gives it a row-vector interpretation and shows that E is a row vector of length 4. Here are a few useful matrix operations: + (addition), − (subtraction), * (multiplication), ^ (power), ' (transpose), \ (left division), and / (right division). You may want to experiment with these operations. Some of these operations deserve our special attention. Suppose that you enter the following three matrices

---

```
A=[1 2 3;  4 5 6];   B=[2 5; 5 8; 9 4];  C=[4 4 4; ...
5 5 5];
```

Note that matrices are entered row by row with semicolons separating the rows. If your input line overflows to the next line, use ellipsis points (...) to indicate it. Now if you do

---

```
D=A*B

D =
    39    33
    87    84
```

---

you see D is the product of A and B. On the other hand, if you enter

---

```
E=A.*C

E =
    4     8    12
    20    25    30
```

---

then you have entry-wise multiplications. The period preceding the * signals the entry-wise operations. To illustrate the use of period once more, you see

```
C.^A

ans =
        4          16          64
      625        3125       15625
```

and also **C^A** will not make any sense. (What about **C^3** or **C.^3**? Try them.) In linear algebra, a system of linear equations is commonly written as $Ax = b$, where x and b are column vectors and A is a square matrix. Its solution is given by $x = A^{-1} b$. In MATLAB, we use **x=A\b** to find the solution. Also, the system can be stated in a row-vector notation as $xA = b$, where x and b are two row vectors. In this case $x = bA^{-1}$ and we use **x=b/A** in MATLAB to solve the problem. When the problem is large, it is more efficient to use the previous approaches to solve linear equations than by first inverting A (say, using the MATLAB function **inv**) and then finding x by a multiplication. When A is not a square matrix, the operations **/** and **** have additional uses in contexts other than those considered in the present text.

Working with Matrices

Since MATLAB is built on matrices and optimized for efficient array operations, we would avoid the use of "do-loops" and work with matrix operations whenever possible. In building a data matrix from its component matrices or vectors, **[C D]** will append matrix D next to matrix C, and **[C;D]** will append D below C. Also, if you enter **H=[ ]**, H is an empty matrix and other matrices of compatible dimension can be appended to it subsequently. To illustrate, you may want to try the following:

```
CD=[C  D]

CD =
        4      4      4     39     33
        5      5      5     87     84
CE=[C;  E]

CE =
        4      4      4
        5      5      5
        4      8     12
       20     25     30
```

Sometimes, you may want to remove some elements from a matrix. As an example, you need to remove the last two rows from the matrix CE. Then you use

```
CE1((3:4),:)=[]

CE1 =
        4      4      4
        5      5      5
```

Thus $CE1$ now only contains the first two rows of CE. Here, we have introduced another useful operation—the colon (:). The operation **1:2:10** will generate a row vector with integers ranging from 1 to 10 in increments of 2. When the middle number of the triplet is omitted, unit increments are assumed. As an example, you see

```
a=1:2:10

a =
     1     3     5     7     9

a=-3:1.25:10

a =
  Columns 1 through 7

   -3.0000   -1.7500   -0.5000    0.7500    2.0000    3.2500    4.5000

  Columns 8 through 11

    5.7500    7.0000    8.2500    9.5000
```

There are many MATLAB functions that will facilitate the building of a data matrix. In this text, we use the following functions frequently: **eye** (an identity matrix), **zeros** (a matrix of zeros), **ones** (a matrix of ones), **diag** (a diagonal matrix). For example, if you need a diagonal matrix K whose diagonal elements are [1 2 3 4] and matrix E of ones of size 3×5, then you enter

```
v=1:4; K=diag(v)

K =
     1     0     0     0
     0     2     0     0
     0     0     3     0
     0     0     0     4

E=ones(3,5)

E =
     1     1     1     1     1
     1     1     1     1     1
     1     1     1     1     1
```

In manipulating data matrices, sometimes you may want to extract a submatrix from an existing matrix. As an example, if you need the 2×2 submatrix located at the southeast corner of the matrix K, then you enter

```
K1=K((3:4),(3:4))

K1 =
     3     0
     0     4
```

Online Help

In addition to the large number of functions available in MATLAB, your installation may also have other MATLAB toolboxes containing additional functions for specialized purposes. By entering **help xyz** where xyz is the name of the function of interest, you will get online help. To illustrate, if you enter

---

help expm

```
EXPM Matrix exponential.
     EXPM(X) is the matrix exponential of X.  EXPM is computed using
     a scaling and squaring algorithm with a Pade approximation.
     Although it is not computed this way, if X has a full set
     of eigenvectors V with corresponding eigenvalues D, then
     [V,D] = EIG(X) and EXPM(X) = V*diag(exp(diag(D)))/V.
     See EXPM1, EXPM2 and EXPM3 for alternative methods.

     EXP(X) (that's without the M) does it element-by-element.
```

---

you get a description of function and its calling protocol. Two examples for using MATLAB in the aforementioned manner (by that, we mean specifically without using user-supplied functions) were given in the appendix of Chapter 1 involving Examples 1.2.7 and 1.2.8.

User-Supplied M-Files

In many applications of MATLAB, you may have to write your own functions, which may invoke MATLAB functions as subroutines, to do the work. When the sequence of MATLAB commands is long or entails many possible changes as you experiment with your computation, writing your own function in a separate file can be more efficient. Such functions can in turn be used in some other functions you may create. Each such function is stored as an ASCII file and has the extension **m**, e.g., **invt_pgf.m** (see Example 1.2.9, Chapter 1 Appendix). These files are called M-files or script files. When an M-file is invoked at a MATLAB prompt, MATLAB opens the file and evaluates all of its commands exactly as it would if you had entered them at the MATLAB prompt (hence the term "script file"). To create an M-file in Microsoft Windows or Macintosh while you are in the MATLAB environment, you choose NEW from the FILE menu and enter your MATLAB commands. You save your M-files using SAVE AS into the directory in which your own M-files reside. (If Windows 95 is your operating system, then you need to configure the Notepad so that your M-files will be saved with the proper file extention **m**. For details about this configuration procedure, please contact The MathWorks.) It is important that the directory containing these M-files is on the MATLAB search path. To accomplish this, you will go to the MATLAB directory and find the M-file **matlabrc.m**. This file is the master start-up M-file. In this file there is a section that starts with **matlabpath) [....** Insert a line underneath it with **'C:\MATLAB\ur_stuff', ...**, where **ur_stuff** is the name of the subdirectory holding your own M-files. If you are working on a UNIX system, typically personal M-files are stored in a subdirectory of your own home directory called **matlab**. Hence your M-files will be directly accessible to MATLAB. In this case, you will create and edit your M-files using a UNIX editor outside of the MATLAB environment (or invoke within MATLAB using ! command). Once these files are created, you can run them from within MATLAB.

Like MATLAB-supplied functions, you may write four types of functions depending on whether a function has a string of input arguments and/or a string of output arguments. As an example, you may have function **xyz, [a,b,c]=function xyz**, function **xyz(d,e, f,g)**, and **[a,b,c]=function xyz(d,e,f,g)** where **xyz** is the name of a function, the sequence of arguments on the left side of each equality are the output arguments and the sequence of arguments on the right side of each equality are the input arguments (the arguments are themselves matrices). The function **invt_pgf** shown in Example 1.2.9 of the Appendix of Chapter 1 is of the first type. In this case, there is no passing of input arguments into the function for use in the computation nor passing of the values of output variables to the MATLAB environment for subsequent uses in MATLAB.

The function **[tx,qx]=invt_lap** shown in Example 1.3.5 of the appendix of Chapter 1 is of the second type. There, we have two vectors **tx** and **qx** as output arguments for use as input arguments in the MATLAB function **plot** for graphing. The values of the two vectors stay intact under the MATLAB environment unless you change them at a MATLAB prompt or until you exit from MATLAB. We note that the MATLAB-supplied function **plot** is of the third type in which there are only input arguments. A function of the fourth type is also shown in Example 1.3.5, namely **[z]=e135(x,y)** (the brackets enclosing z are optional since in this example we have only one output argument). All variables used within a function are local variables. Their names and values disappear when the function call is done. If you need to keep some variables and their values in the MATLAB environment for other uses, then you should return them as arguments or declare them as **global** variables (see the examples under the section Numerical Integration). As an example, the **function [z]=e135(x,y)** is used inside of the function **invt_lap** for passing the needed intermediate values of the Laplace transform. In such a usage, the values of the output variable z along with its name vanish as soon as the computation is completed. To get some idea about the syntax for writing functions, you may review examples shown in the appendices of the various chapters.

Control Flow

To control the flow of computation, the following uses of **for**, **while**, **else**, and **if** occur frequently

```
a=ones(3,2); s=[];
for  i = 1:3
    for  j=1:  2
      if  i*j < 6
        s = [s i*j*a(i,j)];
      else
        s = [s (i+j)*a(i,j)];
      end
    end
end
s

s =
    1    2    2    4    3    5


n=1; sm=0;
while  n < 6
      n=n+1;  sm = sm + n;
end
sm

sm =
    20
```

It is useful to bear in mind that MATLAB is optimized for efficient matrix operations. When possible, you should avoid the use of loops and try to do the computation by equivalent matrix operations. Also, you may have noticed the absence of go-to statements in MATLAB programs (these statements are known for their potential for making trouble anyway).

Numerical Integration

Numerical integration occurs at times in this text. To evaluate a definite integral we use the MATLAB function **c=quad('fname',a,b)** to do $\int_a^b f(x)dx$, where **fname** is either a MATLAB- or user-supplied function. Assuming that you need to find $\int_0^\pi \sin(x)dx$, you enter

```
c = quad('sin', 0, pi)

c =
   2.0000
```

To do double integration, the procedure is somewhat more complicated. Here, we give two illustrative examples. The first one finds

$$\int_0^1 \int_0^1 \sin(e^{xy})dydx$$

with

```
dblquad('objfun','x',0,1,0,1)

ans =
   0.9174
```

We list the three user-supplied functions as follows:

```
function out=dblquad(fun,outvar,xmin,xmax,ymin,ymax)
global ymin;
global ymax;
global fun;
global outvar;
out=quad('frstfun',xmin,xmax);

function out=frstfun(variable)
global ymin;
global ymax;
global fun;
global outvar;
eval(['global ' outvar]); out=ones(size(variable));
for i=1:length(variable)
    eval([outvar, '=variable(i);']);
    out(i)=quad(fun,ymin,ymax);
end

function out=objfun(y,ymin,ymax)
global x;
out =sin(exp(x*y));
```

The preceding functions for doing the double integration are based on a MATLAB technical note (available from ftp.mathworks.com). To understand its details, you need to consult the technical note. On the other hand, mimicking the given procedure to solve your own problem should be straightforward. We can modify the procedure to do double integration in which the limits of the inner integral are functions of the outer variable. As an example, suppose that you need to evaluate

$$\int_1^2 \int_{1-x}^{\sqrt{x}} yx^2\, dy\, dx.$$

Then you enter

```
dblquad('objfun','x',1,2)

ans =

    1.3583
```

In the following listing of the three user-supplied functions involved, most parts identical to their earlier counterparts are omitted for brevity.

```
function out=dblquad(fun,outvar,xmin,xmax)
. . .
out=quad('frstfun',xmin,xmax);


function out=frstfun(variable)
. . .
for i=1:length(variable)
    eval([outvar, '=variable(i);']);
    ymin=1-x; ymax=sqrt(x);
    out(i)=quad(fun,ymin,ymax);
end


function out=objfun(y,ymin,ymax)
global x;
out=y*x*x;
```

About "feval"

You may wonder about the repeated occurrences of `'something'` (namely, enclosing a statement within a pair of single quotes) in the last few program listings. As it turns out, these are applications of the command `feval('function', x1, ..., xn)`. One way to explain the use of this command is by looking at Example 1.2.9 shown in the Appendix of Chapter 1. There the function `invt_pgf` is designed to invert the probability generating function `e129pgf`. In `invt_pgf`, the function `e129pgf` appears three times. What if we have another generation function called `xyz`? We need to make the three substitutions before we can run the program. A better way is to use the command `feval` to evaluate the input (generating) function in `invt_pgf` whenever the generating function appears and take the name of the generating function as an input argument of `invt_pgf`. This is illustrated a follows:

```
invt_pgx('e129pgf')
```

```
p( 1) =    0.0000
p( 2) =    0.3600
p( 3) =    0.1440
p( 4) =    0.1440
p( 5) =    0.0922
```

We list the parts of the function **invt_pgx** that are different from **invt_pgf** as follows:

```
function  invt_pgx(input)
for k=1:n-1
    . . .
    z=r*exp(i*h*k); sum=sum+((-1)^k)*feval(input,z);
end
pn =2*sum+feval(input,r)+(-1)^n*feval(input,-r); pn=u*pn; pmf=[pmf pn];
    . . .
end
```

The preceding example shows that the command **feval** enables us to code MATLAB more effectively. Another example involving the use of **feval** is given in the Appendix of Chapter 3, where we introduce the function **invt_laq** for inverting the Laplace transform.

Graphing

MATLAB has many graphing routines for two- and three-dimensional plots. A rather primitive illustration is given at the end of Example 1.3.5 in the appendix of Chapter 1. Graphs can be given titles, axis labels, and annotations with commands such as **title**, **xlabel**, **ylabel**, and **gtext**. Again, the online help command is useful for obtaining the needed instructions.

Saving Your Results

After you exit from MATLAB, all variables and their values are lost. If you need to save them for a future MATLAB session, enter the command **SAVE** right before you exit from MAT-LAB. The file **matlab.mat** will be created. Next time when you enter MATLAB, just enter the command **LOAD** and your earlier results will be restored. You can also specify a filename for LOAD and SAVE. Under Microsoft Windows, you can use the mouse to SELECT any part of your results of a MATLAB session and use COPY to put it on the CLIPBOARD, and PASTE it elsewhere (say in a word processing document). If you are in a UNIX system and without the benefit of X-Windows, then you can use the command **diary xyz** (where **xyz** is any filename of your choice) at any point in your MATLAB session to record all the screen outputs that follow and turn off the recording with **diary off**. This process can be repeated many times in a single session and the results produced between these diary on-and-off episodes will be saved as an ASCII file with name **xyz** under your home directory. Finally, if you want to save the values of some variables in an ASCII file for use in other contexts (say for use in your favorite graphing routine or in a statistical package), enter the command **save xyz Q R S -ascii**. Then variables **Q, R,** and **S** are saved under the filename **xyz** in your home directory.